T0217113

Lecture Notes in Computer Science 1138

Edited by G. Goos, J. Hartmanis and J. van Leeuwen

Advisory Board: W. Brauer   D. Gries   J. Stoer

Springer
Berlin
Heidelberg
New York
Barcelona
Budapest
Hong Kong
London
Milan
Paris
Santa Clara
Singapore
Tokyo

Jacques Calmet   John A. Campbell
Jochen Pfalzgraf   (Eds.)

Artificial Intelligence and Symbolic Mathematical Computation

International Conference, AISMC-3
Steyr, Austria, September 23-25, 1996
Proceedings

 Springer

Series Editors

Gerhard Goos, Karlsruhe University, Germany

Juris Hartmanis, Cornell University, NY, USA

Jan van Leeuwen, Utrecht University, The Netherlands

Volume Editors

Jacques Calmet
University of Karlsruhe
Am Fasanengarten 5, D-76128 Karlsruhe, Germany
E-mail: calmet@ira.uka.de

John A. Campbell
University College London
Gower Street, WC1E 6BT London, United Kingdom
E-mail: jac@cs.ucl.ac.uk

Jochen Pfalzgraf
RISC-Linz, Johannes Kepler University
A-4040 Linz, Austria
E-mail: pfalzgraf@risc.uni-linz.ac.at

Cataloging-in-Publication data applied for

Die Deutsche Bibliothek - CIP-Einheitsaufnahme

Artificial intelligence and symbolic mathematical computation :
international conference ; proceedings / AISMC-3, Steyr,
Austria, September 23 - 25, 1996. Jacques Calmet ... (ed.). -
Berlin ; Heidelberg ; New York ; Barcelona ; Budapest ; Hong
Kong ; London ; Milan ; Paris ; Santa Clara ; Singapore ;
Tokyo : Springer, 1996
  (Lecture notes in computer science ; Vol. 1138)
  ISBN 3-540-61732-9
NE: Calmet, Jacques [Hrsg.]; AISMC <3, 1996, Steyr>; GT

CR Subject Classification (1991): I.1-2, G.1-2,F.4.1

ISSN 0302-9743
ISBN 3-540-61732-9 Springer-Verlag Berlin Heidelberg New York

Typesetting: Camera-ready by author
SPIN 10513681   06/3142 – 5 4 3 2 1 0   Printed on acid-free paper

Foreword

This volume of the Lecture Notes in Computer Science presents the proceedings of the Third International Conference on Artificial Intelligence and Symbolic Mathematical Computation (AISMC-3), held in Steyr (Austria), September 23–25, 1996, and organised by RISC-Linz (Research Institute for Symbolic Computation, Universtiy of Linz) with support from ProFactor (Steyr).

The AISMC initiative is an interdisciplinary forum which aims at bringing together people from different areas of research and application fields. Emphasis is on the interaction of methods and problem solving approaches from AI and symbolic mathematical computations in a wide sense. The originators of the AISMC initiative are Jacques Calmet and John A. Campbell. AISMC-3 continues the successful events AISMC-1, organised by J.Calmet (Karlsruhe, August 1992), and AISMC-2, organised by J.A.Campbell (Cambridge, August 1994). The proceedings of these conferences have been published as Springer LNCS volumes 737 and 958, respectively. It is intended to continue these meetings biannually. An introductory overview of the basic ideas and intentions behind AISMC can be found in the opening paper by the first two editors in the proceedings of AIMSC-1.

To stress applications and to establish links to engineering disciplines we incorporated in AISMC-3 the branch "Engineering and Industrial Applications". At the end of the first conference day a panel discussion was devoted to this topic.

The conference site in Steyr is a restored old factory building which is now a museum called "Museum industrielle Arbeitswelt". In this sense we consider the conference as an event embedded in the local "cooperation triangle" Linz – Hagenberg – Steyr. This triangle was set up to integrate the cooperation of university institutes and companies. Bruno Buchberger, the director and founder of RISC, has been and still is one of the main propagators and initiators of this idea.

The papers in the proceedings are listed according to the schedule of talks at the conference. As one can see, the titles of the articles reflect the interdisciplinary character of the meeting. Four invited talks are devoted to different areas. We thank the invited speakers for timely sending the full manuscripts of their contribution.

The conference is sponsored by AAAI, ProFactor (Steyr), RISC-Linz and several other institutions and companies (which cannot be listed here as these proceedings go to print). We are grateful to all of them. Last not least we would like to thank all the program committee members and the referees for their valuable support.

June 1996   Jacques Calmet, John A. Campbell, Jochen Pfalzgraf

Conference Committee: Jacques Calmet (Karlsruhe, Germany)
John A. Campbell (London, UK)
Jochen Pfalzgraf (Linz, Austria – Conference Chairman)

Program Committee:
L. Aiello (Rome)
F. Arlabosse (Paris)
B. Buchberger (Linz)
G. Butler (Montreal)
R. Caferra (Grenoble)
J. Calmet (Karlsruhe)
J.A. Campbell (London)
H. Clausen (Salzburg)
A.M. Cohen (Eindhoven)
J. Cunningham (London)
H. Geiger (Munich)
R. Goebl (Vienna)
K. Hingerl (Steyr)
D. Kapur (Albany NY)
L. Kerschberg (Fairfax VA)
H. Kobayashi (Tokyo)
R. Leisen (Bonn)
A. Miola (Rome)
E. Orlowska (Warsaw)
J. Pfalzgraf (Linz)
F. Pfenning (Pittsburgh PA)
G. Reihart (Munich)
M. Rigg (Bracknell)
W. Roque (Porto Alegra)
J. Rosicky (Brno)
E. Sandewall (Linköping)
K.U. Schulz (Munich)
A. Semenov (Novosibirsk)
T. Takeshima (Shizuoka)
T. Wilson (Ithaca NY)

Organized by: RISC-Linz, Johannes-Kepler-Universität Linz
Local Organizers: M. Meisinger
M. Schleicher
V. Sofronie
K. Stokkermans

Table of Contents

Symbolic Computation and Teaching

Dana S. Scott
School of Computer Science
Carnegie Mellon University
Pittsburgh, Pennsylvania 15213, USA

e-mail: dana.scott@cs.cmu.edu

Abstract. Since 1989, the author has tried to show that it is possible to put a complete semester-long course in machine-held form. Several examples have been carried out (in *Mathematica*), and the paper reports on the experience, on the problems encountered, and on some suggestions for future developments.

■ 1. Twenty Questions

In his keynote address at the **Second Annual Conference on Technology in Collegiate Mathematics** in November 1989 at The Ohio State University (see the newsletter UME TRENDS, for January 1990), Professor Lynn Steen effectively spoke for parents and students, scientists and engineers, colleagues and administrators by raising the following *twenty questions for calculus reformers.* Steen suggested that responding to these questions could form an agenda for the current work of people exploring the use of computers in curricular reform. Though several years have passed since Steen wrote these words, the questions remain highly relevant, since satisfactory conclusions about the use of computers have still not been reached.

In looking at the questions, remember they are addressed to mathematics departments, not to computer science departments. Calculus reform is still today a most controversial topic because so many students are required to take calculus, but many mathematicians (in the United States) feel that "reform" has meant "dumbing-down" to a level where the preparation of students in mathematics — both at school and college — is being positively harmed. This is a very broad topic, however, and we are only going to address here some of the problems about using computers and symbolic computation in various other courses with mathematical content, and we cannot enter here into the continuing battle over the future of the Calculus. All the issues are connected, however.

Steen's questions can be broken into five sections:

□ **Learning**
1. Can computers help students understand mathematics?
2. Can students develop mathematical intuition without performing extensive mathematical manipulations?
3. Do the mechanics of computing obscure mathematical insight?
4. Will using computers reduce students' facility to compute by hand?

❑ Curriculum
    5. How does computing change what students should know about mathematics?

    6. How does computing change what students can learn about mathematics?

    7. Where in the curriculum is computing most appropriate?

    8. Will use of computers reduce the need for remediation?

❑ Resources
    9. Can colleges afford computers for all mathematics students?

    10. How much time and distraction is computing worth?

    11. When will there be good software and compatible hardware?

    12. Can textbooks ever reflect contemporary computer examples?

❑ Teaching
    13. How much programming should be taught in mathematics courses?

    14. Can pure mathematicians convey an appropriate computational perspective?

    15. How will new faculty fit into computer-enhanced programs?

    16. Will use of computers improve teaching of mathematics?

❑ Dilemmas
    17. Won't computer packages for calculus lead, as they have in statistics, to much meaningless calculation?

    18. If computers handle routine calculations, what will students do instead?

    19. What are appropriate prerequisites for computer-based calculus courses?

    20. Should mathematics be a lab science?

The author has many answers to and opinions on these questions. Providing some answers and comments will form the main theme of this paper. The bottom line is that, yes, I personally believe that mathematics *should be in part* a laboratory science — the problem in the past has been that we have not had sufficiently powerful tools available to do the necessary experimentation on a large scale. Of course, the computing machine does not replace *imagination* — nor does the laboratory in any other science. Neither does the machine replace the standard means of *exposition* — but it can make the composition and presentation of books and lectures easier and more vivid, and more flexible. As with any tool, considerable effort is required in learning to use it effectively. The financial investment for the institution is considerable as well. The major question to be discussed here is whether the money and effort is worth the gain.

■ 2. The Author's Experience

What follows is a brief chronological survey of the courses in which the author and some of his associates have been involved. Then the syllabi of the *projective geometry* and *discrete mathematics* course will be given later in this section in some detail — especially to emphasize the point that it is possible to produce a substantial semester-long course completely in a computer-based format.

Projective Geometry, Fall, '89, CMU [Scott]
Automata Theory, '90, '91, '93, Stevens Institute for Technology [Sutner]
Topics in Discrete Mathematics, Spr. '91, CMU [Scott]
Projective Geometry, Spr., '93, RISC-Linz [Scott]
Automata Theory, Spring '93, RISC-Linz [Sutner]
Introduction to ModMath, Spr. '94, CMU [Scott, Miller]
Introduction to ModMath, Fall '94, CMU [Scott, Miller]
Problem Solving, Spr. '95 CMU [Scott, Miller, Sleator, Tygar]
Introduction to ModMath, Spr. '95, CMU [Miller, Statman]
Introduction to ModMath, Fall '95, CMU [Albert, Miller, Sutner]
Introduction to ModMath, Spr. '96, CMU [Scott, Miller, Sutner,
Walkington]

As indicated, the first course attempted was a course in projective geometry during the fall semester of 1989 at Carnegie Mellon (CMU). This was followed by a "topics" course, where the students carried out various projects in discrete mathematics and in using computer graphics. During a sabbatical year in Austria at RISC-Linz ('92/'93), the author gave another, improved version of the projective geometry course. That same semester Klaus Sutner also gave at Linz a new version of his course on finite automata, which he had developed over several years at the Stevens Institute for Technology. These courses were presented in lectures with the instructor using a projector from a computer, and sessions were held for students in a computer cluster. We were very indebted to Professor Bruno Buchberger for the opportunity to set up the computer classroom and to deliver these two courses. The very helpful staff and students at RISC-Linz we also essential for making things work.

After returning to the States, Scott began development of an introductory course in discrete mathematics for first-year students with the assistance of Philip L. Miller, head of the Introductory Programming Group at CMU. The computer-based classrooms used for programming were employed for the laboratory sessions, and lectures were given in classrooms with a projector. A more advanced course in discrete mathematics was also tried out with the cooperation of other CS faculty. Sutner then joined the teaching faculty in the Introductory Programming Group in Computer Science at CMU in the fall of 1995. The introductory course had also been rerun twice and a complete revision was made for the Spring Semester 1996, with the cooperation of Miller and Sutner and other CMU faculty.

Some acknowledgments are in order. Over the years, Scott has been very ably assisted at various times by four of his former students: Jean-Philippe Vidal (Paris), Marko Petkovsek (Ljubljana, Slovenia), and J. Todd. Wilson (Cornell), Drew Dean (Princeton), and by one post-doctoral visitor, Dr. Christine Luksch (Darmstadt). He is much indebted to them. Essential advice about *Mathematica* and about computer-based teaching has been obtained at various times from Prof. John Gray (Illinois) and from many people at Wolfram Research, Inc. (Urbana, Illinois), including Stephen Wolfram, Theo Gray, Roman Maeder, Igor Rivin, Henry Cejtin, Cameron Smith, and Nancy Blachman. However, without the extensive involvement, both in planning and in execution, by Philip Miller and Klaus Sutner, the later courses could never have been concieved, mounted or completed. The two of them have been wonderful collaborators and friends, and we hope to continue joint work with new teaching developments in the future.

□ Syllabus for Projective Geometry with *Mathematica*

The course was divided into six chapters, with the objective to end up with the projective classification (over the complex numbers) of *cubic curves*. Starting from scratch, this is a heavy job for one semester!

Chapter 1: A review of some basic algebra. Complex numbers and Argand diagrams; roots of an *n*-th order polynomial equation and the roots of unity; determinants, the line joining two points, and areas of polygons; points in space, the intersection of two lines, and projections between intersecting planes; vector spaces and polynomial algebra.

Chapter 2: Homogeneous coordinates. Homogeneous coordinates for points and lines; projective representations of points and lines; dual spaces and bilinear paring; linear forms and the dual paring; exterior products of forms; meets and joins: computations and geometric interpretation; points on a line and concurrent lines; Desargues' Theorem and computing the intersection of lines determined by points; Theorems of Menelaus, Ceva and Pappus; the uniqueness of the 4th harmonic and grids of squares.

Chapter 3: Projective transformations. Matrices for transformations and point transformations; transformation by substitution and composition of transformations; line transformations; transforming quadrangles and quadrilaterals, and mapping triangles, quadrangles and quadrilaterals; the notion of crossratio and positioning points on a line; preservation of crossratios by transformation; perspective correspondences and compositions of perspectives; classifying transformations by fixed points and fixed lines.

Chapter 4: Polynomial duality. Curves, envelopes, and intersections; conic sections and quadratic equations; intersecting lines with curves and curves with curves; polydifferentiation and the ring of differential operators; evaluation through polydifferentiation; duality at higher degree and the dual vector spaces of polynomials; linear combinations of powers of linear forms; contact, multiplicity, and tangency; polydifferential graded algebra for "general" polynomials; degenerate curves; the Hessian, and testing conics for degeneracy; the tangents at a point on a curve; intersecting lines with a curve from an $(n-1)$-tuple point.

Chapter 5: Conics. Poles and polars with respect to a conic; degenerate and nondegenerate conics and the relation to the Hessian; preservation of crossratios and the crossratio of four points on a conic; meets of conjugate lines, the envelope of a conic; the conic determined by five points; Pascal's Theorem; projectivities on a conic; conics determined by points and tangents; conics through four points; harmonic properties of conjugate points; involutions on a conic; the eight tangents at the intersection of two conics; pencils of conics.

Chapter 6: Cubics and some higher-degree curves. Special points on cubics; independence of points; nodes and cusps; the invariance of the Hessian determinant; preserving polydifferentiation by projective transformations; inflection points of cubics; cubics intersecting at eight points; normal form for cuspidal cubics, for nodal cubics, and classifying non-singular cubics; the configuration of inflection points of a non-singular cubic; graphing and parameterizing curves in general, and the parameterization of the non-singular cubic; the maximum number of double points on a curve; cataloging degenerate cubics; parametrization of a quartic curve; a few examples of other curves; a rational curve with positive defficiency; finding the tangential equation of a curve; analyzing rational functions.

□ Commentary

As this was the first course attempted, and as the number of students that have actually taken it so far (about 12) has been very small, it is as of yet not fully developed as the equivalent of a standard text book. The topics and methods of symbolic computation have been well tested, but far more commentary needs to be written and a better development of graphical illustrations to accompany the examples is needed. More exercises also have to be developed. The files that exist now, unfortunately, could not be used for self-study. If there is ever another opportunity to present this course, the author hopes to put it into publishable form, as he feels the mathematical approached adopted making use of computer algebra in several different ways has the potential for further development. We do not have the space to argue this point in any detail here, however.

The topic of the course is that of algebraic curves in the classical complex projective plane. Algebraic geometry in projective form standardly employs homogeneous polynomials. A polynomial in variables x y z is said to be *homogeneous* if, when multiplied out as a sum of monomials, all the monomials have the *same degree*. In the plane, the zeros of a homogeneous x y z-polynomial represent an algebraic curve. The dual, u v w-polynomials, in a dual set of variables, represent the tangential form of curves, that is to say envelopes.

How can the full impact of duality — familiar from elementary projective geometry as the duality between points and lines — be extended to curves of higher degree? In order to derive the necessary formulae it is most convenient to introduce the ring of *differential operators*. If we consider the space of all x y z-polynomials as a space of (continuous) infinitely differentiable multivariate functions, then we know that on such functions, f, the operators of partial differentiation, $\partial/\partial x$, $\partial/\partial y$, and $\partial/\partial z$, which are written as D[f, x], D[f, y], and D[f, z] in *Mathematica* notation, are associative and commutative (under composition of operators). As is well known, these compositions are also linear operators, and any linear combination of linear operators is again a linear operator. Moreover, composition of linear operators distributes over all linear combinations of operators; hence, the differential operators generate a *commutative ring of operators*.

In the light of these considerations, it is possible to define a ring homomorphism from the (free) ring of u v w-polynomials to these operators. (This homomorphism is actually an isomorphism.) In particular, if f is an x y z-polynomial and g is a u v w-polynomial of *lower* degree, then we can notate by PD[f, g] the result of letting g operate on f as a differential operator by calling up this homomorphism. However, differentiation between x y z-polynomials and u v w-polynomials can be made *symmetric* or *dual* in specific sense. Indeed, if f is a homogeneous x y z-polynomial, and g is a homogeneous u v w-polynomial, then PD[f,g], an operation the author calls *polydifferentiation*, may be defined as the result of allowing the polynomial of lower degree to operate on the polynomial of higher degree as a differential operator. In other words, the symbol u can be regarded as the operator $\partial/\partial x$, or the symbol x can be regarded as the operator $\partial/\partial u$, and similarly for other variables. In the case of polynomials of *equal degree*, by reference to monomials, it is easy to see that in evaluating PD[f, g] it makes no difference which is taken to be the operator and which is the operand.

The *linear   homogeneous* xyz-polynomials, f, can be regarded as *lines*; while the linear uvw-polynomials, g, can be taken as points (up to a constant, non-zero factor, as with homogeneous coordinates). Then the equation PD[f, g] == 0 means that the point g *lies on* the line f. (In the linear case, partial differentiation works just like a dot product between vectors.) The equation also means the line f *passes through* the point g, and this remark is the basis for the explanation of the duality between points and lines. What about higher degrees? What we need is a special case of a theorem of Euler which tell us that, if f is a homogeneous xyz-polynomial of degree n, and if g is taken as the *linear form* a u + b v + c w, then we have the equation PD[f, g^n] == n! f[a, b, c], where f[a, b, c] is the evaluation of the polynomial as a function of xyz at abc. Among other consequences, this theorem tells us that that the equation PD[f, g^n] == 0 means that the point g *lies on the curve* f. But there are many algebraic consequences. For example, by a simple use of the Binomial Theorem, we can argue that if g represents an m-tuple point of the curve f, then the polynomial P[f, g^{n-m}] of degree m factors into the m linear factors representing the *tangents* to the curve f at the point g. The trick here is that pairing PD[f, g] is a bilinear paring between the vector spaces of n-degree homogeneous xyz-polynomials and n-degree homogeneous uvw-polynomials that makes then *dual vactor spaces*. But the polynomials are not just elements of vector spaces, since they are parts of a pair of commutative rings, and PD has many simple properties with respect to the ring structure. A suitable axiomatization of these properties makes the ring of xyz-polynomials fully dual to the ring of uvw-polynomials, and the formal operations — as indicated — can have geometric interpretations.

Mathematica well implements commutative algebra (and many algebraic algorithms) and of course it implements partial differentiation. It does not implement a algebra of differential operators, however. The author was able to do that by defining the function DegP for computing the degree of homogeneous polynomials, and by transcribing the axiomatic characterization of PD directly into rewrite rules between the two kinds of polynomials using *Mathematica* rule sets. These rules incorporate in symbolic form the rules of partial differentiation, and so *Mathematica*'s built-in partial differentiation operator never had to be invoked. Additionally some ideas of exterior algebra had to be represented and some polynomial invariants using determinants. Otherwise, polynomial simplification, factorization, and root finding (all built into *Mathematica*) were the only other constructs needed. One big advantage discovered by using symbolic computation in this way was that *formulae* for the solution of geometric problems could be developed and then used in numerical computation (for example, in creating graphical illustrations).

This approach to classical, plane projective geometry is quite satisfactory, as the symbolic polynomial algebra really gives a middle way between synthetic and analytic geometry. Several difficult problems remain to be investagated, nevertheless:

(1) How to extend this approach to fields of *characteristic* p?
(2) How to make good use of symbolic computation with *algebraic   numbers*?
(3) How to refine the methods to apply to the *real   projective   plane*?
(4) How to move to *higher   dimensions* beyond plane geometry?
(5) How to use of modern methods of symbolic computation in *ideal   theory*?
(6) How to connect with *automated   theorem   proving*?

As regards that last point, *Mathematica* — by doing algebraic transformations and symbolic rewriting — ***does*** actually prove theorems. However, these proofs are not fully automatic, as the user of the system has to issue the commands to be carried out. The user also has to supply the interpretation of the results, often resulting in a non-automated search for significant output. Nor does the system formulate conclusions by itself. Still, the experience was encouraging in that the algebra generated could not be carried out with any pleasure ***by hand***, and the use of the computer definitely helped the author to ***think***. The developments by Edmund M. Clarke (CMU) and his associates on the implementation in *Mathematica* of the ***Analytica*** program indicate that a certain kind of marrage between computer algebra and theorem proving can be very effective. More developments of this kind are needed, however, before courseware can be written that uses theorem proving.

□ Schedule and Syllabus for Introduction to Modern Mathematics with *Mathematica*

The bread-and-butter, service course for first-year students, blandly called **Introduction to Modern Mathematics**, is really a review of some basic algebra (not often well taught anymore in high schools in the USA), an introduction to number theory, a brief introduction to logic, and a glimpse of some topics of discrete mathematics *via* some problems of counting — which as well provide many good examples for doing proofs by mathematical induction. The major objective, of course, is to encourage abstract thinking. Unfortunately, the course curriculum is not well integrated with other courses in Mathematics and Computer Science, in the author's opinion, and so it does not fulfill these purposes as well as it should in the present curriculum.

The reason for attempting a *Mathematica*-based version of the course was to take advantage of elementary programming in problem solving, to allow for the exploration of many examples, and to illustrate results with computer graphics. The author feels he can argue that these objectives were achieved. There is also a hope that students, after becoming familiar with *Mathematica*, will use the software in future courses.

During the two semesters of the academic year '95/'96, the course was organized as two large lectures per week for all the students and then two labs in groups of about 20 students, each with their own computer. It was very helpful that the University and the School of Computer Science remodelled four computer classrooms in the summer of 1995 which could be used for this course (but they are primarily for the teaching of introductory programming, though they are also used for other courses). Further teaching facilites are being prepared now during the summer of 1996.

The outline of the course as delivered by the author as Lectruer in the last spring term is as shown below. The laboratories we handled by a group of five Section Leaders. There was also a voluntary Help Session run by Dr. Miller every Friday afternoon to assist with homework. The computer clusters were open evenings and weekends for students to work individually. (A few students had their own computers.) The examinations were administered in the clusters.

Class organization:

NUMBER OF STUDENTS (Spring '96) Approx. 100.
LECTURES (50 min.) Twice a week.
COMPUTER LABORATORIES (1 hr. 20 min.) Twice a week.
HOMEWORK ASSIGNMENTSS (14) and **QUIZES** (12) Once a week.
MIDTERMS (1 hr. 20 min.) Three Midterms.
FINAL EXAMINATION (3 hr.) At end of course.

Grading:

HW	MTs	Qu	Final	Total
25%	30%	20%	25%	100%

Lecture schedule:

WEEK 1. 2 Lect. [Mathematica, Lists, and Primes]
WEEK 2. 2 Lect. [Quotients, Remainders, and Digits]
WEEK 3. 2 Lect. [Recursion, Factorials, Permutations]
WEEK 4. 2 Lect. [Induction and Well Ordering]
WEEK 5. 2 Lect. [Structural Induction]
WEEK 6. 2 Lect. [Iteration, Periodicity, and Rationals]
WEEK 7. 2 Lect. [Irrationals and Cycles]
WEEK 8. 2 Lect. [Iteration and Chaos]
WEEK 9. 2 Lect. [Boolean Functions and Automata]
WEEK 10. 1 Lect. [Closure, Periodicity, Iteration]
WEEK 11. 2 Lect. [Relations and Functions]
WEEK 12. 2 Lect. [Relational Operations and Permutations]
WEEK 13. 1 Lect. [Structure of Permutations]
WEEK 14. 2 Lect. [Equivalence and Congruence Relations]
WEEK 15. 2 Lect. [GCD, LCM, and the Euclidean Algorithm]
TOTAL 28 Lectures

Course Syllabus:

WEEK 1. **[Mathematica, Lists, and Primes]** Demonstration of the *Mathematica* Frontend and the use of Notebooks; examples of numeric and algebraic expressions, and other details of syntax; introduction to lists and arrays and some basic programming constructs through examples; introduction to prime numbers.

WEEK 2. **[Quotients, Remainders, and Digits]** More on lists; integer and fractional parts; powers and positional notation; the division algorithm; the radix representation of integers; conversion between systems; introduction to recursive definitions; introduction to divisibility and modular arithmetic.

WEEK 3. **[Recursion, Factorials, and Permutations]** Review of recursive definitions by example; generalizing the radix reprsentation; the factorial representation; enumerating permutations; implementing digital arithmetic.

WEEK 4. **[Induction and Well Ordering]** Mathematical Induction; least-element and well ordering principles; other forms of induction; review of theorems requiring inductive proofs; the connection between induction and recursion; finite sets and cardinality; disjoint sums and cartesian products; counting principles. [To First Midterm]

WEEK 5. **[Structural Induction]** Integer lists and the word algebra; principles of induction and recursion; trees as iterated lists; inductive and recursive principles associated with them; Konig's Lemma.

WEEK 6. **[Iteration, Periodicity, and Rationals]** Doing iterations; the Accumulation Laws; periodic and eventually periodic functions in general; the radix representation of fractions; the theorem on periodic representations.

WEEK 7. **[Irrationals and Cycles]** Discussion of rationals vs. irrationals and of proofs of irrationality; a brief look at continued fractions; finding cycles; examples of periodic functions.

WEEK 8. **[Iteration and Chaos]** Finding fixed points and roots by iteration; uses of fixed-point operators; chaos via iteration; the example of the Logistic Function. [To Second Midterm]

WEEK 9. **[Boolean Functions and Automata]** Representation of Boolean functions by polynomials; the enumeration of Boolean functions; sets, subsets, and indicator functions; Boolean set operations; vectorized Boolean functions; the application to cellular automata.

WEEK 10. **[Closure and Periodicity under Iteration]** Closure of a set under a function of list of functions; finite closures and cycles; examples of number-theoretic functions producing cycles; the notorious 3 x + 1 problem; cycles for automata iterations; demonstration of multi-state automata.

WEEK 11. **[Relations and Functions]** The idea of a (binary) relation; properties of relations; functions as special cases of relations; one-one vs. many-one functions; relation tables and Boolean matrices; picturing relations.

WEEK 12. **[Relational Operations and Permutations]** Relational product; representation of relational operations by matrices; transitive relations and the transitive closure; representing and computing permutations; isomorphism of relations. [To Third Midterm]

WEEK 13. [Structure of Permutations] Compositions and inverses of permutations; defining the order of a permutation; finding the cycle structure; computing the order of a permutation; isomorphic permutations.

WEEK 14. [Equivalence and Congruence Relations] Equivalence relations; equivalence classes; equivalential closure; congruence relations in the integers; combinations of congruences.

WEEK 15. [The GCD, LCM, and the Euclidean Algorithm] Defining the GCD and the LCM; relatively prime pairs of integers; properties of divisibility; the Extended Euclidean Algorithm; the prime factorization of integers; the infinitude of the primes.

□ Commentary

The ordering of the topic was not really satisfactory in the Spring '96 version of the course. At the start of the class the students must not only hear about new ideas, they must also become familiar with some basic features of the *Mathematica* language. It is not easy quickly to achieve both aims. The author attempted an introduction via a large suite of examples. Many of the students were confused and kept asking what was the "theory" behind the examples. The meant, apparently, that they did not have enough instructions about what to do with the examples and sufficiently many reasons why they should remember them. However, *number representations* (standard radix, factorial, Fibonacci, and binomial) and discussions of rationals and irrationals and elementary properties of continued fractions provide a very fertile field to connect number-theoretic and algebraic properties with questions of algorithmic computation and the use of data structures. The author feels very firmly that the topics related to representations provide a suitable and sound entry point to discrete mathematics — provided that the examples and exercises are interesting and comprehensible.

Because the students became very restless with the introduction last spring, the author moved perhaps too quickly to the discussion of the various forms of *proofs by mathematical induction* and *definitions by recursion*. It would have been better to combine some of this with the topics of *congruence relations* and the *Euclidean algorithm*, which in one way or the other were postponed to the end of the course. Congruences and properties of divisors give excellent occasions for proofs, and the properties are important for the understanding of many mathematical examples concerning finite structures as well as computational algorithms. Thus, these sections should definitely have entered the lecture program earlier.

The author finds that the topics of *iteration* and the finding of *cycles* in iterative processes give a wide scope not only for proofs and algorithms but also a good way to introduce ideas about functions (injective, surjective, and so on). Also, when applied to complex data structures, there are many interesting questions about counting that come up. Iteration is a basic concept both for mathematics and computer science, and it should have a key role in such a course.

The problems of transforming vectors leads very naturally to discussions of *Boolean functions* and *set operations via* Boolean vectors used as *indicator functions* of subsets. *Mathematica* allows a very easy vectorization of Boolean functions if one uses standard polynomials. One may define a Boolean polynomial as a (multivariate) polynomial that takes on the the values 0 or 1 when 0s and 1s are substituted for the variables. Then, there is an easy discussion of which algebraic operations take Boolean polynomials to Boolean polynomials. (For example, multiplication f g is a good operation, but addition f + g has to be replaced by the compound f + g - f g.) Indeed, all of standard *Boolean algebra* (including valid equations and normal forms) can be treated as a chapter of polynomial algebra; and *Mathematica* can make all the transformations of polynomials needed using only very simple definitions (including a replacement rule for getting rid of unwanted exponents in Boolean polynomials). This use of standard polynomials and polynomial manipulations has three advantages: (a) the operations automatically *vectorize*, (b) the idea *generalizes* to n-valued functions in a natural way, (c) it is not necessary to introduce *abstract algebraic structures* on other sets besides the integers (and rationals and reals) to do Boolean algebra. The author strongly feels that a first-year (even first-semester) course is *not* a good place to introduce abstract algebra. Instead, work with Boolean values and modular arithmetic can be the right *preparation* for the algebra to be studied in a *subsequent* semester. (Provided, of course, that the departmental organization of the curriculum allows for a meaningful linking between course syllabi — something that is not easy to arrange.)

The vectoization of Boolean functions then immediately applies to the implementation of examples of *cellular automata*, which bring up many questions of counting and iteration. *Mathematica* also makes the *graphics* needed for illustrating the histories of cellular automata quite efficient and accessible. Owing to time restrictions, only *one-dimensional* cellular automata were treated in this course; in another course, two- dimensional automata were implemented just as easily. Cellular automata provide many good problems to encourage experimental computing and the making of conjectures.

From vectors and automata it is a natural step to *matrices* and to *permutations*. This is not at all the course to do linear algebra (though it would be reasonable if students took that course at nearly the same time). However, it is not difficult to introduce *Boolean matrices* as *tables for representing relations*. The Boolean polynomials apply to matrices (lists of lists) just as well as they apply to vectors (lists). Thus, the logical operations on (finite) relations become immediately computational (and the understanding of why this is so is a good mathematical exercise).

However, the polynomial $A + B - A\ B$ (where $A\ B$ is the termwise product of matrices) is not the best way to compute the union of relations when the matrices are large (especially for unious of several terms). What the author realized in making up exercises is that computationally (if not algebraically), the compound combination $\text{Sign}[A_1 + A_2 + A_3 + \ldots + A_n]$, where Sign is the standard *signum function* reducing numbers to ± 1 or 0 and working termwise on matrices, is more efficient when many iterations are needed. Moreover, $\text{Sign}[A\ .\ B]$, where $A\ .\ B$ is now the usual matrix multiplication, is exactly the matrix representing the *relational product* of the relations represented by the two given matrices. This idea of relational product and the use of matrix multiplication (including the counting interpretation of $A\ .\ B$ itself) can be explained in elementary terms (and leads naturally to ideas of graph theory — in a later course). And it is quite computationally efficient. Thus, even for quite large matrices, it is reasonable to solve — by fixed-point iteration — the matrix equation $X == \text{Sign}[A + X\ .\ X]$, where A is a given square Boolean matrix, in order to find the matrix representing the *transitive closure* of the relation represented by A.

Relations require many pictures for their understanding: both line (or arrow) digrams and varioous kinds of tables. Of course, *ordering relations* and *equivalence relations* and the basic properties of transitivity and symmetry should be covered through many examples. What of *permutations*? Should these be left to the algebra course? *No, definitely not!* Many students will not take the algebra course, but more positively the idea of a permutation is too basic to be pushed into the future. This introduction is not a group-theory course, but *compositions* and *inverses* of permutations are something to be understood without the axiomatics of groups. In particular, all the properties of relations encountered in the elementary examples immediately bring up the question of *isomorphism* of relations — something to be explained using permutations. Iteration of functions leads to finding permutations. But permutations are functions, and iteration is essential in understanding the cyclic structure of permutations. And, working with permutations brings up many problems to be solved by counting and induction.

In Spring '96 the author did not have time to tie up fully enough the connections between permutations and equivalence relation (*via partitions of sets*) and the very important proof that every permutation is either *even* or *odd*. He strongly believes this course should contain that theorem. And he believes that the integrated use of computing in dealing with (large) permutations makes the concepts and theorems far more meaningful to students than can be done in the limited time available by paper-and-pencil definitions and calculations.

The work in preparing the course and giving it in a computer-based form was very, very painful and time consuming, but the attractiveness and appropriateness of the the topics has convinced us — Scott, Miller, and Sutner — that whatever the future of computer-based teaching at CMU this course should be published quickly in some form involving both print and electronic files. We plan to work together on this project over the coming academic year.

■ 3. The Lessons Learned

The lessons learned by the author over these several years of experience can be conveniently grouped as answers to Professor Steen's queries (for which he no doubt has many answers himself).

□ *Can computers help students understand mathematics? Can students develop mathematical intuition without performing extensive mathematical manipulations? Do the mechanics of computing obscure mathematical insight? Will using computers reduce students' facility to compute by hand?* This is really all one big question. Let us remember that by themselves, computers can do nothing. It is possible we will see in the future self-study tutoring programs which will automatically guide the student through a subject. But the success of tutoring programs for college-level courses will take a terrific amount of development — both in computer software and in course planning. We are not ready to do automatic teaching yet, even though there are some good examples available for some elementary courses.

The present focus of work should be in getting good materials into computer-held form. It is easy to give a single demonstration lecture, but it is quite hard to get 15 weeks of course materials into the computer. Computers can only help students understand mathematics if the mathematics is *in* the computer. And it also takes a lot of thought to construct good examples illustrating the mathematical ideas.

As regards manipulation, what has to be done with the computer is not to use it as a magic, black box (and one using black magic at that), but to break the calculations into phases so the student can see what manipulations are necessary. This should not be a passive activity either, and the student should be given suitable control over choosing manipulations to be done by the computer. The way intuition is gained is through a combination of *success* and *failure* in many examples, where the student is prodded to ask why things are working out — or not working. If this analytical thinking is not encouraged, then the computer will be just a magic instrument provided by the witch doctor. What the author believes should be enhanced is the thinking processes, and whether *extensive* hand calculation is needed for this is a question that only very subtle psychological testing could answer. Success in problem solving is probably a more valuable experience than being able to do hand calculations.

A side comment here may be relevant. Nearly 20 years of using word-processing programs almost daily has improved the author's spelling considerably. Something happened to him in elementary school to make spelling a pain and a chore. However, being able to correct things quickly (not automatically, but through asking the checker to check and the computer-held dictionary to define) has done what the school teachers could not. The key feature is the quick correction. The student has to be able to check answers quickly, and if something is wrong make a change while the problem is uppermost in his mind. This quick response is one of the main advantages of computers — provided the information is there to access. By seeing the correct answers in many contexts, learning takes place. Again, how to assess the extent or efficiency of the learning is again a (difficult) problem for psychologists. The experience the author had in the several versions of the two courses described in the last section, however, convinced him that a program such as *Mathematica* together with preparation of examples can answer questions very quickly and on a scale impossible with hand calculation. Therefore, he is convinced of the value of using computers.

There is a serious danger, nevertheless. The teacher may know very well why things work, and he may call up the computer to give the final answer without showing any *intermediate steps*. It is all too easy to hide things by computer and leave connections invisible. The teacher always has to ask himself whether he is showing enough to make the results understandable and plausible. Computers can be misused in numberless ways, and one must constantly ask whether the answers obtained are reasonable. If there is any doubt, a stepwise analysis must be carried out to be sure that the computer is doing what was intended.

□ *How does computing change what students should know about mathematics? How does computing change what students can learn about mathematics? Where in the curriculum is computing most appropriate? Will use of computers reduce the need for remediation?* Will computers change mathematics? Well, yes and no. Problems are problems, and they must be solved by whatever means. But computing does very much bring home which solutions are *feasible*. For example, a massive search may theoretically solve a problem, but it may be completely infeasible even for moderately sized problems. So the questions of how solutions are carried out and why they are fast or slow is a new aspect that the computers may bring forward and which may have been missing from conventional curricula. The author feels that computers can help make receptive students learn *more mathematics*. And inquisitive student will always want to know why things are true, will always bring up something outside the syllabus, and the computing environment can help answer those questions by letting the student experiment on his own. Moreover, seeing success in one kind of problem suggests at once other problems or further generalizations. The computer often lets one adapt solutions to new cases fairly easily, or at least lets one try to see why a previously usable method fails to be applicable.

Another important feature of the computer is its memory — its *retentiveness*. When paper and pencil homework is returned, does the student ever look at anything other that the mark at the top? Are the papers saved? With computers there is a potential at least for the reuse of work from assignment to assignment — if the teacher is clever enough to ask the right questions in the right way. Certainly the author has found that working out formulae exactly makes it possible to use and reuse the results in new ways. For example, in geometry, the formulae may prove a theorem, but then they may be used in numerical computations to draw pictures. In combinatorics, the result of solving a counting problem results in a formula or program that can be put to work in many other problems. Or, the new problem may require some revisions of something done before, and if the old program in retained, it can then be adapted to meet the new requirements without having to start from scratch. Of course, as with any growing archive — whether electronic or print — some good organization is required for quick recall. The author feels that the software systems designers could make life easier for us on this score by providing better aids to the organization of computer files.

And where in the curriculum should computing enter? Well, should it be excluded from any part of the course of study or area of research? Who does not need to draw a picture or do some algebraic or symbolic simplification? The computer should become the sketchpad, laboratory, archive, and communication device for anyone who is willing to learn how to use it. True, we need some better development of *typesetting* to make the use of the computer less painful for sketching, but the author believes we will see this before the turn of the century. Already in the last two or three years, the computer network facilities have made *communication* much more effective. The author also believes that by the turn of the century we will have seen a big change in mathematical publishing, and this change will have a major impact on teaching as well.

What about remediation? Yes, having materials available in computer-held form will allow the student to review past work and past courses at his own pace. If he is motivated enough, he will find what he needs. Students may also be able to review tests and problem sets more easily in order to understand their mistakes better. E-mail can be effectively used to ask questions and to get answers. The key feature of using the computer will be to get the student to record his question — in whatever form — at the moment at which he is concentrating on the subject. With computers, this can be any time of the day or night; and with portable computers, this can be almost anywhere (except, perhaps in the swimming pool or dance floor). But whether the need for remediation will be *reduced* is not a question we can answer today. We should concentrate in the short term in making remediation easier — by capturing courses in computer-held form. How many times have professors not even been able to find the syllabus for a course from a previous year, let alone copies of the tests and problem sets?

☐ *Can colleges afford computers for all mathematics students? How much time and distraction is computing worth? When will there be good software and compatible hardware? Can textbooks ever reflect contemporary computer examples?* The economic factor is very important. Budgets are tight, and higher education is suffering from a reduction of state funds in every country at the present time. But the author asks: Can colleges and universities afford *not* to supply computers to students (perhaps for purchase at favorable prices)? Can colleges afford *not* to provide the service necessary to make educational computing effective? It is not enough just to buy a bunch of computers and tables and chairs and to plug the machines in. It is not enough just to buy various software packages and to install them. Students and staff have to be trained to use the software, and both the machines and the programs require constant maintenance. Some of the training and care can be done on an institution-wide basis. But there are problems that come up every day that need immediate attention, and departments have to have some dedicated staff to help solve them. Professors cannot be expected to become computer experts capable of doing intensive trouble shooting. Professors who are going to make use of computers — and there will be more and more of them — have to be sufficiently computer literate to make the most of the equipment, and they have to have some technical insight to be able to recommend and oversee the changes that will be necessary nearly every year. But professors change, their assignments change, and their interests change. Therefore, there has to be a kernel staff to keep things running smoothly from year to year. This staff has to understand something of the course requirements and the teaching problems. Different departments probably will have different demands and needs, so some staff has to be attached to each department. Some staff members can be instructors in some courses or course organizers for other courses, but there have also to be technical people who can really care for the machines and software. Not every institution has yet realized this need for staff commitments.

And why can computers be not avoided? Because computers now enter into all parts of business and government — not always very economically to be sure, but they are there to stay. They will come into more and more homes very soon. The truly amazing growth of the Internet (and the crazy "investment bubble" in internet companies!) has made everyone aware of the (some) of the potential of computers for conveying information. Many colleges are already making good use of educational computing. If others do not catch up, they will loose out to competitors. In 1996, private university education now costs something on the order of $1000 per week in the United States. At this price, students will soon realize that they are being cheated if they are not being provided with the opportunity of using computers in their coursework. The question of "time and distraction" can no longer be avoided — colleges have no choice. The problem is how to keep costs in hand. But economy cannot be achieved by minimizing expenditure on staff!

Software, hardware, and textbooks? Everyone is happy to complain about them. But it is in fact the case that more and more textbooks are including computer supplements — either through disks in the books or through Internet archives maintained by the publishers. The author feels that at last the problem of paying for Internet products is being solved, and that publishers will find electronic distribution of certain materials both helpful and appropriate. As far as hardware goes, it is a sad fact that the higher-education market is not large enough to make much impact (witness the awful failure of the NeXT computer and the more recent sad decline of the Apple Macintosh). The author has personally no desire to convert to Microsoft Windows, and only hopes that such operating systems as Linux, which can run on many different machines in a uniform way, will make it possible to develop a sound and reliable educational computing environment. Software suppliers must make their products avaiable for this operating system, however. Whether they will or not remains to be seen.

☐ *How much programming should be taught in mathematics courses? Can pure mathematicians convey an appropriate computational perspective? How will new faculty fit into computer-enhanced programs? Will use of computers improve teaching of mathematics?* The author has heard many mathematics professors say, "I will not teach programming!" Fortunately, not all engineering, physics, and computer science professors are refusing to teach mathematics, since mathematics departments are not always teaching the necessary parts of mathematics in the right way for the more applied areas. It is not just the Calculus: something as basic as *Linear Algebra* can be a very sore point at some institutions.

What the author believes is that if the professors are willing to use programming, then sufficient programming can be learned by students *by example*. The point is that the examples have to be there to be used as models. The people who know how to do it should thus produce more materials for wider distribution. The newer faculty will catch on quickly, but the older faculty may be the ones showing the most resistance. Much of the design and understanding of programming examples is mathematical in nature, and "pure" mathematicians are just as capable of conveying a computing perspective as are the computer science professors. Indeed, the mathematicians may have an advantage in knowing more examples, more proofs, more tricks, and generally more facts than some of the staff that have had to concentrate on technological problems. It takes all kinds, and we should not make more efforts to erect more barriers between subjects and waste time on "protecting" turf. If we do, we are cheating our students in regard to obtaining a wider perspective on the connections among the sciences.

Not all subjects can immediately benefit from the use of computers. But much of algebra, geometry, number theory, numerical analysis, and combinatorics certainly can. And computation, as has been asserted before, often brings up new and mathematically interesting questions (consider cryptography, for example). Would anyone want to teach about differential equations and dynamical systems today without showing what computer simulations and computer graphics can convey about the qualitative descriptions of systems? It is ardently to be hoped not. Perhaps *chaos* has become too much of a popular buzz-word, but it would not have caught the popular imagination without computer graphics. Mathematicians should take more advantage of this publicity, since it is notoriously difficult to give popular descriptions of abstract mathematics. Inasmuch as student numbers are decreasing, why not try to attract more people to the study of mathematics by making it more exciting? For example, the author believes that a course in complex variable theory could be made quite interesting by using computers (say, with *Mathematica*), but unfortunately he has never had the opportunity to try this out. The point is that a textbook can contain only so many pictures. With a computer, hundreds of pictures of difficult transcendental functions can be made. Complex functions seem to be an ideal subject for mathematical lab work.

☐ *Won't computer packages for calculus lead, as they have in statistics, to much meaningless calculation? If computers handle routine calculations, what will students do instead? What are appropriate prerequisites for computer-based calculus courses? Should mathematics be a lab science?* The author does not know how much "meaningless calculation" has gone on in Statistics, but that subject certainly *needs* computers! It is the responsibility of the instructors to provide meaningful problems that take full advantage of computers. It is no longer necessary to "cook the parameters" so the answer turns out to be $1/3$ these days. It is no longer necessary to keep all algebraic expressions down to degree 4 or lower. It is no longer necessary to slog through mammoth tables of integrals. But it is necessary to know what you are doing just to avoid meaningless activities. The elimination of the routine can free the students to explore and to think, but they need to be shown how to make use of the computer as a tool in this way. It takes work to do this, and departments need to find ways to make it possible for interested professors to do this necessary work.

Does a computer-based calculus course need special prerequisites? The author would guess that Bill Davis (Ohio State University), Horacio Porta, and Jerry Uhl (University of Illinois at Urbana-Champaign) who designed over many years a course "Calculus & *Mathematica*" (published in 1994 by Addison-Wesley Publishing Company, Inc.) would say not. It does not take all that long (perhaps about three weeks) to become comfortable with the computer, and then the computer is *transparent* — all the emphasis can be put on the subject matter. Davis, Porta, and Uhl said in one of their introductions to their course:

> "Just as arithmetic is the introduction to *counting*, calculus is the introduction to *measurement* —both exact and approximate. Calculus is a tool-kit of measurement devices: derivatives, differential equations, definite integrals, series expansions and approximations, data fitting, linearizations, vectors, dot products, line integrals, gradients, curls, and divergences. Today's popoular texts present most of the tools but end up focusing too much on the tools and not enough on their *uses*. The successful calculus course produces students who recognize a calculus situtaion and who know what calculus tools to bring to bear on it. This is the goal of Calculus & *Mathematica*."

The author has not taught from their (extensive) materials and can offer no well-founded judgement as to whether they have truly reached their goal. Brief examinations of the text and computer notebooks, and exposure to presentations by those authors about their course suggests that they have. But the only real proof is in actually teaching the course. In any case, they expect that students will be as ready to start their course as any standard calculus course.

And what about *labs*? Davis, Porta, and Uhl also said:

> "Why has the availability of computers and calculators not yet resulted in significant changes in mathematics instruction? Computer advertisements indicate that driving a truck full of computers to school and unloading them at the mathematics classroom door will result in miracles. But, in case after case, high expectations have dissolved when the computers failed to deliver the expected miracles. The fault lies not in the computers themselves. The fault lies in the fact that the already overburdened individual instructor has to figure out what to do with the machines. The expedient and usual response has been to add a lab onto an existing course to allow the students the opportunity to do problems from the text by machine instead of by pencil and paper. This response may stymie the students' thinking process. Students have been taught, in the lecture part of the course, to do the procedures of mathematics by hand. Then, in the lab, they are shown how to do the same mathematics automatically by computer. They may think: Why learn the hand operations when the computer does them automatically? "

What they caution about here is the *disconnecting* of the lab form the main thread of the course. Mathematics can only be a lab science if the lab can be used for exploration — and it has to be set up to do so. In planning a course one also has to ask:

How much lecturing is called for? Davis, Porta, and Uhl have their own answer:

> "In the typical model of C&M instruction, new material is never presented in lecture form. Students meet with their instructor once a week for a discussion of the week's assignment and the relevant Literacy Sheet. There is an implementation of C&M at sixteen rural Illinois high schools in which the only discussions are among the students themselves, because the schools cannot fund full-time instruction in calculus. Anecdotal observations indicate that C&M instructors who insist on introducing ideas via lecture find that difficulties increase in direct proportion to the size of the lecture component of their course. Perhaps C&M students have a lot in common with Winston Churchill who said: *Personally I am always ready to learn, although I do not always like being taught.*"

In other words for, those authors it is *all labs! and no lectures!* This professor does not agree. But, giving good lectures is as difficult as giving good labs. Undoubtedly different courses will need different arrangements.

The experience the author had in the Spring of 1996 in giving lectures was, however, *very discouraging*. It took a lot of physical effort to set up the computer for each lecture. The computer projector was not bright enough to use without turning out most of the lights in the large lecture room. When the lights were out, a small desk lamp was necessary to be able even to see the computer keyboard. The class was too large. The air conditioning had been turned off in an attempt by the University to save money (under the assumption that no cool and fresh air was needed during the winter). Lots of students went to sleep during lectures. The geometry of the hall made it difficult to have contact with the students and to answer questions. It was impossible to write on a blackboard or use an overhead projector while the computer projector was on. As a result of this torture, the author will refuse to lecture in that room ever again if a computer is to be used! Institutions must give more attention to the design of classrooms.

The students complained that the lecturer from time to time was "just showing them how to use *Mathematica*." This was not true. What the lecturer was trying to show then was that it was *possible* to use *Mathematica* — in hopes that the students would do it themselves. Some did; others copied homework from the more enterprising. Since the material for the course was continuously undergoing a complete revision during the term, the lecturer made the mistake of putting *too much material* into the lectures. This was the real reason, he thinks, why the students were not getting the ideas: there was too much to take in every session. Now that the course has been captured in a computer-held form, it would be possible to segment the topics better to make the work of each week more focussed and more enjoyable. But there is still a nagging question about lectures.

It is quite possible that a solid 50-minute period for a lecture to first-year students on mathematics is not advisable. The lecture time should perhaps be broken up into various activities. Or a schedule of three 80-minute directed lab sessions (in smaller groups) including some lecturing, some question answering, some individual problem solving, and some exam post morta would be more effective. That requires more instructor manpower, however. Or perhaps the Illinois model of no lectures whatsoever is the best plan. How can we decide? In the first place the class materials needs to be prepared and tested by use. Then, some kind of controlled experiments have to be run, putting groups of students into different environments. But who is going to do this, and who is going to evaluate the results? What are the criteria for evaluation. Most professors and most departments do not have time (or energy) for elaborate experiments. Even worse, experiments in teaching are not projects leading to tenure for younger staff — who are often the most lively and with the best rapport with the first-year students. And so we bumble on from year to year, sometimes passing through a crisis of "reform" and often letting the pendulum swing back and forth under its own momentum. But the computer is catching up to us, and many of us will soon be out of date and out of place in a new environment of communication.

■ 4. Some Conclusions and Suggestions

The author has often been asked: *Why use Mathematica? Is it not too expensive and too unreliable? And as a programming language is it not too unstructured and inefficient?* The exageration of its reputation for unreliability is overdone in the author's opinion; however, when the new version 3.0 is released this year some very serious reviews must take place, since Wolfram Research has now had a substantial period to correct and refine the program. The expense of the software is indeed a problem, and again a review should be carried out to determine whether the investment is worth the results. But many books now use Mathematica programs and examples, and one has to ask whether a college can afford not to subscribe.

The author found the value of the software to be in the combination of these features: *Mathematica* is at the same time

A programming language
An interactive computing system
A symbolic-numeric-computation engine
A text-processing program
A presentation manager

Perhaps **Maple** could fulfill all these requirements, but the author found that the ease of implementing *rule-rewriting systems* in *Mathematica* led to many applications that students could be taught to do for themselves. But again, some criteria for evaluation of features is needed.

Wolfram Research is promising a revolution in typesetting that will bring *two-dimensional formulae* to interactive computing. We shall have to see how it works. What is definitely needed for the future is better *file management.* If one has only five files (of modest size), once can remember fairly well what is in them. However, when the number of files gets up to 25, and when several of the files are changing every day, it becomes impossible to keep track of what is where. *Mathematica* 3.0 will make notebooks *Mathematica objects*, and so it will become possible to write *Mathematica* programs that process notebooks. This should make dynamic indexing feasible, but it may take some third-party developments to create environments in which courseware development is less time consuming and vexing. As it is, the text-processing capabilities of the current version of *Mathematica* (Version 2.2.2) — which on the Macintosh we already avaiable in 1.0 — were absolutely essential to the author for writing course notes and lectures notebooks. It takes a lot of writing to put a course in shape, and the computer has to make the writing as easy as possible. We can thank Theo Gray for doing an excellent job with the *Mathematica* Front End from the start, and we can look forward to improvements and enhancements. And, to be sure, improvements (especially in printing) are to be desired.

As for programming, there are many dilemmas. *Mathematica* is an interpreted language, and so it can become very slow on larger problems. There is a facility to link Mathematica to compiled programs in the C language, and Klaus Sutner, for one, has made very good use of this facility in implementing algorithms for processing finite automata with large numbers of states. Unfortunately, there are still some problems in using the linking facility with some operating systems, and more implementational problems have to be solved. Also, the creation of environments that can be used by less expert programmers has to be faced. This is one sort of programming problem. Another problem is that Mathematica is an untyped language. This makes some programming easier, but it makes other kinds of programs that require the dynamic checking of features of data structures harder. Some thought should be given to finding a middle way of constructing programs. Or, perhaps, another generation of implementations of symbolic computation software will allow for better compatibility with "industrial grade" programming languages. In the meantime, the author has found that *Mathematica* is a satisfactory language for *prototyping* algoritms for didactic purposes, and so he hopes that he is making students think. If we can give them better tools in the future, so much the better.

Another criticism often heard in using software is: *The medium is becoming the message.* The author both resents and rejects this criticism as regards his own course developments as explained earlier. For him the medium is definitely not the message! But we have to remember that any medium should be well chosen to get the message over well. If one does not understand the medium, however, the message may be wrong. And if one does not understand the *scope* of the medium, the message may be too simple. Therefore, it is necessary to become fairly expert in the medium in order to get the best results. And the students must gain suitable expertise as well to be able to do their work. The makers of software need to keep this problem in mind and provide suitable guides, so that new users can obtain the necessary level of expertise as quickly as possible. One guide (or handbook) may not be suitable for all possible users.

As a medium for mathematics, the computer can truly become *microscope* and a *telescope* to see mathematical phenomena invisible by ordinary means. But the equipment and methods needs good "tuning" so that the "pictures" are neither blurred, distorted, or irrelevant. When the "focus" is good, the coursework and publications already available in many different mathematical subjects do — in the author's considered opinion — show that the money and effort is worth the gain. But we need to do more with the use of computers, and we need to do it better.

Analytica — An Experiment in Combining Theorem Proving and Symbolic Computation *

Andrej Bauer, Edmund Clarke and Xudong Zhao

Carnegie Mellon University
School of Computer Science
5000 Forbes Avenue
Pittsburgh, PA 15213, USA

Abstract Analytica is an automatic theorem prover for theorems in elementary analysis. The prover is written in Mathematica language and runs in the Mathematica environment. The goal of the project is to use a powerful symbolic computation system to prove theorems that are beyond the scope of previous automatic theorem provers. The theorem prover is also able to guarantee the correctness of certain steps that are made by the symbolic computation system and therefore prevent common errors like division by a symbolic expression that could be zero.

We describe the structure of Analytica and explain the main techniques that it uses to construct proofs. Analytica has been able to prove several non-trivial theorems. In this paper, we show how it can prove a series of lemmas that lead to Bernstein approximation theorem.

1 Introduction

Current automatic theorem provers, particularly those based on some variant of resolution, have concentrated on obtaining ever higher inference rates by using clever programming techniques, parallelism, etc. We believe that this approach is unlikely to lead to a useful system for actually doing mathematics. The main problem is the large amount of domain knowledge that is required for even the simplest proofs. In this paper, we describe an alternative approach that involves combining an automatic theorem prover with a symbolic computation system. The theorem prover, which we call *Analytica*, is able to exploit the mathematical knowledge that is built into this symbolic computation system. In addition, it can guarantee the correctness of certain steps that are made by the symbolic computation system and, therefore, prevent common errors like division by an expression that may be zero.

---
* This research was sponsored in part by the National Science Foundation under grant no. CCR-8722633, by the Semiconductor Research Corporation under contract 92-DJ-294, and by the Wright Laboratory, Aeronautical Systems Center, Air Force Materiel Command, USAF, and the Advanced Research Projects Agency (ARPA) under grant F33615-93-1-1330.

Analytica is written in the Mathematica programming language and runs in the interactive environment provided by this system [19]. Since we wanted to generate proofs that were similar to proofs constructed by humans, we have used a variant of the sequent calculus [10, 11] in the inference phase of our theorem prover. However, quantifiers are handled by skolemization instead of explicit quantifier introduction and elimination rules. Although inequalities play a key role in all of analysis, Mathematica is only able to handle very simple numeric inequalities. We have developed a technique that is complete for linear inequalities and is able to handle a large class of non-linear inequalities as well. This technique is more closely related to the BOUNDER system developed at MIT [17] than to the traditional SUP-INF method of Bledsoe [5]. Another important component of Analytica deals with expressions involving summation and product operators. A large number of rules are devoted to the basic properties of these operators. We have also integrated Gosper's algorithm for hypergeometric sums with the other summation rules, since it can be used to find closed form representations for a wide class of summations that occur in practice.

Analytica is able to prove several non-trivial examples, such as the basic properties of the stereographic projection [8], the theorems in the second chapter of Ramanujan's Collected Works [2], and a series of three lemmas that lead to a proof of Weierstrass's example of a continuous nowhere differentiable function [8]. In the appendix we show how Analytica can prove four lemmas from which the Bernstein approximation theorem follows.

There has been relatively little research on theorem proving in analysis. Bledsoe's work in this area [3, 4] is certainly the best known. Analytica has been heavily influenced by his research. More recently, Farmer, Guttman, and Thayer at Mitre Corporation [9] have developed an interactive theorem prover for analysis proofs that is based on a simple type theory. Neither of these uses a symbolic computation system for manipulating mathematical formulas, however. Suppes and Takahashi [18] have combined a resolution theorem prover with the *Reduce* system, but their prover is only able to check very small steps and does not appear to have been able to handle very complicated proofs. London and Musser [15] have also experimented with the use of Reduce for program verification.

2 An overview of Analytica

Analytica consists of four different phases: skolemization, simplification, inference, and rewriting. When a new formula is submitted to Analytica for proof, it is first skolemized to a quantifier free form. Then it is simplified using a collection of algebraic and logical reduction rules. If the formula reduces to true, the current branch of the inference tree terminates with success. If not, the theorem prover checks to see if the formula matches the conclusion of some inference rule. If a match is found, Analytica will try to establish the hypothesis of the rule. If the hypothesis consists of a single formula, then it will try to prove that formula. If the hypothesis consists of a series of formulas, then Analytica will attempt to prove each of the formulas in sequential order. If no inference rule is applicable, then various rewrite rules are used attempting to convert the formula to another equivalent form. If the rewriting phase is unsuccessful, the search terminates in failure; otherwise the simplification, inference and rewriting phases will repeat with the new formula. Backtracking will cause the entire inference tree to be searched before the proof of the original goal formula terminates with failure.

2.1 Skolemization phase

In Analytica (as in Bledsoe's *UT Prover* [3]), we use skolemization to deal with the quantifiers that occur in the formula to be proved. Initially, quantified variables are standardized so that each has a unique name. We define the *position* of a quantifier within a formula as *positive* if it is in the scope of even number of negations, and *negative* otherwise. *Skolemization* consists of the following procedure: Replace $(\exists x.\Psi(x))$ at positive positions or $(\forall x.\Psi(x))$ at negative positions by $(\Psi(f(y_1, y_2, ..., y_n)))$ where $x, y_1, y_2, ..., y_n$ are all the free variables in $\Psi(x)$ and f is a new function symbol, called a *skolem function*. The original formula is satisfiable if and only if its skolemized form is satisfiable. Thus, X is valid if and only if X' is valid where $\neg X'$ is the skolemized form of $\neg X$ [10]. We call $\neg skolemize(\neg f)$ the *negatively skolemized form* of f . A formula is valid if and only its negatively skolemized form is valid. When a negatively skolemized formula is put in prefix form, all quantifiers are existential. These quantifiers are implicitly represented by marking the corresponding quantified variables. The marked variables introduced by this process are called *skolem variables*. The resulting formula will be quantifier-free. For example, the skolemized form of the formula

$$(\exists x.\forall y.P(x,y)) \rightarrow (\exists u.\forall v.Q(u,v))$$

is given by

$$P(x, y_0(x)) \rightarrow Q(u_0(), v),$$

while its negatively skolemized form is

$$P(x_0(), y) \rightarrow Q(u, v_0(u)).$$

where x, y, u and v are skolem variables, and u_0, v_0, x_0, y_0 are skolem functions. Although formulas are represented internally in skolemized form without quantifiers, quantifiers are added when a formula is displayed so that proofs will be easier to read.

2.2 Simplification phase

Simplification is the key phase of Analytica. A formula is simplified with respect to its *proof context*. Intuitively, the proof context consists of the formulas that may be assumed true when the formula is encountered in the proof. The formula that results from simplifying f under context C is denoted by $simplify(f, C)$. In order for the simplification procedure to be sound, $simplify(f, C)$ must always satisfy the the following condition

$$C \models simplify(f, C) \leftrightarrow f.$$

The initial context C_0 in each simplification phase is a conjunction of all of the *given* properties of the variables and constants in the theorem. The initial formula in each simplification phase is the current goal of the theorem prover. In the first simplification phase it is the result of the skolemization phase. In each subsequent simplification phase it is the result of the previous rewriting phase. The simplification procedure for composite formulas is given by the following rules:

1. $simplify(f) = simplify(f, C_0)$
2. $simplify(f_1 \wedge f_2, C) = f_1' \wedge simplify(f_2, C \wedge f_1')$
   where $f_1' = simplify(f_1, C \wedge f_2)$
3. $simplify(f_1 \vee f_2, C) = f_1' \vee simplify(f_2, C \wedge \neg f_1')$
   where $f_1' = simplify(f_1, C \wedge \neg f_2)$
4. $simplify(f_1 \to f_2, C) = f_1' \to simplify(f_2, C \wedge f_1')$
   where $f_1' = simplify(f_1, C \wedge \neg f_2)$
5. $simplify(\neg f, C) = \neg simplify(f, C)$

The soundness of these rules can be easily established by structural induction. For example, if the soundness condition holds for f_1 and f_2, it will also hold for $f_1 \wedge f_2$, etc.

A large number of rules are provided for simplifying atomic formulas (i.e., equations and inequalities) using context information. Some examples of rules for simplifying inequalities are given in Section 5. In addition to the equation and inequality rules, special simplification rules are included to handle functions that are frequently used, such as Abs, Min, Max, Sum, Product, Limit, etc. The simplification of summations and products is discussed in detail in Section 3.

2.3 Inference phase

The inference phase is based on the *sequent calculus* [11]. We selected this approach because we wanted our proofs to be readable. Suppose that f is the formula that we want to prove. In this phase we attempt to find an instantiation for the skolem variables that makes f a valid ground formula. In order to accomplish this, f is decomposed into a set of *sequents* using rules of the sequent calculus. Each sequent has the form $\Gamma \vdash \Delta$, where Γ and Δ are initially sets of subformulas of f. The formula f will be proved, if substitution can be found that makes all of the sequents valid. A sequent $\Gamma \vdash \Delta$ is valid if it is impossible to make all of the elements of Γ true and all of the elements of Δ false.

In Analytica, the function $FindSubstitution(f)$ is used to determine the appropriate substitution for f. If f is not provable, $FindSubstitution(f)$ will return *Fail*. *FindSubstitution* has rules corresponding to each of the rules of the sequent calculus except those concerning quantifiers. The two rules for implication are given as examples:

1. Implication on the left:
   $FindSubstitution(\Gamma, A \to B, \Delta \vdash \Lambda) = \sigma_1 \sigma_2$ where
   $\quad \sigma_1 = FindSubstitution(\Gamma, \Delta \vdash A, \Lambda)$, and
   $\quad \sigma_2 = FindSubstitution(\Gamma \sigma_1, B \sigma_1, \Delta \sigma_1 \vdash \Lambda \sigma_1)$.

2. Implication on the right:
   $FindSubstitution(\Gamma \vdash \Delta, A \rightarrow B, \Lambda) = FindSubstitution(\Gamma, A \vdash \Delta, B, \Lambda)$

Rules are also needed for atomic formulas. The three below are typical.

1. Equation: $FindSubstitution(\Gamma \vdash \Delta, a = b, \Lambda) = \sigma$ where $a\sigma = b\sigma$.
2. Inequality: $FindSubstitution(\Gamma, a < b, \Delta \vdash \Lambda) = \sigma$ where $a\sigma = b\sigma$.
3. Matching: $FindSubstitution(\Gamma, A, \Delta \vdash \Lambda, B, \Theta) = \sigma$ where $A\sigma = B\sigma$.

Backtracking is often necessary in the inference phase when there are multiple subgoals, because a substitution that makes one subgoal valid may not make another subgoal valid. When this happens it is necessary to find another substitution for the first subgoal. In order to restart the inference phase at the correct point, a stack is added to the procedure described above. When a rule is applied that may generate several subgoals, one subgoal is selected as the current goal and the others are saved on the stack. If some substitution σ makes the current subgoal valid, then σ is applied to the other subgoals on the stack and Analytica attempts to prove them. If the other subgoals are not valid under σ, then Analytica returns to the previous goal and tries to find another substitution that makes it valid.

Special tactics are included in the inference phase for handling inequalities and constructing inductive proofs. The tactic that is used for inequalities is described in detail in Section 5 and will not be discussed further here. The induction tactic enables Analytica to select a suitable induction scheme for the formula to be proved and attempts to establish the basis and induction steps. A typical induction scheme is

$$f(n_0) \wedge \forall n(n \geq n_0 \wedge f(n) \rightarrow f(n+1)) \Rightarrow \forall n(n \geq n_0 \rightarrow f(n))$$

In this case, we need only to identify the induction variable n and determine the base value for n. In order to find a suitable induction variable for formula f, we list all variables that appear in f and select those that have type integer. To reduce the search space, we would like to make sure that our choice of the induction variable is a good one. The choice is good if the induction hypothesis is useful for proving the induction conclusion. This will be more likely if the terms that appear in the induction conclusion appear either in the induction hypothesis or in the current context. Hence, we arrive at the following heuristic for selecting the induction variable: Use n as the induction variable to prove $f(n)$ provided that $f(n+1)$ only contains terms that already appear in $f(n)$ or in the current context. Once the induction variable n has been selected, a base value for that variable must be found in order to start the induction. In Analytica, a suitable base value may be determined by calculating the set of lower bounds of n as described in Section 5 and choosing the simplest element of this set. If the basis case fails for this value, Analytica will choose another base value and try again until the basis is proven or no other choice is available. In the former case, the induction step is tried; otherwise the induction scheme fails and Analytica will try other techniques like those in the rewriting phase. This strategy is used in the constructing the induction proof for the second example in Section 2.

2.4 Rewrite phase

Five rewriting tactics are used in Analytica:

1. When the left hand side of an equation in the hypothesis appears in the sequent, it is replaced by the right hand side of the equation. For example,

$$\sum_{k=0}^{n} \frac{2^k}{1+m^{2^k}} = \frac{1}{-1+m} + \frac{2 \cdot 2^n}{1-m^{2 \cdot 2^n}} \implies$$

$$\frac{2 \cdot 2^n}{1+m^{2 \cdot 2^n}} + \left(\sum_{k=0}^{n} \frac{2^k}{1+m^{2^k}}\right) = \frac{1}{-1+m} + \frac{4 \cdot 2^n}{1-m^{4 \cdot 2^n}}$$

substitute using equation

$$\sum_{k=0}^{n} \frac{2^k}{1+m^{2^k}} = \frac{1}{-1+m} + \frac{2 \cdot 2^n}{1-m^{2 \cdot 2^n}} \implies$$

$$\frac{2 \cdot 2^n}{1+m^{2 \cdot 2^n}} + \frac{1}{-1+m} + \frac{2 \cdot 2^n}{1-m^{2 \cdot 2^n}} = \frac{1}{-1+m} + \frac{4 \cdot 2^n}{1-m^{4 \cdot 2^n}}$$

2. Rewrite a trigonometric expression to an equivalent form.
   Given that a is an odd integer, k, m, n are integers, $m \leq n$,

   $$- \cos(\pi a^n x) + (-1)^k \cos(\pi a^{-m+n}(a^m x - k)) = 0$$

   rewrite trigonometric expressions

   True

3. Move all terms in equations or inequalities to left hand side and factor the expression.

   $$\frac{(-1+x_3)^2 (-1+y_2^2 + y_3^2)}{(-1+y_3)^2} = -1 + x_3^2 + \frac{(-1+x_3)^2 y_2^2}{(-1+y_3)^2}$$

   rewrite as

   $$\frac{2(-1+x_3)(x_3 - y_3)}{-1+y_3} = 0$$

4. Solve linear equations.

   $$c + bx + ax^2 = 0 \wedge c + by + ay^2 = 0 \implies x - y = 0 \vee b + a(x+y) = 0$$

   solve linear equation

   $$c = -(x(b+ax)) \wedge c = -(y(b+ay)) \implies x - y = 0 \vee b + a(x+y) = 0$$

5. Replace a user defined function by its definition. In the example below the user defined function S is expanded.

   $$0 < \pi a^m b^m + (1 - ab)\,\mathrm{Abs}(S(m))$$

   expand definition

   $$0 < \pi a^m b^m + (1 - ab)\,\mathrm{Abs}\left(\sum_{n=0}^{-1+m} \frac{b^n(-\cos(\pi a^n x) + \cos(\pi a^n(x+h)))}{h}\right)$$

3 Summation

Summations play an important role in symbolic computation. Nevertheless, Mathematica's ability to handle summations is very limited. A summation with range from n_1 to n_2, where n_1 and n_2 are integers and $n_1 \leq n_2$, is explicitly expanded into a sum with $n_2 - n_1 + 1$ terms. However, a summation with a symbolic range will not be simplified. Consequently, we have introduced a large number of special rules for dealing with summations. Although most of the rules are based on simple identities, Analytica is able to handle a large range of summations in example proofs. Analogous rules for products are also included in Analytica. A few of the rules for summation are listed below. The rules are partitioned into three sets.

1. The first set of rules reduces the number of summations occurring in the expression to be simplified.

   $\sum_{n=n_1}^{n_2} c = c(n_2 - n_1 + 1)$ where c is a constant

   $\sum_{n=n_1}^{n_2} f_1(n) + \sum_{n=n_1}^{n_2} f_2(n) = \sum_{n=n_1}^{n_2} (f_1(n) + f_2(n))$

   $\sum_{n=n_1}^{n_2} f(n) + \sum_{n=n_2+1}^{n_3} f(n) = \sum_{n=n_1}^{n_3} f(n)$

   $\sum_{n=n_1}^{n_3} f(n) - \sum_{n=n_1}^{n_2} f(n) = \sum_{n=n_2+1}^{n_3} f(n)$

   $\sum_{n=n_1}^{n_3} f(n) - \sum_{n=n_2}^{n_3} f(n) = \sum_{n=n_1}^{n_2-1} f(n)$

2. The second set does not change the number of summations, but simplifies summands.

   $\sum_{n=n_1}^{n_2} cf(n) = c \sum_{n=n_1}^{n_2} f(n)$ where c is a constant

   $\sum_{k=n_1}^{n_2} f(k+1) = \sum_{k=n_1+1}^{n_2+1} f(k)$

   $\sum_{k=n_1}^{n_2} f(k-1) = \sum_{k=n_1-1}^{n_2-1} f(k)$

3. The third set does not change the number of summations or the summands, but simplifies the ranges.

   $\sum_{n=n_1}^{n_2} f(n) = - \sum_{n=n_2+1}^{n_1-1} f(n)$ if $n_1 > n_2$

   $\sum_{n=n_1}^{n_2+N} f(n) = \left(\sum_{n=n_1}^{n_2} f(n) \right) + f(n_2 + 1) + ... + f(n_2 + N)$

   $\sum_{n=n_1}^{n_2-N} f(n) = \left(\sum_{n=n_1}^{n_2} f(n) \right) - f(n_2) - ... - f(n_2 - N + 1)$

   where N is positive integer

In many examples, it would be helpful if we could obtain a closed form representation for some summation. *Gosper's algorithm* is able to compute such a representation for a large class of summations. Consequently, we have also integrated this method into our theorem prover. A function g is said to be a *hypergeometric function* if $g(n+1)/g(n)$ is a rational function of n. Gosper's algorithm is able to find a closed form for the series $\sum_{k=1}^{n} a_k$ when each a_k is a hypergeometric function and there is a hypergeometric function g that satisfies $g(n) = \sum_{k=1}^{n} a_k + g(0)$ [14].

4 Inequalities

Inequalities play a key role in all areas of analysis. Since Mathematica does not provide any facility for handling inequalities, we have built several techniques into Analytica for reasoning about them.

4.1 Simplification of inequalities

There are many rules that simplify atomic formulas involving inequalities. However, we only include four examples.

1. $simplify(0 \leq a^n, C) = True$ if $simplify(0 < a, C) = True$
2. $simplify(0 < a^n, C) = True$ if $simplify(0 < a, C) = True$
3. $simplify(a^n \leq 0, C) = False$ if $simplify(0 < a, C) = True$
4. $simplify(a^n < 0, C) = False$ if $simplify(0 < a, C) = True$

There are also rules that use upper and lower bound information to simplify inequalities. If a has a negative upper bound, then $a < 0$ is true, while $a > 0$ and $a = 0$ are both false. The function *Lower(Upper)* gives a set of lower(upper) bounds for its argument and will be discussed in Section 5.3. The set of lower(upper) bounds is calculated in the current context.

1. $simplify(f_1 \leq f_2, C) = False$ if $\exists x[x \in Lower(f_1 - f_2, C) \wedge x > 0]$.
2. $simplify(f_1 \leq f_2, C) = True$ if $\exists x[x \in Lower(f_2 - f_1, C) \wedge x \geq 0]$.
3. $simplify(f_1 < f_2, C) = True$ if $\exists x[x \in Lower(f_2 - f_1, C) \wedge x > 0]$.
4. $simplify(f_1 < f_2, C) = False$ if $\exists x[x \in Lower(f_1 - f_2, C) \wedge x \geq 0]$.

4.2 Proof strategy for Inequalities

Although many inequality formulas can be handled in the simplification phase, some valid inequality formulas cannot be reduced to true in this phase. For example, $(a \leq 0 \wedge b \leq a) \rightarrow b \leq 0$ cannot be proved by the technique used in simplification phase alone. Other more powerful techniques for deciding satisfiability of inequality formulas must be used in addition. If the inequality $a \leq b$ is not directly provable using the techniques in the simplification phase, then Analytica will try to find a term c, such that $a \leq c$ and $c \leq b$ are both provable in the current context. In order to find such a term c, we compute a set of upper bounds for a and a set of lower bounds for b by using information provided by the current context. The sets computed are denoted by $Upper(a)$ and $Lower(b)$, respectively. A term x will be in $Upper(a)$ only if $a <= x$ is true in the current context. Likewise, x will be in $Lower(b)$ only if $x <= b$ is true in the current context. To prove $a \leq b$, it is sufficient to prove that there is some $c \in Upper(a)$ such that $c \leq b$ is true or that there is some $c \in Lower(b)$ such that $a \leq c$ is true.

In order to deal with strict inequalities, we introduce a new symbol S such that both $S_L(a) \leq b$ and $a \leq S_U(b)$ are equivalent to $a < b$. Hence, $S_U(x) \in Upper(a)$ only if $a < x$ is true in the current context, and $S_L(x) \in Lower(a)$ only if $x < a$ is true in the current context. $S_U(a) + b = S_U(a + b)$ because $c \leq S_U(a + b)$ iff $c < a + b$ iff $c - b < a$ iff $c - b \leq S_U(a)$ iff $c \leq S_U(a) + b$. Similarly, $S_L(a) + b = S_L(a + b)$, $-S_L(a) = S_U(-a)$ and $-S_U(a) = S_L(-a)$, etc. This convention permits both strict inequalities and nonstrict inequalities to be handled by the same method.

The technique is complete for linear inequalities, and it can also prove many of the nonlinear inequalities that arise in practice. The technique is not complete for nonlinear inequalities, however.

4.3 Calculating upper and lower bounds for expressions

There are three main ways of obtaining upper and lower bounds for expressions.

1. Obtain bounds from context information:
   Upper and lower bounds for an expression are calculated in the current context. For example, when proving $(a \leq b) \vee c$, the upper bounds of a and the lower bounds of b are calculated under the context of $\neg c$. In general, If $a \leq b$ is a conjunct of the current context, we have

   $$a \in Lower(b), \quad b \in Upper(a),$$

   and if $a < b$ is a conjunct of the current context, we have

   $$S_L(a) \in Lower(b), \quad S_U(b) \in Upper(a).$$

2. Obtain bounds from the monotonicity of some function:
   If f is a monotonically increasing function, and a' is an upper(lower) bound of a, $f(a')$ is an upper(lower) bound of $f(a)$; if f is a monotonically decreasing function and a' is an upper(lower) bound of a, $f(a')$ is a lower(upper) bound of $f(a)$. For example:

   $$\{cx | x \in Upper(a)\} \subseteq Lower(ca), \text{ if } c \leq 0$$

3. Use some known bound on the value of a function:
   If f is bounded, i.e. for all x, $f(x) \leq h(x)$, or $f(x) \geq g(x)$, $h(x)$ is an upper bound for $f(x)$ and $g(x)$ a lower bound for $f(x)$. For example:

   $$x + \frac{1}{2} \in Upper(round(x))$$

   $$x - \frac{1}{2} \in Lower(round(x))$$

5 The Bernstein Approximation Theorem

In this section we show how a non-trivial theorem from real analysis can be proved with Analytica. The proof needs to be broken down into several lemmas. First we state the theorem and provide a manual proof. Then we look at how Analytica proves the same theorem.

Theorem: Let $I = [0,1]$ be the closed unit interval and f a real continuous function on I. Define the n-th Bernstein polynomial for f as

$$B_n(x, f) = \sum_{k=0}^{n} \binom{n}{k} f(k/n) \, x^k \, (1-x)^{n-k}.$$

On the interval I, the sequence of Bernstein polynomials for f converges uniformly to f.

Proof: It follows from the binomial theorem that

$$f(x) = \sum_{k=0}^{n} \binom{n}{k} f(x) x^k (1-x)^{n-k},$$

hence

$$|f(x) - B_n(x)| \le \sum_{k=0}^{n} \binom{n}{k} |f(x) - f(k/n)| x^k (1-x)^{n-k}. \tag{1}$$

Since f is continuous on a closed interval, it is uniformly continuous and bounded. Let $M = \sup\{|f(x)| \; ; \; x \in I\}$.

For every $\epsilon > 0$, there is $\delta(\epsilon) > 0$ such that $|x - y| < \delta(\epsilon)$ implies $|f(x) - f(y)| < \epsilon$ for all $x, y \in I$.

Let $\epsilon > 0$ be arbitrary. Choose n larger than $\delta(\epsilon/2)^{-4}$ and $(M/\epsilon)^2$. We show that the sum on the right side of (1) is less than ϵ. In order to do this, split the sum into two parts:

- *Near* is the sum of those terms in (1) for which $|x - k/n| < n^{-1/4}$, i.e.

$$Near = \sum_{\substack{0 \le k \le n \\ |x-k/n| < n^{-1/4}}} \binom{n}{k} |f(x) - f(k/n)| x^k (1-x)^{n-k}$$

- *Far* is the sum of those terms in (1) for which $|x - k/n| \ge n^{-1/4}$, i.e.

$$Far = \sum_{\substack{0 \le k \le n \\ |x-k/n| \ge n^{-1/4}}} \binom{n}{k} |f(x) - f(k/n)| x^k (1-x)^{n-k}$$

We bound *Near* and *Far* above by $\epsilon/2$. Then the result follows.

For showing that $Near \le \epsilon/2$, note that since the sum involves those terms for which $|x - k/n| < n^{-1/4} \le \delta(\epsilon/2)$, we have $|f(x) - f(k/n)| \le \epsilon/2$, and hence

$$Near \le \frac{\epsilon}{2} \sum_{\substack{0 \le k \le n \\ |x-k/n| < n^{-1/4}}} \binom{n}{k} x^k (1-x)^{n-k} \le \frac{\epsilon}{2} \sum_{k=0}^{n} \binom{n}{k} x^k (1-x)^{n-k} = \frac{\epsilon}{2}$$

To show that $Far \le \epsilon/2$, bound $|f(x) - f(y)| \le 2M$ and note that in *Far* we have $\sqrt{n}\,(x - k/n)^2 \ge 1$. Hence, in *Far*, the inequality

$$|f(x) - f(y)| \le 2M\sqrt{n}(x - k/n)^2$$

is valid. We also need the identity

$$\sum_{k=0}^{n} \binom{n}{k} \left(x - \frac{k}{n}\right)^2 x^k (1-x)^{n-k} = \frac{x(1-x)}{n}. \tag{2}$$

Now we can bound *Far*:

$$Far \leq 2M\sqrt{n} \sum_{\substack{0 \leq k \leq n \\ |x-k/n| < n^{-1/4}}} (x - \frac{k}{n})^2 \binom{n}{k} x^k (1-x)^{n-k}$$

$$\leq 2M\sqrt{n} \sum_{k=0}^{n} (x - \frac{k}{n})^2 \binom{n}{k} x^k (1-x)^{n-k}$$

$$= 2M\sqrt{n}\frac{x(1-x)}{n} \leq \frac{M}{2\sqrt{n}} \leq \epsilon/2$$

This completes the proof.
□

We help Analytica by specifying the following facts as "given":

$$0 < \epsilon$$
$$0 < m$$
$$0 < x \wedge x < 1$$
$$m \leq \delta(\epsilon/2)$$
$$m^2 \leq \frac{\epsilon}{M}$$

Here m is the quantity $n^{-1/4}$ from the above proof. We only considered the cases that $x \neq 0$ and $x \neq 1$ in the proof. Since $B(n,0) \equiv f(0)$ and $B(n,1) \equiv f(1)$, the other two cases are trivial.

We also provide the definitions of $B(n,x)$, the sums *Near* and *Far*, which are called $near(x,n)$ and $far(x,n)$, and the intermediate sum $far1(x,n)$ that appears in the proof of $Far \leq \epsilon_2$. The proof is broken down into five lemmas:

1. $near(x,n) \leq \epsilon/2$
2. $far1(x,n) \leq \epsilon/2$
3. $far(x,n) \leq far1(x,n)$
4. $|f(x) - B(n,x)| \leq far(x,n) + near(x,n)$
5. $|f(x) - B(n,x)| \leq \epsilon$

The input to Analytica for the theorem and the automatically produced proof of lemma 1 and lemma 4 are shown in the appendices.

6   Conclusion

In a related project that we plan to describe in a forthcoming paper, we have managed to prove all of the theorems and examples in Chapter 2 of Ramanujan's Collected Works[2] completely automatically. The techniques that we use are similar to those described in this paper. We believe that the examples that we have been able to prove provide convincing justification for combining powerful symbolic computation techniques with theorem provers.

Nevertheless, there are many ways to improve Analytica. One direction is to add powerful algorithmic techniques for simplifying particular classes of formulas (like extensions of Gosper's algorithm for summations). The difficulty with adding such techniques is that a proof obtained in this manner may be virtually impossible for a human to follow.

Another direction is to strengthen the ability of Analytica to do inductive proofs. The technique that Analytica currently uses for generating induction schemes is quite simple. More research is needed on the generation of complex induction schemes and the identification of sufficiently general hypotheses for inductive proofs. There has been a fair amount of research on this problem [6, 7], but more work should be done in the context of inductive proofs in analysis.

Most proofs in modern analysis are based on set theory and many use topological concepts. Clearly, the extension of Analytica to handle such proofs is critical. Although theorem proving in set theory has been an important problem for a long time, there is no generally accepted technique for constructing such proofs. The most successful work on set theory so far is probably that of Quaife [16]. His work, however, uses a theorem prover based on hyper-resolution and may not produce proofs that are readable.

Better methods for managing hypotheses and previously proved lemmas and theorems are also needed. Techniques developed for proof checking systems like LCF [13] and HOL [12] may be adequate in the short run, but some type of higher-order unification or matching will probably be necessary in the majority of cases. In general, deciding when to use a hypothesis or previous result is a very difficult problem. Every student of elementary calculus learns the mean value theorem by heart, but it is not easy to give a good set of rules for determining when to apply this theorem in order to obtain a simpler bound on some complicated expression.

Certainly, some type of higher order logic would be more appropriate than the first order logic we currently use. The ability to state higher-order lemmas would be an additional advantage of basing the prover on a higher order logic and might help solve the problem described in the last paragraph. We intend to experiment with combining ideas from this paper with Andrews' theorem prover for higher order logic [1] in the near future.

Perhaps, the most serious problem in building a theorem prover like Analytica is the soundness of the underlying symbolic computation system. Mathematica (as well as Macsyma, Reduce, and Maple) has some rules that lead to correct results in most cases but do not lead to correct results all the time. We believe the solution to the soundness problem is to develop the theorem prover and the symbolic computation system together so that each simplification step can be rigorously justified.

Appendices: The Bernstein Approximation Theorem
— Example Proof of Analytica

A   Input To Analytica

```
integer[n] = True;
integer[k] = True;

sum1[f_] := sum[f,   {k,0,n,Abs[k/n-x]<n^(-1/4)}];
sum2[f_] := sum[f,   {k,0,n,n^(-1/4)<=Abs[k/n-x]}];

(* the bernstein polynomial for "f" *)
AddDefinition[B[n_,x_] ==
   sum[f[k/n] Binomial[n, k] x^k (1-x)^(n-k), {k, 0, n}]];

(* Several auxiliary functions *)
AddDefinition[near[x_, n_] ==
   sum1[Abs[f[x]-f[k/n]] Binomial[n,k] x^k (1-x)^(n-k)]];
AddDefinition[far[x_, n_] ==
   sum2[Abs[f[x]-f[k/n]] Binomial[n,k] x^k (1-x)^(n-k)]];
AddDefinition[far1[x_, n_] ==
   2 M Sqrt[n] sum2[(x-k/n)^2 Binomial[n,k] x^k (1-x)^(n-k)]];

(* f is a continuous function on [0, 1] and therefore bounded *)
ContinuousFunction[f] := True;
Domain[f] := ClosedInterval[0, 1];
M=Bound[f];

(* n >= max(delta[f][epsilon/2]^(-4), (M/epsilon)^2) *)
n = m^(-4);
Given[m<= delta[f][epsilon/2]];
Given[m^2 <= epsilon/M];

Given[m>0, epsilon>0, x>0, x<1];

(* Lemmas *)
ProveAndSave[near[x, n]<=epsilon/2];
ProveAndSave[far1[x, n]<=epsilon/2];
ProveAndSave[far[x, n]<=far1[x, n]];
ProveAndSave[Abs[f[x]-B[n,x]]<=far[x, n]+near[x, n]];

(* Theorem *)
Prove[Abs[f[x]-B[n,x]] <= epsilon];
```

B The Proof of Two Lemmas

Theorem :

$$near(x, n) \leq \frac{\epsilon}{2}$$

Proof :

$$near(x, n) \leq \frac{\epsilon}{2}$$

reduces to

$$\frac{-\epsilon}{2} + near(x, n) \leq 0$$

rewrite as

$$\frac{-\epsilon + 2\,near(x, n)}{2} \leq 0$$

reduces to

$$-\epsilon + 2\,near(x, n) \leq 0$$

open definition

$$-\epsilon + 2\,(1-x)^n \sum_{0 \leq k \leq n}^{\left|\frac{k}{n}-x\right| < m} \frac{x^k \left|-f(\frac{k}{n}) + f(x)\right| \binom{n}{k}}{(1-x)^k} \leq 0$$

reduces to

$$-\epsilon + 2\,(1-x)^n \sum_{0 \leq k \leq n}^{-m + \left|\frac{k}{n}-x\right| < 0} \frac{x^k \left|f(\frac{k}{n}) - f(x)\right| \binom{n}{k}}{(1-x)^k} \leq 0$$

replace expression with its lower or upper bounds

$$\epsilon \left(-1 + (1-x)^n \sum_{0 \leq k \leq n}^{-m + \left|\frac{k}{n}-x\right| < 0} \frac{x^k \binom{n}{k}}{(1-x)^k} \right) \leq 0$$

reduces to

$$-1 + (1-x)^n \sum_{0 \leq k \leq n}^{-m + \left|\frac{k}{n}-x\right| < 0} \frac{x^k \binom{n}{k}}{(1-x)^k} \leq 0$$

replace expression with its lower or upper bounds

$$-1 + (1-x)^n \left(\sum_{k=0}^{n} \frac{x^k \binom{n}{k}}{(1-x)^k} \right) \leq 0$$

calculate summations

$$-1 + \frac{(-1)^n (1-x)^n}{(-1+x)^n} \leq 0$$

reduces to

$$True$$

$\square$

Theorem :

$$|f(x) - B(n, x)| \leq far(x, n) + near(x, n)$$

Proof :

$$|-B(n, x) + f(x)| \leq far(x, n) + near(x, n)$$

reduces to

$$|B(n, x) - f(x)| - far(x, n) - near(x, n) \leq 0$$

open definition

$$\left| -f(x) + (1-x)^n \left(\sum_{k=0}^{n} \frac{x^k \binom{n}{k} f(\frac{k}{n})}{(1-x)^k} \right) \right|$$

$$- \left((1-x)^n \sum_{\substack{0 \leq k \leq n}}^{\left| \frac{k}{n} - x \right| < m} \frac{x^k \left| -f(\frac{k}{n}) + f(x) \right| \binom{n}{k}}{(1-x)^k} \right)$$

$$- \left((1-x)^n \sum_{\substack{0 \leq k \leq n}}^{m \leq \left| \frac{k}{n} - x \right|} \frac{x^k \left| -f(\frac{k}{n}) + f(x) \right| \binom{n}{k}}{(1-x)^k} \right) \leq 0$$

reduces to

$$\left| f(x) - (1-x)^n \left(\sum_{k=0}^{n} \frac{x^k \binom{n}{k} f(\frac{k}{n})}{(1-x)^k} \right) \right|$$

$$-(1-x)^n \left(\sum_{\substack{0 \leq k \leq n}}^{-m + \left| \frac{k}{n} - x \right| < 0} \frac{x^k \left| f(\frac{k}{n}) - f(x) \right| \binom{n}{k}}{(1-x)^k} \right.$$

$$\left. + \sum_{\substack{0 \leq k \leq n}}^{m - \left| \frac{k}{n} - x \right| \leq 0} \frac{x^k \left| f(\frac{k}{n}) - f(x) \right| \binom{n}{k}}{(1-x)^k} \right) \leq 0$$

simplify summations

$$\left| f(x) - (1-x)^n \left(\sum_{k=0}^{n} \frac{x^k \binom{n}{k} f(\frac{k}{n})}{(1-x)^k} \right) \right|$$

$$- \left((1-x)^n \left(\sum_{k=0}^{n} \frac{x^k \left| f(\frac{k}{n}) - f(x) \right| \binom{n}{k}}{(1-x)^k} \right) \right) \leq 0$$

replace expression with its lower or upper bounds

$$\left| f(x) - (1-x)^n \left(\sum_{k=0}^{n} \frac{x^k \binom{n}{k} f(\frac{k}{n})}{(1-x)^k} \right) \right|$$
$$- \left((1-x)^n \left| \sum_{k=0}^{n} \frac{x^k \binom{n}{k} \left(f(\frac{k}{n}) - f(x) \right)}{(1-x)^k} \right| \right) \leq 0$$

calculate summations

$$-(1-x)^n \left| -\frac{(-1)^n f(x)}{(-1+x)^n} + \sum_{k=0}^{n} \frac{x^k \binom{n}{k} f(\frac{k}{n})}{(1-x)^k} \right|$$
$$+ \left| f(x) - (1-x)^n \left(\sum_{k=0}^{n} \frac{x^k \binom{n}{k} f(\frac{k}{n})}{(1-x)^k} \right) \right| \leq 0$$

reduces to

$$True$$

☐

References

1. P. B. Andrews. On connections and higher-order logic. *Journal of Automated Reasoning*, 5:257–291, 1989.
2. B. C. Berndt. *Ramanujan's Notebooks, Part I*. Springer-Verlag, 1985.
3. W. W. Bledsoe. The ut natural deduction prover. Technical Report ATP-17B, Mathematical Dept., University of Texas at Austin, 1983.
4. W. W. Bledsoe. Some automatic proofs in analysis. In *Automated Theorem Proving : After 25 Years*. American Mathematical Society, 1984.
5. W. W. Bledsoe, P. Bruell, and R. E. Shostak. A prover for general inequalities. Technical Report ATP-40A, Mathematical Dept., University of Texas at Austin, 1979.
6. R. S. Boyer and J. S. Moore. *A Computational Logic*. Academic Press, 1979.
7. A. Bundy, F. van Harmelen, J. Hesketh, and A. Smaill. Experiments with proof plans for induction. Technical report, Department of Artificial Intelligence, University of Edinburgh, 1988.
8. E. M. Clarke and X. Zhao. Analytica: A theorem prover for mathematica. *The Journal of Mathematica*, 3(1), 1993.
9. W. M. Farmer, J. D. Guttman, and F. J. Thayer. Imps: An interactive mathematical proof system. Technical report, The MITRE Corporation, 1990.
10. M. Fitting. *First-order Logic and Automated Theorem Proving*. Springer-Verlag, 1990.
11. J. H. Gallier. *Logic for Computer Science: Foundations of Automatic Theorem Proving*. Harper & Row, 1986.
12. M. Gorden. Hol: A machine oriented formulation of higher order logic. Technical report, Computer Laboratory, University of Cambridge, 1985.

13. M. Gorden, R. Milner, and C. Wadsworth. *Edinburgh LCF: A Mechanised logic of computation*, volume 131 of *Lecture Notes in Computer Science*. Springer-Verlag, 1979.

14. R. W. Gosper. Indefinite hypergeometric sums in macsyma. In *Proceedings of the MACSYMA Users Conference*, pages 237–252, 1977.

15. R. L. London and D. R. Musser. The application of a symbolic mathematical system to program verification. Technical report, USC Information Science Institute, 1975.

16. A. Quaife. Automated deduction in von neumann-bernays-godel set theory. Technical report, Dept. of Mathematics, Univ. of California at Berkeley, 1989.

17. E. Sacks. Hierarchical inequality reasoning. Technical report, MIT Laboratory for Computer Science, 1987.

18. P. Suppes and S. Takahashi. An interactive calculus theorem-prover for continuity properties. *Journal of Symbolic Computation*, 7:573–590, 1989.

19. S. Wolfram. *Mathematica: A System for Doing Mathematics by Computer*. Wolfram Research Inc., 1988.

Document Recognition, Semantics, and Symbolic Reasoning in Reverse Engineering of Software

G. Butler, P. Grogono, R. Shinghal, I. Tjandra *

Department of Computer Science
Concordia University
1455 de Maisonneuve Blvd. West
Montreal, Quebec, H3G 1M8 Canada
{gregb,grogono,shinghal,ono}@cs.concordia.ca

Abstract. The SOFTDOCS project at Concordia University investigates knowledge acquisition from software documents and the analysis of that knowledge for reverse engineering of legacy systems. It focusses on the recognition and analysis of diagrams rather than natural language processing of textual components of a software document. Rigorous analysis of diagrams requires a formal semantics for them, and utilises tools for symbolic reasoning.

Data flow diagrams (DFDs) are one of many kinds of diagrams that software engineers use to help them understand complex systems. A data flow diagram represents a system as a network of processes connected by data flows. DFDs provide a useful and intuitive way of representing a system and they are easily interpreted by people. Without a formal semantics, however, DFDs cannot be used for automatic software understanding or reverse engineering. The goal of our research is to abstract meaning from existing diagrams, thereby enabling software tools, such as reverse engineering tools, to make use of existing diagrams.

We have previously described a formal semantics for DFDs based on Milner's Calculus of Communicating Systems (CCS). The resulting formal description of a DFD can be analyzed with the aid of the Edinburgh Concurrency Workbench (CWB).

A prototype tool, Π-DFD, hides the details of the formal semantic notation and the commands of CWB. Π-DFD provides engineers with the capability to analyze the structure and semantics of a DFD, to run simulations of the behavior of a DFD, and to display the results graphically. Semantic analysis includes the computation of a DFD's state space; finding a minimal representation for the state space; deciding whether two DFDs are equivalent; and whether one DFD is an abstraction of another, more detailed, DFD. All of these operations are potentially useful in reverse engineering and software understanding.

---

* Present address of I. Tjandra is Department of Computer Science, University of Windsor, Windsor, Ontario, Canada. Email: ono@cs.uwindsor.ca

1 Introduction

Classical methods of structured analysis and design commonly use data flow diagrams (DFDs) to describe a software system [8, 9, 22]. A data flow diagram represents a system as a network of processes connected by data flows. A hierarchy of DFDs may be developed by using a DFD to explain the workings of a higher-level process at increasing levels of detail. DFDs provide a useful and intuitive way of representing a system and they are easily interpreted by people. Many legacy systems from the 70s and 80s were developed using a structured approach, and modern object oriented methodologies also find uses for DFDs [19]. CASE tools provide users with the tools to draw, check, and revise DFDs. Most CASE tools are capable of performing a structural analysis based on a set of connectivity rules but cannot perform semantic processing of DFDs [5].

Understanding (code, designs, requirements) is a critical problem in software engineering, especially in reverse engineering and reuse. About 30 – 35% of total life-cycle costs are consumed in trying to understand software after it has been delivered, in order to make changes [11]. To alleviate this situation tools and methods, which are known generically as *reverse engineering*, have been developed [6]. Basically, the methods aim at recovering knowledge or information involved in the software documents, such as the behavior specification, the design specification, and the program specification of a system, in such a way that knowledge about the software development can be reused. Software development *with reuse* [10, 14, 15] involves using (or adapting) existing software components (which includes code, designs, specifications, requirements). A key requirement is to understand what the component does, and what context constrains the use of the component.

Our work stems from a desire for more capable CASE tools and from our interest in the application of document recognition and understanding techniques to reverse engineering of legacy systems. Paper software documents contain many types of information, including text that is often constrained to a particular domain of discourse, diagrammatic notations, and mathematical notations. One conclusion of our survey [4] is that the state-of-the-art in document recognition is sufficient to recognise line diagrams, and printed text, but not mathematical notations. However, being able to understand and apply the information recognised, seems beyond the state-of-the-art. Highly constrained text can be understood, but no examples from the domain of software documents are known. The geometry of line diagrams can be recognised, but from a reverse engineer's point of view such a level of understanding is insufficient for useful analysis. The SOFTDOCS project has concentrated on understanding data flow diagrams. To "understand" a DFD in a software document requires a representation of a DFD with which one can reason. Hence, a formal semantics of DFDs is of practical interest.

The work of Kasturi [13] allows one to recognise the geometry of a line diagram, such as a DFD. In earlier work [1, 3], we provide a formal semantics for DFDs using Milner's Calculus of Communicating Systems (CCS) [16, 17] and show how to translate from the geometrical structure of a DFD to its formal

Pi-DFD

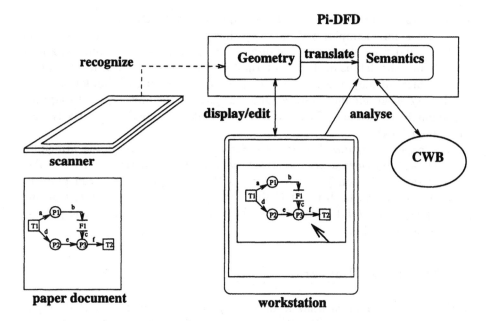

Fig. 1. Overview of the SOFTDOCS Project

CCS description [2]. The Edinburgh Concurrency Workbench (CWB) [7, 18] allows us to reason with this formal representation of a DFD. Figure 1 illustrates this diagrammatically and shows the role of our prototype, Π-DFD.

This paper contains a description of the kinds of analysis that can be supported by the formalism and its tool, CWB, for data flow diagrams from the perspective of a software engineer or reverse engineer. This includes the following kinds of analysis:

Structural queries check the well-formedness of the DFD at each level of the hierarchy and across levels of the hierarchy. These are essentially graph-theoretic properties, however, they can be checked against the CCS formalism as well.

Simulations illustrate operational behaviour of the DFD and allow selective exploration of the state space. Possibilities include step-by-step exploration, random sequences of transitions, enumeration of those states reachable from a given state, and recording and re-running of simulation histories.

Semantic queries address properties of the state space of the DFD such as size of the state space, the minimal equivalent state space, reachability, and deadlock.

Comparative queries allow the comparison of two DFDs, or of a DFD with a higher-level specification (written in μ-logic, a form of modal logic). This includes deciding whether two DFDs are equivalent; and whether one DFD is an abstraction of another, more detailed, DFD. Differences between DFDs can be identified when they are not equivalent.

We have run instances of each of these queries for DFDs under CWB: these
have been small examples with no more than ten nodes. A prototype tool, Π-
DFD, is under construction to support these activities of a reverse engineer
while hiding the details of the CCS formalism and CWB. At present it supports
the graphical display of a data flow diagrams, the running of simulations, and
the display of the simulation state. Our next goal for the prototype is to support
the hierarchical nature of DFDs, which would allow Π-DFD to effectively deal
with larger DFDs.

The paper is organized as follows. Section 2 describes the basic principles
of DFDs and provides a brief overview of a formal semantics for DFDs. The
main section is Section 3, which describes the analysis that one can perform
on DFDs using the facilities of CWB. Section 4 presents an overview of our
prototype system, Π-DFD. We discuss the related work in Section 5 and present
our conclusions in Section 6.

2   Data Flow Diagrams

A data flow diagram represents a software system as a labelled, directed graph.
Figure 2 shows a simple DFD. There is little uniformity in the literature concerning DFD notation; in this paper, we confine ourselves to notations that are
widely used in industry and textbooks. As the name "data flow diagram" suggests, each arc (or edge) of the graph represents a transfer of data. A DFD has
several kinds of node:

- A *rectangle* represents a terminator that may be data source or a data sink.
- A *circle* represents a process.
- A *pair of horizontal lines* represents a data store.

In Figure 2, T_1 and T_2 are terminators; P_1, P_2, and P_3 are processes; and F_1 is
a data store. In an actual DFD, the nodes and edges would have descriptive labels
such as *Keyboard*, *CheckPassword*, and *UserId*, rather than symbolic labels such
as T_1, P_1, and a.

A DFD must satisfy some syntactic requirements. For example:

- each edge must join two nodes;
- a terminator may have input edges or output edges, but not both;
- a process must have at least one input edge and at least one output edge.

2.1   A Semantic Overview of Data Flow Diagrams

Data flow diagrams do not fully capture the behaviour of a system since they
typically omit control and timing information. Furthermore, there is ambiguity
and non-determinism in their normal intuitive meaning as regards the precise
dependency of output data on input data of a process. Here we outline one
semantics; the variations can easily be defined using CCS.

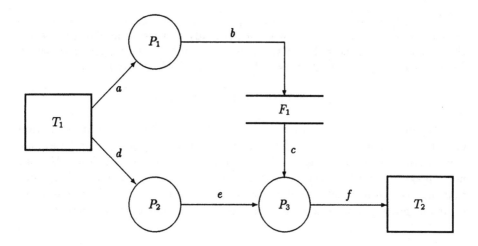

Fig. 2. A Data Flow Diagram

We interpret each node of a DFD as an *agent*. An agent is an active process that can perform a sequence of actions indefinitely. We use an upper case letter, such as P, to represent an agent. We use a lower case letter, such as a, or a lower case letter with a bar, such as $\bar{b}$, to represent an *event*. The bar indicates that the event is an *output event*. If there is no bar, the event is an *input event*. For any x, the events x and $\bar{x}$ are *complementary*. Complementary events allow two agents to interact.

We use CCS to formalize the interactions of agents. Four CCS functions are needed to describe the principal features of DFDs: sequencing, alternative, composition, and restriction.

The infix operator ".". denotes sequencing. For example, the term $x \cdot y$ denotes "the event x followed by the event y". A sequence of events has behaviour and is therefore considered to be an agent. An agent name within a sequence stands for the entire sequence of events in which the agent can participate.

An agent $a \cdot P$ undergoes a *transition* to become the agent P. In symbols:

$$a \cdot P \xrightarrow{a} P.$$

The infix operator "+" expresses alternatives: $X + Y$ denotes an agent that may behave like either X or Y.

To represent the interaction of two agents, we use the infix composition operator "|". The agent $X \mid Y$ can behave in two ways:

- the agents X and Y may proceed independently; or
- if the agents X and Y can experience complementary events, then these events may take place simultaneously, expressing an interaction between X and Y.

For example, in the agent $(\overline{e} \cdot P_2) \mid (e \cdot P_3)$, the agent $\overline{e} \cdot P_2$ is ready to emit the datum e, and the agent $e \cdot P_3$ is ready to accept the datum e. After these events have occurred, the agent becomes $P_2 \mid P_3$. This behaviour, which has no externally visible effect, is called a τ-transition. We write:

$$(\overline{e} \cdot P_2) \mid (e \cdot P_3) \xrightarrow{\tau} (P_2 \mid P_3).$$

We can hide the internal actions of an agent by using the infix restriction operator "\\". The right operand of \\ is a set of complementary pairs of events. The interpretation of the agent $P \setminus \{ x_1, \overline{x}_1, \ldots, x_n, \overline{x}_n \}$ is as follows: "P may not engage in any behaviour involving $x_1, \overline{x}_1, \ldots, x_n, \overline{x}_n$ except τ-transitions that involve complementary pairs x_i and $\overline{x}_i$". Since the set of events consists of complementary pairs, we write $P \setminus \{ e \}$ as an abbreviation for $P \setminus \{ e, \overline{e} \}$.

3 Analyzing Data Flow Diagrams

The Edinburgh Concurrency Workbench (CWB) was designed specifically to automate reasoning with CCS-style semantics. We use CWB interactively to build semantic representations of DFDs, to construct the corresponding state spaces by enumerating transitions, to minimize the state space, and to perform other mechanical tasks.

Figure 3 illustrates how the DFD of our example can be represented in CWB. The event $\overline{a}$ is entered as 'a. In steps (1) through (7), we use the command bi to bind identifiers to their defining terms as in Section 2.1. The final binding, at step (7), binds the identifier Dfd, representing the entire DFD. At step (8), we use the command pa to print the agent Dfd.

At step (9), we ask CWB to compute the size of Dfd: it has 48 states. The command min, used at step (10), instructs CWB to derive a minimal system that is observationally equivalent to Dfd. The minimal system has 9 states and is saved as DfdMin. Step (11) confirms that Dfd and DfdMin are indeed equivalent.

The tables in the appendix summarise the CWB commands. More detail can be found in [18].

4 An Overview of the Π-DFD Prototype

A prototype tool, Π-DFD, is under construction to support the activities of a reverse engineer while hiding the details of the CCS formalism and CWB. All CWB commands can be driven by selection from menus with dialogs of available options. Eventually, all the commands will have a more intuitive graphical interface. At present the interface supports the graphical display of a data flow diagrams, the display of the simulation state, and one can select the steps in the simulation by clicking on the edges of the DFD. Figure 4 shows the interface of the Π-DFD prototype. It is implemented using the ET++ framework [21] which simplifies the construction of graphical user interfaces.

The Edinburgh Concurrency Workbench
(Version 6.12, April 15, 1993)

```
Command: bi T1 input.'a.T1 + input.'d.T1                        (1)
Command: bi P1 a.'b.P1                                          (2)
Command: bi P2 d.'e.P2                                          (3)
Command: bi F1 b.F1 + 'c.F1                                     (4)
Command: bi P3 c.'f.P3 + e.'f.P3                                (5)
Command: bi T2 f.'output.T2                                     (6)
Command: bi Dfd (T1 | P1 | P2 | P3 | F1 | T2)\{a,b,c,d,e,f}     (7)
Command: pa Dfd                                                 (8)
  Dfd = (T1 | P1 | P2 | P3 | F1 | T2)\{a,b,c,d,e,f}
  F1 = b.F1 + 'c.F1
  P1 = a.'b.P1
  P2 = d.'e.P2
  P3 = c.'f.P3 + e.'f.P3
  T1 = input.'a.T1 + input.'d.T1
  T2 = f.'output.T2
Command: size Dfd                                               (9)
 Dfd has 48 states.
Command: min Dfd                                                (10)
 Save result in identifier: DfdMin
 DfdMin has 9 states.
Command: eq                                                     (11)
 Agent: Dfd
 Agent: DfdMin
 true
```

Fig. 3. Using CWB to Analyze the DFD of Figure 2

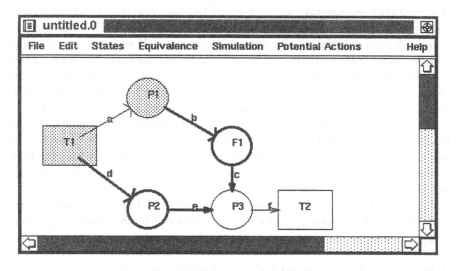

Fig. 4. Interface of the Π-DFD Prototype

The **States** menu provides the functionality of the commands in Table 3 and Table 4. The **Equivalence** menu provides the basic functionality of the commands in Table 2. The **Simulation** menu provides the functionality of the commands in Table 1, as an alternative to selecting the steps in the simulation by clicking on the edges of the DFD. The **Potential Actions** menu provides the functionality of the commands in Table 5.

Our next goal for the prototype is to support graphically the hierarchical nature of DFDs, which would allow Π-DFD to effectively deal with larger DFDs. The semantics and the translator to CWB formalism already support hierarchical DFDs.

5 Related Work

In our survey [4], the only literature we found dealing specifically with the recognition of software documents were two papers on flow charts. However, the work of Kasturi [13] handles the recognition of line drawings in general, including those for software such as flow charts and DFDs.

The literature reports only a small number of attempts to formalize DFDs and other diagrams used in software development. Extended Petri nets have been proposed as a formal foundation for modeling a DFD [20]. Although they introduce a formal semantics, the analysis capability is limited to structural (syntactic) properties. The CASE tool STATEMATE [12] uses temporal logic to provide a formal model for a combination of models, one of which is a DFD.

6 Conclusion

Diagrammatic representations, such as DFDs, have traditionally been used informally. The programmers who used the DFDs were capable of translating them into the formal notation of a programming language. If we are to extract information from DFDs in legacy documents, or to incorporate DFDs into formal software development, an informal interpretation is not sufficient. This overview of the SOFTDOCS project shows that it is possible to use the formal CCS description of a DFD for the rigorous analysis and understanding of a DFD.

The analysis capabilities of CWB deal with properties of the state space, simulations of behaviour, and issues of equivalence. These are reused in our work. The Π-DFD prototype provides a graphical interface to the simulation facilities of CWB, and the menus provide access to the other analysis methods in CWB.

The SOFTDOCS project combines document recognition, formal semantics of CCS, and symbolic reasoning of CWB to contribute to the reverse engineering of legacy software systems developed using Structured Analysis.

Acknowledgements The research described in this paper was funded by the Natural Sciences and Engineering Research Council of Canada, *Fonds pour la Formation de Chercheurs et l'Aide a la Recherche*, and the Institute of Robotics and Intelligent Systems, Canada. Alison Greig and Valerie Large implemented the first version of the Π-DFD prototype.

References

1. G. Butler, P.D. Grogono, R. Shinghal, and I.A. Tjandra. Retrieving information from data flow diagrams. In *Proceedings of Second Working Conference on Reverse Engineering*, (Toronto, July 14–16, 1995). Linda Wills, Philip Newcomb, Elliot Chikofsky (eds), IEEE Computer Society Press, Los Alamitos, CA, 1995,pp. 22–29.
2. G. Butler, P.D. Grogono, R. Shinghal, and I.A. Tjandra. Analyzing the logical structure of data flow diagrams. In *Third International Conference on Document Analysis and Recognition*, August 1995. Poster session.
3. G. Butler, P.D. Grogono, R. Shinghal, and I.A. Tjandra. A Semantics of Data Flow Diagrams. Journal article in preparation.
4. G. Butler, P.D. Grogono, R. Shinghal, and I.A. Tjandra. Knowledge and the recognition and understanding of software documents. Department of Computer Science, Concordia University, February 1995, 47 pages.
5. M.J. Chen and C.G Chung. Preventive structural analysis of dataflow diagrams. *Information and Software Technology*, 34(2):117 – 130, 1992.
6. E. Chikofsky and J. Cross. Reverse engineering and design recovery: A taxonomy. *IEEE Software*, 7(1):13 – 17. 1990.
7. R. Cleaveland, J. Parrow, and B. Steffen. The concurrency workbench. In J. Sifakis, editor, *Automatic Verification Methods for Finite State Systems*, Lecture Notes in Computer Science vol. 407, pages 24–37. Springer-Verlag, 1987.
8. F. DeMarco. *Structured Analysis and System Specification*. Englewood Cliffs, N.J., Yourdon Press, 1978.
9. C. Gane and T. Sarson. *Structured Systems Analysis*. Englewood Cliffs, N.J., Prentice Hall, 1979.
10. M.L. Griss, Software reuse: From library to factory. *IBM Systems Journal* **32**, 4 (1993) 548–566.
11. Hall, P. Overview of reverse engineering and reuse research. *Information and Software Technology*, 34(4):239 – 249. 1992.
12. D. Harel, H. Lachover, A Naamad, A. Pnueli, M. Politi, R. Sherman, and A. Shtul-Trauring. STATEMATE: A working environment for the development of complex reactive systems. In *Proceedings of 10th International Conference on Software Engineering*, pages 396 – 406. IEEE Press, 1988.
13. R. Kasturi, S.T. Bow, W. El-Masri, J. Shah, J.R. Gattiker, U.B. Mokate, A system for interpretation of line drawings. *IEEE Transaction on Pattern Analysis and Machine Intelligence* **12**, 10 (1994) 978 – 992.
14. Charles W. Krueger, Software reuse. *ACM Computing Surveys* **24**, 2 (June 1992) 131–183.
15. H. Mili, F. Mili, A. Mili, Reusing software: Issues and research directions. *IEEE Trans. Software Eng.* **21**, 6 (June 1995) 528–562.
16. R. Milner. *A Calculus of Communicating Systems*, volume 92 of *Lecture Notes in Computer Science*. Springer-Verlag, Berlin-New York, 1980.
17. R. Milner. *Communication and Concurrency*. Prentice-Hall, Englewood Cliffs, N.J., 1989.

18. Faron Moller, *The Edinburgh Concurrency Workbench (Version 6.1)*. Department of Computer Science, University of Edinburgh, October 1992.
19. J. Rumbaugh, M. Blaha, W. Premerlani, F. Eddy, and W. Lorensen. *Object-oriented Modelling and Design*. Englewood Cliffs, N.J., Prentice Hall, 1991.
20. T.H. Tse and L. Pong. Towards a formal foundation for DeMarco data flow. *The Computer Journal*, 32(1):1 – 12, 1989.
21. A. Weinand, E. Gamma, R. Marty, Design and implementation of ET++, a seamless object-oriented application framework. *Structured Programming*, **10**, 2 (1989) 63–87.
22. E. Yourdon and L.L Constantine. *Structured Design: Fundamental of a Discipline of Computer Program and Systems Design*. Englewood Cliffs, N.J., Prentice Hall, 1979.

A CWB Commands

sim	reset the simulation agent, deleting all breakpoints
menu	list the one step transitions from the agent
random	transition at random for a given number of steps
history	list transitions and states encountered
return	reset to a point in the simulation history
break, lb, db	set, list and delete a breakpoint

Table 1. CWB Simulation Commands

eq, cong	observational equivalence
decomp	verification via decomposition
strongeq	strong bisimulation equivalence
strongpre	strong bisimulation preorder relation
pre, precong	weak bisimulation preorder and precongruence
diveq	observational equivalence that respects divergence
maypre, mustpre, testpre	standard testing procedures
mayeq, musteq, testeq	associated testing procedures for equivalence
2/3pre, 2/3eq	2/3 preorder and associated equivalence
branchingeq	branching bisimulation
contraction	contraction preorder
min	binds an identifier to the smallest state space observationally equivalent to the given agent

Table 2. CWB Relational Commands for Pairs of Agents

size	number of states for an agent
states	list all states for an agent
statesexp	list all states, including observable actions and tau transitions needed to reach each state
statesobs	list all states, including observable actions needed to reach each state

Table 3. CWB State Space Commands

stable	can only observable actions be performed initially
fd	list all deadlock states, including observable actions and tau transitions needed to reach each state
fdobs	list all deadlock states, including observable actions needed to reach each state

Table 4. CWB Static Analysis Commands

tr	list all one-step transitions
actders	list all one-step transitions using given action
initobs	initial observable actions
actcl	list of reachable agents by observation of given action
vs	list of possible observation sequences of given length
obsder	list of reachable agents after applying list of observations

Table 5. CWB Derivation Commands

dfobs	generate a weak modality HML formula for two agents that are not observationally equivalent

Table 6. CWB Model Checking Command for Pairs of Agents

Compromised Updates in Labelled Databases

Fátima C. C. Dargam

e-mail: fccd@doc.ic.ac.uk

http://theory.doc.ic.ac.uk/~fccd/index.html

Department of Computing
Imperial College of Science, Technology and Medicine
180 Queens Gate, London SW7 2BZ, UK

Abstract. This paper presents a logical system, CIU_{LDS}, as a labelled realization to our approach of *Compromising Interfering Updates*. The approach proposes a method for handling logically conflicting inputs into knowledge bases, via restricting their consequences. The main idea is to update the database with as many consistent consequences of the inputs as possible, in the case that the inputs themselves are not allowed to be kept in it. And in the case that a revision applies, the idea is to keep as many as possible of the consistent consequences of the retracted sentences as a compromise.[1] The reconciliation of conflicting inputs follows some specified postulates for compromised revision.

CIU_{LDS} is based on the Labelled Deductive Systems framework (LDS). In CIU_{LDS} we take advantage of LDS's labelling facility, to control the derivation process of the compromised consequences. We embed in the labelling propagation conditions, which act on the inference rules, part of the control mechanism for the compromised approach. This mechanism helps the update operations to perform the reconciliation of conflicting inputs. The update operations invoke a specific revision method, which applies some compromising criteria for achieving the revised labelled database, whenever conflicts arise.

In this paper, we present briefly our main motivations and we introduce the specification of our approach, for the case of database updates. We present CIU_{LDS}, by decribing informally its main features and definitions. Finally, we summarize the system's main properties.

1 Introduction

Resolving conflicting updates in dynamic databases, or conflicting actions in planning applications, for instance, are frequent and critically important problems of real applications. Such problems require the revision of theories and knowledge bases. As pointed out by Winslett in [36], it is not realistic to aim for a generic approach in those cases, since theory revision is fundamentally dependent on application-specific mechanisms, principles and heuristics. The approach

---

[1] The idea of reconciling conflicting inputs by compromising on their consequences was proposed by Dov Gabbay, who supervised this research work. See [11].

we propose in this paper, caters for the specific case where compromised solutions apply when conflicts occur. It supports a compromised way of handling conflicting updates for revising databases.

In more practical terms, consider the situation where DB is a database and A an input. Assume that A is inconsistent with DB. Current belief revision/update approaches will keep A and mantain consistency by selecting some element from DB to form a revised database, usually denoted as $DB * A$. There is a lot of research in this area, both theoretical, e.g.: the AGM theory of belief revision[2], and algorithmic research, e.g.: Reason Maintenance Systems[3]. Our aim is to offer an alternative approach, which is flexible enough to keep more data in DB, in the case of conflicts. We view the above situation as a *conflict* between two inputs (DB and A) into an empty database, and we tackle the problem of reconciling these inputs. Under our approach, the conflicting input A is kept in DB only in the case that A generates inconsistency to DB indirectly,[4] in which case a revision also applies in order to restore consistency. However, in the case that A is not allowed to be kept in DB, its consistent consequences, w.r.t. the existing data of DB, are added to the database under the compromised policy of our approach. This way, instead of preventing updates to be performed, when they introduce inconsistency to the system, our approach reconciles the conflicting inputs by compromising on their consequences. We propose to generate the consequences of the conflicting inputs, and get rid of the inconsistency, via a minimal number of retraction of those consequences. We expect the resulting database to be consistent w.r.t. the integrity constraints, and to retain a *safe maximal subset* of the consistent consequences of those updates. This reconciliation of conflicting inputs follows some specified postulates for compromised revision.[5]

1.1 Motivations

As pointed out by Galliers [18], in most of the existing AI research work, conflicts either simply never arise, or are alternatively avoided when they do arise. However, in a constantly changing and unpredictable environment, inconsistencies within the system are most of the times inevitable, and conflict situations do arise. The central interest in [18] is to solve conflicts in cooperative multiagent systems, by facing their positive aspects. They claim that achievement of

---

[2] The AGM theory was first introduced in [1],[2],[3],[4], and since then gained many followers who apply and modify that theory in various ways, see for instance [23], [26], [29],[20],[27],[31],[32], [21],[5],[24],[33] and [14].

[3] Reason Maintenance Systems were initiated in [12] based on justifications, and in [22] based on assumptions. More recent research work following this line have also emerged. Some of them are found in [13],[19],[28],[30] and [35].

[4] By *indirectly* here, we mean that A alone does not violate any of DB's integrity constraints.

[5] In [9] and [10], the compromised revision is defined under a belief revision perspective and some postulates for finite bases with integrity constraints are introduced, as guidelines for the compromised revision function.

cooperation from conflict, among formalized agents may involve the decision of a mutually preferred compromised solution, and/or persuasion to the validity of another position.

Gabbay and Hunter, in [16] and [17], also support that inconsistency should be faced and formalized. They urge for a revision on the way inconsistency is currently being handled in formal logical systems, as opposed to the way it is handled by humans. They claim that there is a need for the development of a framework, in which inconsistency can be viewed in a context-dependent way. As a signal for external and/or internal actions, and not necessarily as a bad element which induces the whole system to collapse. They argue that dealing with inconsistencies is not necessarily a job for restoring consistency, but rather for supplying rules which state how to act in the case of inconsistencies.

We strongly endorse the viewpoints of Galliers and Gabbay & Hunter. Based on the same grounds, we investigate an approach which handles conflicts that introduce inconsistency into a system, and puts forward a compromised reasoning way for dealing with conflicting updates and actions, instead of simply avoiding them.

As in [18], we propose to solve conflicts by facing their positive aspects. We do so, by reconciling the conflicting updates with the underlying knowledge base, and getting as many of their consequences as possible.

We support the point in [16] and [17], that inconsistency (caused by conflicts) should be faced and that we should supply mechanisms for handling situations when they arise. In the current work, we approach such situations by allowing some consequences of the conflicting updates to remain in the database. However, by reconciling the conflicting inputs, we also restore consistency, which does not conform to their view of keeping inconsistency in the system and supplying the appropriate mechanisms to handle it.

Our main motivation in pursuing this approach, comes from the premise that we want to take some advantage of conflicting inputs, and to have some relevant information extracted from them. By reconciling conflicting updates with a knowledge base, our approach allows us to achieve this goal, since it provides us with more informative results than standard approaches. In comparison with approaches that require preference between conflicting inputs, or that simply avoid them by cancelling them out completely, our proposal of compromised revision allows more information to be kept in a theory base. Following our compromised revision approach to conflicts, one will possibly not get all of what he/she originally wanted[6], in the case of conflict. Instead, he/she will get extra data, leading to the direction of the original goal. This is because most of the extra data are related to the goal's consistent consequences.

Via the compromised reasoning strategy of our approach, we are able to keep bits of new information, or a subset of the consequences of updates, which could

---

[6] The idea of this revision approach conforms with the meaning of the word *compromise*. Quoting from the Oxford's Dictionary: *"Compromise" is a settlement of a dispute which each side gives up something it has asked for, and neither gets all it asked for.*

not be totally assimilated, or performed, due to a conflicting situation. These consequences and bits of new information might reveal important facts about the current situation, which then might signal the system to take specific actions, depending on the application context.

Our approach is mainly suitable for applications which allow for compromised solutions, i.e. solutions which present the closest result with relation to the expected one. Some of the application areas which can benefit from our approach are: design processes; resource allocation; and decision making. In these areas, one builds up the goal state of a particular task, via performances of intermediary updates. This procedure allows for compromised results of updates when conflicts arise. As an application example, let us consider a research organization which has the task of deciding the allocation of funds among projects. We assume that it is necessary for the projects to discriminate all the expenses required for each of their phases, allowing for the option of satisfying only partially those phases (compromised solutions). We assume also that the decision makers are not supposed to favour any project in particular. So, if funds are not sufficient to support all the projects' requirements, our approach would be appropriate to be applied in the process of funds allocation. In the sense that it would allow for as many of all the projects' phases as possible, considering the constraints involved in the process.

1.2 Our Approach to Handling Conflicting Inputs

Our approach proposes to reconcile conflicting inputs with respect to the underlying theory, and establishes some policies for dealing with the problem of inconsistency caused by them. The way we approach the problem of conflicting inputs differs from the other existing approaches, in the sense that we allow for a special process of performance of the conflicting updates. A process of *reconciliation of conflicting inputs*, which considers restrictions of the effects of those inputs by compromising on their consequences. We refer to our approach as CIU, meaning *Compromising Interfering Updates*.

By conflicting, or interfering, updates, we mean either simultaneous updates which interfere with each other, generating inconsistencies as part of their combined effects, or updates which are inconsistent to be performed because they conflict with the given database or scenario representation, by violating some of their constraints. Below, we present two examples which illustrate the intuitive notion of our approach with relation to database updates.

Example 1.   *Let us consider the database of formulae as shown below:*

$$(1)\ A \rightarrow B$$
$$(2)\ A \wedge C \rightarrow \perp$$
$$(3)\ A$$

If we want to update this database with the formula (4) C, then, by applying a TMS-like approach [12], for instance, we would force C in, by removing A and all the consequences derived from A, in order to keep consistency, as shown below.

$$(1)\ A\ \rightarrow\ B$$
$$(2)\ A \wedge C\ \rightarrow\ \bot$$
$$(4)\ C$$

In the way we approach this problem, we would also end up with either A or C, but not both. However, we want to be able to keep all the consistent derived consequences of the conflicting update. In this case, we would be able to have B as well in the resulting database, as shown below.

$$(1)\ A\ \rightarrow\ B$$
$$(2)\ A \wedge C\ \rightarrow\ \bot$$
$$(4)\ C$$
$$(5)\ B$$

The example above could be interpreted with the following meanings for the sentences A, B and C: A = "Executive Class Passenger"; B = "Extra baggage allowance"; C = "Economy Class Passenger". Then, we would have that the database update above represents a situation, in which an executive class passenger for some reason has to be moved to the economy class, in a particular flight. However, even in the the economy class, the originally executive class passenger still keeps his/her right of having extra baggage allowance.

CIU can be described as a module of a reasoning system, which is invoked whenever we have conflicting updates, w.r.t. databases and to their sets of integrity constraints. We assume that we have a database module D which is subjected to a module of integrity constraints I. The integrity constraints are assumed to be protected against any update modification, and they restrict the possible transactions on D. The database can be, for instance, a declarative representation of a scenario, in terms of the facts that hold in the current state. A finite set of updates, to be performed on the database, is given as input to the system. We assume that the updates executing module only effectively performs the updates in the case that no database revision is needed. Otherwise, the CIU module is invoked in order to perform the compromised version of the set of updates. In the end, the compromised updated database is supposed to be consistent and to satisfy the constraints in module I. The peculiar characteristic that CIU has in dealing with updates is that, instead of preventing an update to be performed when inconsistency arises,[7] CIU proceeds and generates the consequences of the conflicting updates.

When CIU is invoked, it instigates the compromised performance of the conflicting updates, by firstly generating all their consequences/derivations. Later it takes care of restoring consistency in the database. In order to restore consistency, a special revision procedure takes place. It is based on the minimal elimination of the formulae involved in the generation of inconsistency, and guided by the compromised reasoning policies of the approach.

---

[7] This would make the approach equivalent to many existing ones which do not allow for updates to be performed if they are not consistent with the theory.

Different Kinds of Conflicting Inputs:

Conflicting inputs can be of various kinds. For instance, we can have simultaneous updates which interfere with each other, generating inconsistencies as part of their combined effects, or updates which conflict with the given database, by violating some of its constraints. We can also have the case in which updates are individually consistent to be applied, but if performed in parallel they interfere with each other.[8]

Below, we state clearly all the different kinds of conflicting inputs that we are considering, and we describe how we propose to handle them. We consider that A and B are formulae of the language being considered. We consider classical logic as the underlying logic, including the usual connectives. A database DB is such that $DB = \Delta \bigcup P_\Delta$, where Δ denotes set of formulae which compose the body of the database, and P_Δ denotes the set of integrity constraints which rules Δ.

(a) Conflicting inputs within the update, or within the transaction, e.g. Update $= \{A, \neg A\}$. In this case, the updates are rejected, since one logically cancels the other. However, if within a transaction T we have the following sequence of inputs $T = \{A, \neg A, B\}$, the subset $\{A, \neg A\}$ is removed from T and the remaining inputs in the transaction are still performed. In this case, $T = \{B\}$.

(b) Inputs which conflict directly with some of the integrity constraints which rule the database, e.g. $DB = \Delta \bigcup P_\Delta$, $P_\Delta = \{A \rightarrow \perp\}$, and Update $= \{A\}$. In this case, the input is not allowed to be inserted in the database. However, we allow the consistent consequences of the input to be inserted in DB, with particular status of non-supported data.

(c) Inputs which conflict indirectly with some of the integrity constraints which rule the database, e.g. $DB = \Delta \bigcup P_\Delta$, $P_\Delta = \{A \wedge C \rightarrow \perp\}$, $\Delta = \{C\}$, and Update $= \{A\}$. In this case, the input is inserted in the database and a revision procedure takes place in order to restore consistency and allow the database to accomplish the new update. The revision presents special properties which preserves the consistent consequences of all the retracted formulae from DB.

(d) Inputs which contradict existing data in the database, e.g. $DB = \Delta \bigcup P_\Delta$, $\Delta = \{A\}$, and Update $= \{\neg A\}$. In this case, the input is inserted in the database and a revision on DB takes place, just as described above.

---

[8] A similar motivating approach was pursued by Cholvy [6] in the context of multi-sourced information environment. Cholvy treats the problem of consistency of information provided by different sources, considering the case that the global set of information is inconsistent even if each separate source is consistent. Notice, however, that in this work we propose to deal with inconsistency generated by conflicting updates within the same system, while Cholvy treats the inconsistent information which is due to the combination of different data/knowledge bases. A further analogy between the two approaches requires, at least, a re-definition of the basic conflicting entities, in order to cater for the representation of information sources.

Multiple Updates Case:

In the case of a transaction, which involves a set of single updates, if we have "n" updates which directly conflict with some integrity constraints, the resulting compromised updated database might contain at most "$n-1$" of those updates. Transactions have their consistency initially checked with relation to the three conditions described below. Assume that a transaction $T = \{U_1, U_2, \cdots, U_n\}$, composed of n updates, is to be performed to DB, and I is the set of integrity constraints which rules DB.[9]

1. For any U_i and any U_j in a transaction $T = \{U_1, \cdots, U_n\}$, where $1 \leq i \leq n$; $1 \leq j \leq n$; and $i \neq j$, if U_i expresses a formula which is the complement of the formula expressed by U_j, say A and $\neg A$ [10], then the set $\{U_i, U_j\}$ is retracted from the transaction T.

2. For any U_i in a transaction $T = \{U_1, \cdots, U_n\}$, where $1 \leq i \leq n$, if U_i violates an integrity constraint in I, $\{U_i\} \bigcup I \vdash \perp$, then the update U_i is rejected, however its consistent consequences are allowed to remain as non-supported consequences in the database.

3. For any U_i and U_j in a transaction $T = \{U_1, \cdots, U_n\}$, where $1 \leq i \leq n$; $1 \leq j \leq n$; and $i \neq j$, such that U_i and U_j are not complementary updates in T, and $\{U_i\} \bigcup I \nvdash \perp$ and $\{U_j\} \bigcup I \nvdash \perp$, if $\{U_i, U_j\} \bigcup I \vdash \perp$, then a choice is made between U_i and U_j, according to some meta-level criteria.[11]. In this case, the transaction is then reduced to $T - \{U_k\}$, where k is either i or j, however the consistent consequences of U_k are allowed to remain as non-supported consequences in the database.

In the cases described above, the updates which cause inconsistency are not supposed to be performed, since they are removed from the transaction. However, the transaction is not cancelled due to the fact that some of its updates failed the initial consistency checking phase.[12]

Condition 1 above, ensures that complementary information is cancelled prior to the database transaction performance, in order to avoid redundant update execution.

Condition 2 puts forward that an update U_i which violates directly an integrity constraint cannot be performed. However, under our compromised approach, its consistent consequences can be kept in DB. For instance, if $T =$

---

[9] The formulae considered here are propositional sentences from the system of propositional classical logic.

[10] This would represent adding and deleting the formula A in the same transaction.

[11] This meta-level information is totally context-dependent. We could, for instance, have a total ordering among the updates in T, so that each U_i would be less preferable than U_{i+1}. We will not discuss details about this meta-level based choice in this paper.

[12] Most of the database-update approaches in the literature do not conform with this viewpoint, since they adopt a style denoted sometimes as *all-or-none* updates performance, in the case of inconsistency or integrity constraint violation within a transaction, e.g. [25].

$\{A, B, C\}$, $DB = \{A, D \wedge C \to E\}$ and $I = \{C \to \bot\}$, the transaction would be reduced to $\{A, B\}$, since $\{C\} \bigcup I \vdash \bot$, and the non-supported consequence E would also be added to DB. Eliminating the update that violates the integrity constraint from the transaction, and allowing the other updates in T to be performed is, most of the times, an intuitive procedure which conforms with the compromised philosophy of our approach. Consider the case that A expresses that worker $W1$ gets a raise of 10% on his salary; B expresses that worker $W2$ gets a raise of 30%; and C expresses that worker $W3$ gets a raise of 50% on his salary. Assume that their company has restricted raises of 50% or higher on workers' salaries. Then, update C would not be performed and would have to be negotiated later. However, this would not stop updates A and B from being performed. If we now consider that D expresses that worker $W3$ is a senior professional, and that E expresses that $W3$ becomes a candidate for taking higher supervisory positions in the company. So, by adopting CIU, we would have that worker $W3$ would still be recommended to get a higher position in the company, even though his salary raise is still under negotiations. This is perfectly acceptable, since we would have C as valid, had the company not restricted the raises at that time.

Condition 3 caters for the case when two updates are individually consistent to be performed, but together they violate the set of integrity constraints which rules the database. In the case of two conflicting updates within the same transaction, our approach allows for their consistent consequences to be kept in the database.

2 *CIU* formalized under the *LDS* Framework

Labelled Deductive System (LDS), was introduced by Gabbay in [15]. It is a logical framework which deals with labelled formulae as its basic units of information, where the labels can be of arbitrary form, belonging to a given *labelling algebra*. LDS's derivation rules act on the labels as well as on the formulae. These rules include some prescribed ways, given by the labelling algebra, to propagate the labels. The handling of labelled formulae allows standard proof systems to be extended with non-standard features. Hence, the LDS formalism provides a rich syntactic characterization for a proof-theorectical presentation, which is able to cover a wider operating scope.

By labelling the formulae, we are provided with a way of including in the labels extra information to the system. The original motivation was to be able to code control information in the labels, such as dependencies within a proof, and controlling flags for a derivation process. Nevertheless, structural database information, and network information, among other meta-level pieces of information, can also be incorporated explicitly into the object language via the labels. Moreover, labelling may also be used to facilitate truth maintenance and conflict resolution.

We propose here to build a Labelled Deductive System to formalize our approach. We present the logical system CIU_{LDS} as a labelled realization to our specified system CIU. One of the main motivations for adopting LDS as the underlying framework of this formalization, was to take advantage of its labelling facility, to control the derivation process of the compromised consequences. Actually, we embed in the labelling propagation conditions, which act on the inference rules, part of the control mechanism for the compromised approach. This control mechanism helps the defined update operations to perform the reconciliation of conflicting inputs. The update operations invoke a compromised revision on the labelled database, whenever conflicts arise.[13]

2.1 The System CIU_{LDS}

The Language:

The system's basic units of information are labelled formulae, denoted as declarative units and written as $\gamma : \alpha$, where γ is a label and α is a logical formula. The intended meaning of $\gamma : \alpha$ is that γ indicates the nature of the formula α w.r.t. its data status in the database. The system's language $\mathcal{L}_{CIU}$ is defined as the ordered pair: $\langle \mathcal{L}_\gamma, \mathcal{L} \rangle$, composed of a propositional logical language $\mathcal{L}$, and a distinct language for the labels $\mathcal{L}_\gamma$. The logical language $\mathcal{L}$ provides propositional well formed formulae (wff), considering a countable number of propositional letters, A, B, C, D, $\cdots$, including $\top$ and $\bot$;[14] and the logical connectives $\neg$, $\wedge$, and $\rightarrow$. The labelling language $\mathcal{L}_\gamma$ comprises a finite set of typed constants symbols T, where $T = \{ E, I, P, N \}$, used to qualify the label nature, and the binary function symbols '$\copyright_{\uplus}$', '$\copyright_{\equiv}$', '$\copyright_{\wedge I}$', '$\copyright_{\rightarrow I}$', '$\copyright_{\rightarrow E}$', and '$\copyright_{\bot I}$', which are used to define how labels propagate in relation to the derivation rule being applied. The label types E, I, P, and N express the types of data that they may qualify in the database. Hence, E; I; P; and N, correspond to extensional data; intensional data; protected data; and non-supported data, respectively. The terms of $\mathcal{L}_\gamma$, denoted as t_γ, are such that $t_\gamma \in z$ where z is a type in T; and if $t_{\gamma 1}$ and $t_{\gamma 2}$ are terms of $\mathcal{L}_\gamma$, then $t_{\gamma 1} \copyright_x t_{\gamma 2}$ is also a term of $\mathcal{L}_\gamma$, where $x \in \{\uplus, \equiv, CR, \wedge I, \rightarrow I, \rightarrow E, \bot I\}$. A label γ is a term of $\mathcal{L}_\gamma$. As shown in Figure 1, the functions ' $\copyright_x$ ', combine two label types of $\mathcal{L}_\gamma$ and return another label type as result, in the case that the combination succeeds. For all the other cases not specified in Figure 1, $\gamma_1 \copyright_x \gamma_2$ is not defined.

The Labelled Database:

We consider a labelled database as the tuple $\mathcal{D} = \langle \Delta_{\mathcal{D}}, \preccurlyeq \rangle$, where $\Delta_{\mathcal{D}}$ is a set of declarative units, and $\preccurlyeq$ is an ordering on the declarative units of $\Delta_{\mathcal{D}}$. A declarative unit is a labelled formula of the form $\gamma : \alpha$, where γ is a label of $\mathcal{L}_\gamma$, and α is either a literal formula, or a conjunction of literals, or a clausal formula of $\mathcal{L}$

---

[13] A detailed description of the CIU_{LDS} formal definitions is found in [9].

[14] We consider $\top$ and $\bot$ as distinguished propositions of $\mathcal{L}$, meaning true and false, respectively.

γ_1	γ_2	$\gamma_1 \copyright_\cup \gamma_2$	$\gamma_1 \copyright_\sqsubseteq \gamma_2$	$\gamma_1 \copyright_{\wedge_I} \gamma_2$	$\gamma_1 \copyright_{\to_I} \gamma_2$	$\gamma_1 \copyright_{\to_E} \gamma_2$	$\gamma_1 \copyright_{\perp_I} \gamma_2$
E	E	E	-	E	I	-	I
E	I	-	N	I	I	I	I
E	P	-	-	-	-	I	-
E	N	E	N	N	I	-	I
I	E	-	-	I	I	-	I
I	I	I	-	I	I	I	I
I	P	P	-	-	-	I	-
I	N	-	-	N	I	-	I
P	E	-	-	-	-	-	-
P	I	P	-	-	-	-	-
P	P	P	-	-	-	-	-
P	N	-	-	-	-	-	-
N	E	E	-	N	I	-	I
N	I	-	N	N	I	N	I
N	P	-	-	-	-	I	-
N	N	N	N	N	I	-	I

Fig. 1. Labelling functions.

The intended meaning of using labelled formulae in our database representation is to provide the structural information of the database, to express the nature of each of the formulae available in and from the database. Borrowing the conceptual presentation of deductive databases, we distinguish in $\mathcal{D}$ the explicit facts from the rules. In deductive databases' terms, we distinguish between the extensional and the intensional components of our labelled database $\mathcal{D}$. The extensional data refer to the formulae stored as explicit facts, and the intensional data refer to the deductive rules, in our case the clausal formulae, and the derivable data. Our approach, however, requires that we introduce two more formulae identifications to our database. One which refers to the protected data, and another which addresses the non-supported consequences generated by the compromised solutions of our revision policy. Hence, the possible types of data in $\mathcal{D}$ are: Extensional data, which are declarative units of the form $\gamma : \alpha$, where $\gamma \in$ E,[15] and the formula α is either a literal formula, or a conjunction of literal formulae; Intensional data, which are declarative units of the form $\gamma : \alpha$, where $\gamma \in$ I, and the formula α is either a literal formula, a conjunction of literal formulae, or a clausal formula whose consequence is different from $\bot$; Protected data, or integrity constraints, which are declarative units of the form $\gamma : \alpha$, where $\gamma \in$ P, and α is a clausal formula of the form: $\bigwedge_{i=1}^{n} \beta_i \rightarrow \bot$, where each β_i is a literal formula of $\mathcal{L}$; and Non-supported data, which are declarative units of the form $\gamma : \alpha$, where $\gamma \in$ N, and α is either an atomic formula, or a conjunction of atomic formula of $\mathcal{L}$. This last sort of data will eventually appear after a *compromised revision* takes place on the database.

Example 2. Let us consider the database $\mathcal{D}_1$, where the set $\Delta_{\mathcal{D}_1}$ of declarative units is as shown below:

$$
\begin{aligned}
&(1)\ \text{I}_1\ :\ A\ \rightarrow\ B\\
&(2)\ \text{I}_2\ :\ C\ \rightarrow\ D\\
&(3)\ \text{P}_1\ :\ A \wedge C\ \rightarrow\ \bot\\
&(4)\ \text{E}_1\ :\ A
\end{aligned}
$$

We have that (1) and (2) are the intensional part of the database, (3) represents the protected part of the database, which serves as an integrity constraint; and (4) is the only explicit fact of this database. There is no non-supported data represented in this database.

The ordering relation $\preccurlyeq$ is not part of the language $\mathcal{L}_{CIU}$, but it is used to compare the declarative units of this language on a meta-level. Intuitively, the notion of an ordering, comparing the declarative units of $\Delta_{\mathcal{D}}$, states the meaning of priority or relevance to those elements. In general, such ordering is guided by the requirements of the database application. Among many different meanings that it can give to $\Delta_{\mathcal{D}}$, we can cite that it can express the novelty of the declarative units in the database, or some sort of degree of importance, for instance. Here, we leave the interpretation of the ordering $\preccurlyeq$ open, since we do not specify which database application our formalization is dealing with. However, we assume some basic properties on $\preccurlyeq$. Given a set $\Delta_{\mathcal{D}} = \{\gamma_1 : \alpha_1, \gamma_2 : \alpha_2, \cdots, \gamma_n : \alpha_n\}$, $\preccurlyeq$ is a pre-order, reflexive and transitive, on $\Delta_{\mathcal{D}}$, such that $\forall\, \gamma_i : \alpha_i,\, \gamma_j : \alpha_j \in \Delta_{\mathcal{D}}$, if $\gamma_i \in$ P and $\gamma_j \in$ P, then $\gamma_i : \alpha_i \preccurlyeq \gamma_j : \alpha_j$ and $\gamma_j : \alpha_j \preccurlyeq \gamma_i : \alpha_i$, also written as $\gamma_i : \alpha_i \approx \gamma_j : \alpha_j$. This condition states the natural concept that the protected data in $\Delta_{\mathcal{D}}$ are equivalent in the ordering. This is justified by the fact that protected data are not supposed to be

---

[15] We will use the symbol $\in$ to denote both set-membership and list-membership.

modified or retracted from $\Delta_{\mathcal{D}}$. Hence, there is no need to have them under an ordering, since we will never have to choose any single declarative unit among them to be retracted from $\Delta_{\mathcal{D}}$. Also, $\forall \, \gamma_i : \alpha_i, \, \gamma_j : \alpha_j \in \Delta_{\mathcal{D}}$, if $\gamma_i \in \mathrm{P}$ and $\gamma_j \notin \mathrm{P}$, then $\gamma_i : \alpha_i$ is not comparable to $\gamma_j : \alpha_j$ via $\preccurlyeq$. This last condition places the set of protected formulae in $\Delta_{\mathcal{D}}$ as a distinguished non-related one. The ordering $\preccurlyeq$ propagates to newly inserted declarative units in $\Delta_{\mathcal{D}}$ such that $\Delta_{\mathcal{D}} \bigcup \gamma_i : \alpha_i$ is obtained satisfying one of the following conditions: If $Z \vdash \gamma_i : \alpha_i$, for any set $Z \subseteq \Delta_{\mathcal{D}}$, such that Z is minimal w.r.t. $\subseteq$, then $\gamma_j : \alpha_j \preccurlyeq \gamma_i : \alpha_i$, $\forall \gamma_j : \alpha_j \in Z$. If $\Delta_{\mathcal{D}} \nvdash \gamma_i : \alpha_i$, then $\gamma_j : \alpha_j \preccurlyeq \gamma_i : \alpha_i$, $\forall \gamma_j : \alpha_j \in \Delta_{\mathcal{D}}$. The first condition states the propagation of the ordering $\preccurlyeq$ on the consequences of $\Delta_{\mathcal{D}}$, as a natural dominance dependency. A derived declarative unit has a higher position in the ordering than any of the declarative units involved in its derivation. The second condition states the ordering of the expanded set, by a new declarative unit which is not a consequence of $\Delta_{\mathcal{D}}$. In this case, we assume that the new declarative unit gets the highest priority in the ordering.

2.2 The Derivation Mechanisms

The proof system defined for CIU_{LDS} is given by a set of inference rules; some labelling conditions, which have to be satisfied by the inference rules in order to define the labelling propagation in the derivable declarative units; the notion of proof of a declarative unit; and the notion of the system's consequence relation. The notion of consequence is stated as a binary relation between a database and a declarative unit, denoted as $\mathcal{D} \vdash_{CIU_{LDS}} \gamma : \alpha$, (or $\mathcal{D} \vdash \gamma : \alpha$ for short). The intended meaning is to determine if we can exhibit a proof of $\gamma : \alpha$, denoted as $\rho[\gamma : \alpha]$, from $\mathcal{D}$. $\rho[\gamma : \alpha]$ is a pair $\langle P_\rho, k \rangle$, where P_ρ is a finite sequence of the pairs (or sub-derivations) A/C, $P_\rho = \{A_1/C_1, A_2/C_2, \cdots, A_n/C_n\}$, where $n > 0$, and each A_i, for $1 \leq i \leq n$, is a set of declarative units used as premises by a CIU_{LDS} inference rule IR, in order to reach the consequent C_i which is a single declarative unit. And k is a mapping from the set $\{1, \cdots, n\}$, to the set of inference rules IR_{CIU}, where $IR_{CIU} = \{\mathrm{CR}, \wedge\mathrm{I}, \wedge\mathrm{E}, \rightarrow\mathrm{I}, \rightarrow\mathrm{E}, \neg\mathrm{E}, \perp\mathrm{I}\}$, such that for each i, where $1 \leq i \leq n$, $k(i) = \mathrm{IR}$, for any $\mathrm{IR} \in IR_{CIU}$, and $A_i/C_i = A_{\mathrm{IR}}/C_{\mathrm{IR}}$.

In this formalization, we consider a convenient subset of the set of inference rules relative to each connective defined in the language, presented in the natural deduction style. Given a database $\mathcal{D}$, an inference rule IR is defined as a tuple $\langle A_{\mathrm{IR}}, \varphi_{\mathrm{IR}}, C_{\mathrm{IR}} \rangle$, where A_{IR} indicates a set of declarative units, in or derived by $\mathcal{D}$, used as antecedents, or premises, of the rule; φ_{IR} denotes the labelling condition which needs to be satisfied by the application of the IR inference rule; and C_{IR} represents the declarative unit derived from A_{IR} via the inference rule IR, provided that φ_{IR} holds. Figure 2 shows the inference rules of our system. Namely, conditional reflexivity; $\wedge$ introduction; $\wedge$ elimination; $\rightarrow$ introduction; $\rightarrow$ elimination; $\neg$ elimination; and $\perp$ introduction.

The inference rules, as well as update operations on the database, depend on some labelling propagation conditions, which compose the algebra of labels.

CIU_{LDS} Inference Rules:

Conditional Reflexivity: $\langle A_{\text{CR}}, \varphi_{\text{CR}}, C_{\text{CR}} \rangle$,
where $A_{\text{CR}} = \{ \gamma : \alpha \}$, $C_{\text{CR}} = \gamma : \alpha$, and $\varphi_{\text{CR}}(\gamma, \gamma_i) \in \mathcal{A}$, for any γ_i, such that $\mathcal{D}' \vdash \gamma_i : \alpha$, where $\mathcal{D}' = \mathcal{D} - \{\gamma : \alpha\}$.

$\wedge$ **Introduction:** $\langle A_{\wedge\text{I}}, \varphi_{\wedge\text{I}}, C_{\wedge\text{I}} \rangle$, where $\varphi_{\wedge\text{I}}(\gamma_1, \gamma_2) \in \mathcal{A}$,

$$\frac{\gamma_1 : \alpha_1 \qquad \gamma_2 : \alpha_2}{\gamma_3 : \alpha_1 \wedge \alpha_2}$$

$\wedge$ **Elimination:** $\langle A_{\wedge\text{E}}, \varphi_{\wedge\text{E}}, C_{\wedge\text{E}} \rangle$, where $\varphi_{\wedge\text{E}}(\gamma_1, \gamma_2) \in \mathcal{A}$,

$$\frac{\gamma_1 : \alpha_1 \wedge \alpha_2}{\gamma_2 : \alpha_1} \qquad \frac{\gamma_1 : \alpha_1 \wedge \alpha_2}{\gamma_2 : \alpha_2}$$

$\rightarrow$ **Introduction:** $\langle A_{\rightarrow\text{I}}, \varphi_{\rightarrow\text{I}}, C_{\rightarrow\text{I}} \rangle$, where $\varphi_{\rightarrow\text{I}}(\gamma_1, \gamma_2) \in \mathcal{A}$,

$$\frac{\mathcal{D}, \gamma_1 : \alpha_1 \vdash \gamma_2 : \alpha_2}{\gamma_3 : \alpha_1 \rightarrow \alpha_2}$$

$\rightarrow$ **Elimination:** $\langle A_{\rightarrow\text{E}}, \varphi_{\rightarrow\text{E}}, C_{\rightarrow\text{E}} \rangle$, where $\varphi_{\rightarrow\text{E}}(\gamma_1, \gamma_2) \in \mathcal{A}$,

$$\frac{\gamma_1 : \alpha_1 \qquad \gamma_2 : \alpha_1 \rightarrow \alpha_2}{\gamma_3 : \alpha_2}$$

$\neg$ **Elimination:** $\langle A_{\neg\text{E}}, \varphi_{\neg\text{E}}, C_{\neg\text{E}} \rangle$, where $\varphi_{\neg\text{E}}(\gamma_1, \gamma_2) \in \mathcal{A}$,

$$\frac{\gamma_1 : \neg\neg\alpha}{\gamma_2 : \alpha}$$

$\perp$ **Introduction:** $\langle A_{\perp\text{I}}, \varphi_{\perp\text{I}}, C_{\perp\text{I}} \rangle$, where $\varphi_{\perp\text{I}}(\gamma_1, \gamma_2) \in \mathcal{A}$,

$$\frac{\gamma_1 : \neg\alpha_1 \qquad \gamma_2 : \alpha_1}{\gamma_3 : \perp}$$

Fig. 2. Inference Rules.

These conditions are used to monitor the application of inference rules and the basic update operations. We denote a labelling condition with $\varphi_x(\gamma_1, \gamma_2)$, where x identifies the inference rule or the database operation that it stands for, and γ_1 and γ_2 are the labels which should be combined to satisfy the condition φ_x. Given any labels γ_1 and γ_2, the labelling propagation conditions $\varphi_x(\gamma_1, \gamma_2)$, where $x \in IR_{CIU} \bigcup UP$, where UP is the set of the update operations of CIU_{LDS}, $UP = \{\uplus, \Xi\}$, satisfy the following: For $x \in \{\uplus, \Xi, \wedge I, \rightarrow I, \rightarrow E, \bot I\}$, $\varphi_x(\gamma_1, \gamma_2)$ holds if $\gamma_1 \copyright_x \gamma_2$ returns a label γ_3. For $x = CR$, $\varphi_{CR}(\gamma_1, \gamma_2)$ holds if $\gamma_1 \in E$, and $\gamma_2 \in I$. For $x = \wedge E$, $\varphi_{\wedge E}(\gamma_1, \gamma_2)$ holds if $\gamma_1 \in E \cup I \cup N$ and γ_2 belongs to the same label type as γ_1; For $x = \neg E$, $\varphi_{\neg E}(\gamma_1, \gamma_2)$ holds if γ_1 and γ_2 belong to the same label type z, where $z \in T$ in $\mathcal{L}_\gamma$. And for any other case not specified above, φ_x does not hold. The algebra of labels is simply defined as the set of all the labelling conditions of the form $\varphi_x(\gamma_1, \gamma_2)$, which hold in our system. $\mathcal{A} = \{\varphi_x(\gamma_1, \gamma_2) \mid \varphi_x(\gamma_1, \gamma_2) \text{ holds}\}$. A Labelled Deductive System CIU_{LDS} is a tuple $\langle \mathcal{A}, \mathcal{L}_{CIU}, M_{CIU} \rangle$, where $\mathcal{A}$ is the algebra of labels and $\mathcal{L}_{CIU}$ is the system language, and M_{CIU} represents the possible deduction and change mechanisms of the system. M_{CIU} includes the set of all inference rules and the set of update operations of CIU_{LDS}.

2.3   The Inputs

The inputs of the system are update requests, which invoke an update function of the form $U(\mathcal{D}, \sigma, \delta) \Rightarrow \mathcal{D}'$, where $\mathcal{D}$ and $\mathcal{D}'$ are labelled databases before and after the update, respectively. σ is the type of update to be performed, and δ is the data involved in the update. $\sigma = \{U_+, U_-\}$, where ' U_+ ' implies that the update requests an addition of the argument δ , and ' U_- ' implies that the update operation requested is a deletion. The δ argument denotes a declarative unit $\gamma : \alpha$.

Basically, the allowed updates in CIU_{LDS} w.r.t. declarative units, are the ones involving addition or deletion of declarative units which are either extensional data or intensional data. Atomic formulae of $\mathcal{L}$, can be used in the update only as explicit facts. The intensional data involved in the updates are not supposed to be atomic formulae of $\mathcal{L}$, since these forms of intensional data represent the derivable data from the database. However, derivable formulae are allowed to be added to the database, when they are introduced as explicit facts, using extensional labels. These restrictions avoid the manipulation of non-supported data, and of protected data in the updates, as expected. Non-supported data can only be derived or added to the database by the system, as a compromised solution. And protected formulae cannot be modified by means of updates to the database. Example 3 illustrates some allowed updates in CIU_{LDS}.

Example 3. According to the database $\mathcal{D}_1$ of example 2, the declarative unit $I_3 : B$ is not allowed to be involved in an update request, however it would be allowed if it were given the form of an explicit fact: $E_3 : B$. Hence, $U(\mathcal{D}_1, U_+, E_3 : B)$ is considered as a valid update request, as well as $U(\mathcal{D}_1, U_-, E_1 : A)$, and

$U(\mathcal{D}_1, \; \mathrm{U}_-, \; \mathrm{I}_2 : C \rightarrow D)$. *On the other hand,* $U(\mathcal{D}_1, \; \mathrm{U}_+, \; \mathrm{P}_2 : E \rightarrow \perp)$, *and* $U(\mathcal{D}_1, \; \mathrm{U}_+, \; \mathrm{N}_1 : D)$ *would not be valid update requests.*

The update function invokes the update operations of conditional addition $\uplus$, and compromised retraction, Ξ, such that: $Up(\mathcal{D}, \; \mathrm{U}_+, \; \gamma : \alpha) = \mathcal{D} \uplus \gamma : \alpha$; and $Up(\mathcal{D}, \; \mathrm{U}_-, \; \gamma : \alpha) = \mathcal{D} \; \Xi \; \gamma : \alpha$. These basic update operations carry the reconciling flavour of our compromised approach to conflicting inputs. The operation of conditional inclusion $\uplus$, invokes the compromised revision function $\odot$, when a compromised solution for the update applies. The compromised retraction operation Ξ, already embeds in its definition the mechanism for allowing consequences of retracted declarative units to be added to the database as non-supported data. This operation uses the notion of safe-maximality,[16] when a choice is needed among the compromised consequences.

2.4 The Reconciling Notion

In CIU_{LDS}, the notion of reconciling conflicting updates with the underlying database is mainly represented by the compromised revision function $\odot$, and by the compromised contraction function Ξ. When we get an input request $U(\mathcal{D}, \; \mathrm{U}_+, \; \gamma : \alpha)$, the basic operation of conditional inclusion of a declarative unit into a labelled database is invoked. This function is denoted by the operation $\uplus$. We define that $\mathcal{D} \uplus \gamma : \alpha = \mathcal{D}'$, such that $\mathcal{D}' = \mathcal{D} \bigcup \{\gamma : \alpha\}$, if $\Delta_{\mathcal{D}}, \; \gamma : \alpha \nvdash \perp$; and $\mathcal{D} \odot \gamma : \alpha$, otherwise. In the case that a revision applies within the operation $\uplus$, we have the two following possibilities:

1. $I, \gamma : \alpha \vdash \perp$. This means that the declarative unit violates the integrity constraints in $\mathcal{D}$ directly. In this case, $\gamma : \alpha$ is not allowed to be inserted in $\Delta_{\mathcal{D}}$. However, the compromised revision function takes care of including in the revised database, all the consistent consequences of $\gamma : \alpha$ w.r.t. $\mathcal{D}$, as a compromised solution.

2. $\Delta_{\mathcal{D}}, \gamma : \alpha \vdash \perp$ when $I, \gamma : \alpha \nvdash \perp$. In this case, a revision applies in order to accomodate $\gamma : \alpha$ in the resulting database and to preserve consistency. As defined in the policies of our approach, the compromised revision allows for the consistent consequences of retracted declarative units to remain available in the resulting database. The ordering $\preccurlyeq$ on $\Delta_{\mathcal{D}}$ facilitates

---

[16] We define the notion of a *safe-maximal subset* of an ordered set X, w.r.t. a condition c which involves X in its premise, is denoted as $Smax(X)_c$. $Smax(X)_c$ is defined with the aim that when it substitutes X in the condition c, c succeeds. The set $Smax(X)_c$ is obtained considering the ordering $\preccurlyeq$, the set-inclusion property of minimality, and some auxiliary sets denoted as $Fail(X)_c$, $min(Fail(X)_c)$, $Min(Fail(X)_c)$, and $RMin(Fail(X)_c)$. The set $Fail(X)_c$ contains all the minimal subsets of X w.r.t. $\subseteq$, such that when they substitute X in the condition c, c fails. The set $min(Fail(X)_c)$ contains the subsets of minimal elements w.r.t. $\preccurlyeq$, of each set belonging to $Fail(X)_c$. The set $Min(Fail(X)_c)$ contains all the elements of each set belonging to $min(Fail(X)_c)$. And the set $RMin(Fail(X)_c)$ is a refined construction of the set $Min(\mathcal{Y})$, so that fewer elements of the original ordered set are removed from it.

the process of choosing which declarative unit to discard, in order to regain consistency.

When the input request is $U(\mathcal{D}, \, \text{U}_-, \, \gamma : \alpha)$, the operation Ξ is invoked. Ξ retracts a declarative unit from a labelled database in a compromised way. That is, the compromised contraction function, written as $\Xi(\mathcal{D}, \gamma : \alpha) = \mathcal{D}'$ or $\mathcal{D}' = \mathcal{D} \, \Xi \, \gamma : \alpha$, retracts the existing declarative unit $\gamma : \alpha$ from $\Delta_\mathcal{D}$, and inserts to it the consequences of $\gamma : \alpha$ w.r.t. $\mathcal{D}$ as a compromised solution, provided that $\mathcal{D}' \not\vdash \gamma : \alpha$. In this case, a revision is not needed, because there is no chance that $\mathcal{D} \, \Xi \, \gamma : \alpha \vdash \bot$, since we always consider that $\mathcal{D}$ is initially consistent. The compromised retraction is a very straightforward operation. Basically, if the declarative unit to be retracted is present in the database, it deletes it and preserves its consistent consequences as non-supported consequences. Otherwise, no operation is performed on the database.

The operations $\uplus$ for data inclusion, and Ξ for data retraction, can be described as algorithms which are defined on top of the notion of the consequence relation. That is, the definition of these operations make use of the proof procedure which presents the consequence relation $\vdash$. The consequences which those operations add to the database, without support from their premises, carry the non-supported label type and are subject to continuous checking by the proof system, as the database is further modified. We can say that the derivation mechanism of our system is sensible to the presence of the non-supported declarative units. By this we mean that it applies some restrictions, in the case that a non-supported declarative unit is involved in a derivation process.

2.5 Revision in CIU_{LDS}

In CIU_{LDS}, given a database $\mathcal{D} = \langle \Delta_\mathcal{D}, \preccurlyeq \rangle$, and a declarative unit $\gamma : \alpha$, such that $\gamma : \alpha \in \Delta_\mathcal{D}$, a revision is invoked by the update $Up(\mathcal{D}, \text{U}_+, \gamma : \alpha)$, when one of the following conditions holds: (1) If $I, \gamma : \alpha \vdash \gamma' : \bot$, for some label $\gamma' \in \mathcal{L}_\gamma$; (2) If $I, \gamma : \alpha \not\vdash \gamma' : \bot$, and $\Delta_\mathcal{D}, \gamma : \alpha \vdash \gamma' : \bot$ for $\gamma' \in \mathcal{L}_\gamma$. Where $I = \{\gamma_i : \alpha_i \mid \forall \gamma_i : \alpha_i \in \Delta_\mathcal{D}, \text{ such that } \gamma_i \in \text{P}\}$.

When Condition (1) applies, we take the following revision steps on $\mathcal{D}$:

- If α contradicts a tautology of the underlying logical system, then we reject the update request, and the result of the compromised revision is the original database, such that $\mathcal{D} \odot \gamma : \alpha = \mathcal{D}$.
- Otherwise, we do not add $\gamma : \alpha$ to $\Delta_\mathcal{D}$, but we generate a set of consequences of $\gamma : \alpha$ with respect to $\mathcal{D}$, and we add to $\Delta_\mathcal{D}$ a safe-maximal subset of this set of consequences, relative to the condition that the consequences in it together with $\Delta_\mathcal{D}$ are not inconsistent and do not derive $\gamma' : \alpha$, for any label γ' of $\mathcal{L}_\gamma$.

When Condition (2) applies, we add the input $\gamma : \alpha$ to $\Delta_\mathcal{D}$, and we reject from $\Delta_\mathcal{D}$ some old declarative units, which are not protected, to regain consistency. We adopt the safe-maximality notion whenever we have to choose among a set of declarative units to be retracted from the ordered database, for not satisfying a particular condition imposed to it. We also take into account the inclusion of the consistent consequences of the retracted sentences, as non-supported data.

Example 4. Let us consider the labelled database $\mathcal{D}_1$ of example 2, and the update request $U(\mathcal{D}_1, \text{U}_+, \text{E}_2 : C)$. When the declarative unit $\text{E}_2 : C$ is added to $\Delta_{\mathcal{D}_1}$ as requested, there is an integrity constraint violation. This invokes the compromised revision function. Since the incoming data has priority over the existing data in the database, we will end up retracting $\text{E}_1 : A$ in order to be able to incorporate $\text{E}_2 : C$ into $\mathcal{D}_1$. Hence, the resulting database after the compromised revision is the following:

$$
\begin{aligned}
&(1)\ \text{I}_1\ :\ A\ \rightarrow\ B\\
&(2)\ \text{I}_2\ :\ C\ \rightarrow\ D\\
&(3)\ \text{P}_1\ :\ A \wedge C\ \rightarrow\ \bot\\
&(4)\ \text{E}_2\ :\ C\\
&(5)\ \text{N}_1\ :\ B
\end{aligned}
$$

It is important to notice in the example above, that the declarative unit (5) would not be derivable from the resulting database, had the update operation been a conventional one, without any embedded reconciling revision notion for conflicting updates.

2.6   Transactions

A transaction in CIU_{LDS} is a sequence of updates to be performed on a given labelled database. Let $UP = Up_1, \cdots, Up_n$ be a sequence of update functions where for each $1 \leq i \leq n$ $Up_i(\mathcal{D}_i, \sigma_i, \delta_i) \rightarrow \mathcal{D}'_i$ is such that $\mathcal{D}_i = \mathcal{D}'_{i-1}$, for $i = 1$ $\mathcal{D}_1 = \mathcal{D}$, and for $i = n$ $\mathcal{D}'_n = \mathcal{D}'$. Let UP_{seq} be the set of all the sequences of updates of this form. A CIU_{LDS} Transaction is a function $Trans(\mathcal{D}, UP) = \mathcal{D}'$. A transaction, then incorporates the results of various single updates in sequence. Within each update request Up_i of a transaction, compromised revision $\odot$ may apply. Since, by the definition of single updates, the resulting updated databases are consistent, then we also have that the result of a transaction is a consistent labelled database.[17]

3   Properties of the System CIU_{LDS}

3.1   Properties of the CIU_{LDS} Consequence Relation

As already expected for a consequence relation which does not follow the classical standards, the CIU_{LDS} consequence relation relaxes both reflexivity and

---

[17] Alternatively, one can also consider a more realistic definition of transaction, which only requires that consistency holds at the end of the sequence of updates.

monotonicity conditions. We consider, however, a restricted version of reflexivity, and a weaker monotonicity condition, to be satisfied by $\vdash_{CIU_{LDS}}$.[18]

The consequence relation $\vdash_{CIU_{LDS}}$ satisfies the property of restricted reflexivity, as long as it satisfies the application of the conditional reflexivity inference rule CR. That is, provided that the condition applied to the CR inference rule holds, when required.

The Restricted Monotonicity notion basically guarantees that the addition of a derivable sentence to a knowledge base does not cause any harm to further inferences of the base. In our system, $\vdash_{CIU_{LDS}}$ satisfies the Restricted Monotonicity property, such that $\dfrac{\mathcal{D} \vdash \gamma_1 : \alpha \quad \mathcal{D} \vdash \gamma_2 : \beta}{\mathcal{D}, \gamma_1 : \alpha \vdash \gamma_2 : \beta}$.

The Restricted Monotonicity property also applies to CIU_{LDS}, when we consider the conditional inclusion operation $\uplus$ instead of set inclusion for adding the derivable sentence to the base. If a declarative unit is a consequence of the database, then if we add it to the database via the conditional inclusion operation, restricted to some labelling conditions, this should not interfere with further inferences of the system. $\dfrac{\mathcal{D} \vdash \gamma_1 : \alpha \quad \mathcal{D} \vdash \gamma_2 : \beta}{\mathcal{D} \uplus \gamma_1' : \alpha \vdash \gamma_2' : \beta}$, where $\mathcal{D} \uplus \gamma_1' : \alpha$ is the conditional addition of the declarative unit $\gamma_1' : \alpha$ to the database $\mathcal{D}$, such that $\gamma_1 \notin$ P, $\gamma_1' \in$ E $\cup$ I such that $\varphi_\uplus(\gamma_1', \gamma_1) \in$ A, where γ_1' is obtained as follows: If $\gamma_1 \in$ I and α is a clausal formula, then $\gamma_1' \in$ I. If $\gamma_1 \in$ I and α is not a clausal formula, then $\gamma_1' \in$ E. If $\gamma_1 \in$ E, then $\gamma_1' \in$ E. If $\gamma_1 \in$ N, then $\gamma_1' \in$ E. And γ_2' is restricted to the following condition: If $\gamma_1 : \alpha = \gamma_2 : \beta$, then $\gamma_2' = \gamma_1'$. Otherwise, $\gamma_2' = \gamma_2$.

CIU_{LDS} satisfies the deduction property, under some labelling restrictions. Given a database $\mathcal{D} = \langle \Delta_\mathcal{D}, \preccurlyeq \rangle$, and the declarative units $\gamma_1 : \alpha \to \beta$ and $\gamma_2 : \beta$, we say that the consequence relation $\vdash_{CIU_{LDS}}$ satisfies the Deduction Property, if we have that: $\mathcal{D} \vdash \gamma_3 : \alpha \to \beta$ iff $\mathcal{D}, \gamma_1 : \alpha \vdash \gamma_2 : \beta$, where γ_3 is given by $\gamma_1 \copyright_{\to_I} \gamma_2$, γ_2 is given by the labelling conditions of the inference rules applied in the derivation of $\gamma_2 : \beta$, and α is not a clausal formula.[19]

CIU_{LDS} also satisfies a restricted version of the Deduction Theorem, which justifies the introduction of the logical connective $\to$, by taking into account the conditional inclusion operation, and some labelling restrictions such that: $\mathcal{D} \vdash \gamma_3 : \alpha \to \beta$ iff $\mathcal{D} \uplus \gamma_1 : \alpha \vdash \gamma_2 : \beta$, where γ_3 is given by $\gamma_1 \copyright_{\to_I} \gamma_2$, $\gamma_1 \in$ E $\cup$ I, $\gamma_2 \notin$ N,[20] and α is not a clausal formula.

In general, the full version of Transitivity, or Cut, is such that for all sentences A, B and sets of sentences Δ and Γ, if $\Delta \vdash A$ and $\Gamma, A \vdash B$, then $\Gamma, \Delta \vdash B$. Strong Transitivity[21] states the following variation of the cut

---

[18] We will also refer to $\vdash_{CIU_{LDS}}$ as $\vdash$, for short, whenever its meaning is not compromised.

[19] α is constrained from being a clausal formula, because the system does not support embedded implications.

[20] In this case, γ_2 is constrained from being of type N, because this would allow for the case in which $\gamma_1 : \alpha \notin \mathcal{D} \uplus \gamma_1 : \alpha$.

[21] Also referred to as Unitary Cut in [15].

rule: if $\Delta \vdash A$ and $\Delta, A \vdash B$, then $\Delta \vdash B$. CIU_{LDS} satisfies Strong Transitivity, provided that some labelling restrictions apply.

$$\frac{\mathcal{D} \vdash \gamma_1 : \alpha \quad \mathcal{D}, \gamma_1 : \alpha \vdash \gamma_2 : \beta}{\mathcal{D} \vdash \gamma_2' : \beta},$$ where γ_2' is given by the labelling condi-

tions of the inference rules used for the derivation of $\gamma_2' : \beta$.

Strong Transitivity also holds for the case where we consider $\gamma_1 : \alpha$ added to $\mathcal{D}$, via the conditional addition operation. That is, when we consider $\mathcal{D} \uplus \gamma_1 : \alpha$, provided that some labelling restrictions apply, such that:

$$\frac{\mathcal{D} \vdash \gamma_1 : \alpha \quad \mathcal{D} \uplus \gamma_1' : \alpha \vdash \gamma_2 : \beta}{\mathcal{D} \vdash \gamma_2' : \beta},$$ where γ_2' is given by the labelling con-

ditions of the inference rules used for the derivation of $\gamma_2' : \beta$, and $\gamma_1' \in E \cup I$, such that: If $\gamma_1 \in I$ and α is a clausal formula, then $\gamma_1' \in I$. If $\gamma_1 \in I$ and α is not a clausal formula, then $\gamma_1' \in E$. If $\gamma_1 \in E \cup N$, then $\gamma_1' \in E$.

Non-explosiveness is a notion considered by some pragmatic formalisms, when dealing with contradictory information and conflicting data in general. In traditional logical approaches, the system is forced to collapse when inconsistency is detected. As pointed out in [34], *"It seems to be an unnatural overreaction to abandon a knowledge base once it is discovered to be inconsistent. Rather, one should accommodate it by means of a logic which continues to function plausibly under inconsistency."* CIU_{LDS} supports this viewpoint, and the consequence relation $\vdash_{CIU_{LDS}}$ satisfies Non-explosiveness, such that when $\mathcal{D}, \gamma : \alpha \vdash \gamma^* : \bot$, we do not have that $\mathcal{D}, \gamma : \alpha \vdash \gamma' : \alpha'$, for any $\gamma' : \alpha' \in \mathcal{L}_{CIU}$.

3.2 Properties of the Compromised Revision $\odot$

The revision function $\odot$ preserves the structural properties of a given initial database $\mathcal{D}$, if by revising $\mathcal{D}$ by an input $\gamma : \alpha$, the resulting database $\mathcal{D} \odot \gamma : \alpha$ presents the same structural organization of the initial one. $\mathcal{D} = \langle \Delta_{\mathcal{D}}, \preccurlyeq \rangle$ and $\mathcal{D} \odot \gamma : \alpha = \langle \Delta_{\mathcal{D} \odot \gamma : \alpha}, \preccurlyeq \rangle$.

The revision function $\odot$ is consistent. For any initial consistent database $\mathcal{D}$, and any input $\gamma : \alpha$, if we revise $\mathcal{D}$ by the input, the resulting database $\mathcal{D} \odot \gamma : \alpha$ is consistent. $\mathcal{D} \odot \gamma : \alpha \nvdash \gamma' : \bot$, for any label γ' of $\mathcal{L}_\gamma$.

Compromised revision satisfies a restricted version of the persistence property, which we called *Compromised Persistence*. Basically, the persistence notion states that as much of the original base should survive a revision as possible. Hence, by revising a database $\mathcal{D}$ with a sentence $\gamma : \alpha$ and then retracting $\gamma : \alpha$, we should be able to derive from the resulting database, all the consequences of $\mathcal{D}$ that do not directly contradict $\gamma : \alpha$. The compromised version of this persistence notion, considers that we should be able to derive from the resulting base all the consequences that do not directly contradict $\gamma : \alpha$, and also that do not violate integrity contraints in $\mathcal{D}$. $\forall \gamma' : \alpha'$ such that $\mathcal{D} \vdash \gamma' : \alpha'$, $\gamma' : \alpha' \neq \gamma : \alpha$ and $\gamma' : \alpha' \notin R_{\gamma:\alpha}^*$, $\mathcal{D}' \vdash \gamma' : \alpha'$, where $\mathcal{D}' = Up(\mathcal{D} \odot \gamma : \alpha, U_-, \gamma : \alpha)$.

3.3 Properties of the CIU_{LDS} Updates

The update function preserves the structural properties of a given database $\mathcal{D}$, if by updating $\mathcal{D}$ with an input $\gamma : \alpha$, the resulting database $\mathcal{D}'$ presents the same structural organization of $\mathcal{D}$. That is, $Up(\mathcal{D}, \sigma, \gamma : \alpha) = \mathcal{D}'$, where $\sigma = \{U_{+}, U_{-}\}$, and $\mathcal{D}' = \langle \Delta_{\mathcal{D}'}, \preccurlyeq \rangle$.

The update function $Up(\mathcal{D}, \sigma, \delta)$ is consistent. For any initial consistent database $\mathcal{D}$, and any declarative unit $\gamma : \alpha$, if we update $\mathcal{D}$ with $\gamma : \alpha$, the resulting database $\mathcal{D}'$ is always consistent. $Up(\mathcal{D}, \sigma, \gamma : \alpha) = \mathcal{D}'$, and $\mathcal{D}' \nvdash \gamma' : \perp$, for any label γ' of $\mathcal{L}_{\gamma}$.

3.4 Properties of the CIU_{LDS} Transactions

Since a transaction is defined as a sequence of updates, and we have already proved that the update function preserves the structural properties of a given database $\mathcal{D}$, then it is trivially guaranteed that a transaction $Trans(\mathcal{D}, UP)$, in CIU_{LDS}, returns a resulting database $\mathcal{D}'$, which presents the same structural organization of $\mathcal{D}$.

It is also trivial to show that a transaction $Trans(\mathcal{D}, UP)$ is consistent, since we have already shown that each update operation, in its sequence of updates, returns a resulting database $\mathcal{D}'$ which is always consistent.

4 Final Remarks

In this paper, we have introduced our approach of *Compromised Interfering Updates*, and we have presented the logical system CIU_{LDS}, as a labelled realization to the approach, within the Labelled Deductive Systems framework (LDS). Some relevant properties were presented, establishing the main results of the system.[22] The revision method embedded in the system CIU_{LDS}, is defined for the specific case where compromised criteria of non-supported consequences apply. Thus, within this system, we re-enforce the importance of having different theory change operators available for specific applications, in order to construct more realistic update systems.

The notion of compromised reasoning for revision in knowledge bases, can be expanded in many different directions. Each one concerning a way of taking the most out of what has to be given up in the revision process, for restoring consistency. It is then important to notice that in this work, we have only explored the notion of taking advantage of consistent consequences of the given up elements as a compromise. However, various other ways within this reasoning notion, still remain to be studied.

Under the compromised philosophy of the approach presented here, some other pieces of work were developed by the author. In [7], the problem of dealing with inconsistency after the performance of a database transaction is addressed, within the context of deductive databases. [9] and [10] present more details about

---

[22] Full proofs of the properties cited in this paper are available in [8] and [10].

compromised reasoning and the compromised revision within a belief revision perspective. [10] presents a brief comparison with Truth Maintenance Systems. A more detailed study on their differences, advantages and limitations, is currently under development. As further work, we plan to investigate compromised solutions for modelling simultaneous occurrence of actions, where we have to tackle problems which arise when reasoning about possible conflicts and combined effects of those actions. We believe that this area can benefit much from compromising on solutions.

Acknowledgements

This work was sponsored by the Brazilian Research Council (CNPq), under the grant 202078/90.6. The author is very grateful to Dov Gabbay for his support as a research supervisor. Special thanks to Doris Aragon, Krysia Broda, Sanjay Modgil, Hans Jürgen Ohlbach, Odinaldo Rodrigues, and Alessandra Russo, who carefully read and commented different versions of the document which originated this work. Thanks also to anonymous referees, for their constructive comments and suggestions on an earlier version of this paper.

References

1. Alchourrón, C., Makinson, D.: The Logic of Theory Change: Contraction functions and their associated functions. Theoria. **48** (1982).
2. Alchourrón, C., Makinson, D.: On the Logic of Theory Change: Safe Contractions. Studia Logica. **44** (1985).
3. Alchourrón, C., Makinson, D.: Maps between some different kinds of contractions functions. Studia Logica. **45** (1986).
4. Alchourrón, C., Gärdenfors, P., Makinson, D.: On the Logic of Theory Change: partial meet functions for contraction and revision. Journal of Symbolic Logic. **50** (1985).
5. Boutilier, C., Goldszmidt, M.: Revision by Conditionals Beliefs. Proc. AAAI Conference. (1993).
6. Cholvy, L.: Proving Theorems in a Multi-Source Environment. Proc. IJCAI-93, Chambery. Vol.1 (1993) 66–71.
7. Dargam, F.C.C.: Compromised Updates in Deductive Databases. Research Report, Imperial College, UK. (1996).
8. Dargam, F.C.C.: On Compromising Updates in Labelled Databases. Imperial College - Department of Computing Research Report DoC-96/2. (1996).
9. Dargam, F.C.C.: A Compromised Characterization to Belief Revision. Imperial College - Department of Computing Research Report DoC-96/1. (1996).
10. Dargam, F.C.C.: On Reconciling Conflicting Updates: A Compromised Revision Approach. PhD. Thesis. Department of Computing, Imperial College, UK. (1996 - to appear).
11. Dargam, F.C.C., Gabbay, D.M.: Resolving Conflicting Actions and Updates. (extended abstract). Proc. Compulog Net Meeting on Knowledge Representation and Reasoning CNKRR'93. Lisbon. (1993).
12. Doyle, J.: A Truth Maintenance System. Artificial Intelligence. **12** (1979) 231–272.

13. Elkan, C.: A Rational Reconstruction of Non-monotonic Truth Maintenance Systems. Artificial Intelligence. **43(2)** (1990) 219–234.
14. Freund, M., Lehmann,D.: Belief Revision and Rational Inference. Technical Report TR94-1. Institute of Computer Science, The Hebrew University of Jerusalem. Israel (1994).
15. Gabbay, D.M.: LDS - Labelled Deductive Systems - Volume I Foundations. (Intermediate draft of a forthcoming book by Oxford University Press). MPI-I-94-223. Max-Planck-Institut fur Informatik. Saarbrucken, Germany.(1994).
16. Gabbay, D.M., Hunter, A.: Making Inconsistency Respectable - Part 1: A Logical Framework for Inconsistency in Reasoning. Fundamentals of AI Research. Springer-Verlag, LNCS **535** (1991).
17. Gabbay, D.M., Hunter, A.: Making Inconsistency Respectable - Part 2: Meta-level handling of inconsistency. Springer-Verlag, LNCS **747** (1993).
18. Galliers, J.: The Positive Role of Conflict in Cooperative Multiagent Systems. Decentralized AI. Demazeau, Y., Muller, J.P., eds. North-Holland. (1990) 33–46.
19. Giordano, L., Martelli, A.: An Abductive Characterization of the TMS. Springer-Verlag, LNAI **515**. Proc. TMS ECAI-90. Martins, J.P., Reinfrank, M., eds. (1990).
20. Jackson,P., Pais, J.: Semantic Accounts of Belief Revision. Springer-Verlag, LNAI **515**. Proc. TMS ECAI-90. Martins, J.P., Reinfrank, M., eds. (1990).
21. Katsuno, H., Mendelzon, A.O.: On the Difference between Updating a Knowledge Base and Revising it. Belief Revision. Gärdenfors, P., editor. Cambridge University Press. (1992) 183–203.
22. Kleer, J.d.: An Assumption-based TMS. Artificial Intelligence. **28** (1986) 127–162.
23. Makinson, D.: How to give it up. Synthese. **62** (1985).
24. Makinson, D.: Five Faces of Minimality. Studia Logica. **53(3)** (1993).
25. Monteiro, D., Bertino, E., Martelli, M.: Transactions and Updates in Deductive Databases. Imperial College - Department of Computing Research Report DoC-95/2. (1995).
26. Nebel, B.: A Knowledge level Analysis of Belief Revision. Proc. 1st. Conference on Principles of Knowledge Representation and Reasoning. (1989).
27. Nebel, B.: Reasoning and Revision in Hybrid Representation Systems. Chapter 6 - Belief Revision. Springer-Verlag, LNAI **422** (1990).
28. A Truth Maintenance System Based on Stable Models. Pimentel, S. G., Cuadrado, J. L.: Proc. North American Conference on Logic Programming NACLP. Cleveland, USA. (1989) 274–290.
29. Rao, A., Foo, N.: Minimal Change and Maximal Coherence: A Basis for Belief Revision and Reasoning about Actions. Proc. 11th International Joint Conference on Artificial Intelligence. (1989) 966–971.
30. Rodi, W., Pimentel, S.: A Nonmonotonic Assumption-Based TMS using Stable Bases. Proc. KR'91. (1991) 485–495.
31. Rott, H.: Two Methods of Constructing Contractions and Revisions of Knowledge Systems. Journal of Philosophical Logic. **20** (1991).
32. Rott, H.: On the Logic of Theory Change: more maps between different kinds of contractions functions. Belief Revision. Gärdenfors, P., editor. Cambridge University Press. (1992) 122–141.
33. Sripada, S.: A Temporal Approach to Belief Revision in Knowledge Bases. Proc. 9th. IEEE Conference on AI for Applications CAIA-93. Florida, USA. (1993).
34. Wagner, G.: Vivid Logic. Springer Verlag, LNAI **764** (1994).
35. Wang, X., Chen, H.: On the Semantics of TMS. Proc. IJCAI '91. (1991) 306–309.
36. Winslett, M.: Updating Logical Databases. Cambridge University Press. (1990).

An Inference Engine for Propositional Two-valued Logic Based on the Radical Membership Problem *

Eugenio Roanes-Lozano[1], Luis M. Laita[2],
and Eugenio Roanes-Macías[1]

[1] Universidad Complutense de Madrid, Dept. Algebra (Fac. Educación)
Paseo Juan XXIII s/n, 28040-Madrid, Spain
[2] Universidad Politécnica de Madrid, Dept. I.A. (Fac. Informática)
Campus de Montegancedo, Boadilla del Monte, 28660-Madrid, Spain

Abstract. In this paper, the well-known Radical Membership Problem of Commutative Algebra[3] is adapted to develop an implementation of the inference processes in Knowledge Based Systems.

Let Σ be a set of propositions, Γ a proposition, and let us denote their images, in a certain isomorphism, by σ and γ respectively. It can be established whether or not Γ follows from Σ ($\Sigma \vdash \Gamma$) by checking the equality of ideals $< \sigma \cup \{1 - t \cdot \gamma\} >=< 1 >$ in the polynomial model (t is a new variable). As a consequence, a criterion for consistency is obtained.

1 About This Approach

1.1 Previous Steps

In a previous paper [LR1], a certain polynomial quotient ring, R/I, is constructed as a model for propositional Boolean algebras. Properties of ideals in both structures are detailed. The advantage compared with respect to work with rewriting rules is that a nice structure is provided.

At the end of the article, it is suggested that consistency of KBSs[4] could be studied in the polynomial model by moving to the quotient ring over a certain ideal J: $(R/I)/J$.

In [LR2] and [LR3], the suggestion mentioned above is developed. Consistency is studied by checking if the polynomial forms of the tautology and the

---

* Work partially supported by DGICYT (Spain), project PB-94-0424.

[3] Usually checked with Gröbner Basis, although they are not needed with this approach.

[4] The Knowledge Based Systems to which this article refers are rule-based, that is, they contain knowledge in the form of

   i) A set of production rules of the form $\alpha \wedge ... \wedge \gamma \rightarrow \delta \vee ... \vee \eta$.

   ii) A set of facts, which are some of the literals that appear in the rules.

   iii) (possibly) Some constraints of the form $\nu \wedge ... \wedge \sigma \rightarrow False$.

contradiction are equivalent or not in $(R/I)/J$ (what can be translated into the degeneracy of a certain ideal). The first pays more attention to the algebraic details meanwhile the second one pays more attention to the logical aspects. They include, implementations in Maple V and REDUCE (respectively).

Finally, [LR4] is a resume of [LR2] and [LR3] with some comments about the complexity of the process.

In none of the mentioned papers the process of reasoning is studied. Only the study of equivalence of propositions, which is easier in the above mentioned quotient ring.

1.2 The Development

There are three natural directions for the evolution of this line of research:

i) Analyze in detail the polynomial translations of ⊢ (or, what is equivalent in classical bivalued logic: ⊨), that is the topic of this paper.
ii) Generalize the study of inconsistency to KBSs based on multi-valued logics (a paper for the case of KBSs based on Lukasiewicz three-valued Logic is admitted to be presented at ECAI-96, and the generalization to multi-valued logics is under development now by the authors).
iii) Generalize i) to multi-valued logics (project).

1.3 About the Aim of this Article

In the present article, KBSs (forward and backward) reasoning is studied. Such a reasoning is performed in KBSs by what is known as an *inference engine*. The aim of this paper is to provide a theory and an implementation of the processes that the inference engine of a KBS performs.

As an important consequence of the theory, the article also presents an automatic method for detecting inconsistency in the reasoning processes of KBSs.

Two implementations, one using elementary ideals theory and another one using Gröbner Bases are included.

The advantage of this approach to the study of consistency is that the methods are effective and simple from the mathematical point of view. Other methods for checking inconsistency are hard to implement (like those based on graphs or Petri-nets), are not completely automatic (like truth maintenance systems), or do not provide a theoretical model, what makes extensions more difficult [An]. A comprehensive study of the state of the art in KBS verification can be found in [LL].

1.4 Forward and backward reasoning

There are two standard methods used to obtain consequences in KBSs (really, in the Lindenbaum algebra associated to the KBS).

In *forward reasoning*, the starting is the set of given facts, and the target is the proposition to be proved. By applying Modus Ponens and the rule of And

Introduction, the proposition is tried to be reached.

In *backward reasoning*, the starting is the proposition to be proved. A binary tree which nodes are the logic connectives, is constructed, in order to decompose the proposition to be proved into leaves that are the given facts or unnecessary branches.

As a consequence of the analysis done in this paper, Modus Ponens, Modus Tollens and the rule of And Introduction, are *fired* in the polynomial model when checking ⊢. Therefore, it follows from this paper that both methods can be substituted by this automatic approach.

1.5 The Need of Inference Engines for KBSs

An inference engine is needed both for automatic reasoning in the KBS and for debugging a KBS.

For instance, a flexible model (not dependent on the topology of the station) for interlockings in railway networks has been developed by this group (it will be presented at IMACS-ACA 96). It is based in matrix calculations obtained from two oriented graphs. We are working in the substitution of the model with a rule-based one, using the inference engine detailed in this paper (the reason is that the matrices are very sparse, and that suggests that a rule-based approach will be more efficient).

Moreover, when debugging a KBS, it is very convenient to have an inference engine. The reason is that it can help to make it clear where the inconsistency takes place.

2 How the Membership Problem is Studied in Commutative Algebra

Let R be a polynomial ring over a field and I an ideal of R. A common problem in Ideals Theory is to decide, given an ideal I and a polynomial p , whether or not $p \in Radical(I)$ (Radical Membership Problem)[5].

In general, this is a hard problem to solve, unless Computer Algebra techniques are used. The following theorem is really useful (see [CLO] for details and 4.2.1 in this paper).

Theorem 1. *Let R be a polynomial ring, $u \in R$ and I an ideal of R. Let t be a variable, $t \notin R$, and let us consider the transcendental extension $R[t]$. Then*

---

[5] The problem has a geometric origin. Let us denote by $v(I)$ the algebraic variety of the ideal I and $i(V)$ the ideal of the variety V. Whether or not $v(< p >) \supseteq v(I)$ is equivalent to: $iv(< p >) \subseteq iv(I)$. According to Hilbert's Nullstellensatz, if the base field is algebraically closed, this problem is equivalent to $Radical(< p >) \subseteq Radical(I)$, that is clearly equivalent to $p \in Radical(I)$.

Anyway, this is not the case in the following sections, where the base field is $\mathbb{Z}/2\mathbb{Z}$, that is not algebraically closed (for instance, the polynomial $x^2 + x + 1 \in \mathbb{Z}/2\mathbb{Z}[x]$ has no root).

$$u \in Radical(I) \ (in \ R) \Leftrightarrow I+ <1-t \cdot u>=<1> \ (in \ R[t])$$

Corollary 2. *If the reduced Gröbner Base of the ideal J is denoted $GB(J)$, the problem is equivalent to $GB(I+ <1-t \cdot p>) = 1$.*

A similar problem is the Ideal Membership Problem: does $p \in I$? It can also be checked using reduced Gröbner basis by testing if

$$GB(I+ <p>) = GB(I)$$

due to the uniqueness of these basis (once the order for the monomials and variables is fixed). Another way to check the ideal membership is to check if the Normal Form of p modulo the ideal I is 0 or not.

3 A Polynomial Model for the Propositional Boolean Algebra Associated to a KBS

3.1 The Polynomial Algebra Isomorphic to a Propositional Boolean Algebra

Let $\mathcal{A} = R/I = (\mathbb{Z}/2\mathbb{Z})[x, y, ..., z]/<x^2 - x, y^2 - y, ..., z^2 - z>$. Then $(\mathcal{A}, +, \cdot)$ is a polynomial class ring[6].

Proposition 3. *Let $(\mathcal{C}, \vee, \wedge, \neg, \rightarrow)$ be a propositional Boolean algebra, for $\underline{0}$ and $\underline{1}$ respectively denoting contradiction and tautology and $P, Q, ..., R$ denoting the propositional variables.*
Let us consider the Boolean algebra $(\mathcal{A}, \tilde{+}, \cdot, 1+, \text{"is a multiple"})$, where

$$\mathcal{A} = (\mathbb{Z}/2\mathbb{Z})[p, q, ..., r]/<p^2 - p, q^2 - q, ..., r^2 - r>$$

and $\tilde{+}$ is defined as follows:

$$if \ a, b \in \mathcal{A}, \ then: \ a\tilde{+}b = a + b - ab \ .$$

Let us denote by φ the function

$$\varphi : (\mathcal{C}, \vee, \wedge, \neg, \rightarrow) \longrightarrow (\mathcal{A}, \tilde{+}, \cdot, 1+, \text{"is a multiple"})$$

defined as follows: $\varphi(P) = p, \varphi(Q) = q, ..., \varphi(R) = r$ and for any $A, B \in \mathcal{C}$, if $\varphi(A) = a$ and $\varphi(B) = b$ then $\varphi(A \vee B) = a\tilde{+}b$, and $\varphi(\neg A) = 1 + a$ (as a consequence of the De Morgan laws, $\varphi(A \wedge B) = ab$). Then φ is an order preserving isomorphism.

Corollary 4. *$A \rightarrow B$ corresponds by φ with $b|a$.*

Remark. Within a Knowledge Based System (KBS), $\mathcal{C}$ is the Lindenbaum algebra constructed by introducing the equivalence relation $\leftrightarrow$ into the set of all well formed formulas that can be built using as language the propositional variables that appear in the KBS.

---

[6] This subsection is detailed in [LR1].

3.2 Ideals and Filters of a Boolean Algebra

Definition 5. The principal ideal (respectively principal filter) of the Boolean algebra $(\mathcal{C}, \vee, \wedge, \neg, \rightarrow)$ generated by $B \in \mathcal{C}$ is $E_B = \{X \in \mathcal{C} : X \rightarrow B\}$ (respectively $E^B = \{X \in \mathcal{C} : B \rightarrow X\}$) .

Proposition 6. *The ideals and filters of $\mathcal{C}$ correspond by φ with the ideals and filters of the Boolean algebra $\mathcal{A}$, and these ideals are exactly the same as the ideals of the polynomial class ring $\mathcal{A}$.*

Theorem 7. *$(\mathcal{A}, +, \cdot)$ is a principal ideal ring and all filters are also principal. Moreover, the ideal generated by $S = \{s_1, s_2, ..., s_n\} \subseteq \mathcal{A}$, is also generated by $s_1 \tilde{+} s_2 \tilde{+} ... \tilde{+} s_n$. The filter generated by S is generated by $s_1 \cdot s_2 \cdot ... \cdot s_n$. As a consequence, $\mathcal{A}$ is an unique factorization ring (not a UFD).*

3.3 The Polynomial Algebra Isomorphic to the Boolean Algebra Associated with a KBS

Definition 8. The "Boolean algebra associated to the KBS" is a structure $(\mathcal{C}^*, \vee, \wedge, \neg, \rightarrow)$ where $\rightarrow$ is the relation obtained applying the rules of logical deduction to the implications of $\mathcal{C}$ and the rules, facts and constraints of the KBS (consequently, $\leftrightarrow$ is not the usual implication), and $\mathcal{C}^*$ is the set of equivalence classes defined by this enlarged equivalence relation $\leftrightarrow$ in $\mathcal{C}$.

Proposition 9. *If the rules $Rule_1, ..., Rule_n$, the facts $Fact_1, ..., Fact_m$ and the constraints $Cons_1, ..., Cons_u$ of a KBS are added as true to the Boolean Algebra $\mathcal{C}$ of section 2.1, the structure obtained is*

$$\mathcal{C}^* = \mathcal{C}/E_{\neg Rule_1 \vee ... \vee \neg Rule_n \vee \neg Fact_1 \vee ... \vee \neg Fact_m \vee \neg Cons_1 \vee ... \vee \neg Cons_u}$$

and it is isomorphic to the image of $\mathcal{A}$ in the natural surjective homomorphism[7]

$$\psi : \mathcal{A} \longrightarrow \mathcal{A}/J$$

where J is the ideal

$$< \varphi(Rule_1) + 1, ..., \varphi(Rule_n) + 1, \varphi(Fact_1) + 1, ...,$$
$$\varphi(Fact_m) + 1, \varphi(Cons_1) + 1, ..., \varphi(Cons_u) + 1 >$$

3.4 Forward Reasoning Inconsistency

Remark. Let b be the conjunction of rules, facts and integrity constraints of a KBS. E^b (respectively $E_{\neg b}$) is the set of all elements of $\mathcal{C}^*$ that are implied by b (respectively imply $\neg b$). If $E^b = \mathcal{C}^*$ (resp. $E_{\neg b} = \mathcal{C}^*$) any formula is implied by b (resp. imply $\neg b$). The idea of inconsistency can be expressed in both of these ways.

---

[7] This subsection and the following one are detailed in [LR2] and [LR3].

Theorem 10. *The equality* $J = A$ *translates the KBS concept of forward reasoning inconsistency.*

Remark. In such case of degeneracy, the whole class ring collapses to a single element, and therefore $\underline{0} \leftrightarrow \underline{1}$ in C^*. This translates the fact that inconsistency takes place when the above mentioned Lindenbaum algebra C is equal to its ideal $E_{\neg Rule_1 \vee \ldots \vee \neg Rule_n \vee \neg Fact_1 \vee \ldots \vee \neg Fact_m \vee \neg Cons_1 \vee \ldots \vee \neg Cons_u}$.

3.5   First Step of the Implementation in Maple V

The infix operators "logic or" ($\vee$), "logic and" ($\wedge$), "implies" ($\rightarrow$) and "if and only if" ($\leftrightarrow$) are denoted, respectively, as &or, &and, &implies and &iff. The prefix operator "negation" ($\neg$) is denoted as neg.

Maple is "case sensitive", so we identify the upper case and lower case that are going to be used as names of propositions.

```
▷ P:=p;
▷ Q:=q;
▷ .....
▷ '&or':=proc(alfa,beta)
▷    convert(expand(alfa+beta+alfa*beta),mod2);
▷ end;
▷ '&and':=proc(alfa,beta)
▷    convert(expand(alfa*beta),mod2);
▷ end;
▷ '&implies':=proc(alfa,beta)
▷    convert(expand(1+alfa*(1+beta)),mod2);
▷ end;
▷ '&iff':=proc(alfa,beta)
▷    (alfa &implies beta) &and (beta &implies alfa);
▷ end;
▷ neg:=proc(alfa)
▷    convert(expand(1+alfa),mod2);
▷ end;
```

The element that generates the ideal which is generated by the elements in the list S , can be calculated using the following procedure

```
▷ baseideal:=proc(S : list)
▷    local i,gen;
▷    gen := 0;
▷    for i to nops(S) do gen := convert(expand(gen &or op(i,S)),mod2) od;
▷    gen;
▷ end;
```

4 Inference Engine

In this section, an interpretation of the inference rules Modus Ponens, Modus Tollens and And Introduction (which are precisely those used in the firing of rules in propositional KBSs), is given in terms of the Radical Membership Problem.

4.1 Formal Deduction

Lemma 11. *If a formula B is the conclusion by Modus Ponens ($\{A, A \rightarrow B\} \vdash B$), then the image of the negation of B (thesis) by φ belongs to the ideal of A generated by the images by φ of the negation of both A and $A \rightarrow B$ (the hypotheses).*

Polynomial interpretation: $a = 1 \wedge b|a \Rightarrow b = 1$.

Polynomial proof: If A and $A \rightarrow B$ (i.e. $\neg A \vee B$) are true, their images by φ are equal to one:

$$a = 1 \quad ; \quad (1 + a) + b + (1 + a) \cdot b = 1$$

From the first equation, $a + 1 = 0$ and substituting this value in the second equation: $b = 1$ (i.e. B is true in the propositional Boolean algebra).

Proof using ideals: $\varphi(A \rightarrow B) = \varphi(\neg A \vee B) = (1 + a) + b + (1 + a) \cdot b$. Instead of checking if

$$a = 1 \wedge (1 + a) + b + (1 + a) \cdot b = 1 \Rightarrow b = 1$$

it can be checked if

$$a + 1 = 0 \wedge (1 + a) + b + (1 + a) \cdot b + 1 = 0 \Rightarrow 1 + b = 0$$

i.e. if

$$1 + b \in < a + 1, (1 + a) + b + (1 + a) \cdot b + 1 > = < 1 + a \cdot b >$$

what clearly holds

$$1 + b = (1 + b) \cdot (1 + a \cdot b) \ .$$

Let us note that what we have proved is

$$\varphi(\neg B) \in < \varphi(\neg A), \varphi(\neg(A \rightarrow B)) > \ .$$

Lemma 12. *If $\neg A$ is the conclusion by Modus Tollens ($\{\neg B, A \rightarrow B\} \vdash \neg A$), then the image by φ of the negation of $\neg A$ belongs to the ideal of A generated by the images of the negation of both $\neg B$ and $A \rightarrow B$ (the hypotheses).*

Polynomial interpretation: $1 + (1 + b) = b = 0 \land b|a \Rightarrow a = 0$.

Polynomial proof: If $\neg B$ and $A \rightarrow B$ (i.e. $\neg A \lor B$) are true, their images by φ are equal to one:

$$1 + b = 1 \quad ; \quad (1 + a) + b + (1 + a) \cdot b = 1$$

From the first equation, $b = 0$; and substituting that value in the second equation: $1 + a = 1 \Leftrightarrow a = 0$ (i.e. $\neg A$ is True in the propositional Boolean algebra).

Proof using ideals: It has to be checked if

$$\varphi(\neg\neg A) \in < \varphi(\neg\neg B), \varphi(\neg(A \rightarrow B)) >$$

that is if

$$a \in < b, (1 + a) + b + (1 + a) \cdot b + 1 >=< a + b + a \cdot b >$$

which is clear, because

$$a = a \cdot (a + b + a \cdot b) \ .$$

Lemma 13. *If a formula $A \land B$ is the conclusion by the rule of And Introduction ($\{A, B\} \vdash A \land B$), then the image by φ of the negation of $A \land B$ belongs to the ideal of A generated by the image by φ of the negation of the hypotheses (A and B).*

Polynomial interpretation (and proof): $a = 1 \land b = 1 \Rightarrow a \cdot b = 1$.

Proof using ideals: It has to be checked if

$$\varphi(\neg(A \land B)) \in < \varphi(\neg A), \varphi(\neg B) >$$

i.e. if

$$1 + a \cdot b \in < 1 + a, 1 + b >=< 1 + a \cdot b > \ .$$

Theorem 14. *Formal deduction of a thesis from a set of hypotheses by Modus Ponens, Modus Tollens and the rule of And Introduction, can be performed by checking whether or not the image by φ of the negation of the thesis belongs to the ideal (of the polynomial ring A) generated by the images by φ of the negation of the hypotheses. i.e.:*

$$\{A, B, ..., C\} \vdash D \Leftrightarrow 1 + d \in < 1 + a, 1 + b, ..., 1 + c >$$

Proof: $\Rightarrow$) Has been proven in the three previous lemmas.
$\Leftarrow$) The elements of an ideal of a ring can be expressed as an algebraic combination of the generators of the ideal. Therefore, in our case, they can be written as combinations using the operations $\tilde{+}$ and $\cdot$ (as $+$ and $\cdot$ can be expressed using only $\tilde{+}$ and $\cdot$), that respectively correspond by φ^{-1} with $\lor$ and $\land$.

Finally, the rule of And Introduction is supposed to hold. And, as $\alpha \land \beta \rightarrow \alpha \lor \beta$, it follows from Modus Ponens that $\lor$ can also be introduced. Therefore, anything that can be constructed from the generators of the ideal using $\land$ and $\lor$ can be deduced using Modus Ponens and the rule of And Introduction.

4.2 The Ideal Membership Problem in the Ring $\mathcal{A}$

Idempotency holds in $\mathcal{A}$ (the ring defined in 3.1), and therefore all ideals are radical. So, the Ideal Membership Problem is equivalent in this ring to the Radical Membership Problem. A theorem similar to Th. 1 will be proved for $\mathcal{A}$ below.

Theorem 15. *Let $\mathcal{A}$ be the polynomial ring, $u \in \mathcal{A}$ and J an ideal of $\mathcal{A}$. Let t be a variable, $t \notin \mathcal{A}$, and let us consider the transcendental extension $\mathcal{A}[t]$ with idempotency (i.e. $\mathcal{A}[t]/< t^2 - t >$). Then*

$$u \in Radical(J) \ \ (in \ \mathcal{A}) \ \ \Leftrightarrow \ \ J + <1 - t \cdot u> = <1> \ \ (in \ \mathcal{A}[t]/< t^2 - t >) \ .$$

Note 16. To clarify the ring considered in each case, let us denote $J_\mathcal{A}$ the ideal in $\mathcal{A}$ and $J_{\mathcal{A}[t]}$ its extension to $\mathcal{A}[t]/< t^2 - t >$.

Proof: $\Rightarrow$) This implication is always straight forward:

$$u \in J_\mathcal{A} \Rightarrow t \cdot u \in J_{\mathcal{A}[t]} \Rightarrow 1 = (t \cdot u) + (1 - t \cdot u) \in J_{\mathcal{A}[t]} + <1 - t \cdot u>_{\mathcal{A}[t]} \ .$$

$\Leftarrow$) This other implication is easy to prove in this particular ring. If:

$$J_{\mathcal{A}[t]} + <1 - t \cdot u>_{\mathcal{A}[t]} = <1>_{\mathcal{A}[t]}$$

then

$$\exists i \in J_{\mathcal{A}[t]}, \exists a \in \mathcal{A}[t] : \quad i + a \cdot (1 - t \cdot u) = 1 \ .$$

and multiplying by $u \cdot t$ (idempotency holds):

$$i \cdot u \cdot t = u \cdot t \Rightarrow u \cdot t \in J_{\mathcal{A}[t]} \Rightarrow u \in J_\mathcal{A} \ .$$

4.3 A Straight Forward Implementation

The procedure baseideal has already been implemented in 2.5. Therefore, a procedure isDeducible, that checks whether or not thesis can be deduced from the propositions in the list S , can be implemented as follows [8]:

```
▷ isDeducible:=proc(S : list , thesis : polynom)
▷    if baseideal([op(map(neg,S)) , 1-t*neg(thesis)]) = 1
▷       then print('YES')
▷       else print('NO')
▷    fi;
▷ end;
```

This procedure produces a list with the negations of the elements of the list S and $1 - t \cdot negation(thesis)$. Then it computes the element that generates such ideal and checks if it is 1 or not.

---

[8] It should be noted that the variable t must not be used for any other purpose.

Example 1. Circularity: $\{B, A \rightarrow B, B \rightarrow C, C \rightarrow A\} \vdash A$
(a formula in a circularity can be obtained from one of the others) but: $\{B, A \rightarrow B, B \rightarrow C, C \rightarrow A\} \not\vdash \underline{0}$ (there is no formal contradiction in a circularity).

▷ isDeducible([B, A &implies B, B &implies C, C &implies A], A);
  YES
▷ isDeducible([B, A &implies B, B &implies C, C &implies A], 0);
  NO

(the element that generates the ideal is: $1 + abct$). These results are obtained immediately in a 8MB RAM 486 running Maple V.3.

5 Extension of the Inference Engine to the Boolean Algebra Associated with a KBS

Let us observe that all the previous reasoning can be extended to the model of the Boolean algebra associated with a KBS developed in [LR2] and [LR3]. Therefore, an inference engine for the Boolean algebra associated with a KBS is also available.

Example 2. Two streets cross at a certain place. The access is controlled by four traffic lights (one for each direction in each street): A, B in one street and C, G in the other. If one traffic light is green, the two in the perpendicular street should be red (green will be assigned to true and red to false). If no turning left is allowed, it is enough to consider the rules:

▷ R1:= A &implies (neg(C) &and neg(G));
▷ R2:= B &implies (neg(C) &and neg(G));
▷ R3:= C &implies (neg(A) &and neg(B));
▷ R4:= G &implies (neg(A) &and neg(B));

If the traffic light A is green, can we deduce that C is red?

▷ isDeducible([A,R1,R2,R3,R4],neg(C));
  YES

Can traffic lights A and B be green simultaneously? (If contradiction can not be deduced, then they can be green simultaneously).

▷ isDeducible([A,B,R1,R2,R3,R4],0);
  NO

(the element that generates the ideal is: $1 + abt + gabt + abct + abcgt$).

Can traffic lights A and C be green simultaneously?

▷ isDeducible([A,C,R1,R2,R3,R4],0);
  YES

As contradiction can be deduced, they can not be green simultaneously. These results are obtained in a few seconds in a 8MB RAM 486 running Maple V.3.

Remark. The forward reasoning consistency of a KBS is a special case. It has to be established whether or not the contradiction can be deduced from any of the maximal sets of potential facts, by applying forward reasoning to the rules and constraints.

6   A not so Simple but Faster Implementation

The degeneracy of the ideal in Theorem 17 can be checked in other ways. Although the approach shown above is simpler, from the point of view of speed, very good results have been obtained calculating the Gröbner Basis of the ideal using CoCoA (CoCoA [CN] is a special purpose CAS, oriented to Gröbner Bases calculations in polynomial rings over finite fields)[9] .

This result is somehow surprising, as the method in subsection 4.3 directly calculates the generator (and consequently a minimal base, as the ideals are principal in $(R/I)/J$), although it has to be simplified.

CoCoA also provides a Normal Form command (NF). Therefore, the inference engine could have also been constructed using Theorem 14 (instead of Theorem 15) and NF.

Below there is an example of a KBS with 100 propositional variables, 100 rules and 10 facts. Times to check a tautological consequence are around 15 seconds in a 16MB RAM Pentium PC.

This size of KBS seems to be more o less the upper boundary in this kind of computer. Times are sensibly lower for a KBS half the size of this.

Example 3. The following code is given to CoCoA:
i) Polynomial ring to be used:

```
W ::= Z/(2)[x[1..100],t];
USE W;
```

ii) The ideal I (that introduces idempotency):

```
I := Ideal( [x[K_]^2-x[K_] — K_ In 1..100] ) + Ideal(t^2-t);
```

iii) Polynomial definition of the bivalued connectives (O stands for ∨ and Y stands for ∧, to avoid conflicts with the existing CoCoA commands OR and AND). Let us observe that CoCoA does not allow the user to define infix operators; therefore the logical operators are now prefix. The ideal I is defined in ii) above.

```
NEG(M):=NF(1-M,I);
O(M,N):=NF(M+N+M*N,I);
```

---

[9] Information can be obtained at cooca@dima.unige.it

```
Y(M,N):=NF(M*N,I);
IMP(M,N):=NF(1+M+M*N,I);
IFF(M,N):=NF(1+M+N,I);
```

iv) Translation of the procedure isDeducible of section 4.3 to CoCoA (this one uses Gröbner Bases instead of calculating a generator of the principal ideal):

```
Define ISDEDUCIBLE(S_,TH_)
  If GBasis(I + Ideal( Concat( [NEG(J_) | J_ In S_] , [1-t*NEG(TH_)] ) )) = [1]
    Then Print 'YES'
    Else Print 'NO'
  End;
End;
```

v) The complete KBS with 100 rules, 10 facts and 1 integrity constraint:

```
R1:=NF(IMP( Y(Y(Y(Y(x[1],x[8]),NEG(x[16])),x[17]),x[18]) , O(x[2],x[13])), I );
R2:=NF(IMP( Y(Y(Y(NEG(x[3]),x[9]),x[19]),NEG(x[22])) , NEG(x[13])),I );
R3:=NF(IMP( Y(x[19],Y(x[20],x[21])) , O(x[10],x[31])),I );
R4:=NF(IMP( Y(NEG(x[16]),x[17]) , NEG(x[31])),I );
R5:=NF(IMP( Y(x[1],x[19]) , NEG(x[23])), I );
R6:=NF(IMP( Y(x[20],x[21]) , x[24]), I );
R7:=NF(IMP( Y(x[1],Y(NEG(x[3]),x[5])) , O(NEG(x[16]),O(x[26],x[27]))), I );
R8:=NF(IMP( Y(NEG(x[16]),x[17]) , x[18]),I );
R9:=NF(IMP( Y(x[17],x[19]) , x[26]), I );
R10:=NF(IMP( Y(x[5],x[9]) , NEG(x[27])), I );
R11:=NF(IMP( Y(x[5],x[8]) , O(NEG(x[22]),x[25])), I );
R12:=NF(IMP( Y(x[9],x[20]) , NEG(x[25])), I );
R13:=NF(IMP( Y(Y(NEG(x[25]),x[26]),NEG(x[28])) , O(x[29],x[30])), I );
R14:=NF(IMP( Y(Y(x[5],x[8]),x[9]) , NEG(x[30])), I );
R15:=NF(IMP( Y(x[19],NEG(x[25])) , NEG(x[28])), I );
R16:=NF(IMP( Y(Y(Y(Y(x[2],NEG(x[3])),x[9]),x[32]),x[33]) , O(x[4],x[12])), I );
R17:=NF(IMP( Y(x[20],x[21]) , O(x[32],x[34])), I );
R18:=NF(IMP( Y(x[19],x[20]) , NEG(x[34])),I );
R19:=NF(IMP( Y(x[21],Y(x[37],x[21])) , O(O(x[51],x[52]),x[53])),I );
R20:=NF(IMP( Y(Y(NEG(x[3]),x[5]),x[8]) , O(x[33],x[35])), I );
R21:=NF(IMP( Y(x[17],x[19]) , NEG(x[35])), I );
R22:=NF(IMP( Y(x[5],x[9]) , NEG(x[12])),I );
R23:=NF(IMP( Y(Y(Y(x[2],NEG(x[3])),x[5]),x[36]) , O(x[6],x[11])), I );
R24:=NF(IMP( Y(Y(NEG(x[3]),x[5]),x[9]) , O(NEG(x[11]),x[38])),I );
R25:=NF(IMP( Y(NEG(x[16]),x[17]) , O(x[36],NEG(x[37]))),I );
R26:=NF(IMP( Y(Y(Y(Y(x[1],x[5]),x[8]),x[9]),x[39]) , O(O(x[7],x[10]),x[40])), I );
R27:=NF(IMP( Y(x[17],NEG(x[40])) , x[39]),I );
R28:=NF(IMP( Y(x[1],NEG(x[3])) , NEG(x[10])), I );
R29:=NF(IMP( Y(x[21],x[37]) , NEG(x[38])), I );
R30:=NF(IMP( Y(Y(x[2],x[4]),x[6]) , O(O(NEG(x[1]),x[14]),x[15])), I );
R31:=NF(IMP( Y(x[2],x[5]) , NEG(x[14])), I );
```

R32:=NF(IMP(Y(NEG(x[3]),x[8]) , NEG(x[15])), I);
R33:=NF(IMP(Y(NEG(x[13]),NEG(x[14])) , x[41]), I);
R34:=NF(IMP(Y(x[39],x[41]) , O(x[42],NEG(x[43]))),I);
R35:=NF(IMP(Y(Y(x[5],x[8]),x[9]) , NEG(x[42])), I);
R36:=NF(IMP(Y(NEG(x[43]),x[41]) , O(x[44],x[45])),I);
R37:=NF(IMP(Y(x[5],NEG(x[16])) , NEG(x[44])),I);
R38:=NF(IMP(Y(x[19],x[45]) , NEG(x[38])), I);
R39:=NF(IMP(Y(NEG(x[23]),x[24]) , x[46]), I);
R40:=NF(IMP(Y(Y(x[37],x[46]),NEG(x[31])) , O(x[47],x[48])), I);
R41:=NF(IMP(Y(Y(x[8],NEG(x[40])),NEG(x[11])) , NEG(x[47])), I);
R42:=NF(IMP(Y(x[48],x[24]) , O(x[49],x[50])), I);
R43:=NF(IMP(Y(NEG(x[23]),x[2]) ,O(NEG(x[50]),NEG(x[37]))), I);
R44:=NF(IMP(Y(NEG(x[23]),NEG(x[50])) , x[11]), I);
R45:=NF(IMP(Y(x[1],x[2]) , NEG(x[3])), I);
R46:=NF(IMP(Y(x[39],Y(NEG(x[10]),x[8])) , NEG(x[52])), I);
R47:=NF(IMP(Y(x[5],Y(x[9],x[6])) , NEG(x[53])), I);
R48:=NF(IMP(Y(NEG(x[52]),NEG(x[53])) ,O(O(x[54],x[56]), NEG(x[57]))), I);
R49:=NF(IMP(Y(NEG(x[47]),NEG(x[16])) ,NEG(x[52])), I);
R50:=NF(IMP(Y(NEG(x[35]),NEG(x[57])) ,NEG(x[56])),I);
R51:=NF(IMP(Y(NEG(x[16]),NEG(x[47])) , O(x[57], NEG(x[58]))),I);
R52:=NF(IMP(Y(x[1], NEG(x[3])) , NEG(x[57])),I);
R53:=NF(IMP(Y(NEG(x[58]),x[17]) , O(x[60],x[59])), I);
R54:=NF(IMP(Y(x[19],Y(NEG(x[14]),NEG(x[35]))) , NEG(x[59])),I);
R55:=NF(IMP(Y(x[2],Y(x[4],x[6])) , O(O(NEG(x[5]),x[14]), x[61])), I);
R56:=NF(IMP(Y(x[8],x[61]) , O(x[62],x[63])),I);
R57:=NF(IMP(Y(x[19],NEG(x[40])) , NEG(x[62])),I);
R58:=NF(IMP(Y(x[1],Y(NEG(x[3]),x[63])) , O(O(x[64],NEG(x[65])),x[66])),I);
R59:=NF(IMP(x[17] , NEG(x[64])),I);
R60:=NF(IMP(Y(x[20],x[21]) , x[65]),I);
R61:=NF(IMP(Y(x[1], x[66]) , O(O(x[67],x[68]), x[69])),I);
R62:=NF(IMP(NEG(x[53]) , NEG(x[67])),I);
R63:=NF(IMP(Y(x[37], x[17]) , NEG(x[68])),I);
R64:=NF(IMP(Y(NEG(x[3]) ,x[69]) , x[70]),I);
R65:=NF(IMP(Y(x[1],x[5]) , NEG(x[3])),I);
R66:=NF(IMP(Y(NEG(x[3]), x[8]) , x[9]),I);
R67:=NF(IMP(Y(x[8],x[5]) , NEG(x[16])),I);
R68:=NF(IMP(Y(x[17],x[19]) , x[20]),I);
R69:=NF(IMP(Y(x[20],x[8]) , x[21]),I);
R70:=NF(IMP(x[21] , x[37]),I);
R71:=NF(IMP(NEG(x[3]) , x[25]),I);
R72:=NF(IMP(Y(NEG(x[62]), x[8]) , O(NEG(x[7]),x[72])),I);
R73:=NF(IMP(Y(NEG(x[64]), x[8]) , NEG(x[72])),I);
R74:=NF(IMP(Y(NEG(x[72]), x[17]) , O(O(x[73],x[74]), NEG(x[1]))),I);
R75:=NF(IMP(x[17] , NEG(x[73])),I);
R76:=NF(IMP(Y(x[1], NEG(x[3])) , NEG(x[74])),I);

R77:=NF(IMP(Y(x[19], NEG(x[73])) , x[75]),I);
R78:=NF(IMP(Y(x[75], NEG(x[74])) , NEG(x[18])),I);
R79:=NF(IMP(Y(x[1], x[9]) , NEG(x[18])),I);
R80:=NF(IMP(NEG(x[16]) , x[21]),I);
R81:=NF(IMP(Y(NEG(x[31]),x[41]) ,O(O(x[82],NEG(x[83])),x[84])),I);
R82:=NF(IMP(Y(x[5],x[41]) , O(x[83],x[85])),I);
R83:=NF(IMP(Y(x[1], NEG(x[43])) , NEG(x[85])),I);
R84:=NF(IMP(x[5] , NEG(x[82])),I);
R85:=NF(IMP(Y(x[84],x[81]) ,x[86]),I);
R86:=NF(IMP(Y(x[86],NEG(x[85])) , O(O(x[87],x[88]),x[89])),I);
R87:=NF(IMP(x[1], NEG(x[88])),I);
R88:=NF(IMP(x[81] , NEG(x[89])),I);
R89:=NF(IMP(x[87] , O(x[88],x[90])),I);
R90:=NF(IMP(NEG(x[89]),NEG(x[88])),I);
R91:=NF(IMP(Y(x[8],x[90]) , O(x[91],x[92])),I);
R92:=NF(IMP(x[5],NEG(x[85])),I);
R93:=NF(IMP(Y(x[17],x[19]) , NEG(x[92])),I);
R94:=NF(IMP(NEG(x[92]) , NEG(x[94])),I);
R95:=NF(IMP(Y(x[90],x[91]) , O(O(x[93],x[94]),x[95])),I);
R96:=NF(IMP(Y(x[93], NEG(x[88])) , O(O(x[96],NEG(x[97])),x[98])),I);
R97:=NF(IMP(NEG(x[88]) , NEG(x[95])),I);
R98:=NF(IMP(Y(NEG(x[43]),NEG(x[95])) , NEG(x[96])),I);
R99:=NF(IMP(Y(x[94],x[46]) , O(x[97],x[99])),I);
R100:=NF(IMP(x[100] , NEG(x[99])),I);
H1:=x[1];
H2:=x[5];
H3:=x[8];
H4:=x[17];
H5:=x[19];
H6:=NEG(x[40]);
H7:=x[41];
H8:=NEG(x[43]);
H9:=x[81];
H10:=x[100];
TIC:=NF(NEG(Y(NEG(x[44]),x[41])), I);

The generators of the ideal J:

GJ:= [H1,H2,H3,H4,H5,H6,H7,H8,H9,H10,R1,R2,R3,R4,R5,R6,R7,R8,R9,R10,
R11,R12,R13,R14,R15,R16,R17,R18,R19,R20,R21,R22,R23,R24,R25,
R26,R27,R28,R29,R30,R31,R32,R33,R34,R35,R36,R37,R38,R39,R40,
R41,R42,R43,R44,R45,R46,R47,R48,R49,R50,R51,R52,R53,R54,R55,
R56,R57,R58,R59,R60,R61,R62,R63,R64,R65,R66,R67,R68,R69,R70,
R71,R72,R73,R74,R75,R76,R77,R78,R79,R80,R81,R82,R83,R84,R85,
R86,R87,R88,R89,R90,R91,R92,R93,R94,R95,R96,R97,R98,R99,R100,TIC];

Let us ask the system:

```
ISDEDUCIBLE(GJ , x[24] );
YES
ISDEDUCIBLE(GJ , x[27] );
YES
ISDEDUCIBLE(GJ , 0 );
YES
```

Not only x[24] and x[27] can be obtained. As 0 can be deduced, anything can be proved (there is inconsistency).

Removing rules 28, 30, 44, 71, 76, 75, 74, 78, 79 and TIC, the KBS obtained is consistent. Let us call H the correspondent ideal and GH a list of its generators, which are omitted for the sake of brevity. Then

```
ISDEDUCIBLE(GH , x[24] );
YES
ISDEDUCIBLE(GH , x[27] );
NO
ISDEDUCIBLE(GH , 0 );
NO
```

i.e. x[24] can be proved, but x[27] can not. Consequently, there is no inconsistency in this case.

All the calculations in this example take less than 2 minutes in a 16 MByte RAM Pentium PC.

7   Conclusions

This article deals with a problem similar to that of [AB] and [KN]. But in this article, instead of considering a polynomial ring with simplification rules, a class ring is considered as the algebraic model. This is what makes it possible to obtain a model of the Boolean algebra associated to a KBS (as the quotient of the previous quotient ring over a certain ideal). Therefore, inference and consistency can be treated in KBSs.

Moreover, in [AB] and [KN], Gröbner Basis are used. Whereas, in this paper, the Radical Membership Criterion is adapted, and it has been shown how the use of Gröbner Basis can be avoided in this particular context (although it is generally faster to use them).

This pure algebraic approach allows the application of algebraic techniques and all the power of Computer Algebra Languages. Both facts make the implementation original and extremely simple.

Related Literature

References

[An]   A. de Antonio: Una Interpretación Algebraica de la Verificación de Sistemas Basados en el Conocimiento. Univ. Politécnica de Madrid (Ph.D. Thesis) (1994).

[AB]   J.A. Alonso, E. Briales: Lógicas Polivalentes y Bases de Gröbner. In: C. Martin (editor): Actas del V Congreso de Lenguajes Naturales y Lenguajes Formales (1995) 307-315.

[Bu]   B. Buchberger: Applications of Gröbner Bases in non-linear Computational Geometry. In: J.R. Rice (editor): Mathematical Aspects of Scientific Software. IMA Volumes in Math. and its Applications, vol. 14. Springer-Verlag (1988).

[CLO]   D. Cox, J. Little, D. O'Shea: Ideals, Varieties, and Algorithms. Springer-Verlag (1991).

[CN]   A. Capani, G. Niesi: CoCoA User's Manual. University of Genova (1996).

[En]   H.B. Enderton: A Mathematical Introduction to Logic. Academic Press (1972).

[Ha]   P.R. Halmos: Lectures on Boolean Algebras. Springer-Verlag (1974).

[He]   A. Heck: Introduction to Maple. Springer-Verlag (1993).

[Hs]   J. Hsiang: Refutational Theorem Proving using Term-Rewriting Systems. Artificial Intelligence **25** (1985) 255-300.

[KN]   D. Kapur, P. Narendran: An Equational Approach to Theorem Proving in First-Order Predicate Calculus. In: Proceedings of the 9th International Joint Conference on Artificial Intelligence (IJCAI-85), vol. 2 (1985) pages 1146-1153.

[LL]   L. M. Laita, L. de Ledesma: Knowledge-Based Systems Verification. In: J.G. Williams, A. Kent (editors): Encyclopedia of Computer Science and Technology. M. Dekker (to appear).

[LR]   L. M. Laita, B. Ramírez, L. de Ledesma A. Riscos: A Formal Model for Verification of Dynamic Consistency of KBSs. Computers & Mathematics **29-5** (1995) 81-96.

[LR1]   L. M. Laita, L. de Ledesma, E. Roanes L., E . Roanes M.: An Interpretation of the Propositional Boolean Algebra as a k-algebra. Effective Calculus. In: J. Campbell, J. Calmet (editors): Proceedings of the Second International Workshop/Conference on Artificial Intelligence and Symbolic Mathematical Computing (AISMC-2). Lecture Notes in Computer Science **958**. Springer-Verlag (1995) 255-263.

[LR2]   E. Roanes L., L. M. Laita, E. Roanes M.: Maple V in A.I.: The Boolean Algebra Associated to a KBS. CAN Nieuwsbrief **14**, (1995) 65-70.

[LR3]   E. Roanes L., L.M. Laita: Verification of Knowledge Based Systems with Commutative Algebra and Computer Algebra Techniques. Proceedings of the 1st International Conference on Applications of Computer Algebra (IMACS). New Mexico University (USA) (1995) (electronic book).

[LR4]   L.M. Laita, E. Roanes L.: Verification of Knowledge Based Systems: An Algebraic Interpretation. In: Proceedings of the International Joint Conference on Artificial Intelligence (IJCAI-95) (Workshop on Verification and Validation of Knowledge Based Systems). McGill University, Montreal (Canadá), (1995) 91-95.

[Me]   E. Mendelson: Boolean Algebra. McGraw-Hill (1970).

[Sh]   J.R. Shoenfield: Mathematical Logic. Addisson-Wesley (1967).

Programming by Demonstration: A Machine Learning Approach to Support Skill Acquision for Robots

R. Dillmann and H. Friedrich

University of Karlsruhe, Institute for Real-Time Computer Systems and Robotics, Kaiserstr. 12, D-76128 Karlsruhe, Germany
dillmann@ira.uka.de

Abstract

Programming by Demonstration (PbD) is a programming method that allows to add new functionalities to a system by simply showing the desired task or skill in form of few examples. In the domain of robotics this paradigm offers the potential to reduce the complexity of robot task programming and to make programming more "natural". In case of programming an assembly task PbD allows with the help of a video or a laser camera and a data glove the automatic generation the necessary robot program for the assembly task. In addition, the demonstration of the task with few different assembly situations and strategies may achieve a generalized assembly function for all possible variants of the class. In order to realize such a PbD system at least two major problems have to be solved. First, the sensor data trace of a demonstration has to be interpreted and transformed into a high-level situation-action representation. This task is not yet well understood nor solved in general. Second, if a generalization is required, induction algorithms must be applied to the sensor data trace, to find the most general user-intended robot function from only few examples. In this paper mainly the second problem is focused. The described experimental PbD environment consists of an industrial robot, a 6D space mouse used as input device, and some sensors. Various data can be recorded during a demonstration for further processing in the PbD system implemented on a workstation. The objective is to exploit the possibilities of integrating learning and clustering algorithms for automated robot programming. In particular it is investigated how human interaction with the PbD system as well as user-initiated dialogs can support inductive learning to acquire generalized assembly programs and skills.

Keywords: Programming by Demonstration (PbD), Robot Programming by Demonstration (RPD), Teaching, Robots, Manipulators, User Intentions, Dialog-Based Learning (DBL).

1 Introduction

Programming by Demonstration (PbD) is an approach for fast software development in application domains where the user works interactively with a computer or some machinery. Such a domain may be programming of a graphical user interface as well as graphics and text editors where the user interacts with the editor program by using a keyboard, a mouse, a pen, or other input devices. Another important area is the domain of robot application programming and of other machines where the user moves and manipulates the device by remote controls. In all applications the goal of PbD is to "show" the system a new functionality by demonstrating the desired behavior, rather than coding it in a programming language. What has been demonstrated is stored in form of traces of sensor data, which is then transformed into generalized sensor-action sequences. If the learned sequence is analyzed and verified to be correct, a computer program can be generated that allows a robot to repeat the demonstrated task. To achieve higher efficiency of the learned program, additional generalization procedures may be applied to the learned robot function.

Given the problem of an assembling task to be executed by a robot manipulator. For programming such a task, classical teach-in methods may be used. The programmer directs the robot's end-effector to all necessary locations, where interaction like grasping an object or manipulation of a workpiece may happen. At each relevant position the teacher stores the location as a tuple $(x, y, z$ and the orientation angles), and additional commands like open or close gripper. Replaying the whole sequence of stored "button hits" repeats exactly the demonstrated program. Teach-in programming of manipulators can take a lot of time and is really effective only if large numbers of exactly the same product items shall be assembled.

Programming by demonstration with the capability of learning overcomes the classical teach-in especially for programming the manufacturing task of assemly of small product lines with a high number of product variants. In case of service robots it would be convenient just to demonstrate the service task with the own hands, observed by a video camera and supported by a data glove which provides the data related to the tactile interaction.

Unfortunately, PbD systems as described in this scenario are not yet available even in form of experimental prototypes and some PbD like systems are realized in very specialized domains. The reasons are manifold: Sensors for PbD like haptic and visual sensors are characterized by low resolution and are still too expensive. Sensor processing of natural human demonstration sequences is actually not fast and accurate enough to meet the rigorous needs of a manufacturing process. A human arm and hand with more than 20 DOF for motion has a higher flexibility in performing object manipulations than a standard industrial robot with 6-DOF. Another problem is that a human teacher demonstrates in most cases not optimal and no exact exaples. Thus, a variety of interactions between machines, robots, and humans have the need for additional information like context and background knowledge or a performance criterion. In flexible manufacturing systems, not all steps of a complex assembly task are usually performed at one place. The whole process is characterized by a sequence of actions over time at different stations along a conveyor belt. Therefore, programming the whole assembly task by demonstration can probably not be done giving only one example.

However, the task may be subdivided into subtasks which can be programmed with the help of demonstration techniques -even with generalization procedures-. Even such a strategy may significantly reduce the programming effort. Typical robot subtasks are pick & place operations, compliant operations, loading of pallets, sorting tasks, part feeding tasks, etc. They are far from complex assembly tasks, but still need time-consuming and cumbersome teach-in programming procedures. Graphical off line programming techniques are also in use, but they need also a significant amount of teach-in postprocessing in order to adjust the generated program to the target system [25].

Robot subtasks like pick & place, compliant actions or sorting objects often imply the need for the generalization of geometrical positions, of constants to variables, and of similar objects into classes. There is also a need for finding loops or recursive control schemata in the action traces. Such generalizations are well-known from inductive machine learning, particularly in form of learning from examples. In fact, as in most practical applications (see the discussion in [28]), a combination of analytical and inductive learning is used for robot programming by demonstration (RPD). This means that previous knowledge of the domain and particular knowledge about tasks and environments have to be available. Nevertheless, PbD and RPD cope with inherent problems that do not necessarily exist for all learning-from-examples tasks. In reality usually only a few of examples are available, typically 1 - 5, because nobody wants to demonstrate the same task too often. Thus, there is a great uncertainty about what the user intends with the demonstrated examples, in particular because the hypothesis space is large.

As a solution to overcome this problem we propose dialog-based learning (DBL) [5]. DBL includes the tutor into the induction process with the help of dialogs that may be initiated by the RPD system as well as by the person. The effectiveness of DBL has already been shown in other PbD domains in which—based on only few examples—functions for graphics editors [26] and personal assistance software [4] could be learned (induced).

The structure of this paper is organized as follows. In section 2 the principle of robot programming by demonstration is described in detail and a solution for an adequate system architecture is presented. Section 3 contains the description of a RPD system and describes in detail the two major phases of RPD. In section 4 an example is illustrated where human interaction is used to support an induction process. The presented approach is discussed and compared with other methods in section 5. Finally, section 6 concludes the paper and outlines our future work.

2   Robot Programming by Demonstration

"The key characteristic of robots is their versatility; but it can be exploited only if the robot can be programmed easily. In some cases, the lack of adequate programming tools can make some tasks impossible to perform.

*In other cases, the cost of programming may be a significant fraction of
the total cost of an application. For these reasons, robot programming
systems play a crucial role in robot development".*

(Tomas Lozano-Perez, 1983)

Having in mind Lozano-Perez' demand for an adequate robot programming
environment, a short overview on robot programming methods is given. A de-
scriptione of the principles of the approach of Robot Programming by Demonstration (RPD) follows.

2.1 Basic Robot Programming Methods

Today, many different programming methods and -languages with a large number of modifications exist. In [19], three major categories can be identified:

1. Robot programming by guiding;
2. Explicit or robot-level programming;
3. Implicit or task-level programming.

In robot guiding (also known as teaching by showing, teach-in, or play-back),
the operator moves the manipulator and records explicitly all relevant positions.
The major advantage of this method is that only positions are taught that can
really be reached by the robot and that the explicit record yields to very precise
trajectories. This method is very easy to perform, but, it has some major drawbacks: The robot cannot be used for the production process during the teaching
phase and the stored trajectory cannot be altered; furthermore it is not possible
to include sensor based controls into the generated program. Usually, only the
robot used for teaching can exactly follow the recorded path.

Explicit or off line programming in a high level programming language introduces control structures like loops and branches, and allows to react to external
events by using sensors, thus, very flexible robot programs can be generated. For
the programming process, either special-purpose languages like VAL II (Versa-
tile Assembly Language) or general-purpose languages with special extensions,
like PasRo (Pascal for Robots), are used. These languages offer much more flexibility than teach-in, and the robot can be programmed off line. But, due to
the rich semantics and power of these language compilers, programs are hard to
write by robot users who are not experianced in programming. The most advanced method for programming of robots is implicit programming. This method
requires a set of complex planners, because the system input is neither a trajectory nor a position sequence, but the programmer has to determine the goal and
subgoals of the desired robot task. From the task description the planner derives
an action skeleton plan and expands it with the help of planning knowledge.
Problems with this method arise due to the high complexity of the task which
requires a set of planners which are highly complex in case of assembly tasks.

All three methods can be extended and combined. For example, for programming of robot spot or arc welding tasks, CAD model data is used and adjusted

with teach-in methods to generate the robot trajectory. To simplify model based programming, graphical simulation systems are used in combination with other methods. Off line programming allows testing of alternatives and the visualization of the simulated trajectories or action sequences. Advanced graphical simulation systems are based on virtual environments (VE) (see, e.g., [9]) which include the dynamic behaviour of the robot. With robot programming by demonstration (RPD), a rich, powerful and flexible programming method is proposed, which is easy to learn and to use.

2.2 Principles of Robot Programming by Demonstration

Robot programming by demonstration is an effective method that supports programming of robot applications, but which can also be used to learn skills for a robot system. The idea of RPD is to combine teach-in with learning and to integrate the user in all processing and generalisation steps. Learning in an RPD system can be realized at different levels of abstraction. Therefore, learning of skill as well as knowledge acquisition by the system via human demonstration might be integrated in the system [10].

The two major levels of an integrated RPD system are as follows:

1. Learning on the symbolic level: The system can either generalize robot programs completely or partly, using sequences of elementary operations as input data for the learning process [23, 24]; or the acquisition, extension, and refinement of robot action knowledge and object descriptions may be realized by processing data extracted from experiments [16].
2. Learning on the subsymbolic level: On the lower level, either trajectories can be reconstructed from freehand motions [29, 30] or robot controllers performing basic skills might be generated with the help of artificial neural networks [14]. The method described within this paper, is focusing learning of robot programs on the symbolic level.

Three major processing phases can be identified in a RPD system: First, the demonstration phase in which a solution for the task is demonstrated by the user. All data of this demonstration is segmented and stored in a trace for further processing. Thereafter an analyzing component is applied to transform the stored trace on the symbolic level, namely to the representation given by the planning language used in the succeeding components of the system. Secondly, the induction phase is processed, where the segmented demonstration trace is matched with additional information related to the user's intentions. Finally, the execution phase is performed in case the generated program has to be applied. This includes the selection, sequentialization, and parametrisation of the the program as well as the robot control parameters.

3 The RPD environment

This section describes the function and the architecture of a RPD system which has been realized prototypically at the University of Karlsruhe. The techniques and methods applied in the 3 processing phases are described. Preceding this description, a short overview of the planning language used by the system is given.

The Planning Language The language providing the symbolic representation used for generating and processing programs from demonstrations basically consists of five Elementary Operations (EOs) that are: grip(O), ungrip(O), approach(s, d), depart(s, d), and transfer(s, d). Here, O denotes either an object of the world model or a relational condition that can be evaluated and unified with the world model in order to determine an object. s and d are frames (also called locations) describing positions and orientations of the manipulators tool center point.

Like most procedural programming languages, the planning language includes higher level elements that represent if-then-else branches and loops. Their body holds a sequence of other operators that can be branches, loops, or even whole plans themselves. A plan is an operator containing a sequence of operators and representing the solution to a given task. It can itself be used as part of another plan, branch, or loop. Currently, the loop operators are not used. However, they are already defined for the next development stage of the system.

All operators of the planning language are modeled following the terminology of STRIPS operators (see [8]), i.e. for every operator its application depends on the evaluation of a precondition and after the application a certain postcondition holds.

Conditions themselves are expressed in terms of combinations of disjunctions and conjunctions of relations. These are describing relationships between objects like on(O1, O2) and in(O1, O2) or object-features like obj-type(O, type). Terms are always given and transformed into disjunctive normal form (DNF). They can be evaluated with respect to the world model that is given initially and updated each time one of the operators described above is applied. The implemented variable concept allows to use variables as attributes of relations. Unification of a term holding variable attributes allows to determine all combinations of variable instantiations on the current world model that yield a positive evaluation of the term.

3.1 Demonstration Phase

Demonstrating a task In order to generate a new robot program, the PbD process starts with the demonstration of an example. Several different methods exist for that step, each with particular strengths and drawbacks. In the system presented here, a space mouse is used to control the robot. Only a training cycle is needed to generate data from the demonstration.

Filtering Pauses occurring in demonstrations and unsteady motions due to not perfectly trained users result in noisy data. Therefore, filtering of the recorded trace follows the demonstration in order to remove redundant vectors to reduce the amount of data.

Analysis The next step is the analysis of the filtered trace, which is mainly used to transform the "raw data" into the symbolic form of the planning language. To achieve a correct transformation, i.e. the recognition of the correct sequence of EOs hidden in the recorded vector sequence, two problems have to be solved. Firstly the trace has to be segmented into subsequences, each of them representing a certain EO. Secondly each segment has to be classified and mapped on one of the five EOs. Obviously the two tasks of segmentation and classification depend on each other and therefore have to be done simultaneously. In this sense the transformation task is similar to the problem of continuous speech recognition. Another aspect also common to the speech recognition domain is that of time variations. Just like people speak with different speeds, various users perform demonstration movements with different speeds. Moreover, the duration of a movement usually depends on the distance between the start and the final location.

In the domain of speech recognition Time Delay Neural Networks (TDNNs) have been proved to be powerful tools for various segmentation and classification problems [31], overcoming also the problem of time variations in the input signal. For this reason the application of TDNNs was considered to solve the analysis problem in the RPD system. For each of the BOs a TDNN was generated, consisting of three layers with 5 neurons in the input and exactly one in the output layer. The number of neurons in the hidden layer varies for the different BOs as well as the applied time-delays and training epochs, as can be seen from Table 1.

Motions are restricted to translational movements and each net was trained with nine to ten training sequences (10–20 vectors per training sequence) of pick & place operations. The demonstrations vector sequence is fed simultaneously into each EO TDNN. By studying the outputs produced, i.e. the net activations, a decision about which EO was performed during what time-span of the demonstration is taken.

Table 1. Some parameters of the TDNNs for Elementary Operation recognition

EO Net	Perceptrons in the hidden-layer	Time-delays on 1st layer	Number of training epochs
transfer	9	10	10,000
approach	10	15	12,000
depart	10	15	12,000
grip	8	8	10,000
ungrip	8	8	10,000

Automatic correction of classification errors, like a missing gripper event between a detected approach and depart pair is done by applying some heuristics about possible operation sequences in a post-processing step. Following this stage the transformation from the subsymbolic to the symbolic level takes place. The determined sequence of Elementary Operations of the planning language is generated and instantiated with respect to the segmented vector sequence. The boundaries of the recognized EOs are given. Now the user has the option to correct the EO sequence in order to improve the performance. Stored corrections from multiple demonstrations might be used later to retrain the TDNNs and thus improve the automatic analysis performance of the system over time.

3.2 Induction Phase

Constructing a plan from the EO sequence generated by the analysis module is a task consisting of the following phases that can be performed by the induction unit:

1. Formation of Object Groups (OGs), where each OG contains a subset of the EO sequence that represents the manipulation of exactly one object.
2. Determination of postcondition of each EO sequence of step 1.
3. Acquisition of the user's intention considering the demonstration via man-machine communication.
4. Detection and generation of branches, i.e. conditioned manipulations and elimination of superfluous manipulations.
5. Acquiring the object selection criteria for the manipulated objects and introducing variables based on user interaction.

These steps will be discussed in the following.

Preparing EOs; Grouping & Calculating Conditions For further processing of a given EO sequence, it is segmented into subsets, the Object Groups.

Starting from the robot transfer movement towards an object, proceeding with its grip, the sequence of Elementary Operations is traced until the object is finally ungripped and the manipulator departed from it. Therefore, each Object Group encloses a subpart of the users demonstration that is characterized by the manipulation of one specific object.

The consideration of interactions of objects enables multiple cycles of transfers to other objects and interactions of the gripped object with those. These interactions take place in between the grip and ungrip of the primarily manipulated object.

After the identification of all Object Groups, a forward trace over the EO sequence is performed. In this process the parametrized pre- and postconditions of every EO are determined. Also the postconditions of every Object Group and finally the whole EO sequence, i.e. the demonstration itself, are determined. Every type of EO has certain preconditions and postconditions that have to be valid for its application respectively are valid after application. In addition, there are pre- and postconditions that depend on the environmental context in which the EO is executed. If, for example, the postcondition for the EO approach(s,d) has to bedetermined, the result depends on whether an object has already been grasped or not. If the gripper does not hold an object (i.e., it is free), approach is prior to a gripping action. Therefore, the gripper encloses the object O to be gripped at the end of the approach. This allows the insertion of the enclosed (O,Manipulator) relation to the postcondition of the approach. A special case is given when an object is ungripped. Here, all possible relations to all other objects have to be tested and added to the ungrip EO's postcondition.

While forward tracing through the EO sequence and determining the conditions of the Elementary Operators, the world model is updated by virtually applying the EOs to it. Starting with the first EO and the initial environmental configuration, the world model data is updated sequentially after the determination of each operators conditions with respect to the operators definition.

As soon as the conditions of all EOs related to an OG are determined, they can be fused to the postcondition of the group. Naturally this includes to check for relations built up by one EO possibly beeing destroyed later by another EO related to the same Object Group. The same fusing of postconditions is done for all OGs in order to determine the whole demonstration's postcondition. In other words, the demonstration's postcondition is the accumulation of all changes that occured throughout the demonstration.

But do all these changes really reflect exactly the users intention? Or were some actions simply forced by the environmental conditions present at that moment? Or were some of the manipulations even unnecessary in order to solve the given task?

Acquiring new Knowledge by User Interaction Since the primary focus in the proposed PbD system is not to induce generalizations and/or knowledge from a large number of examples, a different direction than classical ML induction techniques has to be followed to generate real executable programs the user intends the robot to do. In order to keep the number of required demonstrations as small as possible another source of information may be integrated in the induction process—the user itself. The postconditions of the demonstration that are related to the same Object Group are presented to the user successively. Thus he can select the conditions that reflect his intention best and can discard the others. Thereby the system gathers the information necessary to identify actions which are conditioned with respect to the user's intention. With this knowledge, branches can be derived and if-then-else operators can be generated. Parts of the demonstration that where found to be superfluous can even be deleted.

Generalizing and Optimizing; Branching & Garbage Collection While the processing strategy in the phase of the induction process was characerized by forward tracing through the input EO sequence, the sequence is now traced in the reverse direction, i.e. backwards. Tracing backward, branching, determination of preconditions of the Object Groups, and optimization by deletion of superfluous manipulations is done. In the following, the developed branching algorithm is given in pseudo-code in which the initializations and update rules are given precisely. I denotes the users intention, UI unsatisfied intentions, UP-rec unsatisfied preconditions, PostC(ogi) the postcondition and PreC(ogi) the precondition of argument ogi.

```
1.determine-branches (plan, I):
2.  PostC(plan) := UI := I;
3.  UPreC := ;;
4.  loop OGcurrent:= OGlastto OGfirst          /* backward trace */
5.    determine-precondition (OGcurrent);
6.    if direct-contribution (OGcurrent, UI) then
7.      UI := (UI = PostC(OGcurrent));          /* intended manipulation */
8.      UPreC := (UPreC = PostC(OGcurrent)) ^ Prec(OGcurrent);
9.    elseif indirect-contribution (OGcurrent, UpreC) then ;
10.     generate-branch(OGcurrent);              /* conditioned manipulation */
11.       (body branch) := OGcurrent;            /* generate branch */
12.       BC := (equal (PostC(OGcurrent) " UPrec) nil);
13.       UPreC := (UPreC _ ((UPreC = PostC(OGcurrent)) ^ Prec(OGcurrent)));
14.    else                                      /* superfluous manipulation */
15.       dump (OGcurrent);
16.   end-loop;
17.   Prec(plan) := UPreC;
18.   return (plan);
19.end determine-branches.
```

For each Object Group checked during the backward trace three possible reasons for its occurrence exist:

1. The OG was intended. This means to perform the related manipulation is crucial for the achievement of the intended goal of the demonstration, namely the previously acquired user intention.
2. The OG was not intended but is nevertheless crucial for the achievement of the desired goal, since it enables an intended OG following later on.
3. The OG is superfluous and can be omitted without thwarting the achievement of the desired goal configuration.

OGs of the first kind are called to be directly contributing to the demonstrations, i.e. the user's goal. To test whether an Object Group is directly contributing its postcondition is simply matched against the set of conditions that represent the unsatisfied intentions (UI). If there is an intersection the OG creates parts of the user intended goal state—it is directly contributing.

The second type of OGs is called to be indirectly contributing to the demonstration. Here, the postcondition has an empty intersection set with the unsatisfied user intentions, but there is an intersection with the set representing the unsatisfied preconditions (UPreC) of the following OGs. The term "intersection" used in the last phrase is meant in the way that parts of the OGs postcondition enable the positive evaluation of parts of the UPreCs. For example, we assume that as a result of a manipulation free space is provided to be occupied by another object. This might enable a manipulation that requires a subpart of this space for delivering some object. Here the amount and location of newly provided space might not be equivalent to the required space but nevertheless there is a dependency; an intersection. The last type of OGs is fairly easy to handle. It simply does not contribute to the intended goal at all, neither directly nor indirectly.

Now the three different types of Object Groups are defined and it has been shown how a particular OG can be classified as one of these types. The remaining question is: how is an OG of a particular type to be treated during the performed backward trace in order to optimize or generalize the plan? As outlined in the algorithm given above, a directly contributing OG is kept without any changes. Its contribution is deleted from the unsatisfied intentions and its precondition is added to the unsatisfied preconditions. Furthermore, possible indirect contributions are deleted from the UPreCs. In the contrary, OGs that are not contributing at all are simply deleted. This is in fact an optimization since an unnecessary motion sequence is erased. The interesting OGs in terms of generalization are indirectly contributing ones. Each of those is embedded in a newly generated branch. The branching condition (BC) is instantiated with an OGs indirect contribution, namely the intersection of the OGs postcondition with the unsatisfied preconditions. The unsatisfied intentions remain unchanged because there is no direct contribution. However, the unsatisfied preconditions are updated by a disjunction. One disjunctive term being the unchanged former UPreCs, corresponding to the case when the branch is not carried out. The other disjunctive term being the former UPreCs with the OGs indirect contribution deleted and the precondition of the branchs OG added. This corresponds to the case in which the manipulations embedded in the branch are carried out.

Backtracing through the whole demonstration, generating branches when possible, and updating the unsatisfied preconditions leads to a generalized and optimized plan. Its precondition which determines its applicability is equivalent to the set of accumulated unsatisfied preconditions. Obviously the "factor" of generalization corresponds directly to the number of disjunctive terms in the precondition of the resulting plan, which depends on the number of conditioned manipulations that where detected throughout the backward trace.

Further Generalization; Object Selection Criteria & Variables Two methods exist for further user guided generalizations. The first one consists of editing the selection criteria that are determining which objects are chosen for manipulation when applying a generalization algorithme. This allows to match object knowledge related to the objects type (e.g. cube, cylinder, : :) as well as all valid inter object relations (e.g. on, in, : :) with the generated plan. Backpropagating these object attributes to the plans precondition results in a generalization, widening the applicability of the plan.

Second, variables can be introduced by user interaction in the postcondition of the plan which reflects the users intention. For example, we assume that an object O was put into another one (Oc) which is described as being of type case. Furthermore these facts shall be represented in the user's intention as the relations in(O,Oc) and type(Oc,case). Introducing a variable for the type case and backpropagating this change throughout the whole plan widens its applicability since every object able to serve as a container for O can be used to instantiate Oc, not only one of type case.

Finally, the induction component stores the generated plan in a knowledge base. In the storage process all objects are replaced by variables. This enables the unification process necessary for plan selection and instantiation as described below. Thus, generated plans can be applied to given tasks by the system's execution components. The representation of whole plan in terms of STRIPS operators having pre- and postcondition also allows to use them as macro operators in a planning process. This way more complex tasks can be solved by combining plans that where generated from different demonstrations.

3.3 Execution Phase

The third processing phase addresses the problem how to select an applicable plan to a given task among previously generated ones, how to instantiate its variables with respect to the current world model, and finally how to execute it with a given manipulator.

Plan Selection For the selection of a plan that enables the robot system to solve a given task two prerequisites have to be fulfilled: (i) the task's goal specification has to be given as a combination of the predefined relations in disjunctive normal form (DNF); (ii) the current environmental configuration has to be given as a valid world model. Given these two inputs a plan is selected by firstly matching the tasks goal specification with the postconditions of all available plans. All plans that are left after this process serve as candidates for the second selection phase. In this phase, each plan's precondition is matched with the given world model. Plans that "survive" this phase are applicable and do lead to the desired goal. Currently the first plan found to fulfill both selection phases is chosen for execution. However, this is a greedy method leading mostly to a suboptimal selection. Future extensions will be added to assess applicable appropriate plans in terms of execution time, complexity of required motions etc. to enable a better selection.

Preparing the Execution; Plan Instantiation and Sequencing At first glance the problem of how to prepare the selected plan for execution, namely how to instantiate its free variables and how to select the appropriate objects in the current world configuration, seems to be difficult to solve. Surprisingly, to determine the instantiations is not difficult at all. In fact, these instantiations are provided as a side effect of the selection phase. The matching processes required for plan selection include unifications of the objects in the world model with the variables in the plan's conditions, followed by the evaluation of these instantiated conditions on the world model. Therefore the determination of an applicable and appropriate plan includes one or more sets of possible and valid instantiations from which a random one is chosen. As in the selection process, the optimization of the selection of a valid instantiation set will be a future system's extension. Based on the world model, the instantiated plan is transformed into a sequence of Basic Operations, i.e. branching conditions are tested and the bodies of the corresponding branches are either omitted or prepared for execution.

Solving the Task; Plan Execution The last step is the execution of the generated program. Usually, the same robot that was used for the demonstration will be used for the execution, too. But, due to the flexibility of the program, it is also possible to execute the instantiated plan with another robot or with a simulation system. Of course, the basic capabilities (degrees of freedom, sensory equipment, work space) of the system used for the execution must not differ from those supported by the system used for the demonstration. In our system, the generated program that consists of a set of basic operations can either be transformed into VAL II commands (the programming language of the controller used with PUMA 260 robot) or into commands suitable for a graphical simulation interface. A transformation into other robot programming or simulation system languages is no principle problem, since only the module performing the transformation from the planning BOs to the desired language has to be added.

4 Interaction with the User—an Example

In this section, the process of programming a task by demonstration, applying the methods described in section 3, is illustrated by following an example through the phases of demonstration and induction. The example is demonstrated and the generated program executed with a real industrial robot. Thereby the system overcomes one of the most criticized aspects of former Robot PbD systems which have solved problems in simulated environments only [27] or have treated the robot as a dimension-less point in space [1].

Setting up the Scene; The Example The example task chosen for illustrating the methods used in the developed PbD system is as follows: Two cylinders shall be placed in two defined locations.

The user performs four manipulations while demonstrating the task. Two of those are required in any case, namely the manipulations of the cylinders.

Demonstration Phase The data recorded during the demonstration are the position of the tool center point (TCP) of the robot, the distance between the gripper jaws, and the values of the space mouse moved by the user. More data available like forces, torques, and orientations can be taken into account . A snapshot of the raw trace recorded during the demonstration and its filtered

Table 2. Recorded trace data containing raw unfiltered data. Left part (from left to right): position & gripper values of the spave mouse position of the TCP, and gripper width. Right part: (xt; xt1), gripper event & distance(xt, xt1).

((0 0 0 0)	(219.86 222.71 428.43)	(4939))						
((0 0 -5 0)	(219.76 222.55 429.18)	(4939))						
((0 0 0 0)	(219.76 222.55 429.18)	(4939))						
((0 0 0 2)	(219.74 222.51 429.18)	(2692))	) 0	0	0	2	0	
((0 0 0 0)	(219.74 222.51 429.18)	(2692))						
((0 0 40 0)	(219.79 222.74 424.78)	(2692))	) 0	0	4	0	4	
((0 0 42 0)	(219.96 223.07 394.19)	(2692))	) 0	0	31	0	31	
((-1 0 0 0)	(219.86 253.47 375.73)	(2692))	) 0	30	18	0	35	
((0 0 37 0)	(219.15 265.87 362.32)	(2692))	) 1	1	13	14	0	19
((0 0 37 0)	(219.23 265.68 336.63)	(2692))	) 0	0	25	0	25	
((0 0 0 0)	(219.34 266.01 320.16)	(2692))	) 0	0	17	0	17	
((0 0 0 0)	(219.36 266.08 320.13)	(2692))						
((0 0 0 0)	(219.37 266.11 320.10)	(2692))						

representation is shown in Table 2. The sequence shows a grip and part of a depart operation.

As can be seen from the data in Table 2 (left side), sometimes the operator does not move the robot's arm at all. These standstills or "zero-movements" are eliminated by the filtering process whose parameters can be modified by the user if desired. After the trace has been filtered, the TDNNs are applied to segment the trace and classify the segments into the five classes of Basic Operations. Part of the result of this step is shown in Table 3.

Table 3. Analysis results applying TDNNs; from left to right: EO identifier, start vector in filtered trace, and symbolic EO representation. grip & depart correspond to the data in Table 2

```
(0 2)      ) transfer ((220, 92, 378),  (220, 223, 402))
(1 14)     ) approach((220, 223, 402), (220, 223, 429))
(3 17)     ) grip    ((220, 223, 429), (2692))
(2 21)     ) depart  ((220, 223, 425), (202, 255, 320))
(0 2)      ) transfer ((202, 255, 320), (-69, 270, 334))
```

After the analysis the trace is presented to the user by a graphical tool. This allows to check the results of the previously performed automatic segmentation and classification. For illustrating the accuracy of the TDNN based classification, two segments that were classified to represent approach EOs were highlighted. The high classification accuracy can be evaluated by comparing the approach segments with the lightest curve which represents the z-values of the trace. This shows the high performance of the automatic analysis since an approach operation in the experiment is mainly characterized by negative translational z-values of the TCP.

The demonstration phase ends with the storage of the determined EO sequence.

Induction Phase As described in section 3, Object Groups are formed by the induction module and the demonstrations postcondition is determined (see Table 4).

Table 4. The determined postcondition of the demonstration.

(position(cyl1,(-68 240 530)) ∧ in(cyl1,shelf9) ∧ type(cyl1,cylinder) ∧
space(cylinder,(220 225 460)) ∧ position(cub1,(225 205 460)) ∧ on(cub1,tray) ∧
type(cub1,cube) ∧ position(cub2,(125 245 460)) ∧ on(cub2,tray)
∧ type(cub2,cube) ∧ space(cube,(-160 130 530)) ∧ position(cyl2,(70 180 530))
∧ in(cyl1,shelf10) ∧ type(cyl1,cylinder) ∧ space(cylinder,(307 65 460)))

In the next step, the user has to select the intended subparts of the postcondition determined by the system. In our example, the manipulations of the two cylinders were intended which is reflected by the specified user intention shown in Table 5.

Table 5. The subpart of the plan's postcondition reflecting the intention of the user.

(position(cyl1,(-68 240 530)) ∧ in(cyl1,shelf9) ∧ type(cyl1,cylinder) ∧
position(cyl2,(-70 180 530)) ∧ in(cyl2,shelf10) ∧ type(cyl2,cylinder))

Applying the generalization procedure leads to the generalized plan shown in Table 6. The superfluous manipulation of the cube has been eliminated and one branch has been generated reflecting the manipulation of the other cube, which was contributing only indirectly to the solution of the example task.

By specifying the initial positions of the two cylinders on the tray as variables, further generalization is achieved. Thus the precondition of the final plan shown

Table 6. The generalized plan (manipulations are abbreviated by their Object Groups).

```
OG1                  ;; put first cylinder in the shelf
If (equal (space(cylinder, (-70 180 530))) nil)
    OG2              ;; manipulate cube if necessary
OG4                  ;; put second cylinder in the shelf
```

in Table 7 determines the plan to be applicable when there are two cylinders on the tray, there is space in the first target shelf, and either also space in the second target shelf or a cube to be removed from it.

Table 7. The precondition of the generated generalized program, determining its applicability.

((on(cyl1,tray) ∧ type(cyl1,cylinder) ∧ space(cylinder,(-68 240 530)) ∧
on(cyl2,tray) ∧ type(cyl2,cylinder) ∧ space(cylinder,(-70 180 530))) ∧
(on(cyl1,tray) ∧ type(cyl1,cylinder) ∧ space(cylinder,(-68 240 530)) ∧
on(cyl2,tray) ∧ type(cyl2,cylinder) ∧ in(cub1,shelf10) ∧ type(cub1,cube) ∧
position(cub1,(-72 170 530)) ∧ space(cub1,(225 205 460)))))

5   Related Work

In this section, the relationship of RPD to other research work in this area will be discussed. Due to the number of existing approaches, it is neither possible to discuss every system nor to describe each selected approach in detail. Since the RPD approach presented here is more focussed on the acquisition of procedural knowledge, the work which is more related to adaptive control method as e.g. in [18] or [2] will not be reviewed. In the following, two different "classes" of RPD will be distinguished: RPD with direct repetition and RPD with generalized repetition.

5.1   RPD with direct repetition

The term RPD with direct repetition is expressing that the demonstrated sequence of robot actions is used to exactly repeat the same task. The generated program cannot be applied to even slightly modified products, environments, or manufacturing situations (see [21]). In particular, RPD with direct repetition is interesting for manufacturers in order to shorten programming cycles, provided that humans could demonstrate a task simply with their hands, i.e. in the most "natural way", while video and/or laser range cameras watch the scene. Two research projects explore exactly this technique. Kuniyoshi et al. [17] have called their RPD approach with direct repetition Teaching by Showing. A human worker builds various structures with different blocks using her or his hands while being observed by a stereo vision system. With several interpretation modules (like the movement detector, the hand tracker, the object finder, the edge junction classifier), all information necessary for the action recognizer are extracted from the image sequences. The recognizer generates a script-like action model that symbolically describes the extracted motion sequence. The recognition is model-based, i.e., it uses predefined action templates for large transfer motions as well as local motions like approach and depart movements. The recognizer instantiates those templates with the names of the identified objects, their coordinates, and spatial relation information (like alignments between objects). Finally, LISP operators like place-on-table, pick or reach-with-obj are created which can be sent to the robot control in order to repeat the task.

Research similar to that of Kuniyoshi et al. is performed by Ikeuchi and Suehiro [13]. They call their approach Assembly Plan from Observation or APO. In contrast to Kuniyoshi, they use a video camera and a laser range finder. Furthermore, they have defined a more elaborated and complete model. The basic idea in APO is the detection of the so-called face contact relations between all objects and use them to derive unambiguous interpretations and generalizations of EOs. Since the model has to be able to completely explain every possible action only small problems can be solved with the APO approach.

5.2 RPD with generalized repetition

RPD with the possibility of generalizing the demonstrated programs has two advantages over RPD with direct repetition. First, RPD with generalized repetition could shorten programming times if large repetitive robot tasks like loading a pallet can be programmed by detecting loops from few example demonstrations of the repetitive steps. Second, programs with generalized repetitions can easily be adapted to new situations and, therefore, are interesting for the aspect of programming reusable code.

An application of RPD with generalized repetition was reported by Heise [11] from the University of Calgary. Her system Etar (Example-Based Task Acquisition in Robots) is able to generate programs for blocksworld constructions and, more recently, also programs for placing objects on pallets or a conveyor belt [12]. The programs contain variables, loops, and branches. Similar to our demonstration interface, Heise also does not operate with complex vision sensors but uses the master-slave robot setup Excalibur that records 10 robot locations per second.

In order to extract the important data from the stream of robot locations, Heise defined a heuristic called "focus of attention" (FOA). This focus contains all location data in the vicinity of the objects to be manipulated (vicinity is the distance of the robot gripper to an object, a configurable value, typically three inch). All location data on a path between two vicinities are thrown away and replaced by one transport-motion command. Although the FOA criterion considers the context of actions to segment traces, it does not guarantee a user-intended segmentation.

Generalization in Etar is realized in two steps. In the first step, the "outer generalization" derives the program structure by grouping actions and searching for loops. Subsequences enclosed by gripper operations are identified in the action sequence. In each case, they define one semantical unit that describe the transfer of objects within the scene. After the subsequences have been identified, similar subsequences will be used to induce loops. Each loop generalizes a set of similar subsequences. The method used is similar to those applied in [7] or [1]. Structures within the subsequences are determined using the same method and induce further inner loops or branches. Etar provides no mechanism to deal with ambiguities while clustering similar subsequences. Therefore, the derived program structure might not meet the user's intention. The RPD system is using interaction (see section 4) with the user to ensure that the user-intended loops and branches are induced. The only other approaches to use additional interactions to ensure a program structure that meets the user's intention are implemented as part of the ProDeGE+ graphics editor [26] or Metamouse [20].

In the second step, Etar performs the "inner induction" that generalizes the parameters of the EOs as well as the determination of the loop and branch conditions. This is done using an expensive generate & test algorithm which determines exactly one solution. Nevertheless, this method does not ensure that the solution meets the user's intention. The RPD system presented does not only provide automated inner induction, but supports these steps by user interaction. This is similar to approaches in the graphics editor domain [20, 6, 26] or learning agents [5, 4]. It results in safer robot programs that more probably meet the user's intention. Furthermore, the RPD approach can be applied to a larger variety of problems because it is not restricted to a fixed set of basic functions as this is necessary for Etar's inner induction. Another possible approach to cope with very few examples is explanation-based learning (EBL). It was explored by Segre in his Arms system (Acquiring Robotic Manufacturing Schemata), a graphical simulation testbed for blocksworld and some peg-in-hole problems [27]. Arms uses EBL to derive a function schema from one example that allows him the application to situations similar with respect to the object types and spatial relations, but different with respect to geometric locations and distances. Arms is based on a sophisticated geometric reasoner that analyzes the start and goal states of a certain assembly.

6 Conclusions and future work

In this paper, an application of Programming by Demonstration (PbD) in the robotics domain was presented. The objective of this research is the exploration of possibilities to use machine learning for automated or semi-automated robot programming. The need for such programming environments actually comes from the industrial demand to lower the costs of production processes. With respect to the robot programming task within such processes the costs can be reduced significantly by incorporating PbD through:

- shortening programming and test cycles,
- enabling workers without or with little programming skills to perform the programming, and
- making robot programs (or program fragments) reusable for various applications.

In order to support this effort an industrial PUMA 260b manipulator with some sensors and a space mouse was used as input device. Using this demonstrational interface all robot locations as well as some sensor data are recorded. The trace is first filtered to get rid of unnecessary intermediate locations. Then a segmentation takes place to find out meaningful subsequences which are classified into five classes of symbolic Elementary Operations (EOs). For solving this segmentation and classification task on the continuous demonstration data, subsymbolic machine learning techniques (Time Delay Neural Networks) are used.

After the transformation of the demonstration representation on the symbolic level the induction component is started to generalize and optimize the determined sequence of Basic Operations. Therefore, the planning language contains operations for branches and loops besides the EOs, all of those modeled as STRIPS-like operators. Moreover a variable concept is integrated. The postcondition for each operator and the whole demonstration is determined. By incorporating a man-machine interface, the user is asked questions about the intentions related to the demonstration. By analyzing the demonstration with respect to the user's intentions superfluous and conditioned subparts are detected. Superfluous manipulations are eliminated, whereas branches are generated whenever conditioned manipulations are encountered. With this method, supported by introducing variables by the user in a dialogue, optimization and generalization takes place. The resulting plan is stored for later usage in a knowledge base.

The execution module selects an appropriate and applicable plan to a given task and environment from the available ones. The selected plan is prepared for execution by instantiating its variables with respect to the given environmental configuration and is transformed into a sequence of executable operations. Finally, the prepared plan is translated to the specific robot control language of the desired robot system and executed.

A critical drawback of the system is the use of STRIPS-like operators in the planning language. One might ask: "What is the benefit of RPD when the STRIPS operators, a world model, and a goal specification are already available? Simply apply a standard planning algorithm, and sooner or later it will produce a plan that solves the task!" Although this is true in principle, planning is an NP-hard problem which means that—given a sufficiently complex world model and goal specification—the off line planning will take very long. Therefore, in many cases it is much more comfortable and faster for the user to give a demonstration. Moreover, it is also reasonable to integrate a standard planner into the RPD system, i.e. to combine PbD and off line planning in order to make the best use of both methods.

6.1 Future Research

To extend the presented approach to more complex applications including interactions of grasped parts with other objects is the challenge for future work. Particular extensions are:

- Reactive behavior: Industrial robots can operate in a static, well-structured and well-known environment only if nothing unforeseen happens. In the future, on line planning and an exception handling which provides reactive behavior to the robot are important additionals to the system. This requires learning of sensor guided control strategies.
- New applications: If the robot is able to react to specific actions in the world by using external sensors, new applications become possible, e.g. manufacturing and assembly operations. Moreover, new operators representing pre-learned low-level skills (see [3, 15]) will be integrated in the planning language. This will enlarge the basic manipulation capabilities of the system allowing compliant motions, e.g. insertion operations.
- Induction of loops: Some of the systems presented in section 5 do support the detection of repetitively occurring demonstration parts and to induce loop structures from them. An extension that will enable the system presented here to induce loop structures, too, is currently under development. Again, the focus has been set on the generation of user intended loops, i.e. loops that do not only reflect a repetitively occurring structure, but also the user's intention.

Adding these extensions to the RPD system will improve its applicability and planning performance significantly. Finally, the result will be a system which might serve as a programming system in real-world domains like manufacturing or service environments (e.g. households etc.), supporting the user in programming even complex tasks easy and comfortable by demonstration.

References

1. Andrea, P.M., *Justified Generalization: Acquiring Procedures from Examples*, Technical Report AI-TR-834. Artificial Intelligence Laboratory, MIT, 1985.
2. Atkeson, C.G., Aboaf, E.W., McIntyre, J., and Reinkensmeyer, D.J., *Model-based robot learning*, in Proceedings of the 4th International Symposium on Robotics Research, 1987.
3. Baroglio, C., Giordana, A., Kaiser, M., Nuttin, M., and Piola, R., *Learning Controllers for Industrial Robots, Machine Learning*, 1996.
4. Bocionek, S., *Agent systems that negotiate and learn, International Journal Human-Computer Studies*, **42**, pp. 265–288, 1995.
5. Bocionek, S. and Sassin, M., *Dialog-Based Learning (DBL) for Adaptive Interface Agents and Programming-by-Demonstration Systems*, Technical Report CMU-CS-93-175. Carnegie Mellon University, Pittsburgh, 1993.
6. Cypher, A., *EAGER: Programming Repetitive Tasks by Example*, in CHI '91 Conference Proceedings (pp. 33–39). New Orleans, Louisiana: ACM Press, 1991.

7. Dufay, B. and Latombe, J.-C., *An approach to automatic robot programming based on inductive learning*, International Journal of Robotics Research, **3**, pp. 97–115, 1984.

8. Fikes, R. E. and Nilsson, N. J., *Strips: A new approach to the application of theorem proving to problem solving*, Artificial Intelligence, **2**, pp. 189–208, 1971.

9. Flaig, T., Neugebauer, J.-G., and Wapler, M., *VR4RobotS: a New Off-line Programming System Based on Virtual Reality Techniques*, in Proceedings of the 25th International Symposium on Industrial Robots (pp. 671–678). Hannover, Germany, 1994.

10. Friedrich, H. and Kaiser, M., *What can Robots learn from Humans?*, in IFAC Workshop on Human-Oriented Design of Advanced Robotic Systems (pp. 1–6). Vienna, Austria, 1995.

11. Heise, R., *Demonstration Instead of Programming: Focussing Attention in Robot Task Acquisition*, Research Report 89/360/22. University of Calgary, 1989.

12. Heise, R., *Programming Robots by Example*, Research Report 92/476/14. University of Calgary, 1992.

13. Ikeuchi, K., Kawade, M., and Suehiro, T., *Towards Assembly Plan from Observation: Task Recognition with Planar, Curved and Mechanical Contacts*, in Proceedings of the IEEE/RJS International Conference on Intelligent Robots and Systems (pp. 2294–2301). Yokohama, Japan, 1993.

14. Kaiser, M., Giordana, A., and Nuttin, M., *Integrated Acquisition, Execution, Evaluation and Tuning of Elementary Skills for Intelligent Robots*, in Proceedings of the IFAC Symposium on Artificial Intelligence in Real Time Control (pp. 145–150). Valencia, Spain, 1994.

15. Kaiser, M., Retey, A., and Dillmann, R. (1995), *Robot skill acquisition via human demonstration*, in Proceedings of the International Conference on Advanced Robotics (pp. 763–768), Barcelona, Spain, 1995.

16. Kreuziger, J. and Hauser, M., *A New System Architecture for Applying Symbolic Learning Techniques to Robot Manipulation*, in Proceedings of the IEEE/RSJ International Conference on Intelligent Robots and Systems. Yokohama, Japan, 1993.

17. Kuniyoshi, Y., Masayuki, I., and Inoue, H., *Learning by watching: Reusable task knowledge from visual observation of human performance. IEEE Transactions on Robotics and Automation*, **10**, pp. 799–822, 1994.

18. Liu, S. and Asada, H., *Teaching and training of deburring robots using neural networks*, in IEEE International Conference on Robotics and Automation (pp. 339–345), 1993.

19. Lozano-Perez, T., *Robot Programming*, in *Proceedings of the IEEE*, **71**, (pp. 821–841), 1983.

20. Maulsby, D.L. and Witten, I.H., *Metamouse: An Instructable Agent for Programming by Demonstration*, in A. Cypher (ed.), *Watch What I Do: Programming by Demonstration*. MIT Press, 1993.

21. McKerrow, P.J., *Introduction to Robotics*, in *Electronic Systems Engineering*. Addison-Wesley, 1991.

22. Milne, R., *Building Successful Applications: The Wrong Way and the Right Way*, in G. Barth et al. (eds.), *KI-94 – Anwendungen der Künstlichen Intelligenz*. Springer, 1991.

23. Münch, S., Kreuziger, J., Kaiser, M., and Dillmann, R., *Robot Programming by Demonstration (RPD) – Using Machine Learning and User Interaction Methods for the Development of Easy and Comfortable Robot Programming Systems*, in

Proceedings of the 25th International Symposium on Industrial Robots (pp. 685–693). Hannover, Germany, 1994.

24. Münch, S., Sassin, M., and Bocionek, S., *The Application of PbD Methods to Real-World Domains: Two Case Studies*, in Proceedings of the 7th Australian Joint Conference on Artificial Intelligence (pp. 92–99). Armidale, Australia, 1994.

25. Neubauer, W., Bocionek, S., Möller, M., and Rencken, W., *Learning Systems Behavior for the Automatic Correction and Optimization of Off-line Robot Programs*, in Proceedings of the IEEE International Conference on Intelligent Robots and Systems, Raleigh, 1992.

26. Sassin, M., *Creating user-intended programs with programming by demonstration*, in Proceedings of the IEEE/CS Symposium on Visual Languages (pp. 153–160). St. Louis, Missouri, 1994.

27. Segre, A.M., *Machine Learning of Robot Assembly Plans*. Kluwer Academic Publishers, 1988.

28. Thrun, S.B. and Mitchell, T.M., *Integrating Inductive Neural Network Learning and Explanation-based Learning*, in Proceedings of the 13th International Joint Conference on AI (pp. 930–936). Chambery, France, 1993.

29. Ude, A., *Trajectory Generation from Noisy Positions of Object Features for Teaching Robot Paths. Robotics and Autonomous Systems*, 11, pp. 113–127, 1993.

30. Ude, A., Bröde, H., and Dillmann, R., *Object Localization Using Perceptual Organization and Structural Stereopsis*, in Proceedings of the 3rd International Conference on Automation, Robotics and Computer Vision, Singapore, 1994.

31. Waibel, A., Hanazawa, T., Hinton, G., Shikano, K., and Lang, K., *Phoneme recognition using time-delay neural networks. IEEE Transactions on acoustics, speech and signal processing*, pp. 328–339, 1989.

Knowledge-Based Information Processing in Manufacturing Cells - The Present and the Future

Prof. Dr.-Ing. Gunther Reinhart
Dipl.-Ing. Rolf Diesch
Dr.-Ing. Michael R. Koch

Institute for Machine Tools and Industrial Management (iwb)
Technical University of Munich
Germany

Abstract. Existing concepts for control and malfunction handling in flexible manufacturing systems are mostly centrally structured. Accordingly, the existing knowledge-based systems for diagnosis and quality assurance have a central structure. These approaches don't fulfill future requirements towards increased availability and malfunction tolerance. At the iwb a new hierarchic control concept for autonomous manufacturing cells was developed and implemented. The approach enables an independent handling of occuring malfunctions, relieving the operators from routine interventions. The efficient use of decentral degrees of freedom to react autonomously to occuring malfunctions requires knowledge-based approaches. An analysis of the distributed knowledge bases and the different representations of knowledge shows, that an integrated design of the knowledge-based systems is vital for the successful development of future autonomous manufacturing cells.

1   Introduction

The increasing competition which manufacturing companies have faced on international markets in recent years has led to extended requirements being imposed on manufacturing systems and flexible manufacturing cells. Existing concepts and solutions no longer satisfy these requirements. In this situation, autonomous manufacturing cells open up new opportunities. The development from traditional flexible manufacturing cells towards the autonomous manufacturing cells of the future is described below, with the focus on knowledge-based information processing.

2   Flexible manufacturing cells

2.1   Initial information

In order to ensure a high level of availability, flexible manufacturing systems are increasingly often structured into individual, decoupled part-systems ([14]), which are described as "cells" *(Fig. 1)*. The dividing-up principle is based on the production of an order in a number of stages on several machines ([5]).

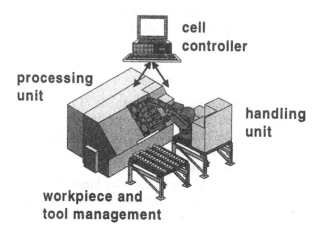

Fig. 1. Elements of a flexible manufacturing cell

2.2 Control of flexible manufacturing cells

Centralistic systems, described as "cell controllers" (e.g. [5]), are used in most cases to control flexible manufacturing cells. The cell controllers control and monitor the sequence within the manufacturing cell, that is to say coordination of the individual elements in the cell including the machine peripherals. Freely programmable sequence controllers - usually based on Petri networks - are used for this purpose.

With the aid of the cell controller, no problems are encountered in initiating fault-free sequences. Problems occur, however, in the response to malfunctions and unexpected events, since the control system is based on a purely deterministic description of the sequence. For this reason, separate systems for dealing with malfunctions were developed subsequently.

2.3 Dealing with malfunctions in flexible manufacturing cells

A "malfunction" is understood here as a deviation from the desired or nominal behavior. Dealing with malfunctions in manufacturing cells encompasses every aspect of the identification, localization and rectification of the malfunctions. If dealing with the malfunction relates to the system and the process, that is to say sequences within the system, we speak of diagnosis ([14]). If dealing with the malfunction relates to the product and the process, that is to say machining of the product, we speak of quality assurance.

Knowledge-based malfunction treatment methods have been developed for both diagnosis and quality assurance ([14]; [7]). It is characteristic of both cases that in accordance with the centralistic structure, central malfunction treatment systems have also grown up, for example for the central diagnosis of components in a manufacturing cell ([14]) or even of all the components in a complete manufacturing system ([13]).

2.4 Evaluation

Their centralistic structure results in the methods of control and malfunction treatment in flexible manufacturing cells referred to here suffering from significant shortcomings: the centralization of information, knowledge on how to deal with malfunctions and decision-taking competence lead to decentral opportunities and potentials remaining unused ([2]).

Taking these shortcomings and the demand for higher manufacturing-system availability as the starting-points, procedures which could be applicable to future manufacturing systems have been developed at the iwb.

3 Autonomous manufacturing cells

3.1 Autonomy as a guiding principle

For the further development of flexible manufacturing concepts, the term "autonomy" was selected as a guiding principle.

Five aspects of autonomy can be used to identify the principal characteristics of autonomous systems in the manufacturing area: "malfunction tolerance", "task orientation", "clarity", "adaptation" and "reflection" ([9]).

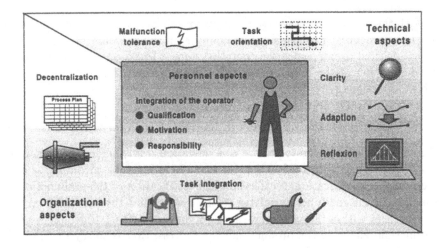

Fig. 2. Organizational, personnel and technical aspects of autonomy

By "malfunction tolerance" we mean the ability to carry out predetermined tasks within limited periods of time and without human intervention, despite the occurrence of malfunctions. "Task orientation" means the resolution and processing of task- related objects on the basis of decentral degrees of freedom and competences, again without human intervention. "Clarity" means the ability to observe the system's actions and to comprehend the decisions that are taken. "Adaptation refers to the system's ability to conform by itself to changes related to the surroundings, objectives and the system itself, for instance if the system's configuration is changed. "Reflection" is used to denote a system's ability to observe and assess its own actions, so that it can steadily improve its ability to perform the tasks it has been set. "Reflection" is therefore the basis for the system's ability to learn *(Fig. 2)*.

In addition to the technical aspects of autonomy, it is necessary above all to consider interaction with organizational aspects of autonomy, that is to say task integration and decentralization, and also personnel aspects of autonomy, that is to say integration of the operator. Task integration calls for the operator's knowledge and skills to be extended so that additional tasks can be solved, relating for example to quality assurance. At the same time, the decentralization of tasks and functions presupposes that mental activity will be transferred back from the office to the workshop, for example so that NC programming can be decentralized. A major precondition for the success of the measures described above is the operator's motivation to learn and to accept responsibility. However, he or she must also be granted additional decision-taking competence in the relevant task area in parallel with the greater degree of responsibility deriving from this extended task content.

3.2   Malfunction-tolerant control of autonomous manufacturing cells

The results of analyzing the technical control tasks in autonomous manufacturing cells are presented below. Three different categories of technical control tasks are found to result if sequences are to be free from malfunctions. The first comprises tasks which involve order processing. Secondly, during order processing the interaction of cell components and cooperation with external components, for example transport systems, must be coordinated. Thirdly, the cell components must perform the tasks relating to machining, handling or measuring the workpieces. *(Fig. 3)*

In accordance with the structuring of technical control tasks, [8] developed a malfunction-tolerant control concept based on the hierarchical structuring of control and malfunction treatment functions on three levels *(Fig. 4)*. These levels can be referred to as the organization, coordination and execution levels.

The organization level

Based on the cell's requirements, the organization level lays down the overall behavior of the cell during processing of the order. Planning the processing of the order is performed, taking the required logistical target values into account. As part of internal cell requirement control and with the agreed deadlines taken into account, the planning process determines the optimum order sequences and makes the orders to be processed available to the coordination level.

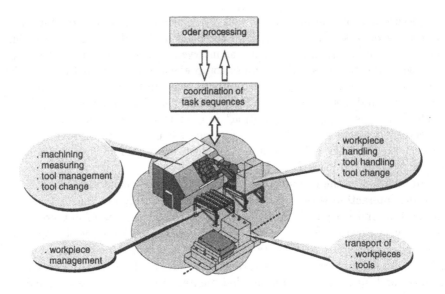

Fig. 3. Technical control tasks in autonomous manufacturing cells

The agents in the execution level, which represent the cell components, report any changes in their condition to the organization level. In this way, the organization level can take the effects of malfunctions on orders in the cell into account quickly, and react accordingly ([8]).

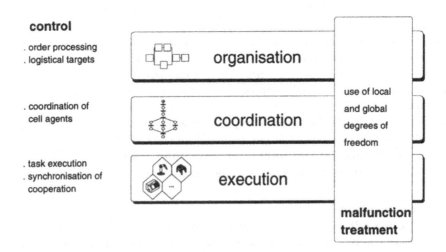

Fig. 4. Control and malfunction treatment hierarchy levels

The coordination level

The coordination level's task is to coordinate the agents in the manufacturing cell during the preparation, machining and rectification of the orders. The coordination level receives the orders for processing from the organization level. The sequence which has to be initiated in the cell is described with the aid of single-mark Petri networks. If two or more agents are available to perform a function, the coordination can select the most suitable one according to the actual situation. If cooperation between agents is needed in order to carry out a task, the coordination level informs the affected agents of this as a coordinated cooperation procedure.

In addition to initiating sequences, the coordination level performs malfunction treatment tasks. These include for example dealing with malfunctions which occur when agents are cooperating, initiating avoidance actions or utilizing resources from outside the cells when dealing with malfunctions ([8]).

The execution level

On the execution level, the components in the cells and their functions are represented by agents; these are only present on the execution level. For the preparation and subsequent performance of tasks, the agents make what are known as "agent functions" available, for example "Machine workpiece" or "Handle workpiece". The tasks received from the coordination level are first checked for plausibility. If this is not established, the task is rejected. During performance of the task, the agent functions can be divided up into elementary operations which represent the smallest possible breakdown of a task. Agents can perform tasks independently, but can also cooperate with other agents, for instance when workpieces are transferred. Cooperation takes place in accordance with a rigidly defined synchronization pattern, that is to say no agent hierarchies are formed on the execution level.

In accordance with the malfunction tolerance principle, an agent attempts initially to solve its task independently, even if malfunctions have occurred, in other words with no outside help. After this, either the occurrence of the malfunction or the successful completion of the task is reported to the coordination level ([8]).

A brief evaluation of the concept developed for malfunction-tolerant control is given below. Dividing the control system up into three levels reduces its complexity and improves the user's ability to comprehend system actions, in other words enhances clarity. Degrees of freedom are available on all levels for control and for dealing with malfunctions; they can be used independently by the system on the basis of available competences. At the same time, the agents' decentral degrees of freedom permit a transition from a sequence-oriented to a behavior-oriented control paradigm.

4 Knowledge-based approaches in conventional flexible manufacturing cells

4.1 Overview

In the remarks below, a distinction should be made between the adoption of knowledge-based methods in a single manufacturing cell and their adoption in the entire manufacturing surroundings.

In the manufacturing surroundings, artificial-intelligence methods are used for a variety of applications. For example, they support production planning and control ([16]), are used for the classification of production parts ([4]) or employed for process monitoring ([12]). Of the many and varied uses of expert systems in technical manufacturing control, [10] is one example.

Within manufacturing cells, the fields of application for artificial-intelligence methods today still remain largely limited to the quality assurance and diagnosis areas. The features and shortcomings established in respect of these methods are described below.

4.2 Diagnosis

Various methods of using knowledge-based systems for dealing with malfunctions in a diagnosis context in manufacturing cells have been developed. Examples of the varied concepts in this area are [14] and [17].

A significant aspect in the development of diagnosis in manufacturing cells is the drawing up of a basic structure which in general reflects the largest possible part of the knowledge relevant to malfunction treatment. [14] developed for example a diagnosis system which maintains and updates knowledge relevant to diagnosis centrally for all the components of a manufacturing cell. When a malfunction has to be rectified within the cell, this central component is accessed.

4.3 Quality assurance

In a close parallel with diagnosis, knowledge-based systems are used in flexible manufacturing-cell quality assurance to depict malfunction knowledge with reference to the product ([7]; [11]).

During conventional malfunction treatment processes, quality-relevant malfunctions, if they occur, are analyzed with the aid of the machine operating personnel's experience and expert knowledge. In order to change over to computer-aided treatment of malfunctions, this knowledge is presented completely and in a manner suitable for processing in a knowledge base. Basic malfunction knowledge is used for the general procedure independent of the malfunction if one should occur, but complemented by malfunction-specific knowledge for the relevant application.

4.4 Analysis and evaluation

When the knowledge stored in manufacturing cells for quality assurance and for diagnosis is compared, it is evident that in both cases malfunction causalities and cell or machining sequences are stored in knowledge bases *(Fig. 5)*. In addition, knowledge about the system, product, process and environment is needed for quality assurance. Knowledge about the system is also needed for diagnosis, but differently structured, that is to say in state transitions and structural "trees". This indicates that for both quality assurance and diagnosis, partly the same knowledge is presented redundantly in differing structures. A clear allocation of knowledge exclusively to quality assurance or diagnosis as the case may be is not possible. The two areas cannot be kept apart in some cases.

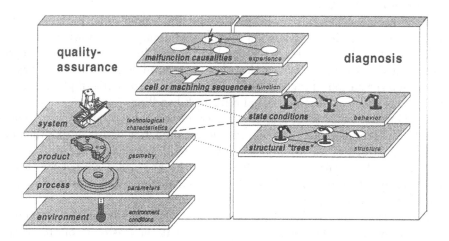

Fig. 5. Structure of the knowledge presented for quality assurance and for diagnosis

The redundancies determined here and the overlaps discovered between quality-assurance and diagnosis knowledge suggest that there is a need for restructuring of the knowledge in manufacturing cells.

The methods of knowledge-based diagnosis and quality assurance which have been referred to exhibit deficits in the knowledge processing area. One major deficit for both of them is in the acquisition of knowledge. This is a very complex process, which entails high costs ([6]). Changes in the cell configuration have to be shown in the knowledge base. This adaptation of knowledge bases until now frequently had to be undertaken by artificial- intelligence experts, since an easily handled user interface is often lacking from current systems. Concepts for automating the acquisition of knowledge in manufacturing cells do exist ([1]), but have not yet been accepted.

Furthermore, the concepts and approaches drawn up so far tend for the most part to address the operation of manufacturing cells. A complete approach to knowledge-based diagnosis and quality assurance, taking into account at the same time the design, start-up and continuous operation of manufacturing cells, does not seem to be in the offing at the time of writing.

Based on these investigations, the following section deals in detail with knowledge processing in autonomous manufacturing cells.

5 Knowledge-based approaches in future autonomous manufacturing cells

5.1 The current position of knowledge processing in autonomous manufacturing cells

The organization level

At the organization level, knowledge is needed for regular order processing and also for the replanning of orders if a malfunction occurs. The knowledge needed for regular task handling is on the one hand stored implicitly in the sequence determination algorithms, and on the other memorized in the form of fuzzy rules for the interpretation of logistic tasks. Knowledge permits the implementation of logistical targets in the individual cell's task handling.

Replanning strategies for individual orders permit the continuous, autonomous operation of the manufacturing cell, even if a malfunction occurs during the machining of an order. An attempt is made to replace damaged orders in order control by orders which can be carried out. The knowledge needed for this is permanently integrated into the order control algorithms, and additions are not possible unless the program is modified.

In addition, current knowledge concerning the abilities of the individual agents is present on the organization level, so that the feasibility of orders can be continuously monitored.

The coordination level

On the coordination level, knowledge concerning the coordination of the individual cell agents for order processing is stored in the form of Petri networks. Knowledge concerning the treatment of malfunctions is also stored in these networks. For example, a malfunction knowledge base is used to describe the relationships between malfunction reports, malfunction causes and the corresponding rectification measures. In this knowledge base, various causes of malfunctions including their probabilities, together with elimination strategies and their probability of success are memorized. Inputs to this malfunction knowledge base are only undertaken by the developer.

The execution level

The execution level is formed from individual agents, which from the knowledge processing standpoint are all built up according to the same pattern. Important elements in the agents are the condition model and the behavior model.

The condition model contains the actual condition of the agent concerned. By means of the knowledge memorized in the condition model, the agents can check the feasibility of agent functions independently. This condition modeling process forms the basis for agent behavior modeling.

Memorized in the behavior model is the way in which the individual agent must react to external demands when fulfilling a task. For cooperation between several agents in carrying out a task, the individual agents' actions must be synchronized, that is to say precisely coordinated. For this purpose, a permanent synchronizing routine is stored in every agent. With the aid of this knowledge, each agent can interact with every other agent without difficulties. Neural Networks are used to analyse and evaluate sensor data, which is recieved by the individual agents ([15]).

(Fig. 6) summarizes the form of presentation for knowledge and its incorporation and utilization in the three levels.

As *(Fig. 6)* shows, the knowledge present in the various levels of cell control is memorized in extremely heterogeneous forms of presentation. Knowledge was programmed into the individual levels either implicitly, that is to say directly, or explicitly, that is to say in the form of fuzzy rules, neural networks, lists, Petri networks or state-transition graphs.

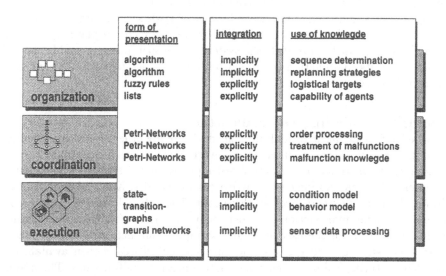

	form of presentation	integration	use of knowlegde
organization	algorithm	implicitly	sequence determination
	algorithm	implicitly	replanning strategies
	fuzzy rules	explicitly	logistical targets
	lists	explicitly	capability of agents
coordination	Petri-Networks	explicitly	order processing
	Petri-Networks	explicitly	treatment of malfunctions
	Petri-Networks	explicitly	malfunction knowlegde
execution	state-transition graphs	implicitly implicitly	condition model behavior model
	neural networks	implicitly	sensor data processing

Fig. 6. Existing knowledge processing structure in autonomous manufacturing cells

An analysis of the current status of knowledge processing in autonomous manufacturing cells reveals that knowledge is always input when degrees of freedom exist in the system. In this case the system can take decisions of its own, which are made possible and influenced by the incorporation of knowledge. The resulting decentral knowledge structure reflects the deliberate use of decentral potentials in the decision-making process.

Current operating areas for artificial intelligence methods arose directly from the corresponding requirements on the individual levels. For this reason, a complete view of the control structure with regard to the introduction of knowledge-based systems is lacking.

5.2 Future knowledge needed for autonomous manufacturing cells

In future manufacturing cells, extended knowledge built up on the basis of the knowledge already present must be available. It must be possible for machine operating personnel to change and extend this more comprehensive knowledge. The types of knowledge which have to be examined in this context are the knowledge needed for the planning of tasks in the sense of task orientation and the knowledge needed for dealing with malfunctions.

Knowledge for planning

The heading "knowledge for the planning of tasks" groups together the knowledge needed for agent action planning, sequence planning on the coordination level and the control of orders on the organization level. Preparation of low-personnel operating phases is of particular importance.

By means of additional product and system knowledge, the individual manufacturing cell can select suitable strategies for optimizing order planning. Knowledge of the process, the product, the system or the surroundings enable the cell to determine its malfunction-treatment optimization strategies and to improve them still further.

Knowledge for dealing with malfunctions

Knowledge for dealing with malfunctions deals among other things with replanning strategies if orders are interrupted, evasion operations as a means of counteracting malfunctions during actual order processing or strategies for passing on orders which can no longer be carried out punctually. Knowledge on dealing with malfunctions extends over all three control-structure levels, in order to ensure a consistent malfunction-treatment sequence. It is of decisive importance in this connection for the user to be able to input malfunction treatment mechanisms and strategies to the system. This new knowledge is then available to the manufacturing cell for independent action against malfunctions. The manufacturing cell's knowledge base must also store details of which actions can be taken at which times and in which circumstances. This enables a mechanism to be created which protects the cell against damage.

The planning and malfunction treatment knowledge referred to here is available firstly in the manufacturing cell in the form of a basic configuration, but can also be modified or added to by the machine's operating personnel.

In future manufacturing cells, the role to be played by the machine user will acquire even more prominence, since the latter will undertake extended tasks at the cells, so that the individual user's specific knowledge will be of increasing importance. This specialist knowledge is essential when dealing with malfunctions, for quality assurance measures or for optimum order control. It must be made possible for the user to contribute specialist knowledge to the manufacturing cell.

Based on the attempts already undertaken to use knowledge-based systems in autonomous manufacturing cells, and with future operating areas in mind, the section below examines the requirements which knowledge-based information processing in autonomous manufacturing cells must satisfy.

5.3 Requirements to be satisfied by knowledge processing in autonomous manufacturing cells

For the continued development of autonomous manufacturing cells it is necessary to examine the control structure which has been developed as a whole, with reference to the potential applications of knowledge-based systems. The demands which will be made of future manufacturing cells must of course be taken into account.

If knowledge-based systems are to be operated in future manufacturing cells, it is of central importance for knowledge input for manufacturing-cell development not to be solely in the hands of AI experts or application specialists. On the contrary, users must be able to modify and extend this knowledge so that their own specific expert knowledge can be input to the manufacturing process. For this reason, knowledge-based systems in manufacturing cells must have an easily operated user interface, by means of which the memorized knowledge can be changed and added to. At the same time, it should be possible to record user actions automatically, so that they can be included in the knowledge base subsequently. The modification of existing knowledge and the integration of additional knowledge makes the manufacturing cells capable of "learning".

When the user introduces new knowledge to the manufacturing cells, care must be taken to ensure that this new knowledge does not contradict existing knowledge. The new knowledge input must therefore be checked for consistency with the knowledge already available. This ensures that knowledge-based decisions at the various levels lead to consistent decisions.

New knowledge for a manufacturing cell must not result exclusively from user actions. Data obtained during start-up must also be absorbed reliably into the knowledge base.

As the next step following the introduction of new knowledge, a check can be made as to whether this has consequences for the existing knowledge. First of all, the ability to transfer knowledge to other areas must be checked.

The aim of adopting artificial-intelligence methods is to permit knowledge to be re- used throughout and beyond the manufacturing cell's life-cycle. In this way, knowledge obtained from an existing manufacturing cell can also be used by new generations of cells. The feedback of knowledge from the operating phase to the development phase thus becomes possible, to the benefit of a new generation of manufacturing cells. For this reason the knowledge must be presented and structured appropriately, so that it can satisfy this future manufacturing-cell design requirement. According to [3], however, the problems arising in connection with the re-usability of knowledge bases are still severe.

Newly acquired knowledge must also be accessible quickly and easily in the documentation applicable to a manufacturing cell. It must be possible to utilize a change in the knowledge base with a minimum of effort for a documentation update. In this way, the documentation which the user is able to access is maintained at the newest status.

In all cases the knowledge available in an autonomous manufacturing cell must be stored explicitly, which will normally mean in rules or networks. This is the only way of ensuring that the knowledge concerned can also be modified and extended without undue effort during operation of the manufacturing cell.

5.4   Evaluation and the resulting need for action

Extending the existing areas of operation for knowledge-based systems to permit a complete approach procedure in the context of an overall malfunction-tolerant control concept and the requirements arrived at above have consequences for the further development of autonomous manufacturing cells.

For developing future manufacturing cells a complete approach to knowledge-based systems is necessary. It is furthermore important to investigate whether one central or several decentral knowledge bases can satisfy the requirement as stated above more effectively. The decentralization of the control structure undertaken as part of the development of autonomous manufacturing cells has proved successful in terms of control and malfunction treatment. It will also be capable of satisfying the demands made of future manufacturing cells. For this reason, although there seem to be cogent reason for the appropriate decentralization of the necessary knowledge, this needs to be investigated further.

Whether it would be possible or desirable to reduce the heterogeneity of the forms in which existing knowledge is presented in autonomous manufacturing cells should also be investigated. In order to permit interaction between various decentral knowledge bases, knowledge must be stored in a more homogeneous form. The manner in which distributed knowledge from a manufacturing cell can be used in other manufacturing cells will have to be determined, as well as the manner in which this knowledge can be integrated both syntactically and semantically into manufacturing cells.

6   Summary

Taking deficits in existing centralistic systems for the control of manufacturing cells and the treatment of malfunctions as a starting point, new opportunities are offered by the progress being made towards autonomous manufacturing cells. A major precondition for their successful introduction, however, is the systematic utilization of decentral degrees of freedom within the systems. For this, knowledge is needed which will enable the systems to take the correct decisions quickly and consistently. The demands which must be made of knowledge processing in future autonomous manufacturing cells are therefore correspondingly high. Accordingly, the knowledge processing aspect is of decisive importance in the successful design of future autonomous manufacturing cells.

References

1. Birkel, G: Aufwandsminierter Wissenserwerb für die Diagnose in flexiblen Produktionszellen (Knowledge acquisition at minimum complexity for diagnosis in flexible production cells). Berlin, publ. Springer, 1995 (iwb Research Reports, 1984).
2. Duffie, N.A., Chitturi, R. and Mou, J.: Fault-tolerant Heterarchichal Control of Heterogeneous Manufacturing System Entities. Journal of Manufacturing Systems 7 (1988) 4, pp. 315-328.
3. Fredrich, H.: Verteiltes Assistenzsystem zur Fehlersuche an Fertigungsanlagen (Distributed assistance system for troubleshooting on manufacturing plant). Munich, publ. Hanser, 1995 (Produktionstechnik, Berlin, 171).
4. Greska, W.: Wissensbasierte Analyse und Klassifizierung von Blechteilen (Knowledge-based analysis and classification of sheet-metal parts). Munich, publ. Hanser 1995 (Fertigungstechnik, Erlangen, 49).
5. Groha, A.: Universelles Zellenrechnerkonzept für flexible Fertigungssysteme (Universal cell-controller concept for flexible manufacturing systems). Berlin, publ. Springer, 1988 (iwb Research Reports 14).
6. Huber, K.-P. and Nakhaeizadeh, G.: Maschinelle Lernverfahren als Unterstützung beim Wissenserwerb von Diagnose (Machine learning processes as support for knowledge acquisition by expert diagnosis systems). In: Puppe, F. and Günter, A.: Expertensysteme 1993. Berlin, publ. Springer, 1993 (Informatik aktuell).
7. Kahlenberg, R.: Integrierte Qualitätssicherung in flexiblen Fertigungszellen (Integrated quality assurance in flexible manufacturing cells). Berlin, publ. Springer, 1995 (iwb Research Reports 82).
8. Koch, M.R.: Autonome Fertigungszellen - Gestaltung, Steuerung und integrierte Störungsbehandlung (Autonomous manufacturing cells - design, control and integrated treatment of malfunctions). Berlin, publ. Springer, 1996 (iwb Research Reports 98).
9. Koch, M.R.: Von flexiblen zu autonomen Systemen. Höhere Verfügbarkeit durch beherrschte Komplexität bei autonomen Fertigungssystemen (From flexible to autonomous systems. Improved availability through mastery of complexity on autonomous manufacturing systems). TECHNICA 43 (1994) 20, pp. 14-19.
10. Kupec, T.: Wissensbasiertes Leitsystem zur Steuerung flexibler Fertigungssysteme (Knowledge-based control system for flexible manufacturing systems). Berlin, publ. Springer, 1991 (iwb Research Reports 37).
11. Pfeifer, T., Grob, R. and Klonaris, P.: Erfahrungen mit dem wissensbasierten Fehleranalysesystem CAFA (Experience obtained from the CAFA knowledge-based malfunction analysis system). Technisches Messen 62 (1995), pp. 380-384.
12. Reinhart, G. and Löffler, T.: Signalklassifikation im Rahmen der akustischen Fügeprozessüberwachung (Signal classification in the course of acoustic joining- process monitoring). Technisches Messen 62 (1995), pp. 370-374.
13. Reuschenbach, W.: Entwicklung und Einsatz eines universellen Stördatenerfassungssystems mit wissensbasierter Diagnose für Produktionseinrichtungen (Devel-

opment and introduction of a universal malfunction data recording system with knowledge-based diagnosis for production facilities). Aachen and Mainz, 1992.

14. Schönecker, W.: Integrierte Diagnose in Produktionszellen (Integrated diagnosis in production cells). Berlin, publ. Springer, 1992 (iwb Research Reports 45).

15. Wagner, M.: Fehlertolerante Steuerung maschinennaher Abläufe (Fault-tolerant control of sequences in tool machines). Berlin, publ. Springer, 1995 (iwb Research Reports 100).

16. Weigelt, M. and Mertens, P.: Produktionsplanung und - steuerung mit verteilten wissensbasierten Systemen (Production planning and control with distributed knowledge-based systems). In: Brauer, W. and Hernàndez, D. (Eds.), Verteilte Künstliche Intelligenz und kooperatives Arbeiten. Berlin, publ. Springer, 1991 (Fachberichte Informatik 291)

17. Wiedmann, H.: Objektorientierte Wissensrepräsentation fr die modellbasierte Diagnose an Fertigungseinrichtungen (Object-oriented knowledge presentation for model-based diagnosis on manufacturing plant). Berlin, publ. Springer, 1993 (ISW Forschung und Praxis 84)

Calculi for Qualitative Spatial Reasoning

A G Cohn

Division of Artificial Intelligence,
School of Computer Studies,
University of Leeds,Leeds LS2 JT, UK.
Telephone: +44 113 233 5482.
Email: agc@scs.leeds.ac.uk WWW: http://www.scs.leeds.ac.uk/

Abstract. Although Qualitative Reasoning has been a lively subfield of AI for many years now, it is only comparatively recently that substantial work has been done on qualitative *spatial* reasoning; this paper lays out a guide to the issues involved and surveys what has been achieved. The papers is generally informal and discursive, providing pointers to the literature where full technical details may be found.

1 What is Qualitative Reasoning?

The principal goal of Qualitative Reasoning (QR) [86] is to represent not only our everyday commonsense knowledge about the physical world, but also the underlying abstractions used by engineers and scientists when they create quantitative models. Endowed with such knowledge, and appropriate reasoning methods, a computer could make predictions, diagnoses and explain the behaviour of physical systems in a qualitative manner, even when a precise quantitative description is not available[1] or is computationally intractable. The key to a qualitative representation is not simply that it is symbolic, and utilises discrete quantity spaces, but that the distinctions made in these discretisations are *relevant* to the behaviour being modelled – i.e. distinctions are only introduced if they *necessary* to model some particular aspect of the domain with respect to the task in hand. Even very simple quantity spaces can be very useful, e.g. the quantity space consisting just of $\{-, 0, +\}$, representing the two semi-open intervals of the real number line, and their dividing point, is widely used in the literature, e.g. [86]. Given such a quantity space, one then wants to be able to compute with it. There is normally a natural ordering (either partial or total) associated with a quantity space, and one form of simple but effective inference is to exploit the transitivity

---

[1] Note that although one use for qualitative reasoning is that it allows inferences to be made in the absence of complete knowledge, it does this not by probabilistic or fuzzy techniques (which may rely on arbitrarily assigned probabilities or membership values) but by refusing to differentiate between quantities unless there is sufficient evidence to do so; this is achieved essentially by collapsing 'indistinguishable' values into an equivalence class which becomes a qualitative quantity. (The case where the indistinguishability relation is not an equivalence relation has not been much considered, except by [55, 54].)

of the ordering relation. More interestingly, one can also devise qualitative arithmetic algebras [86] to achieve this; for example figure 1 is a table for addition on the above quantity space. Note that certain entries are ambiguous: this is a recurring feature of Qualitative Reasoning – not surprisingly, reducing the precision of the measuring scale decreases the accuracy of the answer. Much research in the Qualitative Reasoning literature is devoted to overcoming the detrimental effects on the search space resulting from this ambiguity, though there is not space here to delve into this work[2]. However one other aspect of the work in traditional Qualitative Reasoning is worth noting here: a standard assumption is made that change is continuous; thus, for example, in the quantity space mentioned above, a variable cannot transition from - to + without first taking the value 0. We shall see this idea recurring in the work on qualitative spatial reasoning described below.

	+	0	-
+	+	+	?
0	+	0	-
-	?	-	-

Fig. 1. The addition table for the qualitative quantity space $\{+,0,-\}$. Note the ambiguity present in two of the entries marked with a '?'.

2 What is Qualitative Spatial Reasoning?

QR has now become a mature subfield of AI as evidenced by its 10th annual international workshop, several books (e.g. [86] [33],[57]) and a wealth of conference and journal publications. Although the field has broadened to become more than just Qualitative Physics (as it was first known), the bulk of the work has dealt with reasoning about scalar quantities, whether they denote the level of a liquid in a tank, the operating region of a transistor or the amount of unemployment in a model of an economy.

Space, which is multidimensional and not adequately represented by single scalar quantities, has only a recently become a significant research area within the field of QR, and, more generally, in the Knowledge Representation community. In part, this may be due to the *Poverty Conjecture* promulgated by Forbus, Nielsen and Faltings [86]: "there is no purely qualitative, general purpose kinematics". Of course, qualitative spatial reasoning (QSR) is more than just kinematics, but it is instructive to recall their third (and strongest) argument for the conjecture – "No total order: Quantity spaces don't work in more than one dimension, leaving little hope for concluding much about combining weak information about

---

[2] In the present context it is perhaps worth while pointing out the interesting work of [56] which shows that in the presence of such ambiguity, humans usually have a preferred model, at least for Allen's qualitative temporal interval calculus [1].

spatial properties". They correctly identify transitivity of values as a key feature of a qualitative quantity space but doubt that this can be exploited much in higher dimensions and conclude: "we suspect the space of representations in higher dimensions is sparse; that for spatial reasoning almost nothing weaker than numbers will do".

The challenge of QSR then is to provide calculi which allow a machine to represent and reason with spatial entities of higher dimension, without resorting to the traditional quantitative techniques prevalent in, for example, the computer graphics or computer vision communities.

Happily, over the last few years there has been an increasing amount of research which tends to refute, or at least weaken the 'poverty conjecture'. There is a surprisingly rich diversity of qualitative spatial representations addressing many different aspects of space including topology, orientation, shape, size and distance; moreover, these can exploit transitivity as demonstrated by the relatively sparse transitivity tables (cf the well known table for Allen's interval temporal logic [86]) which have been built for these representations (actually 'composition tables' is a better name for these structures).

3 Possible applications of qualitative spatial reasoning

Researchers in qualitative spatial reasoning are motivated by a wide variety of possible application areas, including: Geographical Information Systems (GIS), robotic navigation, high level vision, the semantics of spatial prepositions in natural languages, engineering design, commonsense reasoning about physical situations, and specifying visual language syntax and semantics. Below I will briefly discuss each of these areas, arguing the need for some kind qualitative spatial representation. Other application areas include document-type recognition [37] and domains where space is used as a metaphor, e.g. [59].

GIS are now commonplace, but a major problem is how to interact with these systems: typically, gigabytes of information are stored, whether in vector or raster format, but users often want to abstract away from this mass of numerical data, and obtain a high level symbolic description of the data or want to specify a query in a way which is essentially, or at least largely, qualitative. Arguably, the next generation of GIS will be built on concepts arising from *Naive Geography* [30] which requires a theory of qualitative spatial reasoning.

Although robotic navigation ultimately requires numerically specified directions to the robot to move or turn, this is not usually the best way to plan a route or other spatially oriented task: the AI planning literature [82] has long shown the effectiveness of hierarchical planning with detailed decisions (e.g. about how or exactly where to move) being delayed until a high level plan has been achieved; moreover the robot's model of its environment may be imperfect (either because of inaccurate sensors or because of lack of information), leading to an inability to use more standard robot navigation techniques. A qualitative model of space would facilitate planning in such situations. One example of this kind of work is [58]; another, solving the well known 'piano mover's problem' is [32].

While computer vision has made great progress in recent years in developing low level techniques to process image data, there is now a movement back [34] to try to find more symbolic techniques to take the results of these low level computations and produce higher level descriptions of the scene or video input; often (part of) what is required is a description of the spatial relationship between the various objects or regions found in the scene; however the predicates used to describe these relationships must be sufficiently high level, or qualitative, in order to ensure that scenes which are semantically close have identical or at least very similar descriptions.

Perhaps one of the most obvious domains requiring some kind of theory of qualitative spatial representation is the task of finding some formal way of describing the meaning of natural language spatial prepositions such as "inside", "through", "to the left of" etc. This is a difficult task, not least because of the multiple ways in which such prepositions can be used (e.g. [53] cites many different meanings of "in"); however at least having a formal language at the right conceptual level enables these different meanings to be properly distinguished. Examples of research in this area include [5, 85].

Engineering design, like robotic navigation, ultimately usually requires a fully metric description; however, at the early stages of the design process, it is often better to concentrate on the high level design, which can often be expressed qualitatively. The field of qualitative kinematics (e.g. [31]) is largely concerned with supporting this kind of activity.

The fields of qualitative physics and naive physics [86] have concerned themselves with trying to represent and reason about a wide variety of physical situations, given only qualitative information. Much of the motivation for this was given above in the section on qualitative reasoning; however traditionally these fields, in particular qualitative physics, have had a rather impoverished spatial capacity in their representations, typically restricting information to that which can be captured along a single dimension; adding a richer theory of qualitative spatial reasoning to these fields would increase the class of problems they could tackle.

Finally, the study and design of visual languages, either visual programming languages or some kind of representation language, perhaps as part of a user interface, has become rather fashionable; however, many of these languages lack a formal specification of the kind that is normally expected of a textual programming or representation language. Although some of these visual languages make metric distinctions, often they are predominantly qualitative in the sense that the exact shape, size, length etc. of the various components of the diagram or picture are unimportant – rather, what is important is the topological relationship between these components and thus a theory of qualitative spatial representation may be applicable in specifying such languages [42].

4   Aspects of qualitative spatial representation

There are many different aspects to space and therefore to its representation: not only do we have to decide on what kinds of spatial entity we will admit (i.e. commit to a particular ontology of space), but also we can consider developing different kinds of ways of describing the relationship between these kinds of spatial entity; for example we may consider just their topology, or their sizes or the distance between them, or their shape. Of course, these notions are not entirely independent as we shall see below.

4.1   Ontology

In developing a theory of space, one can either decide that one will create a *pure* theory of space, or an *applied* one, situated in the intended domain of application; the question is whether one considers aspects of the domain, such as rigidity of objects, which would prevent certain spatial relationships, such as interpenetration, from holding. In order to simplify matters in this paper, we shall concentrate mainly on pure spatial theories – one could very well argue that such a theory should necessarily precede an applied one which would be obtained by extending a purely spatial theory.

Traditionally, in mathematical theories of space, points are considered as primary primitive spatial entities (or perhaps points and lines), and extended spatial entities such as regions are defined, if necessary, as sets of points. However, within the QSR community, there has been a strong tendency to take regions of space as the primitive spatial entity. There are several reasons for this. If one is interested in using the spatial theory for reasoning about physical objects, then one might argue that the spatial extension of any actual physical object must be region-like rather than a lower dimensional entity. Similarly, most natural language (non mathematical) uses of the word "point" do not refer to a mathematical point: consider sentences such as "the point of pencil is blunt". Moreover, it turns out that one can define points, if required, from regions (e.g. [15]). Another reason against taking points as primitive is that many people find it counterintuitive that extended regions can be composed entirely of dimensionless points occupying no space! However, it must be admitted that sometimes it is useful to make an abstraction and view a 3D physical entity such as a potholed road as a 2D or even 1D entity. Of course, once entities of different dimensions are admitted, a further question arises as to whether mixed dimension entities are to be allowed. Further discussion of this issue can be found in [18, 48].

Another ontological question is what is the nature of the embedding space, i.e. the universal spatial entity? Conventionally, one might take this to be R^n for some n, but one can imagine applications where discrete, finite or non convex (e.g. non connected) universes might be useful.

Once one has decided on these ontological questions, there are further issues: in particular, what primitive "computations" will be allowed? In a logical theory, this amounts to deciding what primitive non logical symbols one will admit without definition, only being constrained by some set of axioms. One could argue that this set of primitives should be small, not only for mathematical elegance and to make it perhaps easier to assess the consistency of the theory, but also because this will simplify the interface of the symbolic system to a perceptual component resulting in fewer primitives to be implemented; the converse argument might be that the resulting symbolic inferences may be more complicated (and thus perhaps slower) and for the kinds of reasons argued for in [50].

One final ontological question is how to model the multi dimensionality of space: one approach (which might appear superficially attractive) is to attempt to model space by considering each dimension separately, projecting each region to each of the dimensions and reasoning along each dimension separately; however, this is easily seen to be inadequate: e.g. two individuals may overlap when projected to both the x and y axes individually, when in fact they do not overlap at all: consider figure 2.

Fig. 2. Projecting regions to dimensions may give misleading information, e.g. about disjointness of these two regions.

4.2 Topology

Topology is perhaps the most fundamental aspect of space and certainly one that has been studied extensively within the mathematical literature. It is often described informally as "rubber sheet geometry", although this is not quite accurate. However, it is clear that topology must form a fundamental aspect of qualitative spatial reasoning since topology certainly can only make qualitative distinctions; the question then arises: can one not simply import a traditional mathematical topological theory wholesale into a qualitative spatial representation? Although various qualitative spatial theories have been influenced by mathematical topology, there are a number of reasons why such a wholesale importation seems undesirable in general [48]; not only does traditional topology deal with much more abstract spaces that pertain in physical space or the space to be found in the kinds of applications mentioned above, but also we are interested in qualitative spatial *reasoning* not just representation, and this has been paid little attention in mathematics and indeed since typical formulations involve higher order logic, no reasonable computational mechanism would seem to be immediately obvious.

One exception to the disregard of earlier topological theories by the QSR community, is the tradition of work to be found in the philosophical logic literature, e.g. [87, 25, 88, 14, 15, 10]. This work has built axiomatic theories of space which are predominantly topological in nature, and which are based on taking regions rather than points as primitive – indeed, this tradition has been described as "pointless geometries" [40]. In particular the work of Clarke [14, 15] has lead to the development of the so called RCC systems [73, 72, 71, 69, 23, 19, 7, 45, 17, 48] and has also been developed further by [85, 4].

Clarke took as his primitive notion the idea of two regions x and y being connected (sharing a point, if one wants to think of regions as consisting of sets of points): $C(x, y)$. In the RCC system this interpretation is slightly changed to the closures of the regions sharing a point[3] – this has the effect of collapsing the distinction between a region, its closure and its interior, which it is argued has no relevance for the kinds of domain with which QSR is concerned (another reason for abandoning traditional mathematical topology). This primitive is surprisingly powerful: it is possible to define many predicates and functions which capture interesting and useful topological distinctions. The set of eight jointly exhaustive and pairwise disjoint (JEPD) relations illustrated in figure 3 are one particularly useful set (often known as the RCC8 calculus)and indeed have been defined in an entirely different way by [27]: three sets of points can be associated with each region – its interior, boundary and complement; the relationship between two regions can be characterized by a 3x3 matrix each of whose elements denotes whether the intersection of the corresponding sets from each region are empty or not. After taking into account the physical reality of space and some specific assumptions about the nature of regions, which can then be translated into constraints between the matrix values, it turns out that there are exactly 8 possible matrices, corresponding to the eight RCC8 relations. By changing the underlying assumptions about what a region is, and by allowing the matrix to represent the codimension of the intersection, different calculi with more JEPD relations can be derived, for example, one with enumerates all the relations between areas, lines and points.

The RCC calculus assumes that regions are all of the same dimension in any interpretation, though the dimension is left open. Although it seems possible [7, 44] to determine the dimensionality syntactically, this is quite cumbersome and a more direct approach may perhaps be preferable – thus an alternative calculus [46] is being developed, named the INCH calculus, based on a primitive $INCH(x, y)$, whose intended interpretation is that the extended spatial entity x includes a chunck of y – i.e. if x and y are the same dimension then $INCH(x, y)$ is the same as $C(x, y)$, otherwise x must be of higher dimension than y and they must not be disjoint.

4.3 Orientation

Orientation is not a binary relation – at least three elements need to be specified to give an orientation between two of them (and possibly more in dimensions higher than 2D). If we want to specify the orientation of a *primary object* (PO)

---

[3] Actually, given the disdain of the RCC theory as presented in [72] for points, a better interpretation, given some suitable distance metric, would be that $C(x, y)$ means that the distance between x and y is zero.

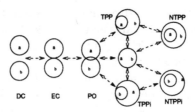

Fig. 3. 2D illustrations of of the relations of the RCC8 calculus and their continuous transitions (*conceptual neighbourhood*).

with respect to a *reference object* (RO), then we need some kind of *frame of reference* (FofR). An *extrinsic* frame of reference imposes an external, immutable orientation: e.g. gravitation, a fixed coordinate system, or an third object (such as the North pole). A *deictic* frame of reference is with respect to the "speaker" or some other internal observer. Finally, an *intrinsic* frame of reference exploits some inherent property of the RO – many objects have a natural "front", e.g. humans, buildings and boats. This categorization manifests itself in the display of qualitative orientation calculi to be found in the literature: certain calculi have an explicit triadic relation while others presuppose an extrinsic frame of reference and, for example, use compass directions [35, 51]. Of those with explicit triadic relations is it especially worth mentioning the work of Schlieder [76] who develops a calculus based on a function which maps triples of points to one of three qualitative values, -, 0 or +, denoting anticlockwise, collinear and clockwise orientations respectively. This can be used for reasoning about visible locations in qualitative navigation tasks, or for shape description [78] or to develop a calculus for reasoning about the relative orientation of pairs of line segments [77] – see figure 4. Another important triadic orientation calculus is that of Roehrig [75]; this calculus is based on a relation $CYCORD(x, y, z)$ which is true (in 2D) when x, y, z are in clockwise orientation. Roehrig shows how a number of qualitative calculi (not only orientation calculi) can be translated into the CYCORD system (whose reasoning system can then be exploited).

4.4   Distance and size

Distance and size are related in the sense that traditionally we use a linear scale to measure each of these aspects, even though distance is normally thought of as being a one dimensional concept, whilst size is usually associated with higher dimensional measurements such as area or volume. The domain can influence distance measurements, as we shall see below, but first I will discuss pure spatial representations. These can be divided into two main groups: those which measure on some "absolute" scale, and those which provide some kind of relative measurement. Of course, since traditional Qualitative Reasoning is primarily concerned with dealing with linear quantity spaces, the qualitative algebras and the transitivity of such quantity spaces mentioned earlier can be used as a distance or size measuring representation.

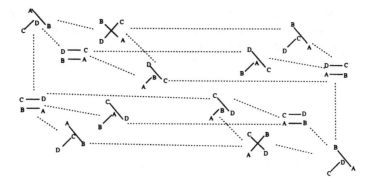

Fig. 4. The 14 JEPD relations of Schlieder's oriented line segment calculus and their *conceptual neighbourhood.*

Also of interest in this context are the order of magnitude calculi [64, 68] developed in the QR community. These calculi introduce measuring scales which allow one quantity to be described as being *much larger* than another, with the consequence that it requires summing many (in some formulations even an infinite number) of the former quantities in order to surpass the second, "much larger" quantity. All these "traditional QR" formalisms are of the "absolute" kind of representations mentioned above as is the Delta calculus [90] which introduces a triadic relation, $x(>, d)y$: x is larger/bigger than y by amount d; terms such as $x(>, y)y$ mean that x is more than twice as big as y.

Of the 'relative' representations specifically developed within the spatial reasoning community, perhaps the first is the calculus proposed by de Laguna [25], which introduces a triadic $CanConnect(x, y, z)$ primitive, which is true if the body[4] x can connect y and z by simple translation (i.e. without scaling, rotation or shape change). From this primitive it is quite easy to define notions such as equidistance, nearer than, and farther than. Also note that this primitive allows a simple size metric on regions to be defined: one region is larger than another if it can connect regions that the other cannot. Another technique to determine the relative size of two objects was proposed by Mukerjee and Joe [66] and relies on being able to translate regions (assumed to be shape and size invariant) and then exploit topological relationships – if a translation is possible so that one region becomes a proper part of another, then it must be smaller. Interestingly, these seem to be about the only proposals which are grounded in a region based theory – all the other representations mentioned in this section take points as their primitive spatial entity.

Distance is closely related to the notion of orientation: e.g. distances cannot usually be added unless they are in the same direction, and the distance between a point and region may vary depending on the orientation, see figure 5. Thus it is

---

[4] Technically, this perhaps is not a pure spatial representation in the way Laguna presented it, but it is close enough to be presented in this group of representations I think.

perhaps not surprising that there have been a number of calculi which are based on a primitive which combines distance and orientation information. Arguably, unless both of these aspects are represented then the calculus is not really a calculus of distance, though it might be said that this is a calculus of position rather than mere distance.

Fig. 5. Distance may vary with orientation.

One straightforward idea [35] is to combine directions as represented by segments of the compass with a simple distance metric (*far*, *close*). A slightly more sophisticated idea is to introduce a primitive which defines the position of a third point with respect to a directed line segment between two other points [91] – see figure 6. A calculus which combines the Delta calculus and orientation is presented in [89].

Fig. 6. There are 15 qualitatively different positions a point c (denoted by the shaded circles) can be with respect to a vector from point a to point b. Some distance information is represented, for example the darker shaded circles are in the same orientation but at different distances from ab.

The most sophisticated qualitative distance calculus to date is the framework for representing distances [52] which is currently being extended to include orientation. In this framework a distance is expressed in a particular *frame of reference* (FofR) between a *primary object* (PO) and a *reference object* (RO). A distance system is composed of an ordered sequence of *distance relations* (between a PO and an RO), and a set of *structure relations* which give additional information about how the distance relations relate to each other (apart from their distance ordering given implicitly by the ordered sequence). Each distance has an *acceptance area* (which in the case of an isotropic space will be a region the same shape as the PO, concentrically located around the PO); the distance between successive acceptance areas defines a sequence of intervals:$\delta_1, \delta_2, ...$ The structure relations define relationships between these δ_i. Typical structure relations might specify a monotonicity property (the δ_i are increasing), or that each δ_i is greater than the sum of all the preceding δ_i. The structure relationships can also be used to specify order of magnitude relationships, e.g. that $\delta_i + \delta_j \widetilde{} \delta_i$ for $j < i$. The structure relationships are important in refining the *composition tables* (see below). In a *homogeneous* distance system all the distance relations have the same structure relations; however this need not be the case in a *heterogeneous* distance system. The proposed system also allows for the fact that the context may affect the distance relationships; this is handled by having different frames of reference, each with its own distance system and with inferences in different frames of reference being composed using *articulation rules* (cf. [54]). Analogously to orientation calculi, intrinsic, extrinsic and deictic frames of reference can be distinguished.

It is possible that different qualitative distance calculi (or FofR) might be needed for different scale spaces – Montello [65] suggests that there are four main kinds of scale of space, relative to the human body: *figural space* pertains to distances smaller than the human body and which thus can be perceived without movement (e.g. table top space and pictures); *vista space* is similar but pertains to spaces larger than the human body, making distortions more likely; *environmental space* cannot be perceived without moving from one location to another; finally, *geographic space* cannot be properly apprehended by moving – rather it requires indirect perception by a (figural space) map. One obvious effect of moving from one scale, or context to another, is that qualitative distance terms such as "close" will vary greatly; more subtly, distances can behave in various "non mathematical" ways in some contexts or spaces: e.g. distances may not be symmetrical – e.g. because distances are sometimes measured by time taken to travel, and an uphill journey may take longer than the return downhill journey. Another "mathematical aberration" is that in some domains the shortest distance between two points may not be a straight line (e.g. because a lake or a building might be in the way,).

4.5 Shape

One can think of theories of space as forming a hierarchy ordered by expressiveness (in terms of the spatial distinctions made possible) with topology at the top and a fully metric/geometric theory at the bottom. Clearly in a purely topological theory only very limited statements can be made about the shape of a region: whether it is has holes (in the sense that a torus has a hole), or interior voids, or whether it is in one piece or not. A series of papers have explored exactly what it is possible to represent, both within the RCC calculus and, to a lesser extent, using the nine intersection model, e.g.: [21, 20, 47, 16, 29, 26].

However, if one's application demands finer grained distinctions than these, then some kind of semi-metric information has to be introduced[5]; there is a huge choice of extending topology with some kind of shape primitives whilst still retaining a qualitative representation (i.e. not becoming fully metric). The QSR community has only just started exploring the various possibilities; below we briefly describe some of the approaches.

---

[5] Of course, the orientation and distance primitives discussed above already add something to pure topology, but as already mentioned these are largely point based and thus not directly applicable to describing region shape.

There are a number of ways to classify these approaches; one distinction is between those techniques which constrain the possible shapes of a region and those that construct a more complex shaped region out of simpler ones (e.g. along the lines of constructive solid geometry [74], but perhaps starting from a more qualitative set of primitives). An alternative dichotomy can be drawn between representations which primarily describe the boundary of an object (e.g. the sequence of different kinds of curvature extrema[60] along its contour) compared to those which represent its interior (e.g. symmetry based techniques). Arguably [12], the former techniques are preferable since shape is inherently not a one dimensional concept.

One natural distinction to make is between convex and concave regions (note that topology only allows certain rather special kind of non convex regions to be distinguished, and in any case does not allow the concavities to be explicitly referred to – it is a theory of 'holed regions', rather than of holes per se). The RCC theory has shown that many interesting predicates can be defined once one takes the notion of a convex hull of a region (or equivalently, a predicate to test convexity). For example one can distinguish all the different kinds of shape displayed in figure 7. In fact, it has recently been shown [24], that this system essentially is equivalent to an affine geometry: any two compact planar shapes not related by an affine transformation can be distinguished.

Fig. 7. All these shapes and many more can be distinguished using a convex hull primitive.

Various different notions of the inside of a region can be distinguished using a convex hull primitive [17, 22] – these can all be viewed as different kinds of hole. A very interesting line of research [13, 84] has investigated exactly what holes are and proposes an axiomatisation of holes based on a new primitive: Hosts(x, y) – which is true if the *body* x hosts hole y; note that this is not a theory of pure space: holes cannot host other holes, only physical objects can act as hosts.

Another recent proposal [11] is to take the notion of two regions being congruent as primitive; from this it is possible to define the notion of a sphere, and then import Tarski's theory of spheres and related definitions such as 'betweenness' [81]. That this theory is more powerful than one just with convex hull is shown by the fact convexity can now be defined in a congruence based system.

Some representations naturally lend themselves to a hierarchical description; for example, if we take the convex hull of a (2D) shape, then we can easily define its concavities and describe their relationships to each other and the original shape; however these concavities may not be convex, and therefore we can take their convex hulls and recursively describe the relationship between the concavities of each concavity [80, 17].

4.6 Uncertainty

Even though a qualitative calculus already makes some attempt to represent and reason about uncertainty as discussed above, sometimes some extra mechanism may be required. Of course, it is always possible to glue on some standard numerical technique for reasoning about uncertainty (e.g. [38]), but there has also been some research on extending existing qualitative spatial reasoning techniques to explicitly represent and reason about uncertain information. For example, a recent GISDATA workshop on representing and reasoning about regions with indeterminate boundaries generated two papers [21, 16] which extended the RCC calculus and the 9-intersection in very similar ways to handle these kind of regions. Another interesting line of research is to work with an indistinguishability relation which is not transitive and thus fails to generate equivalence classes [83, 55]. Yet another line of research is the development of nonmonotonic spatial logics [79, 3].

5 Qualitative spatial reasoning

Although much of the work in QSR has concentrated on representational aspects, various computational paradigms are being investigated including constraint based reasoning (e.g. [51]). However, the most prevalent form of qualitative spatial reasoning is based on the composition table (originally known as a transitivity table [2], but now renamed since more than one relation is involved and thus it is relation composition rather than transitivity which is being represented). Given a set of n JEPD relations, the $n \times n$ composition table specifies for each pair of relations R1, and R2 such that R1(a,b) and R2(b,c) hold, what the possible relationships between a and c could be. In general, there will be a disjunction of entries, as a consequence of the qualitative nature of the calculus. Most of the calculi mentioned in this paper have had composition tables constructed for them, though this has sometimes posed something of a challenge [70]. One approach to the automatic generation of composition tables has been to try to reduce each calculus to a simple ordering relation [75]. Another, perhaps more general approach, is to formulate the calculus as a decidable theory (many calculi, e.g. the original RCC system, are presented as first order theories), ideally even as a tractable theory, and then use exhaustive theorem proving techniques to analyze and thus generate each composition table entry; a reformulation of the RCC first order theory in a zero order intuitionistic logic [6] [7] was able to generate the appropriate composition tables automatically. Another approach would have been been use a zero order modal logic [9].

5.1 Spatio-temporal reasoning

So far we have been concerned purely with static spatial calculi, so that we can only represent and reason about snapshots of a changing world. It is thus import-

---

[6] This reformulation is interesting in that it becomes a true spatial logic, rather than a theory of space: the "logical symbols" have spatial interpretations, e.g. implication is interpreted as parthood and disjunction as the sum of two regions.

ant to develop calculi which combine space and time in an integrated fashion. Assuming that change is continuous, as is the case in standard qualitative reasoning, it important to know which qualitative values or relations are neighbours in the sense that if a value or predicate holds at one time, then there is some continuous change possible such that the next value or predicate to hold will be a neighbour. Continuity networks defining such neighbours are often called *conceptual neighbourhoods* in the literature following the use of the term [36] to describe the of structure Allen's 13 JEPD temporal relations [1] according to their conceptual closeness (e.g. *meets* is a neighbour of both *overlaps* and *before*). Most of the qualitative spatial calculi reported in this paper have had conceptual neighbourhoods constructed for them; see figures 3 and 4 for example[7].

Perhaps the most common form of computation in the QR literature is qualitative simulation and using conceptual neighborhood diagrams is quite easy to build a qualitative *spatial* simulator [23]. Such a simulator takes a set of ground atomic statements describing an initial state[8] and constructs a tree of future possible states – the branching of the tree results from the ambiguity of the qualitative calculus. Of course, continuity alone does not provide sufficient constraints to restrict the generation of next possible states to a reasonably small set in general – domain specific constraints are required in addition. These may be of two kinds: intra state constraints restrict the spatial relationships that may hold within any state whilst inter state constraints restrict what can hold between adjacent states (or in general, across a sequence of states). Both of these constraint types can be used to prune otherwise acceptable next states from the simulation tree. Additional pruning is required to make sure that each state is consistent with respect to the semantics of the calculus (e.g. that there is no cycle of proper part relationships) – the composition table can be used for this purpose. A desirable extension, by analogy with earlier QR work, would be to incorporate a proper theory of spatial processes couched in a language of QSR. ; some work in this direction is reported in [62].

One problem is that the conceptual neighbourhood is usually built manually for each new calculus – a sometimes arduous and error prone operation if there are many relations; techniques to derive these automatically would be very useful; one attempt in this direction [19] tried to exploit the interesting observation that every entry in most [9] composition tables form a connected subgraph of the conceptual neighbourhood diagram though was not sufficiently powerful to generate the complete conceptual neighbourhood diagram from the composition table in general. An analysis of the structure of conceptual neighbourhoods is reported by Ligozat [61]. A more foundational approach which exploits the continuity of the

---

[7] A close related notion is that of "closest topological distance" [28] – two predicates are neighbours if their respective 9-intersection matrices differ by fewer entries than any other predicates; however the resulting neighbourhood graph is not identical to the true conceptual neighbourhood or continuity graph – some links are missing.

[8] The construction of an envisioner [86] rather than a simulator would also be possible of course.

[9] There are the odd exceptions [6].

underlying semantic spaces has been investigated by Galton [39] but the process is still far from automatic.

5.2 Complexity and Completeness

Metatheoretic results such as completeness and complexity results are still quite sparse in the QSR literature. The intuitionistic formulation of RCC8 has been shown to provide a polynomial procedure to test the consistency of sets of ground atoms [67] but for other topological calculi or tasks the results are not so favourable [49, 43]. Related to the question of decidability is the question of the completeness of the theory (are there no contingent formulae?) and the stronger result of ℵ0-categoricity (all models are isomorphic) – some investigations into providing sufficient extra existential axioms for RCC8 to ensure this are reported in [8].

6 Final comments

An issue which has not been much addressed yet in the QSR literature is the issue of cognitive validity – claims are often made that qualitative reasoning is akin to human reasoning, but with little or no empirical justification; one exception to this work is the study made of a calculus for representing topological relations between regions and lines [63] where native speakers of several different languages were asked to perform tasks in which they correlated spatial expressions such as "the road goes through the park" with a variety of diagrams which depicted a line and a region which the subjects were told to interpret as as road and a park.

As in so many other fields of knowledge representation it is unlikely that a single universal spatial representation language will emerge – rather, the best we can hope for is that the field will develop a library of representational and reasoning devices and some criteria for their most successful application. Finally, as in the case of non spatial qualitative reasoning, quantitative knowledge and reasoning must not be ignored – qualitative and quantitative reasoning are complementary techniques and research is needed to ensure they can be integrated, for example by developing reliable ways of translating between the two kinds of formalism. Equally, interfacing symbolic QSR to the techniques being developed by the diagrammatic reasoning community [41] is an interesting and important challenge.

In this paper I have tried to provide an overview of the field of qualitative spatial reasoning; the field active and there has not been space to cover everything (for example qualitative kinematics [31]). A European funded *Human Capital and Mobility* Network, Spacenet, has recently been initiated across eleven sites and the web page (http://www.scs.leeds.ac.uk/spacenet/) provides an entry to point to the ongoing work at these sites and elsewhere.

7 Acknowledgments

The support of the EPSRC under grants GR/H/78955 and GR/K65041, and also the CEC under the HCM network SPACENET is gratefully acknowledged.

In writing this paper I have been greatly influenced not only by my colleagues in the qualitative spatial reasoning group here at Leeds (in particular Brandon Bennett, John Gooday and Nick Gotts), but also by the many discussions I have participated in at the Spacenet workshops with the participants from the other 10 sites – my sincere thanks to them all. The responsibility for any errors in this paper rests entirely with me of course.

References

1. J F Allen. An interval-based representation of temporal knowledge. In *Proceedings 7th IJCAI*, pages 221–226, 1981.
2. J F Allen. Maintaining knowledge about temporal intervals. *Communications of the ACM*, 26(11):832–843, 1983.
3. N Asher and J Lang. When nonmonotonicity comes from distance. In L Nebel, B amd Dreschler-Fischer, editor, *KI-94: Advances in Artificial Intelligence*, pages 308–318. Springer-Verlag, 1994.
4. N Asher and L Vieu. Toward a geometry of common sense: A semantics and a complete axiomatization of mereotopology. In *Proceedings of the International Joint Conference on Artificial Intelligence (IJCAI-95), Montreal*, 1995.
5. M Aurnague and L Vieu. A three-level approach to the semantics of space. In C Zelinsky-Wibbelt, editor, *The semantics of prepositions - from mental processing to natural language processing*, Berlin, 1993. Mouton de Gruyter.
6. B. Bennett. Some observations and puzzles about composing spatial and temporal relations. In R Rodríguez, editor, *Proceedings ECAI-94 Workshop on Spatial and Temporal Reasoning*, 1994.
7. B. Bennett. Spatial reasoning with propositional logics. In J Doyle, E Sandewall, and P Torasso, editors, *Principles of Knowledge Representation and Reasoning: Proceedings of the 4th International Conference (KR94)*, San Francisco, CA., 1994. Morgan Kaufmann.
8. B Bennett. Carving up space: steps towards construction of an absolutely complete theory of spatial regions. In *Proc. JELIA96*, 1996.
9. B Bennett. Modal logics for qualitative spatial reasoning. *Bulletin of the Interest Group in Pure and Applied Logic (IGPL)*, 1996.
10. L. Biacino and G. Gerla. Connection structures. *Notre Dame Journal of Formal Logic*, 32(2), 1991.
11. S Borgo, N Guarino, and C Masolo. A pointless theory of spade based on strong connecction and congruence. Technical report, LADSEB-CNR, Padova, 1996.
12. J M Brady. Criteria for representations of shape. *Human and Machine Vision*, 1993.
13. R Casati and A Varzi. *Holes and Other Superficialities*. MIT Press, Cambridge, MA, 1994.
14. B L Clarke. A calculus of individuals based on 'connection'. *Notre Dame Journal of Formal Logic*, 23(3):204–218, July 1981.
15. B L Clarke. Individuals and points. *Notre Dame Journal of Formal Logic*, 26(1):61–75, 1985.
16. E Clementini and P Di Felice. An algebraic model for spatial objects with undetermined boundaries. In P Burrough and A M Frank, editors, *Proceedings, GISDATA Specialist Meeting on Geographical Entities with Undetermined Boundaries,*. Taylor Francis, 1996.

17. A G Cohn. A hierarchcial representation of qualitative shape based on connection and convexity. In A Frank, editor, *Proc COSIT95*, LNCS, pages 311–326. Springer Verlag, 1995.

18. A G Cohn, B Bennett, J Gooday, and N Gotts. Representing and reasoning with qualitative spatial relations about regions. In O Stock, editor, *Temporal and spatial reasoning*. Kluwer, to appear.

19. A G Cohn, J M Gooday, and B Bennett. A comparison of structures in spatial and temporal logics. In R Casati, B Smith, and G White, editors, *Philosophy and the Cognitive Sciences: Proceedings of the 16th International Wittgenstein Symposium*, Vienna, 1994. Hölder-Pichler-Tempsky.

20. A G Cohn and N M Gotts. A theory of spatial regions with indeterminate boundaries. In C. Eschenbach, C. Habel, and B. Smith, editors, *Topological Foundations of Cognitive Science*, 1994.

21. A G Cohn and N M Gotts. The 'egg-yolk' representation of regions with indeterminate boundaries. In P Burrough and A M Frank, editors, *Proceedings, GISDATA Specialist Meeting on Geographical Objects with Undetermined Boundaries*, pages 171–187. Francis Taylor, 1996.

22. A G Cohn, D A Randell, and Z Cui. Taxonomies of logically defined qualitative spatial relations. *Int. J of Human-Computer Studies*, 43:831–846, 1995.

23. Z Cui, A G Cohn, and D A Randell. Qualitative simulation based on a logical formalism of space and time. In *Proceedings AAAI-92*, pages 679–684, Menlo Park, California, 1992. AAAI Press.

24. E Davis. Personal communication., June 1996.

25. T. de Laguna. Point, line and surface as sets of solids. *The Journal of Philosophy*, 19:449–461, 1922.

26. M Egenhofer. Topological similarity. In *Proc FISI workshop on the Toplogical Foundations of Cognitive Science*, volume 37 of *Reports of the Doctoral Series in Cognitive Science*. University of Hamburg, 1994.

27. M Egenhofer and J Herring. Categorizing topological spatial relationships between point, line and area objects. In *The 9-intersection: formalism and its use for natural language spatial predicates, Technical Report 94-1*. National Center for Geographic Information and Analysis, Santa Barbara, 1994.

28. M J Egenhofer and K K Al-Taha. Reasoning about gradual changes of topological relationships. In A U Frank, I Campari, and U Formentini, editors, *Theories and Methods of Spatio-temporal Reasoning in Geographic Space*, volume 639 of *Lecture Notes in Computer Science*, pages 196–219. Springer-Verlag, Berlin, 1992.

29. M J Egenhofer, E Clementini, and P Di Felice. Toplogical relations between regions with holes. *Int. Journal of Geographical Information Systems*, 8(2):129–144, 1994.

30. M J Egenhofer and D Mark. Naive geography. In A U Frank and W Kuhn, editors, *Spatial Information Theory: a theoretical basis for GIS*, volume 988 of *Lecture Notes in Computer Science*, pages 1–16. Springer-Verlag, Berlin, 1995.

31. B. Faltings. A symbolic approach to qualitative kinematics. *Artificial Intelligence*, 56(2), 1992.

32. B Faltings. Qualitative spatial reaoning using algebraic topology. In A U Frank and W Kuhn, editors, *Spatial Information Theory: a theoretical basis for GIS*, volume 988 of *Lecture Notes in Computer Science*, pages 17–30, Berlin, 1995. Springer-Verlag.

33. B. Faltings and P. Struss, editors. *Recent Advances in Qualitative Physics*. MIT Press, Cambridge, Ma, 1992.

34. J Fernyhough, A G Cohn, and D C Hogg. Real time generation of semantic regions from video sequences. In *Proc. ECCV96*, LNCS. Springer Verlag, 1996.

35. A Frank. Qualitative spatial reasoning with cardinal directions. *Journal of Visual Languages and Computing*, 3:343–371, 1992.

36. C Freksa. Temporal reasoning based on semi-intervals. *Artificial Intelligence*, 54:199–227, 1992.

37. H Fujihara and A Mukerjee. Qualitative reasoning about document design. Technical report, Texas A and M University, 1991.

38. M Gahegan. Proximity operators for qualitative spatial reasoning. In W Kuhn A Frank, editor, *Spatial Information Theory: a theoretical basis for GIS*, number 988 in Lecture Notes in Computer Science, pages 31–44, Berlin, 1995. Springer Verlag.

39. A Galton. Towards a qualitative theory of movement. In W Kuhn A Frank, editor, *Spatial Information Theory: a theoretical basis for GIS*, number 988 in Lecture Notes in Computer Science, pages 377–396, Berlin, 1995. Springer Verlag.

40. G. Gerla. Pointless geometries. In F. Buekenhout, editor, *Handbook of Incidence Geometry*, chapter 18, pages 1015–1031. Eslevier Science B.V., 1995.

41. J et al Glasgow. *Diagrammatic Reasoning*. MIT Press, 1995.

42. J M Gooday and A G Cohn. Using spatial logic to describe visual languages. *Artificial Intelligence Review*, 10(1-2), 1995.

43. N Gotts. Personal communication, June 1996.

44. N M Gotts. Defining a 'doughnut' made difficult. In C. Eschenbach, C. Habel, and B. Smith, editors, *Topological Foundations of Cognitive Science*, volume 37 of *Reports of the Doctoral programme in Cognitive Science*. University of Hamburg, 1994.

45. N M Gotts. How far can we 'C'? defining a 'doughnut' using connection alone. In J Doyle, E Sandewall, and P Torasso, editors, *Principles of Knowledge Representation and Reasoning: Proceedings of the 4th International Conference (KR94)*. Morgan Kaufmann, 1994.

46. N M Gotts. Formalising commonsense topology: The inch calculus. In *Proc. Fourth International Symposium on Artificial Intelligence and Mathematics*, 1996.

47. N M Gotts and A G Cohn. A mereological approach to spatial vagueness. In *Proceedings, Qualitative Reasoning Workshop 1995 (QR-95)*, 1995.

48. N M Gotts, J M Gooday, and A G Cohn. A connection based approach to commonsense topological description and reasoning. *The Monist*, 79(1):51–75, 1996.

49. M. Grigni, D. Papadias, and C. Papadimitriou. Topological inference. In C.S. Mellish, editor, *proceedings of the fourteenth international joint conference on artificial intelligence (IJCAI-95)*, volume I, pages 901–906. Morgan Kaufmann, 1995.

50. P J Hayes. The naive physics manifesto. In D Mitchie, editor, *Expert systems in the micro-electronic age*. Edinburgh University Press, 1979.

51. D Hernández. *Qualitative Representation of Spatial Knowledge*, volume 804 of *Lecture Notes in Artificial Intelligence*. Springer-Verlag, 1994.

52. D Hernandez, E Clementini, and P Di Felice. Qualitative distances. In W Kuhn A Frank, editor, *Spatial Information Theory: a theoretical basis for GIS*, number 988 in Lecture Notes in Computer Science, pages 45–58, Berlin, 1995. Springer Verlag.

53. A Herskovits. *Language and Spatial Cognition. An interdisciplinary study of prepositions in English*. Cambridge University Press, 1986.

54. J Hobbs. Granularity. In *Proceedings IJCAI-85*, pages 432–435, 1985.

55. S Kaufman. A formal theory of spatial reasoning. In *Proc Int. Conf. on Knowledge Representation and Reasoning*, pages 347–356, 1991.

56. M Knauff, R Rauh, and C Schlieder. Preferred mental models in qualitative spatial reasoning: A cognitive assessment of allen's calculus. In *Proc. 17th Annual Conf. of the Cognitive Science Society*, 1995.

57. B Kuipers. *Qualitative Reasoning*. MIT Press, Cambridge, MA., 1994.

58. B J Kuipers and T S Levitt. Navigating and mapping in large-scale space. *AI Magazine*, 9(2):25–43, 1988.

59. F Lehmann and A G Cohn. The EGG/YOLK reliability hierarchy: Semantic data integration using sorts with prototypes. In *Proc. Conf. on Information Knowledge Management*, pages 272–279. ACM Press, 1994.

60. M Leyton. A process grammar for shape. *Artificial Intelligence*, page 34, 1988.

61. G Ligozat. Towards a general characterization of conceptual neighbourhoods in temporal and spatial reasoning. In F D Anger and R Loganantharah, editors, *Proceedings AAAI-94 Workshop on Spatial and Temporal Reasoning*, 1994.

62. M Lundell. A qualitative model of gradient flow in a spatially distributed parameter. In *Proc 9th Int. Workshop on Qualtiative Reasoning, Amsterdam*, 1995.

63. D Mark, D Comas, M Egenhofer, S Freundschuh, J Gould, and J Nunes. Evaluating and refining computational models of spatial relations through cross-linguistic human-subjects testing. In W Kuhn A Frank, editor, *Spatial Information Theory: a theoretical basis for GIS*, number 988 in Lecture Notes in Computer Science, pages 553–568, Berlin, 1995. Springer Verlag.

64. M Mavrovouniotis and G Stephanopoulos. Formal order-of-magnitude reasoning in process engineering. *Computers and Chemical Engineering*, 12:867–881, 1988.

65. D Montello. Scale and multiple pyschologies of space. In I Campari A Frank, editor, *Spatial Information Theory: a theoretical basis for GIS*, number 716 in Lecture Notes in Computer Science, pages 312–321, Berlin, 1993. Springer Verlag.

66. P Mukerjee and G Joe. A qualitative model for space. In *Proceedings AAAI-90*, pages 721–727, Los Altos, 1990. Morgan Kaufmann.

67. B. Nebel. Computational properties of qualitative spatial reasoning: First results. In *Procdings of the 19th German AI Conference*, 1995.

68. O Raiman. Order of magnitude reasoning. In *AAAI-86: Proceedings of the National Conference on AI*, pages 100–104, 1996.

69. D A Randell and A G Cohn. Exploiting lattices in a theory of space and time. *Computers and Mathematics with Applications*, 23(6-9):459–476, 1992. Also appears in "Semantic Networks", ed. F. Lehmann, Pergamon Press, Oxford, pp. 459-476, 1992.

70. D A Randell, A G Cohn, and Z Cui. Computing transitivity tables: A challenge for automated theorem provers. In *Proceedings CADE 11*, Berlin, 1992. Springer Verlag.

71. D A Randell, A G Cohn, and Z Cui. Naive topology: Modelling the force pump. In P Struss and B Faltings, editors, *Advances in Qualitative Physics*, pages 177–192. MIT Press, 1992.

72. D A Randell, Z Cui, and A G Cohn. A spatial logic based on regions and connection. In *Proc. 3rd Int. Conf. on Knowledge Representation and Reasoning*, pages 165–176, San Mateo, 1992. Morgan Kaufmann.

73. D.A. Randell and A.G. Cohn. Modelling topological and metrical properties of physical processes. In R Brachman, H Levesque, and R Reiter, editors, *Proceedings 1st International Conference on the Principles of Knowledge Representation and Reasoning*, pages 55–66, Los Altos, 1989. Morgan Kaufmann.

74. A A G Requicha and H B Boelcke. Solid modelling: a historical summary aand contemporary assessment. *IEEE Computer Graphics and Applications*, 2:9–24, 1992.

75. R Röhrig. A theory for qualitative spatial reasoning based on order relations. In *AAAI-94: Proceedings of the 12th National Conference on AI*, volume 2, pages 1418–1423, Seattle, 1994.

76. C Schlieder. Representing visible locations for qualitative navigation. In N Piera Carreté and M G Singh, editors, *Qualitative Reasoning and Decision Technologies*, pages 523–532, Barcelona, 1993. CIMNE.

77. C Schlieder. Reasoning about ordering. In W Kuhn A Frank, editor, *Spatial Information Theory: a theoretical basis for GIS*, number 988 in Lecture Notes in Computer Science, pages 341–349, Berlin, 1995. Springer Verlag.

78. C Schlieder. Qualitative shape representation. In P Burrough and A M Frank, editors, *Proceedings, GISDATA Specialist Meeting on Geographical Objects with Undetermined Boundaries*. Francis Taylor, 1996.

79. M Shanahan. Default reasoning about spatial occupancy. *Artificial Intelligence*, 1995.

80. J Sklansky. Measuring concavity on a rectangular mosaic. *IEEE Trans. on Computers*, C-21(12):1355–1364, 1972.

81. A. Tarski. Foundations of the geometry of solids. In *Logic, Semantics, Metamathematics*, chapter 2. Oxford Clarendon Press, 1956. trans. J.H. Woodger.

82. A Tate, J Hendler, and M Drummond. A review of AI planning techniques. In J Allen, J Hendler, and A Tate, editors, *Readings in Planning*. Morgan Kaufman, San Mateo, CA, 1990.

83. T Topaloglou. First order theories of approximate space. In F Anger et al., editor, *Working notes of AAAI workshop on spatial and temporal reasoning*, pages 283–296, Seattle, 1994.

84. A C Varzi. Spatial reasonng in a holey world. In *Proceedings of the Spatial and Temporal Reasoning workshop, IJCAI-93*, pages 47–59, 1993.

85. L Vieu. *Sémantique des relations spatiales et inférences spatio-temporelles*. PhD thesis, Université Paul Sabatier, Toulouse, 1991.

86. D S Weld and J De Kleer, editors. *Readings in Qualitative Reasoning About Physical Systems*. Morgan Kaufman, San Mateo, Ca, 1990.

87. A N Whitehead. *Process and reality: corrected edition*. The Free Press, Macmillan Pub. Co., New York, 1978. edited by D.R. Griffin and D.W. Sherburne.

88. J.H. Woodger. *The Axiomatic Method in Biology*. Cambridge University Press, 1937.

89. K Zimmermann. Enhancing qualitative spatial reasoning – combining orientation and distance. In I Campari A Frank, editor, *Spatial Information Theory: a theoretical basis for GIS*, number 716 in Lecture Notes in Computer Science, pages 69–76, Berlin, 1993. Springer Verlag.

90. K Zimmermann. Measuring without distances: the delta calculus. In W Kuhn A Frank, editor, *Spatial Information Theory: a theoretical basis for GIS*, number 988 in Lecture Notes in Computer Science, pages 59–68, Berlin, 1995. Springer Verlag.

91. K Zimmermann and C Freksa. Enhancing spatial reasoning by the concept of motion. In A Sloman, editor, *Prospects for Artificial Intelligence*, pages 140–147. IOS Press, 1993.

Combining Local Consistency, Symbolic Rewriting and Interval Methods

Frédéric Benhamou and Laurent Granvilliers

LIFO, Université d'Orléans,
I.I.I.A., Rue Léonard de Vinci
B.P. 6759 45067 ORLEANS Cedex 2 France
{benhamou,granvil}@lifo.univ-orleans.fr

Abstract. This paper is an attempt to address the processing of non-linear numerical constraints over the Reals by combining three different methods: local consistency techniques, symbolic rewriting and interval methods. To formalize this combination, we define a generic two-step constraint processing technique based on an extension of the Constraint Satisfaction Problem, called Extended Constraint Satisfaction Problem (ECSP). The first step is a rewriting step, in which the initial ECSP is symbolically transformed. The second step, called approximation step, is based on a local consistency notion, called weak arc-consistency, defined over ECSPs in terms of fixed point of contractant monotone operators. This notion is shown to generalize previous local consistency concepts defined over finite domains (arc-consistency) or infinite subsets of the Reals (arc B-consistency and interval, hull and box-consistency). A filtering algorithm, derived from AC-3, is given and is shown to be correct, confluent and to terminate. This framework is illustrated by the combination of Gröbner Bases computations and Interval Newton methods. The computation of Gröbner Bases for subsets of the initial set of constraints is used as a rewriting step and operators based on Interval Newton methods are used together with enumeration techniques to achieve weak arc-consistency on the modified ECSP. Experimental results from a prototype are presented, as well as comparisons with other systems.

Keywords: Constraint Satisfaction Problem, local consistency, arc-consistency, filtering algorithms, non-linear constraint solving, Gröbner bases, Newton methods, interval arithmetic, interval constraints.

1 Introduction

The notion of Constraint Satisfaction Problems (CSP) was introduced in the seventies [30, 23, 19] to address combinatorial problems over finite domains and has been heavily studied since then. A Constraint Satisfaction Problem is given by a set of variables, a set of domains associated to these variables and representing their possible values and a set of relations defined over these variables. Solving of CSPs being a NP-hard problem, several approximations of the solution space, computed by local consistency methods have been proposed, the most famous being arc-consistency [19] and path-consistency [23].

Since the introduction of local consistency in Constraint Logic Programming [29] various extensions have been proposed, among which methods to solve so-called interval constraints [10, 8, 26, 15, 14, 5, 18, 3, 4, 28, 27]. More recently several authors have studied various combinations of solvers in the case of continuous real constraints [20, 22, 13] and, in particular, combinations of techniques from computer algebra (Gröbner bases), Interval Constraint methods and techniques from numerical analysis. Similar ideas for the case of 0/1 linear constraints are suggested in [2] which proposes to compute cutting planes to prune the search space before applying a classical finite domain constraint solver.

We propose in this paper to formalize this combination by defining a generic two-step constraint processing technique based on an extension of the Constraint Satisfaction Problem, called Extended Constraint Satisfaction Problem (ECSP). The first step is a rewriting step, in which the initial ECSP is symbolically transformed. The second step, called approximation step is based on a local consistency notion, called weak arc-consistency, defined over ECSPs in terms of fixed point of contractant monotone operators. This notion is shown to generalize arc-consistency [23, 19], arc B-consistency [18] and interval, hull and box consistency [4]. A filtering algorithm, derived from AC-3 is given and is shown to be correct, confluent and to terminate. This framework is illustrated by the combination of Gröbner Bases computations and Interval Newton methods. In order to maximize the trade-off between pruning and computation time, Gröbner Bases are computed for subsets of the initial set of constraints in the rewriting step. Operators based on Interval Newton methods are used together with enumeration techniques to achieve weak arc-consistency on the modified ECSP. These operators are basically the ones proposed in [4, 28]. Experimental results from a prototype are presented, as well as comparisons with other systems.

With respect to the related papers mentioned above, we believe that this paper generalizes the local consistency notions addressing numerical constraints processing that we are aware of, and proposes a novel combination of the use of Gröbner Bases as a preprocessing step applied to reasonably small subsets of the initial constraint set and of state of the art local consistency techniques based on Interval Newton methods. Finally, preliminary experimental results, detailed at the end of the paper, show significant speed-ups with respect to other recent publications.

The remaining of the paper is organized as follows. Section 2 defines Extended Constraint Satisfaction Problems. Section 3 defines weak arc-consistency and the corresponding filtering algorithm and gives the main theoretical results of the paper. Section 4 introduces a rewriting step used to preprocess the initial CSP and to generate an ECSP preserving the declarative semantics. Section 5 presents an instance of this framework combining Gröbner bases and Interval Newton methods. Section 6 gives experimental results and we conclude in section 7.

2 Extended Constraint Satisfaction Problems

2.1 Definitions

We define an *Extended Constraint Satisfaction Problem* over the Reals[1] as a finite set of constraints (i.e. conjunctions of atomic constraints over the Reals) S over variables $\{x_1, \ldots, x_n\}$ to which are associated contractant monotone operators called *constraint narrowing operators*, defined over subsets of $\mathbb{R}^n$ called *approximate domains* and an n-ary Cartesian product X of elements of $\mathbb{R}$ representing the domains of the variables. Note that to each constraint corresponds a relation over $\mathbb{R}^n$ but that ECSPs are defined over formulae and not relations. This will be used to define symbolic transformations of ECSPs. In the rest of the paper, when no confusion is possible, we will use constraints to denote the relations they represent.

Before giving a more formal definition of ECSPs, we introduce approximate domains and constraint narrowing operators, already discussed in [3] in the framework of interval constraint programming.

Definition 1 *An* approximate domain A *over* $\mathbb{R}$ *is a subset of* $2^{\mathbb{R}}$, *closed under (eventually infinite) intersection, such that* $\mathbb{R} \in A$, *and for which the inclusion is a well-founded ordering (an approximate domain contains no infinite decreasing sequence of elements).*

Approximate domains, partially ordered by set inclusion, constitute complete lattices in which the meet operation is defined by set intersection. The join of sets is defined as the smallest set larger than all of them, i.e. the approximation of their union. (For a discussion on a lattice theoretic view of interval constraints, see [26]).

Given an approximate domain A, and an n-ary relation ρ over $\mathbb{R}$, a *narrowing operator* for the relation ρ is a correct, contractant, monotone, idempotent function $N : A^n \longrightarrow A^n$. More formally,

Definition 2 *Let A be an approximate domain. Let ρ be an n-ary relation over* $\mathbb{R}$. *The function* $N : A^n \longrightarrow A^n$ *is a* constraint narrowing operator *for the relation ρ iff for every $u, v \in A^n$, the four following properties hold:*

$$(1) \qquad N(u) \subset u, \qquad \text{(Contractance)}$$

$$(2) \qquad u \cap \rho \subset N(u), \qquad \text{(Correctness)}$$

$$(3) \; u \subset v \; \text{implies} \; N(u) \subset N(v), \text{(Monotonicity)}$$

$$(4) \qquad N(N(u)) = N(u). \qquad \text{(Idempotence)}$$

We can now introduce formally ECSPs.

---

[1] This definition can be generalized to any set but it is not the topic of this paper.

Definition 3 *Let A be an approximate domain over $\mathbb{R}$.*
An Extended Constraint Satisfaction Problem *is a pair (S, X), where*
$S = \{(C_1, N_1), (C_2, N_2), \ldots, (C_m, N_m)\}$ is a set of pairs made of a real constraint
C_i and of a constraint narrowing operator N_i for C_i, and X is an n-ary Cartesian
product of elements of A.

Declarative and approximate semantics of ECSPs

First, note that classical CSPs, assuming that the relations are given intension-
ally, can be represented as pairs (S, X) where S is a set of constraints and X
the Cartesian product of the domains associated to the variables appearing in
S. It follows that to every ECSP $E = (\{(C_1, N_1), (C_2, N_2), \ldots, (C_m, N_m)\}, X)$
corresponds a CSP $E' = (\{C_1, C_2, \ldots, C_m\}, X)$.

The declarative semantics of the ECSP E, denoted E^*, is then defined as
the declarative semantics of E', that is the set of n-ary tuples $(x_1, \ldots, x_n) \in X$
satisfying the conjunction of the constraints $C_1, C_2, \ldots, C_m$.

Definition 4 (Declarative semantics of ECSPs)
Let $E = (\{(C_1, N_1), (C_2, N_2), \ldots, (C_m, N_m)\}, X)$ be an ECSP. The declarative
semantics of E is defined as

$$E^* = \bigcap_{i=1}^{m} C_i \cap X$$

Unfortunately this set is generally either computationaly intractable or even
uncomputable. This leads to define what we call the approximate semantics of
E, denoted $\overline{E}$, as being the greatest common fixed point of the N_i's included in
X.

Definition 5 (Approximate semantics of ECSPs)
Let $E = (\{(C_1, N_1), (C_2, N_2), \ldots, (C_m, N_m)\}, X)$ be an ECSP. If $fp(N_i)$ denotes
the set of fixed points of N_i, then the approximate semantics of E is defined as

$$\overline{E} = max(\{u \in \bigcap_{i=1}^{m} fp(N_i) \mid u \subseteq X\})$$

The proof of existence of $\overline{E}$ is constructive and based on properties of the
filtering algorithm presented in section 3.1.

Due to the correctness property of narrowing operators with respect to their
associated constraints, we have the following, expected, result concerning the
two semantics:

Property 1 *Let E be an ECSP. Then, $E^* \subseteq \overline{E}$.*

We describe in the next section the local consistency notions related to the
approximate semantics of ECSPs.

3   Weak arc-consistency

As for CSPs, ECSPs are processed by local consistency techniques, that is propagation. The local consistency used in ECSPs is an extension of arc-consistency, called here *weak arc-consistency*.

We recall here first the definition of arc-consistency for Constraint Satisfaction Problems. Intuitively, a CSP is said to be arc-consistent iff for every variable appearing in a constraint C, there is no value of its associated domain which cannot be "completed" with values from the domains of the other variables in order to satisfy C. More formally:

Definition 6 (arc-consistency)
Let $E = (\{C_1, \ldots, C_m\}, X)$ be a CSP. Then, E is arc-consistent iff $\forall i \in \{1, \ldots, m\}, \forall j \in \{1, \ldots, n\}, X_i \subset \pi_i(C_j \cap X)$, where, for every n-ary relation ρ, $\pi_i(\rho)$ denotes the ith projection of ρ.

Moreover a CSP $E = (S, X)$ is said to be *maximally arc-consistent* with respect to a Cartesian product of domains X' iff X is the greatest Cartesian product (with respect to inclusion) such that E is arc-consistent and $X \subseteq X'$.

These definitions are generally used, in a slightly different formalism, for finite sets and binary constraints, but can be extended, at least theoretically, to infinite sets and n-ary constraints. We give below the definition of weak arc-consistency for ECSPs.

Definition 7 (weak arc-consistency)
Let $E = (\{(C_1, N_1), (C_2, N_2), \ldots, (C_m, N_m)\}, X)$ be an ECSP. E is weakly arc-consistent iff $X = \overline{E}$.

The definition for *maximal weak arc-consistency* is expressed as follows:

Definition 8 (maximal weak arc-consistency)
Let $E = (S, X)$ be an ECSP. Let X' be a Cartesian product of domains such that $X \subseteq X'$. Then, if E' is the ECSP (S, X'), E is maximally weak arc-consistent with respect to X' iff E is weakly arc-consistent and $X = \overline{E'}$.

We then state without proof, due to size limitations, the following properties:
Let $E = (\{(C_1, N_1), \ldots, (C_m, N_m)\}, X)$ be an ECSP over an approximate domain A. Let $E' = (\{C_1, \ldots, C_m\}, X)$ be the CSP associated to E.

1. If for all i in $\{1, \ldots, m\}$, and for all u in A^n, $N_i(u)$ is the Cartesian product of the n projections of the relation $u \cap C_i$, then E is weakly arc-consistent iff E' is arc-consistent.
2. If A is made of all intervals whose bounds are floating point numbers, and if for all i in $\{1, \ldots, m\}$, and for all u in A^n, $N(u) = \text{apx}(\rho \cap u)$, where apx is the function that associates to any subset ρ of $\mathbb{R}^n$ the intersection of all the elements of A containing ρ, then E is weakly arc-consistent iff E' is hull-consistent (as defined in [4]) and iff it is arc B-consistent (as defined in [18]).

3. if A is made of all finite unions of intervals whose bounds are floating point numbers and the N_i's are defined as above, then E is weakly arc-consistent iff E' is interval-consistent (as defined in [4]) .

4. weak arc-consistency generalizes also box-consistency [4] and the same type of result as above can be shown as will be developped in section 5.2.

3.1   Filtering algorithm

The algorithm used to compute, from an ECSP E and a Cartesian product of domains X, an equivalent ECSP maximally weak arc-consistent with respect to X is essentially an immediate adaptation of AC-3 [19] and of the filtering algorithms used in interval constraint-based systems like BNR-Prolog [26], CLP(BNR) [5] or Newton [4, 28].

```
filtering( in {(C_1, N_1), ..., (C_m, N_m)} ; inout X = X_1 × ... × X_n )
begin
    S := {C_1, ..., C_m};
    while S ≠ ∅ and X ≠ ∅ do
        C := choose one C_i in S;
        X' := N_i(X);
        if X' ≠ X then
            S := S ∪ {C_j | ∃v_k ∈ var(C_j) ∧ X'_k ≠ X_k};
            X := X'
        endif
        S := S \ {C}
    endwhile
end;
```

Again, we state without proof the main properties of the algorithm:

1. if the computation of the constraint narrowing operators terminates, the algorithm terminates due to the fact that set inclusion is a well-founded ordering for approximate domains.

2. the algorithm is correct in the sense that the solution space defined by the declarative semantics of the initial ECSP is *included* in the computed Cartesian product. In particular, if this Cartesian product is empty, the ECSP is unsatisfiable.

3. the algorithm is confluent in the sense that the output is independant of the order in which the constraints are chosen.

4. given as input E, the output is $\overline{E}$.

4   Symbolic rewriting

Given a CSP over the Reals, it is often very efficient to preprocess it with symbolic rewriting techniques before applying the filtering algorithm, to generate redundant and/or surrogate constraints. Therefore, we introduce a preprocessing step called *constraint rewriting*.

Constraint rewriting consists in two phases:

1. repeated syntactical transformations of the constraints of a CSP preserving its declarative semantics,

2. association of constraint narrowing operators to the generated set of constraints.

Note that the first phase alone can solve the CSP when the rewriting techniques are complete and computationaly tractable w.r.t. the problem at hand (e.g. cylindrical algebraic decomposition for polynomials, syntactic processing of 0/1 constraints, etc.). Another remark is that, in order to compute a usable ECSP, to each generated narrowing operators must correspond a terminating algorithm for the considered constraint.

The second part of this paper is devoted to an instance of this generic framework addressing the approximation of solutions of multivariate polynomials over the Reals.

5 Combining Gröbner bases and interval Newton methods

We describe in this section an instance of the framework introduced in the previous sections whose purpose is to approximate the solutions of systems of nonlinear polynomial equations. Different programming languages and systems have been designed to handle similar problems such as Newton [4, 28] based on interval Newton methods and local consistency techniques, CLP(BNR) [5] implementing interval propagation methods, CoSAc [22] which combines Maple, a Gröbner bases module and a linear solver, the system presented in [20] based on Gröbner bases, the simplex algorithm and interval propagation methods, TKIB [14] introducing an approximation notion called tightening and the Krawckzyk operator [16, 17].

In this paper, we propose to combine Gröbner bases computations over subsets of the initial system, used as a preprocessing step, interval Newton methods to compute weak arc-consistency and enumeration techniques used to separate the solutions. This combination was motivated by three main arguments. First, syntactical transformations of the system by means of Gröbner bases computations may speed up the resolution step by adding redundant constraints used to reduce the search space. Second, computing Gröbner bases over subsets of the initial system still permits to generate redundant constraints while keeping the algorithm memory and time consumption in reasonable bounds. Finally, propagation based on interval Newton methods has been preferred to classical interval propagation algorithms for their superior efficiency.

5.1 Constraint rewriting

As mentioned above, the constraint rewriting step of our system consists in computing Gröbner bases for selected subsets of the initial system. We recall first very briefly some basics on Gröbner bases.

Gröbner bases

The concept of Gröbner bases for polynomial ideals is due to Buchberger [6]. The main idea is to transform a multivariate polynomial set to obtain a new set in a certain normal form which makes possible the resolution of many problems in polynomial ideal theory and in particular the problem of solving polynomial equations (see [7, 9] for more details).

The computation of Gröbner bases essentially depends on the following three notions:

1. Every polynomial is ordered using a *total ordering on monomials* $\succ$ which is Noetherian and compatible with the multiplication on monomials. The maximal monomial of a polynomial p wrt $\succ$ is called the *leading term* of p.
2. The *reduction process* of a polynomial f wrt a set of polynomials F computes an irreducible polynomial obtained as a remainder after iteratively dividing f by the elements of F.
3. The *S-polynomial* of two polynomials f and g, denoted $S(f, g)$, is a combination of two polynomials that cancels their leading terms.

We define Gröbner bases by giving the algorithmic characterization proposed by Buchberger in [7]:

Definition 9 (Gröbner basis)
Let $G = \{f_1, \ldots, f_n\}$ be a set of polynomials. G is a Gröbner basis iff for all pairs $(i, j) \in \{1, \ldots, n\}$ the reduction wrt G of the S-polynomial $S(f_i, f_j)$ is equal to 0.

The basic algorithm computing Gröbner bases naturally follows from the definition. Let F be a set of polynomials. Let F' be a set of polynomials initialized with F. The algorithm consists in successive computations of reduced S-polynomials over F'. These S-polynomials are added to F' at each step. The algorithm stops, after a finite number of steps, when no S-polynomial different from 0 can be computed. The resulting set of polynomials is a Gröbner basis and is denoted GB(F). A Gröbner basis can be further simplified into a *reduced Gröbner basis* (see details in [7]).

Another property which is of interest for this paper is that, given a system of polynomial equations $S = \{f_1 = 0, \ldots, f_n = 0\}$ and $G = \{g_1, \ldots, g_m\}$ a Gröbner basis such that $G = \mathrm{GB}(\{f_1, \ldots, f_n\})$, the systems $\{g_1 = 0, \ldots, g_m = 0\}$ and S have the same solutions.

Rewriting step

The Gröbner bases are used as a preprocessing step to generate redundant equations which may prune the search space. In what follows we choose the total degree ordering for monomials and the usual lexicographic ordering for variables.

Let $S = \{f_1 = 0, \ldots, f_n = 0\}$ be a constraint system. Let $\{F_0, F_1, \ldots, F_m\}$ be a partition of the set $\{f_1, \ldots, f_n\}$. The rewriting step consists in the computation of reduced Gröbner bases G_i such that for all $i \in \{1, \ldots, m\}$ $G_i = \text{GB}(F_i)$. The set F_0 contains the constraints that are not rewritten. The correctness of this computation is guaranteed by Gröbner bases properties (see the previous section). The termination comes from the termination of Gröbner bases algorithms. The order in which Gröbner bases are computed do not change the resulting system due to the independence of the sets F_i.

In some cases, inconsistency can be detected at this stage. By application of the Weak Nullstellensatz theorem, if there exists $i \in \{1, \ldots, p\}$ such that the constant polynomial 1 belongs to G_i then G_i is inconsistent and so is the whole system.

Consider the set $\{f_1 = x^2 + y^2 - 1, f_2 = x^2 - y\}$. The reduction of the S-polynomial $S(f_1, f_2)$ gives the univariate polynomial $f_3 = y^2 + y - 1$ whose roots can be easily extracted. The elimination of x is possible because the leading terms of f_1 and f_2 are equals. This suggests a heuristic to compute the partition in such a way that its elements contain polynomials whose leading terms share variables. In the current implementation the partition is computed by hand however we believe that this process could be automated as suggested by the previous example.

Example

To illustrate this constraint rewriting step, we present an example taken from [14].

$$
S = \begin{cases}
\bullet\ x_1^2 + x_2^2 + x_3^2 + x_4^2 - 2*x_1 - 3 = 0 \\
\bullet\ x_1^2 + x_2^2 + x_3^2 + x_4^2 - 4 = 0 \\
\circ\ x_1 + x_2 + x_3 + x_4 - 1 = 0 \\
\circ\ x_1 + x_2 - x_3 + x_4 - 3 = 0
\end{cases}
\qquad
S' = \begin{cases}
x_2^2 + x_3^2 + x_4^2 - 3.75 = 0 \\
x_1 + x_2 - x_3 + x_4 - 3 = 0 \\
x_3 + 1 = 0 \\
x_1 - 0.5 = 0
\end{cases}
$$

The system S is divided in two subsystems S_1 and S_2, materialized by two different kinds of bullets. The resulting set S' is the union of $\text{GB}(S_1)$ and $\text{GB}(S_2)$. In S' the domains of variables x_1 and x_3, which are reduced to a singleton, can be computed immediately. Moreover we remark that three polynomials from S were removed during the computation of reduced Gröbner bases for each subsystems.

5.2 Constraint solving

The approximate resolution of the system after the rewriting step is basically the one proposed in [28] which corresponds to state of the art techniques in constraint solving. We show here how this method can be expressed in terms of computing weak arc-consistency over a certain ECSP.

Basics

We recall first some basic notions from [4, 28]. Due to space restrictions we do not recall the basics of interval computations which can be found for example in [24, 1]. We overload function symbols over the reals with function symbols over the intervals when no confusion is possible. Let D be a set of floating point intervals.

Given a real expression f, the *natural interval extension* F of f is the interval expression which is a copy of f where operation symbols are replaced by interval operations, variables by interval variables and constants by interval constants. Interval constraints are of the form $F = 0$ where F is an expression over intervals. The semantics of interval functions and constraints is quite natural and follows the definitions given in [4]. An interval I is said to be *canonical* w.r.t D iff $\forall J \in D, J \subseteq I$ implies $I = J$.

Let $C = (F = 0)$ be a constraint where F is an interval expression on variables $X_1, \ldots, X_n$ over D and let $I_1 \times \ldots \times I_n$ be a Cartesian product of elements of D. The *i*th *projection* of C is the constraint $F' = 0$ where F' is obtained from F by replacing the variables X_j by the intervals I_j for all $j \in \{1, \ldots, i-1, i+1, \ldots, n\}$ and is denoted $\pi_i(C)$. We also define the *constraint enclosure* of C, denoted $\text{EN}(C)$, as being the smallest subset of D^n such that, for every n-ary tuple A made of canonical intervals, if A is a solution of C then $A \in \text{EN}(C)$.

Interval Newton methods

Newton methods are numerical algorithms approximating zeros of functions and derived from the mean value theorem. We present here two methods, one based on the Newton interval function and other based on the Taylor series approximation.

Let f be a real function continuously differentiable between x and y and consider that y is a zero of f. It can be deduced from the mean value theorem that $y = x - f(x)/f'(a)$. The Newton method iterates this formula to approximate roots of f. This method has been extended to interval functions [24, 16, 12, 1, 11, 17, 25]. Let X be an interval containing x and y and suppose that F' is the natural interval extension of f', F the natural interval extension of f and $m(X)$ the approximation of the center of X. The Newton interval function is the function

$$N(X) = m(X) - F(m(X))/F'(X)$$

From this definition one can design an *interval Newton method* enclosing roots of interval functions. Given an initial interval X_0 and an interval function F, a sequence of intervals $X_1, X_2, \ldots, X_n$ is computed using the iteration step $X_{i+1} = N(X_i) \cap X_i$. X_n is either empty, which means that X_0 contains no zero of F, or is a fixed point of N.

The centered form is based on the Taylor series approximation. Let $C = (F = 0)$ be an interval constraint, X be a vector of variables, $I = (I_1, \ldots, I_n)$ be a vector of intervals and J be the vector of the continuous partial derivatives of F. We denote by $M(I)$ the midpoint vector $(m(I_1), \ldots, m(I_n))$. The centered form is the interval expression $F(X) = F(M(I)) + J(I)(X - M(I))$. If X is a zero of F it follows

$$F(M(I)) + J(I)(X - M(I)) = 0$$

In what follows this last constraint will be denoted $\text{TAY}(C)$. Let C_i be the ith projection of this constraint. One can remark that C_i is an unary constraint over X_i. After a basic transformation of C_i, X_i can be expressed as a constant interval expression $G(I)$. The method based on the Taylor series approximation computes $X_i = I_i \cap G(I)$.

Generation and resolution of ECSPs

The main idea is to consider, for each non-linear constraint over p variables, $2p$ "copies" of the constraint. To each copy is associated a constraint narrowing operator defined over closed floating point intervals and implementing one of the two methods described above.

More precisely, let $E = (\{C_1, \ldots, C_m)\}, X)$ be a CSP over $\mathbb{R}$. The approximate domain D is the set of all closed intervals whose bounds are floating point intervals.

Then, an ECSP $E' = (S \cup S', X)$ is computed from E by the following algorithm:

```
generation( in E = (S, X) ; out E' = (S' ∪ S'', X) )
begin
     S' := ∅;  S'' := ∅;
     while S ≠ ∅ do
          C := choose one constraint in S;
          P := arity(C);
          for i:=1 to P do
               S'  := S' ∪ {(C, N'_i)};
               S'' := S'' ∪ {(C, N''_i)};
          endfor
          S := S \ {C}
     endwhile
end;
```

First let us remark that, in E', to every constraint narrowing operator N corresponds a variable from the initial set $\{X_1, \ldots, X_n\}$. The index of this variable is denoted $\text{ind}(N)$. Then, for every $(C', N') \in S'$, for every $(C'', N'') \in S''$ and for every n-ary Cartesian product u of elements of D, N' and N'' are defined as follows:

$$N'(u) = \text{EN}(\pi_{\text{ind}(N')}(C))$$
$$N''(u) = \text{EN}(\pi_{\text{ind}(N'')}(\text{TAY}(C)))$$

Finally, given such an ECSP E', the (possible) solutions are computed using a branch-and-prune algorithm [28] which is basically an iteration of two steps:

1. a pruning step which computes weak arc-consistency for E' using the filtering algorithm presented in section 3.1. In this case weak arc-consistency is equivalent to box-consistency.
2. a branching step which generates two subproblems by splitting one non-canonical interval, when possible.

The following example (continuing the example developed in the previous section) describes the constraint resolution process. It will give the underlying intuition of the generic branch-and-prune algorithm. Consider that the initial Cartesian product of domains is $[-10, 10]^4$ and the required precision for the resulting intervals is 10^{-12}. Weak arc-consistency is computed, providing the following intervals:

$$x_1 \in X_1 = [+0.5000000000000, +0.5000000000000]$$
$$x_2 \in X_2 = [-0.1583186821562, +1.6583186821562]$$
$$x_3 \in X_3 = [-1.0000000000000, -1.0000000000000]$$
$$x_4 \in X_4 = [-0.1583186821562, +1.6583186821562]$$

No more pruning can be done and the intervals X_2 and X_4 are not canonicals. A branching step is applied and generates two subintervals (thus two Cartesian products) from X_4 which is split. The pruning step on the first Cartesian product gives immediately the first solution:

$$x_1 \in [+0.5000000000000, +0.5000000000000]$$
$$x_2 \in [+1.6513878188659, +1.6513878188660]$$
$$x_3 \in [-1.0000000000000, -1.0000000000000]$$
$$x_4 \in [-0.1513878188660, -0.1513878188659]$$

Then the second solution is derived from the second Cartesian product after backtracking:

$$x_1 \in [+0.5000000000000, +0.5000000000000]$$
$$x_2 \in [-0.1513878188660, -0.1513878188659]$$
$$x_3 \in [-1.0000000000000, -1.0000000000000]$$
$$x_4 \in [+1.6513878188659, +1.6513878188660]$$

6   Experimental results

We have implemented a prototype of a polynomial constraint solver, which we call here INGB, integrating the interval Newton method called IN described in the previous sections and a Gröbner bases module called GB. In order to compare constraint solvers we propose two parameters. The computation time is the most used computational parameter and permits to compare behaviours of rather different systems. This parameter is used here to compare our system with CoSAc [22] and Newton [28]. The number of branchings is the second parameter and illustrates the use of Gröbner bases as a preprocessing step. In particular, INGB is compared to TKIB [14].

The following computational results are given for a SUN Sparc20 and the width of the resulting intervals are smaller than 10^{-12}. The results of TKIB were computed on a Silicon Graphics workstation with a 50 MHz processor MIPS 4000/4010. The results of CoSAc and Newton were computed on a SUN Sparc10. The Gröbner bases are computed using different orderings for monomials and the usual lexicographic ordering for the variables. Derivatives are computed symbolically.

Example 1 *This example comes from [14] and concerns the intersection of a circle and a parabola. We compare* INGB *with the solver* TKIB *which implements an approximation method called tightening, the Krawczyk operator and enumeration methods. A reduced Gröbner basis S' is computed for the whole system S.*

$$S = \begin{cases} x^2 + y^2 - 1 = 0 \\ x^2 - y = 0 \end{cases} \qquad S' = \begin{cases} x^2 - y = 0 \\ y^2 + y - 1 = 0 \end{cases}$$

The initial intervals for x, y are $[-1.5, 1.5]$. The experimental results show that INGB *has the best possible behaviour on the number of enumeration steps (one branching for two solutions).*

Example 2 *This example was developed in the previous section and is also taken from [14]. The initial intervals for the variables are $[-10, 10]$. The table 1 shows again an improvement w.r.t. the number of branchings and a good runtime behaviour.*

Example 3 *This example is taken from [21, 22] and consists in moving a rectangle through a right angled corridor. The system S is divided in three parts materialized by different bullets for the last two subsystems. GB is applied on each of the last two parts.*

$$S = \begin{cases} y - b = 0 \\ b - r * t = 0 \\ w^2 - 1 + t^2 = 0 \\ \bullet \ x - l * t^3 - L * w = 0 \\ \bullet \ y - L * t - l * w^3 = 0 \\ \bullet \ L - 1 = 0 \\ \bullet \ l - 2 = 0 \\ \circ \ x - a = 0 \\ \circ \ 2 * a - 3 = 0 \end{cases} \qquad S' = \begin{cases} t^2 + w^2 - 1 = 0 \\ -r * t + b = 0 \\ L - 1 = 0 \\ l - 2 = 0 \\ a - x = 0 \\ -b + y = 0 \\ t^3 + 0.5 * w - 0.5 * x = 0 \\ w^3 + 0.5 * t - 0.5 * y = 0 \\ x - 1.5 = 0 \end{cases}$$

The initial intervals are $[-10^6, 10^6]$ for x, y, l, t, L, b, r, a and $[0, 10^6]$ for w. The experimental results show that INGB *has a better behaviour than* CoSAc *which is due for the most part to interval Newton methods and to the partial use of Gröbner bases in our system.*

Example 4 *This example comes from [28] and describes an economics modelling problem for which we consider the case of four variables. A reduced Gröbner basis S' is computed for the whole system S.*

$$S = \begin{cases} \bullet\ x_1 * x_4 + x_1 * x_2 * x_4 - 0.35 = 0 \\ \bullet\ x_2 * x_4 + x_1 * x_3 * x_4 - 1.086 = 0 \\ \bullet\ x_3 * x_4 - 2.05 = 0 \\ \bullet\ x_1 + x_2 + x_3 + 1 = 0 \end{cases}$$

$$S' = \begin{cases} x_1 + x_2 + x_3 + 1 = 0 \\ \\ x_2 + \frac{850}{2593} * x_3 + \frac{500}{2593} * x_4 + \frac{3311}{2593} = 0 \\ \\ x_3 + \frac{10000}{71463} * x_4^2 + \frac{15240}{23821} * x_4 + \frac{514433}{595525} = 0 \\ \\ x_4^3 + \frac{1143}{250} * x_4^2 + \frac{1543299}{250000} * x_4 + \frac{2929983}{200000} = 0 \end{cases}$$

The initial intervals are $[-10^8, 10^8]$ for each variables. The experimental results in the table 1 show that the computation of Gröbner bases speeds up interval Newton methods. The reason is that the roots of the last polynomial can be computed immediately using Newton methods. Then the propagation in S' of the values of x_4 allows to solve the first three equations using the same process.

Benchs	Methods	CPU time (in seconds)			Branching(s)	Solution(s)
		GB	IN	Total		
Ex1	INGB	0.01	0.02	0.03	1	2
	TKIB	-	-	0.03	3	2
Ex2	INGB	0.02	0.11	0.13	1	2
	TKIB	-	-	0.78	10	2
Ex3	INGB	0.05	0.19	0.24	0	1
	CoSAc	2	-	2	-	1
Ex4	INGB	0.13	0.07	0.20	0	1
	Newton	-	0.60	0.60	?	1

Table 1. Experimental results.

7 Conclusion

In this paper, we have proposed an extension of the notion of Constraint Satisfaction Problems, called Extended Constraint Satisfaction Problem, to formalize the collaboration of different solvers over continuous domains. We have defined an extended notion of arc-consistency, called weak arc-consistency and shown that it generalizes previous local consistency notions for finite and continuous CSPs. Weak arc-consistency is characterized by a fixed point semantics over the lattice of approximate domains. To illustrate this framework we have proposed a novel combination of the use of Gröbner Bases as a preprocessing step applied to reasonably small subsets of the initial constraint set and of state of the art local consistency techniques based on Interval Newton methods. Finally, we have reported experimental results and compared with other recent implementations. We intend to continue these experiments and notably to automatize the process which partition the initial system in the preprocessing step. We envisage also to instantiate this framework to other combinations of symbolic and numeric methods.

Acknowledgements

We are grateful to Alain Colmerauer, Gérard Ferrand and Pascal Van Hentenryck for fruitful discussions on these topics.

References

1. G. Alefeld and J. Herzberger. *Introduction to Interval Computations*. Academic Press, 1983.
2. P. Barth and A. Bockmayr. Finite domain and cutting plane techniques in CLP($\mathcal{PB}$). In L. Sterling, editor, *Proceedings of ICLP'95*, pages 133–147, Kanagawa, Japan, 1995. MIT Press.
3. F. Benhamou. Interval Constraint Logic Programming. In A. Podelski, editor, *Constraint Programming: Basics and Trends*, volume 910 of *LNCS*, pages 1–21. Springer-Verlag, 1995.
4. F. Benhamou, D. McAllester, and P. Van Hentenryck. CLP(Intervals) Revisited. In *Proceedings of ILPS'94*, pages 124–138, Ithaca, NY, USA, 1994.
5. F. Benhamou and W. J. Older. Applying Interval Arithmetic to Real, Integer and Boolean Constraints. *Journal of Logic Programming*, 1996. forthcoming.
6. B. Buchberger. *An Algorithm for Finding a Basis for the Residue Class Ring of a Zero-Dimensional Polynomial Ideal*. PhD thesis, University of Innsbruck, 1965. (in German).
7. B. Buchberger. Gröbner Bases: an Algorithmic Method in Polynomial Ideal Theory. In *Multidimensional Systems Theory*, pages 184–232. D. Reidel Publishing Company, 1985.
8. J. G. Cleary. Logical Arithmetic. *Future Computing Systems*, 2(2):125–149, 1987.
9. D. Cox, J. Little, and D. O'Shea. *Ideals, Varieties and Algorithms*. Springer-Verlag, 1992.
10. E. Davis. Constraint Propagation with Interval Labels. *Artificial Intelligence*, (32), 1987.
11. E. R. Hansen and R. I. Greenberg. An Interval Newton Method. *Applied Mathematics and Computation*, 12:89–98, 1983.

12. E. R. Hansen and S. Sengupta. Bounding Solutions of Systems of Equations using Interval Analysis. *BIT*, 21:203–211, 1981.
13. H. Hong. Confluency of Cooperative Constraint Solving. Technical Report 94-08, RISC-Linz, Johannes Kepler University, Linz, Austria, 1994.
14. H. Hong and V. Stahl. Safe Start Region by Fixed points and Tightening. *Computing*, 53(3-4):323–335, 1994.
15. E. Hyvönen. Constraint Reasoning based on Interval Arithmetic. The Tolerance Propagation Approach. *Artificial Intelligence*, 58:71–112, 1992.
16. R. Krawczyk. Newton-Algorithmen zur Bestimmung von Nullstellen mit Fehlerschranken. *Computing*, 4:187–201, 1969.
17. R. Krawczyk. A Class of Interval Newton Operators. *Computing*, 37:179–183, 1986.
18. O. Lhomme. Consistency techniques for numeric CSPs. In R. Bajcsy, editor, *Proceedings of the 13th IJCAI*, pages 232–238, Chambéry, France, 1993. IEEE Computer Society Press.
19. A. Mackworth. Consistency in Networks of Relations. *Artificial Intelligence*, 8(1):99–118, 1977.
20. P. Marti and M. Rueher. A Distributed Cooperating Constraint Solving System. *International Journal on Artificial Intelligence Tools*, 4(1):93–113, 1995.
21. E. Monfroy. Gröbner Bases: Strategies and Applications. In *Proceedings of AISMC'92*, volume 737 of *LNCS*, pages 133–151. Springer-Verlag, 1992.
22. E. Monfroy, M. Rusinowitch, and R. Schott. Implementing Non-Linear Constraints with Cooperative Solvers. In K. George, J. Carroll, D. Oppenheim, and J. Hightower, editors, *Proceedings of ACM Symposium on Applied Computing*, pages 63–72, February 1996.
23. U. Montanari. Networks of Constraints: Fundamental Properties and Applications to Picture Processing. *Information Science*, 7(2):95–132, 1974.
24. R. E. Moore. *Interval Analysis*. Prentice-Hall, Englewood Cliffs, NJ, 1966.
25. A. Neumaier. *Interval Methods for Systems of Equations*. Cambridge University Press, 1990.
26. W. Older and A. Vellino. Constraint Arithmetic on Real Intervals. In F. Benhamou and A. Colmerauer, editors, *Constraint Logic Programming: Selected Research*. MIT Press, 1993.
27. I. Shvetzov, A. Semenov, and V. Telerman. Constraint Programming based on Subdefinite Models and its Applications. In *ILPS'95 post-conference workshop on Interval Constraints*, Portland, Oregon, USA, 1995.
28. P. Van Hentenryck, D. McAllester, and D. Kapur. Solving Polynomial Systems Using a Branch and Prune Approach. *SIAM Journal on Numerical Analysis*. (To appear).
29. P. Van Hentenryck, H. Simonis, and M. Dincbas. Constraint Satisfaction Using Constraint Logic Programming. *Artificial Intelligence*, 58(1-3):113–159, Dec. 1992.
30. D. L. Waltz. Generating Semantic Descriptions from Drawings of Scenes with Shadows. In P. H. Winston, editor, *The Psychology of Computer Vision*. McGraw Hill, 1975.

Proof Transformation for Non-Compatible Rewriting

Reinhard Bündgen

Wilhelm-Schickard-Institut, Universität Tübingen
Sand 13, D-72076 Tübingen
⟨buendgen@informatik.uni-tuebingen.de⟩

Abstract. We present a new inference rule based characterization of abstract completion procedures. The completeness of this inference system is shown by proof transformation techniques using a very powerful new proof ordering. This new ordering improves over previously proposed orderings because (1) it can handle semi-compatible reduction relations and (2) it explains all known complete redundancy criteria in a uniform framework.

1 Introduction

Reduction relations are ubiquitous in symbolic computation. They are in general used to compute in congruence classes defined by a set of equations describing the congruence relation. That is applying arbitrary contexts to the left- and right-hand sides of an equation always yields two equal objects. The task of the reduction relation is to compute a canonical normal form for each object and therefore it should be confluent and terminating.

Given two congruent objects, a (non-failing) completion procedure compiles a set of equations (and/or rules) into a rule set for which the two objects have a common normal form. On termination the resulting rule system describes a confluent reduction relation that defines a decision procedure for that congruence. Otherwise completion is just a semi-decision procedure. There are well-known completion procedures in the area of computer algebra (e. g., Buchberger's algorithm [5]), and in the area of automated deduction (e. g., Knuth-Bendix completion [11]).

Completion procedures are often presented by a set of inference rules that may be applied non-deterministically. The main challenge of deduction processes based on inference rules is proving their completeness. That is showing that all valid theorems (congruences) can eventually be proved. For term rewriting systems and the Knuth-Bendix completion this can be done very elegantly using proof transformation techniques [2].

Reduction relations can be partitioned into two classes: Reduction relations that are compatible w. r. t. the application of contexts (sometimes also called stable or monotonic) and those that are not compatible even though their reflexive-transitive-symmetric closure preserves compatibility w. r. t. arbitrary contexts. Examples of the first class are term rewriting systems (even modulo finite congruence classes like associative-commutative theories) and string rewriting systems. Most interesting reduction relations of the second class are semi-compatible. That is if an object a is in one step reducible to b then the reduction relation preserves that the two objects obtained by applying a common context to a and b have a common normal form. Semi-compatible reduction relations are in general used to work with canonical representatives of possibly infinite congruence classes. Examples of semi-compatible reduction relations are polynomial reduction systems as used for Gröbner base computations and string rewriting systems on reduced words as used with finitely presented groups. Semi-compatible reduction relations can also be defined by the extension rules computed in a symmetrization based completion procedure [8] and in the context of normalized term rewriting [14].

The proof transformation technique of [2] depends on the compatibility of the rewrite relation. Thus it cannot be applied to the completion of semi-compatible reduction relations. It was actually remarked in [16] that the proof transformation technique for term rewriting cannot be applied to Buchberger's algorithm. In this paper we introduce the framework of proof transformations to semi-compatible reduction relations. Therefore we present a new set of inference rules which basically replaces the concept of equations (i. e., critical pairs) by "promises" to consider a reduction ambiguity. In addition, we define a new proof ordering for that system that implies all known complete redundancy criteria including those critical pair criteria that are not covered by the term orderings proposed in [2]. Both our new inference system and the new proof ordering generalize the ones usually used with term rewriting systems. But they are also applicable to algebraic completion procedures like Buchberger's algorithm. A major goal of this work is to design a *generic* completion procedure. Therefore we must determine which properties of the domain, the congruences and the rewrite relations are *essential* to allow for the construction of a completion procedure.

The next section introduces abstract reduction relations. In Section 3, an abstract characterization of context applications and the behavior of relations under context application is given. This culminates in a formulation of an abstract critical pair theorem. Section 4 generalizes the proof transformation method of [2] to abstract compatible reduction relations and Section 5 presents our new inference system and proof ordering used to prove the completeness of a completion procedure for semi-compatible reduction relations. In Section 6, we apply the new proof orderings to describe redundancy criteria. Finally we show that all our results hold for an even more general class of reduction relations which we call weakly compatible. Proofs not presented in this paper can be found in the full version of this paper [6].

2 Abstract Rewriting

Let $\mathcal{D}$ be some set of objects and $\rightarrow_A$ be a binary relation on $\mathcal{D} \times \mathcal{D}$. Then $\leftarrow_A$ is the inverse relation of $\rightarrow_A$. $\leftrightarrow_A$, $\rightarrow_A^+$, $\rightarrow_A^*$ and $\leftrightarrow_A^*$ are the symmetric -, transitive -, transitive and reflexive -, and the symmetric, transitive and reflexive closures of $\rightarrow_A$ respectively. The relation $\rightarrow_A$ is *terminating* if there is no infinite chain $a_1 \rightarrow_A a_2 \rightarrow_A \cdots$.

A situation of the form $b \leftarrow_A a \rightarrow_A c$ is called a *rewrite* or *reduction ambiguity*. Two objects $b, c \in \mathcal{D}$ are *joinable* if there is a $d \in \mathcal{D}$ with $b \rightarrow_A^* d \leftarrow_A^* c$. We then write $b \downarrow_A c$. We say $b \leftarrow_A a \rightarrow_A c$ is *confluent* if b and c are joinable. A reduction relation $\rightarrow_A$ is *confluent* if for all $a, b, c \in \mathcal{D}$ such that $b \leftarrow_A^* a \rightarrow_A^* c$, b and c are joinable. To prove the confluence of a terminating relation it suffices to show a weaker condition. $\to_A$ is *locally confluent* if all reduction ambiguities are confluent. In [15], Newman showed that for all terminating relations, confluence is equivalent to local confluence. Another criterion for confluence was introduced by Winkler and Buchberger in [17]. Let $\succ\ \supseteq\ \to_A$ be a strict ordering on $\mathcal{D}$. $b \leftarrow^*_A a \to^*_A c$ is *subconnected* w. r. t. $\succ$ if there are $d_1,\ldots,d_n \in \mathcal{D}$ with $a \succ d_i$ for $1 \le i \le n$ and $b \leftrightarrow_A d_1 \leftrightarrow_A \ldots \leftrightarrow_A d_n \leftrightarrow_A c$. We then write $b \overset{a}{\leftrightsquigarrow}_A c$.

Lemma 1 (Winkler & Buchberger). *If $\succ\ \supseteq\ \to_A$ is terminating then $\to_A$ is confluent iff all reduction ambiguities are subconnected w. r. t. $\succ$.* □

For the special case of k-algebras this criterion was already known to Bergman [4].

An object $a \in \mathcal{D}$ is *reducible* by $\to_A$ if there is another object $a' \ne a$ such that $a \to_A a'$, otherwise a is *irreducible*. If $\to_A$ is confluent and terminating then $\to_A$ is called *canonical*. Then for each object $a \in \mathcal{D}$ there is an irreducible object $a' \in \mathcal{D}$ with $a \to^*_A a'$ and a' is called the *normal form* of a w. r. t. $\to_A$. We write $a' = a{\downarrow}_A$.

A *rule* has the form $l \to r$ where $l, r \in \mathcal{D}$ describe patterns for a set of objects in $\mathcal{D}$. A rule may be applied to an object which 'contains' a part that 'fits' the pattern of l. This part will then be 'replaced' by a corresponding part which 'fits' the pattern of r. The 'part' to be replaced is called a *redex*. The exact meaning of 'containment', 'fitting' and 'replacement' depends on the domain $\mathcal{D}$. However in any case each term contains itself and fits itself. Thus l can be replaced by r using the rule $l \to r$. A set of rules $\mathcal{R}$ defines a *reduction relation* $\to_{\mathcal{R}}$ in the sense that $a \to_{\mathcal{R}} b$, if b can be obtained from a by a single application of a rule in $\mathcal{R}$. A reduction relation defined by a set of rules is called a *rewrite relation*. We say $\mathcal{R}$ is *terminating, confluent, canonical* etc. if $\to_{\mathcal{R}}$ is so.

An equation $a \leftrightarrow b$ is a rule that can be applied both from left to right as $a \to b$ and from right to left as $b \to a$. The reduction relation induced by a set $\mathcal{E}$ of equations is denoted by $\leftrightarrow_{\mathcal{E}}$. Equations usually describe a congruence.

We denote multisets enumerating their elements within brackets. E. g., $[a, a, b]$ is the multiset containing a twice and b once. By $\succ_{mul}$ we denote the multiset extension of a (strict) ordering $\succ$, i. e., $M_1 \succ_{mul} M_2$ if $M_1 \ne M_2$, $M_2 = (M_1 \setminus X) \cup Y$, and $\forall y \in Y \exists x \in X : x \succ y$ for some subset $X \subseteq M_1$. A multiset ordering is terminating iff the inducing ordering on the domain of elements is terminating [9].

3 Rewriting in Contexts

A congruence is an equivalence relation that is compatible w. r. t. a class of contexts associated with that congruence. That is if a congruent b then for all associated contexts C, a "in context C" is congruent to b in the same context. Reduction relations are normally used to describe congruences. Therefore their behavior in arbitrary contexts associated with the underlying congruence relation is crucial.

3.1 Complete Classes of Contexts

Definition 2. A *context* is a function $C : \mathcal{D} \to \mathcal{D}$. A *context class* $\mathcal{C}$ is a set of contexts such that for every $C_1, C_2 \in \mathcal{C}$ there is a context $C_3 = C_1 \circ C_2 \in \mathcal{C}$.

Definition 3. A context class $\mathcal{C}$ is *complete for a rule* $l \to r \in \mathcal{R}$, and $a \in \mathcal{D}$ that is reducible by $l \to r$, if there is a context $C \in \mathcal{C}$ such that $a = C[l]$ and $a \to_{l \to r} C[r]$. A context class is *complete for a set of rules* $\mathcal{R}$ if it is complete for all rules in $\mathcal{R}$ and all $a \in \mathcal{D}$. It is *complete* if it is complete for all possible rules for the domain $\mathcal{D}$.

Very often a complete set of contexts $\mathcal{C}$ is generated by a proper subset $\mathcal{C}' \subset \mathcal{C}$ such that $\mathcal{C} = \bigcup_{i \geq 0} \mathcal{C}_i$ where $\mathcal{C}_0 = \mathcal{C}'$ and $\mathcal{C}_{i+1} = \{C_1 \circ C_2 \mid C_1, C_2 \in \mathcal{C}_i\}$.

Example 1.   – Substitution applications and replacements into arbitrary terms are contexts that generate a complete context class for terms and term rewriting systems.

- Additions of polynomials and multiplications by monomials generate a complete context class for polynomials and polynomial reduction systems.

- For and string rewriting systems, appending strings as suffixes describes a context class that is not complete.   □

Given that we have a complete context class, all reducible objects can be described as rule left-hand sides augmented by certain contexts. If the generating contexts of the complete context class can be partitioned into a finite set of context classes, this partition of the generating contexts gives raise to some kind of structural inductive reasoning for reducible objects since all reducible objects are either a rule left-hand side or a reducible object to which some context of a generating context partition has been applied.

Note that from $a = C[l]$ does not necessarily follow that a is reducible by the rule $l \to r$. Next we want to characterize contexts that grant reducibility if their argument is reducible.

Definition 4. A context C is *redex preserving* for some $a \in \mathcal{D}$ if for every rule $l \to r$ such that $a = C_l[l]$ for some context C_l and $a \to_{l \to r} C_l[r]$ it follows that $C[a] \to_{l \to r} C[C_l[r]]$.

In other words all reductions possible at a are still possible after applying a redex preserving context.

Definition 5. A rewrite relation on $\mathcal{D}$ definable by a set of rules is *proper* for a complete context class $\mathcal{C}$ if for each rule $l \to r$ and each $C \in \mathcal{C}$ with $C[l] \to_{l \to r} C[r]$, it follows that C is context preserving for l.

Lemma 6. *Given a proper rewrite relation and a context C such that $a = C[l]$ is reducible by $l \to r$ then $a = C_l[l]$ for some context C_l that is redex preserving for l.*   □

Example 2. Term rewriting systems, polynomial reduction systems and string rewriting systems each describe proper rewrite relations.   □

3.2   Compatibility Properties

Let us now define some properties that describe the behavior of relations under context applications.

Definition 7. A binary relation $\triangleright \subseteq \mathcal{D} \times \mathcal{D}$ is *compatible* w. r. t. a context class $\mathcal{C}$ if for all $a, b \in \mathcal{D}$ and all $C \in \mathcal{C}$, $a \triangleright b$ implies $C[a] \triangleright C[b]$.

Definition 8. A binary relation $\triangleright \subseteq \mathcal{D} \times \mathcal{D}$ is *semi-compatible* w. r. t. a context class $\mathcal{C}$ if for all $a, b \in \mathcal{D}$ and all $C \in \mathcal{C}$, $a \triangleright b$ implies $C[a] \triangleright \cdots \triangleright c \triangleleft \cdots \triangleleft C[b]$ for some $c \in \mathcal{D}$.

Definition 9. A relation $\triangleright$ is *essentially compatible* for a set $\mathcal{R}$ of rules w. r. t. a context class $\mathcal{C}$ if for all $l \to r \in \mathcal{R}$ and all $C \in \mathcal{C}$ that are redex preserving for l, $l \triangleright r$ implies that $C[l] \triangleright C[r]$.

Convention: In the sequel we will only consider those contexts w. r. t. which the congruences of interest are compatible and proper rewrite relations for which those contexts form a complete context class.

3.3   Termination of the Rewrite Relation

Lemma 10. *Let $\mathcal{R}$ be a set of rules describing a proper rewrite relation and $\succ$ be an ordering that is essentially compatible for $\mathcal{R}$ and a complete set of contexts. If $l \succ r$ for all $l \to r \in \mathcal{R}$ then $\succ \supseteq \to_{\mathcal{R}}$.*

Proof: If $a \to_{\mathcal{R}} b$ then there is a $l \to r \in \mathcal{R}$ and a context C that is redex preserving for l such that $a = C[l]$ and $b = C[r]$. Thus $a \succ b$ because $\succ$ is essentially compatible. $\square$

Definition 11. An ordering $\succ$ is *admissible* if it is terminating and if for all $a, b \in \mathcal{D}$ with $a \succ b$ and all contexts that are redex preserving for a, $C[a] \succ C[b]$ holds.

Theorem 12. *Let $\succ$ be an admissible ordering and $\mathcal{R}$ be a set of rules describing a proper rewrite relation. If $l \succ r$ for all $l \to r \in \mathcal{R}$ then $\to_{\mathcal{R}}$ is terminating.*

Proof follows directly from Lemma 10 and the definition of admissible orderings. $\square$

3.4   Confluence under Context Application

Now we investigate how reduction ambiguities that are confluent or subconnected by a semi-compatible reduction relation behave under applications of contexts.

Theorem 13. *Let $\to_R$ be a semi-compatible rewrite relation included in an admissible ordering $\succ$. If $b \succ a$, $b \succ c$ and $a \downarrow_R c$ then $C[a] \overset{C[b]}{\leftrightsquigarrow}_R C[c]$ for all contexts C that are redex preserving for b.*

Proof By $a \downarrow_R c$ we have $a = a_0 \to_R a_1 \to_R \ldots \to_R a_m = c_n \leftarrow_R \ldots \leftarrow_R c_1 \leftarrow_R c_0 = c$ and $b \succ a_i$, $b \succ c_j$ for $0 \leq i \leq m, 0 \leq j \leq n$. By semi-compatibility of $\to_R$ we get $C[a] = C[a_0] \downarrow_R C[a_1] \downarrow_R \ldots \downarrow_R C[a_m] = C[c_n] \downarrow_R \ldots \downarrow_R C[c_1] \downarrow_R C[c_0] = C[c]$ and $C[b] \succ C[a_i]$, $C[b] \succ C[c_j]$ for $0 \leq i \leq m, 0 \leq j \leq n$. In particular $C[b] \succ d$ for each d occurring in the above chain connecting $C[a]$ and $C[c]$ because C is redex preserving for b and $\succ$ is admissible. $\qquad \square$

Theorem 14. *Let $\to_R$ be a semi-compatible reduction relation included in an admissible ordering $\succ$. If $a \overset{b}{\leftrightsquigarrow}_R c$ w. r. t. $\succ$ then $C[a] \overset{C[b]}{\leftrightsquigarrow}_R C[c]$ for all contexts C that are redex preserving for b.*

Proof By $a \overset{b}{\leftrightsquigarrow} c$ we have $a = a_0 \downarrow_R a_1 \downarrow_R \ldots \downarrow_R a_m = c_n \downarrow_R \ldots \downarrow_R c_1 \downarrow_R c_0 = c$ and $b \succ a_i$, $b \succ c_j$ for $0 \leq i \leq m, 0 \leq j \leq n$. By Theorem 13, we get $C[a] = C[a_0] \overset{C[b]}{\leftrightsquigarrow}_R C[a_1] \overset{C[b]}{\leftrightsquigarrow}_R \ldots \overset{C[b]}{\leftrightsquigarrow}_R C[a_m] = C[c_n] \overset{C[b]}{\leftrightsquigarrow}_R \ldots \overset{C[b]}{\leftrightsquigarrow}_R C[c_1] \overset{C[b]}{\leftrightsquigarrow}_R C[c_0] = C[c]$ because C is redex preserving for b and $\succ$ is admissible. Hence $C[a] \overset{C[b]}{\leftrightsquigarrow}_R C[c]$. $\qquad \square$

The concept of a *critical reduction ambiguity* is crucial if local confluence is to be implied by the confluence of a finite set of reduction ambiguities.

Definition 15. Let R be a set of rules. A finite set $CA(R)$ of reduction ambiguities of $\to_R$ is called a *set of critical ambiguities* for R if for each rewrite ambiguity $a \leftarrow_R b \to_R c$ that is not confluent there is an ambiguity $s \leftarrow_R t \to_R u \in CA(R)$ and a context C that is redex preserving for t such that $a = C[s]$, $b = C[t]$ and $c = C[u]$.

Now we can prove an abstract critical pair theorem.

Theorem 16. *Let $\to_R$ be a set of rules defining a proper semi-compatible rewrite relation $\to_R$. If R is included in an admissible ordering and $CA(R)$ is a set of critical ambiguities of R then $\to_R$ is confluent iff all critical ambiguities in $CA(R)$ are subconnected.*

Proof By Lemma 1 and Theorem 14. $\qquad \square$

Critical pair theorems are known for term rewriting systems [11] and polynomial reduction systems [5]. An abstract critical pair theorem was presented for the first time in [8].

4   Proof Transformation for Compatible Rewriting

Completion procedures are often described as a set of inference rules transforming a pair of equations $\mathcal{E}$ and rules $\mathcal{R}$. Fig. 1 shows such a set of inference rules. $\succ \supseteq \to_R$ is a

Delete: $\frac{(\mathcal{E} \cup \{s \leftrightarrow s\}; \mathcal{R})}{(\mathcal{E}; \mathcal{R})}$

Compose: $\frac{(\mathcal{E}; \mathcal{R} \cup \{s \to t\})}{(\mathcal{E}; \mathcal{R} \cup \{s \to u\})}$ if $t \to_\mathcal{R} u$

Simplify: $\frac{(\mathcal{E} \cup \{s \leftrightarrow t\}; \mathcal{R})}{(\mathcal{E} \cup \{s \leftrightarrow u\}; \mathcal{R})}$ if $t \to_\mathcal{R} u$

Orient: $\frac{(\mathcal{E} \cup \{s \leftrightarrow t\}; \mathcal{R})}{(\mathcal{E}; \mathcal{R} \cup S)}$ if $\begin{cases} S \text{ is a set of rules such that } s \leftrightarrow_S t \\ \text{and } \forall l \to r \in S : (l \succ r \wedge l \leftrightarrow^*_{\mathcal{R} \cup \{s \leftrightarrow t\}} r) \end{cases}$

Collapse: $\frac{(\mathcal{E}; \mathcal{R} \cup \{s \to t\})}{(\mathcal{E} \cup \{u \leftrightarrow t\}; \mathcal{R})}$ if $\begin{cases} s \to_\mathcal{R} u \text{ by} \\ l \to r \in \mathcal{R} \text{ where } s \to t \rhd l \to r \end{cases}$

Superpose: $\frac{(\mathcal{E}; \mathcal{R})}{(\mathcal{E} \cup \{s \leftrightarrow t\}; \mathcal{R})}$ if $s \leftarrow_\mathcal{R} u \to_\mathcal{R} t$

Fig. 1. Inference rules for abstract completion

terminating ordering. The relation $\rhd$ on $\mathcal{D} \times \mathcal{D}$ used in the collapse step must be terminating and we require that from $l \succeq l'$ and $r \succeq r'$ follows $l \to r \rhd l' \to r'$. We write $(\mathcal{E}, \mathcal{R}) \vdash_I (\mathcal{E}', \mathcal{R}')$ to denote that $(\mathcal{E}', \mathcal{R}')$ has been derived from $(\mathcal{E}, \mathcal{R})$ by the application of an inference rule I.

The soundness of such an inference system follows from the fact that for each inference rule I that transforms $(\mathcal{E}, \mathcal{R})$ to $(\mathcal{E}', \mathcal{R}')$, $\leftrightarrow^*_\mathcal{E} \cup \leftrightarrow^*_\mathcal{R} = \leftrightarrow^*_{\mathcal{E}'} \cup \leftrightarrow^*_{\mathcal{R}'}$ holds. Yet the completeness proof for such a procedure namely the proof that for each $a(\leftrightarrow_\mathcal{E} \cup \leftrightarrow_\mathcal{R})^* b$ we may find an $(\mathcal{E}', \mathcal{R}')$ with $(\mathcal{E}, \mathcal{R}) \vdash^* (\mathcal{E}', \mathcal{R}')$ such that $a \downarrow_{\mathcal{R}'} b$ is much harder.

Completeness is usually proved using a technique called proof transformation. A *proof* of $a_0 = a_n$ in $(\mathcal{E}, \mathcal{R})$ is a chain $\langle a_0 \sim a_1 \sim \cdots \sim a_n \rangle$ where each $\sim$ stands for $\leftrightarrow_{a \leftrightarrow b}$, $\to_{l \to r}$ or $\leftarrow_{l \to r}$ for $a \leftrightarrow b \in \mathcal{E}, l \to r \in \mathcal{R}$. We will omit the indices if they are not required. A proof P of $a = b$ and a proof Q of $b = c$ can be concatenated to a proof $P.Q$ of $a = c$. Two proofs are equivalent if they prove the same equality. A proof of the form $\langle a \to \cdots \to b \leftarrow \cdots \leftarrow c \rangle$ is called a *rewrite* or *reductional proof*.

The relation on proofs that transforms a proof P of $a = b$ in $(\mathcal{E}, \mathcal{R})$ to a proof P' for $a = b$ in $(\mathcal{E}', \mathcal{R}')$ if $(\mathcal{E}, \mathcal{R}) \vdash (\mathcal{E}', \mathcal{R}')$ is written $P \Rightarrow_c P'$. Following [1], a *proof ordering* $\gg_c$ is a terminating ordering on proofs that is

1. compatible w. r. t. a complete context class $\mathcal{C}$, i. e., if $\langle a_0 \sim \cdots \sim a_n \rangle \gg_c \langle b_0 \sim \cdots \sim b_m \rangle$ then $\langle C[a_0] \sim \cdots \sim C[a_n] \rangle \gg_c \langle C[b_0] \sim \cdots \sim C[b_m] \rangle$ for all $C \in \mathcal{C}$,
2. compatible w. r. t. concatenation of proofs, i. e., if $Q \gg_c Q'$ then $P'.Q.P'' \gg_c P'.Q'.P''$ and
3. includes default confluence, i. e., every rewrite proof of $a = b$ is smaller than a proof $\langle a \leftarrow c \to b \rangle$.

Then we can show

Theorem 17. *If the proof transformation relation is included in a proof ordering then the inference rules of Figure 1 form a complete completion procedure.* $\quad\Box$

This view of completion has been proved for the case of term rewriting in [1, 12, 2].

Let us describe an appropriate proof ordering. The ordering we present here is slightly less complicated than the one given in [1]. First we may assign to each elementary proof $\langle a \sim b \rangle$ a cost function $c_c(a \sim b)$ as follows:

$$c_c(a \leftrightarrow_{s\leftrightarrow t} b) = (\{a, b\}, \bot)$$
$$c_c(a \rightarrow_{l\rightarrow r} b) = (\{a\}, l \rightarrow r)$$
$$c_c(a \leftarrow_{l\rightarrow r} b) = (\{b\}, l \rightarrow r).$$

Then $\langle a \sim b \rangle \gg_c \langle d \sim e \rangle$ if $c_c(a \sim b) >_c c_c(d \sim e)$ where $>_c$ compares its components lexicographically. Its first component is compared w. r. t. the multiset extension of $\succ$ and the second component w. r. t. $\rhd$ where $\bot$ is considered minimal w. r. t. $\rhd$. Now $\succ$ is compatible and terminating and $\rhd$ is context independent and terminating thus $\gg_c$ is compatible and terminating. Proofs can be compared by the multiset extension $\ggg_c$ of $\gg_c$ which is then terminating [9], compatible w. r. t. contexts and compatible w. r. t. concatenation of proofs.

Now we must show that $\ggg_c$ includes default confluence and that all elementary proofs in $(\mathcal{E}, \mathcal{R})$ affected by an application of an inference $(\mathcal{E}, \mathcal{R}) \vdash_I (\mathcal{E}', \mathcal{R}')$ result in a smaller proof in $(\mathcal{E}', \mathcal{R}')$. Note in the next and all following analysises of transformations of proof patterns, $\ggg_x$ is associated with the multiset extension of $>_x$. In addition, the right-hand sides of the ordering relations give only an upper bound of the actual proof costs. Underscores and ellipses denote components that do not influence the result. If two transformations are symmetrical only one of them is presented.

default confluence $s \leftarrow t \rightarrow u \Rightarrow_c s \rightarrow^* v \leftarrow^* u$:
$\quad [(\{t\}, \_), (\{t\}, \_)] \ggg_c [(\{s\}, \_), (\{u\}, \_), \dots]$
delete $s \leftrightarrow s \Rightarrow_c s$: $[(\{s, s\}, \_)] \ggg_c [\,]$
orient $s \leftrightarrow t \Rightarrow_c s \rightarrow t$: $[(\{s, t\}, \_)] \ggg_c [(\{s\}, \_)]$
simplify $s \leftrightarrow t \Rightarrow_c s \rightarrow t' \leftrightarrow t$: $[(\{s, t\}, \_)] \ggg_c [(\{s\}, \_), (\{t, t'\}), \_]$
superpose $s \leftarrow t \rightarrow u \Rightarrow_c s \leftrightarrow u$: $[(\{t\}, \_), (\{t\}, \_)] \ggg_c [(\{s, u\}, \_)]$
compose $s \rightarrow_{l\rightarrow r} t \Rightarrow_c s \rightarrow_{l\rightarrow r'} t' \leftarrow t$:
$\quad [(\{s\}, l \rightarrow r)] \ggg_c [(\{s\}, l \rightarrow r'), (\{t\}, \_)]$ because $r \succ r'$
collapse $s \rightarrow_{l\rightarrow r} t \Rightarrow_c s \rightarrow_{u\rightarrow v} t' \leftrightarrow t$:
$\quad [(\{s\}, l \rightarrow r)] \ggg_c [(\{s\}, u \rightarrow v), (\{t, t'\}, \_)]$ because $l \rightarrow r \rhd u \rightarrow v$.

Hence follows

Theorem 18. $\ggg_c$ *is a proof ordering and* $\Rightarrow_c \subseteq \ggg_c$ $\quad\Box$

Corollary 19. *The inference rules of Figure 1 form a complete completion procedure for compatible rewrite relations.* $\quad\Box$

Unfortunately $\gg_c$ is no longer a proof ordering if the rewrite relation is only semi-compatible w. r. t. the set of complete contexts. The reason for this is that the first component of $\gg_c$ depends on the compatibility of $\succ$[1].

Note that the compatibility requirement for proof orderings is only needed to lift the ordering from equations and rules to their application in arbitrary contexts. Thus a proof ordering need not be compatible w. r. t. contexts as long as it can be lifted from rules and equations to their application in arbitrary contexts. This is a weaker requirement than compatibility because there may be contexts that inhibit the application of a certain rule or equation. But even with this weaker definition $\gg_c$ does not qualify as a proof ordering. This will be investigated in the next section.

5 Proof Transformation for Semi-compatible Rewriting

In this section, we assume that the rewrite relation on $\mathcal{D}$ is proper, and that $\succ \supseteq \to_{\mathcal{R}}$ is an admissible ordering and $\to_{\mathcal{R}}$ is semi-compatible for all sets $\mathcal{R}$ of rules.

5.1 The Standard Approach

Let us first investigate how the ordering $\gg_c$ from the last section behaves in this case. Before we start we must recall the following facts. In each proof step $\langle a \to_{l \to r} b \rangle$ and $\langle b \leftarrow_{l \to r} a \rangle$ we have $a = C[l]$ for some context C that is redex preserving for l. Similarly we get that for each $\langle a \leftrightarrow_{s \leftrightarrow t} b \rangle$ $a = C[s]$ or $b = C[s]$ for some C that is context preserving for s or $a = C[t]$ or $b = C[t]$ for some C that is context preserving for t. Further we assume in a first step that the relation $\leftrightarrow_{\mathcal{E}}$ inherits semi-compatibility from $\to_{\mathcal{R}}$ (in later steps we will get rid of this assumption) i. e., if $s \leftrightarrow t$ then for all contexts C, $C[s] \leftrightarrow^* C[t]$. Now we can describe the effect of proof transformation. Question marks describe relations or objects which are unknown.

default confluence $s \leftarrow t \to u \Rightarrow_c s \stackrel{t}{\leftrightsquigarrow} u$:
$$[(\{t\}, \_), (\{t\}, \_)] \gg_c [(\{s\}, \_), (\{u\}, \_), \ldots] \text{ because of Theorem 13}$$
delete $s \leftrightarrow s \Rightarrow_c s$: $[(\{s, s\}, \_)] \gg_c []$
orient $s \leftrightarrow t \Rightarrow_c s \downarrow t$: $[(\{s, t\}, \_)] \gg_c [(\{s\}, \_), (\{t\}, \_), \ldots]$
simplify $s \leftrightarrow t \Rightarrow_c s \to t' \leftrightarrow^* t$: $[(\{s, t\}, \_)] ? [(\{s\}, \_), (\{t, ?\}, \_), \ldots]$
 or $s \leftrightarrow t \Rightarrow_c s \downarrow t' \leftrightarrow t$: $[(\{s, t\}, \_)] ? [(\{s\}, \_), (\{t, t'\}, \_), \ldots]$
superpose $s \leftarrow t \to u \Rightarrow_c s \leftrightarrow^* u$:
$$[(\{t\}, \_), (\{t\}, \_)] ? [(\{s, ?\}, \_), (\{u, ?\}, \_), \ldots]$$
compose $s \to_{l \to r} t \Rightarrow_c s \to_{l \to r} t' \downarrow t$:
$$[(\{s\}, l \to r)] \gg_c [(\{s\}, l \to r'), (\{t\}, \_), (\{t'\}, \_), \ldots]$$
 because $r \succ r'$, $s \succ t$, $s \succ t'$
collapse $s \to_{l \to r} t \Rightarrow_c s \to_{u \to v} t' \leftrightarrow^* t$:
$$[(\{s\}, l \to r)] ? [(\{s\}, u \to v), (\{t, ?\}\_), \ldots].$$

---

[1] An analogous statement holds for the ordering used in [1] that relies on the compatibility of $\succ$, the subterm ordering and the subsumption ordering.

5.2 Promises

The proof transformation above shows two problems. The first one is that we deliberately postulated how equational steps are to be applied in arbitrary contexts. The second more serious problem is that we cannot show that $\Rightarrow_c$ is included in $\ggg_c$. This is in particular the case for the *superpose*, *simplify* and *collapse* inferences. Note that in all these cases the problem was caused by equality applications in contexts. Note also that we have not shown a single witness that $\Rightarrow_c$ is actually not included in $\ggg_c$.

Therefore we will get rid of equations. We will rather use "promises" that a rewrite ambiguity will eventually become subconnected during the completion[2]. A *promise* to eventually subconnect the rewrite ambiguity $s \leftarrow t \rightarrow u$ will be written $s \stackrel{t}{\longleftrightarrow} u$. We can use promises in proofs: $a \leftrightarrow_P b$ if there is a promise $s \stackrel{t}{\longleftrightarrow} u \in P$ and $a = C[s]$ and $b = C[u]$ or $a = C[u]$ and $b = C[s]$ for some context C that is redex preserving for t. Note we do not demand that C is context preserving for s or u. A proof of $a = b$ using a single promise $s \stackrel{t}{\longleftrightarrow} u$ will be denoted by $a \stackrel{C[t]}{{}_s\longleftrightarrow_u} b$. Thus proofs containing promises as elementary steps are actually proofs using reductions only; yet some peaks are labeled as promised. Completion can now be described by inferences on a pair of rules promises. An alternative view is that real inferences are only applied to rules and the promises are some book keeping mechanism organizing a "to-do-list" for the completion process. In the case of (conditional) term rewriting the superposition object in the promises corresponds to the "history dependent complexity" information proposed by Ganzinger [10]. Fig. 2 shows the modified inferences. Note that the inference rules

Delete: $\dfrac{(P \cup \{s \stackrel{t}{\longleftrightarrow} s\}; R)}{(P; R)}$

Compose: $\dfrac{(P; R \cup \{s \rightarrow t\})}{(P; R \cup \{s \rightarrow u\})}$ if $t \rightarrow_R u$

Simplify: $\dfrac{(P \cup \{s \stackrel{t}{\longleftrightarrow} u\}; R)}{(P \cup \{s \stackrel{t}{\longleftrightarrow} v\}; R)}$ if $u \rightarrow_R v$

Orient: $\dfrac{(P \cup \{s \stackrel{t}{\longleftrightarrow} u\}; R)}{(P; R \cup S)}$ if $\begin{cases} S \text{ is a set of rules such that } s \stackrel{t}{\leadsto}_{S \cup R} u \\ \text{and } \forall l \rightarrow r \in S : (l \succ r \wedge l \leftrightarrow^*_{R \cup \{s \leftrightarrow t\}} r) \end{cases}$

Collapse: $\dfrac{(P; R \cup \{s \rightarrow t\})}{(P \cup \{u \stackrel{s}{\longleftrightarrow} t\}; R)}$ if $\begin{cases} s \rightarrow_R u \text{ by} \\ l \rightarrow r \in R \text{ where } s \rightarrow t \triangleright l \rightarrow r \end{cases}$

Superpose: $\dfrac{(P; R)}{(P \cup \{s \stackrel{t}{\longleftrightarrow} u\}; R)}$ if $s \leftarrow_R t \rightarrow_R u$

Fig. 2. Inference rules for abstract completion with promises

in Fig. 1 are special cases of those in Fig. 2 if all promises are treated as equations.

---

[2] Eventually fullfilling a promise implies the fairness of the completion procedure.

The applications of the inference rules in Fig. 2 define again a proof transformation relation. We will denote this proof transformation by $\Rightarrow_s$. Let us now define a proof ordering including $\Rightarrow_s$.

First we define the ordering $\succ_s$ on $\mathcal{D} \cup \mathcal{D}'$ where $\mathcal{D}'$ is a labeled copy of $\mathcal{D}$: $\mathcal{D}' = \{a' \mid a \in \mathcal{D}\}$ that is disjoint from $\mathcal{D}$ such that $a \succ_s a'$, for all $a \in \mathcal{D}$ and $a' \succ_s b$ for all $a, b \in \mathcal{D}$ with $a \succ b$. Loosely speaking we can say that each a' is infinitesimally smaller than a. It is easy to see that $\succ_s$ is essentially compatible and terminating.

Now we define a cost function of elementary proof steps, where a proof step is elementary if it is a rewrite step or a rewrite ambiguity that is labeled by a promise:

$$c_s(a_s \xleftrightarrow{C[t]}_u b) = (C[t]', \perp, s \xleftrightarrow{t} u)$$
$$c_s(a \to_{l \to r} b) = (a, l \to r, \perp)$$
$$c_s(a \leftarrow_{l \to r} b) = (b, l \to r, \perp).$$

Costs computed by c_s can be ordered according to the lexicographic combination $>_s$ of orderings on their components: The first component is ordered w. r. t. $\succ_s$, the second w. r. t. $\triangleright$, the third component orders promises as follows: $s \xleftrightarrow{t} u > v \xleftrightarrow{w} x$ if $t \succ^* w$ or $t = w$ and $[s, u] \succ^*_{mul} [v, x]$ where $\succ^*$ is a terminating ordering including $\succ$. Clearly $>_s$ is terminating. Two elementary proofs can be compared by $\gg_s$ as follows: $\langle a \sim b \rangle \gg_s \langle c \sim d \rangle$ if $c_s(a \sim b) >_s c_s(c \sim d)$. Proofs are ordered w. r. t. $\ggg_s$, the multiset extension of $\gg_s$. Thus it follows

Lemma 20. $\ggg_s$ *is terminating.* □

Now we have to check whether $\ggg_s$ includes default confluence and whether elementary proofs and reduction ambiguities affected by an inference rule are transformed into a smaller proof. Recall again that for all $\langle a \to_{l \to r} b \rangle$ and $\langle b \leftarrow_{l \to r} a \rangle$ there is a context C such that $a = C[l]$ and C is redex preserving for l. In addition for all $\langle a_s \xleftrightarrow{C[t]}_u b \rangle$, C is context preserving for t and $a = C[s]$ and $b = C[u]$ or $a = C[u]$ and $b = C[s]$.

default confluence $a \leftarrow c \to b \Rightarrow_s a \xleftrightarrow{c} b$:
$\quad [(c, \_, \_), (c, \_, \_)] \ggg_s [(a, \_, \_), (b, \_, \_), \ldots]$ because of Theorem 14
delete $a_s \xleftrightarrow{C[t]}_u a \Rightarrow_s a$: $\ [(C[t]', \_, \_)] \ggg_s [\,]$
orient $a_s \xleftrightarrow{C[t]}_u b \Rightarrow_s a \xleftrightarrow{C[t]} b$: $[(C[t]', \_, \_)] \ggg_s [(\{a\}, \_, \_), (\{b\}, \_, \_), \ldots]$
$\quad$ because of Theorem 14. Note that $a \xleftrightarrow{C[t]} b$ is a proof containing objects from $\mathcal{D}$ all of which are smaller than $C[t]$. Thus all of these objects are also smaller than $C[t]'$.
simplify $a_s \xleftrightarrow{C[t]}_u b \Rightarrow_s a \downarrow C[v]_v \xleftrightarrow{C[t]}_u b$:
$\quad [(C[t]', \perp, s \xleftrightarrow{t} u)] \ggg_s [(C[t]', \perp, v \xleftrightarrow{t} u]), (a, \_, \_), (C[v], \_, \_), \ldots]$
$\quad$ because $s \succ v$
superpose $a \leftarrow b \to c \Rightarrow_s a_s \xleftrightarrow{C[t]}_u c$: $[(b, \_, \_), (b, \_, \_)] \ggg_s [(C[t]', \_, \_)]$
$\quad$ because $b = C[t]$ for some context preserving C
compose $a \to_{l \to r} b \Rightarrow_s a \to_{l \to s} C[s] \downarrow b$:
$\quad [(a, l \to r, \_)] \ggg_s [(a, l \to s, \_), (b, \_, \_), (C[s], \_, \_), \ldots]$
$\quad$ because $a = C[l]$ for context preserving C, $r \succ s$, $a \succ b$, $a \succ C[s]$

collapse $a \to_{l \to r} b \Rightarrow_s a \to_{l \to s} C[v]\ {}_v\overset{a}{\longleftrightarrow}_r b$:

$$[(a, l \to r, \_)] \ggg_s [(a, t \to s, \_), (a', \_, \_)]$$

because $a = C[l]$ for context preserving C, $l \to_{t \to s} v$ and $l \to r \rhd t \to s$.

Thus it follows

Lemma 21. *The relation* $\ggg_s$ *is a proof ordering for semi-compatible rewrite relations and* $\Rightarrow_s \subseteq \ggg_s$. □

and we get the main theorem

Theorem 22. *Let $\mathcal{D}$ be a domain with semi-compatible rewrite relation, and let $\succ$ be an admissible ordering on $\mathcal{D}$ that includes $\to_\mathcal{R}$ then the inference rules in Fig. 2 form a complete completion procedure.* □

6 Implications

In this section, we want to relate our results to known completion procedures and additional inference rules.

6.1 Starting Equations

As mentioned in Section 5, promise based completion subsumes standard completion. Yet some (standard) completion procedures start with equations as inputs. For these cases we must answer the question how we want to deal with starting equations in completion procedures that use promises. In such a case we may add for each starting equation $s \leftrightarrow t$ a new (pseudo) element $\top(s, t)$ to $\mathcal{D}$ that is maximal w. r. t. $\succ$ and for which each context is context preserving. Then we can consider each equation $s \leftrightarrow t$ as promise $s \overset{\top(s,t)}{\longleftrightarrow} t$.

6.2 Subsumption of Promises

We can define an additional inference rule *subsume* that transforms a promise into an equivalent but smaller promise

$$\text{Subsume:} \quad \frac{(\mathcal{P} \cup \{s \overset{t}{\longleftrightarrow} u, v \overset{w}{\longleftrightarrow} x\}; \mathcal{R})}{(\mathcal{P} \cup \{v \overset{w}{\longleftrightarrow} x\}; \mathcal{R})} \quad \text{if} \quad \begin{cases} s \overset{t}{\longleftrightarrow} u \succ v \overset{w}{\longleftrightarrow} x \text{ and} \\ s \to_\mathcal{R}^* C[v], t \to_\mathcal{R}^* C[w], u \to_\mathcal{R}^* C[x] \\ \text{and } C \text{ context preserving for } w \end{cases}$$

Then the subsumption inference rule yields in contexts that are redex preserving for t:

subsume $a\ {}_s\overset{C[t]}{\longleftrightarrow}_u b \Rightarrow_{subsume} a\ \overset{C[C_1[w]]}{\longleftrightarrow} C[C_1[v]]\ {}_v\overset{C[C_1[w]]}{\longleftrightarrow}_x C[C_1[x]]\ \overset{C[C_1[w]]}{\longleftrightarrow} b$:

$$[(C[t]', \bot, s \overset{t}{\longleftrightarrow} u)] \ggg_s [(C[t]', \bot, v \overset{w}{\longleftrightarrow} x), \dots]$$ because $t \to_\mathcal{R}^* C_1[w]$ for some context preserving C_1 and all objects in $C[C_1[x]]\ \overset{C[C_1[w]]}{\longleftrightarrow} b$ and in $a\ \overset{C[C_1[w]]}{\longleftrightarrow} C[C_1[v]]$ are smaller than $C[C_1[w]]'$.

Note that this subsumption rule is less general than the subsumption usually defined for critical pairs in the area of term rewriting systems. This is because the above subsumption rule must take the superposition objects t and w into account.

Theorem 23. *Let $\mathcal{D}$ be a domain with a proper and semi-compatible rewrite relation, and let $\succ$ be an admissible ordering including $\rightarrow_\mathcal{R}$ then the inference rules in Fig. 2 together with the subsumption inference rule form a complete completion procedure.* $\square$

6.3 Promise Redundancy Criteria

In the context of promise based completion, critical pair criteria can be characterized very easily as deletion of redundant promises:

$$\text{Delete redundancy: } \frac{(\mathcal{P} \cup \{s \xleftrightarrow{t}_u u\}; \mathcal{R})}{(\mathcal{P}; \mathcal{R})} \text{ if } \begin{cases} P \text{ proves } s = u \text{ in } (\mathcal{P}; \mathcal{R}) \\ \text{and } \langle s\ _s\xleftrightarrow{t}_u u\rangle \ggg_s P \end{cases}$$

The redundancy deletion rule yields in contexts that are redex preserving for t:

delete redundancy $a\ _s\xleftrightarrow{C[t]}_u b \Rightarrow_{delete\_redundancy} C[P]$:

$[(C[t]', \perp, s \xleftrightarrow{t} u)] \ggg_s$ "cost of $C[P]$" because $[(t', \perp, s \xleftrightarrow{t} u)] >_s c_s(d \sim e)$ for every elementary proof step $\langle d \sim e\rangle$ in P. Then $[(C[t]', \perp, s \xleftrightarrow{t} u)] >_s c_s(f \sim g)$ for every elementary proof step $\langle f \sim g\rangle$ in $C[\langle d \sim e\rangle]$ because C is context preserving for t.

Therefore we get

Theorem 24. *Let $\mathcal{D}$ be a domain with a proper and semi-compatible rewrite relation, and let $\succ$ be an admissible ordering including $\rightarrow_\mathcal{R}$ then the inference rules in Fig. 2 together with the redundancy deletion inference rule form a complete completion procedure.* $\square$

Since all proof transformations for inference rules for promise based completion presented so far are included in $\ggg_s$ they are all compatible:

Theorem 25. *Let $\mathcal{D}$ be a domain with a proper and semi-compatible rewrite relation, and let $\succ$ be an admissible ordering including $\rightarrow_\mathcal{R}$ then the inference rules in Fig. 2 together with the subsumption and the redundancy deletion inference rules form a complete completion procedure.* $\square$

In [2] a class of critical pair criteria is described that is sound and preserves the completeness of the completion procedure. Such a critical pair criterion is called *composite*. It allows to delete a critical pair $s_0 \leftrightarrow s_n$ derived from a rewrite ambiguity $s_0 \leftarrow t \rightarrow s_n$ if there are proofs $P_1, ..., P_n$ proving $s_0 = s_1, ..., s_{n-1} = s_n$ respectively such that $t \succ s_i$ and $\langle s_0 \leftarrow t \rightarrow s_n\rangle \ggg_c P_i$ for $1 \le i \le n$ even though the whole proof $P_1.P_2.\cdots.P_n$ may be greater than $\langle s_0 \leftarrow t \rightarrow s_n\rangle$ w.r.t. $\ggg_c$.

This criterion is implicit in our proof ordering $\gg_s$ because if $s_0 \overset{t}{\longleftrightarrow} s_n \gg_s P_i$ for $1 \leq i \leq n$ then $s_0 \overset{t}{\longleftrightarrow} s_n \gg_s P_1.P_2.\cdots.P_n$. This is so by virtue of multiset orderings because a promise is an elementary proof whereas the rewrite ambiguity $\langle s_0 \leftarrow t \rightarrow s_n \rangle$ can be decomposed into two elementary proofs $\langle s_0 \leftarrow t \rangle$ and $\langle t \rightarrow s_n \rangle$.

Note that the formulation of the composite critical pair criterion based on critical pairs (or equations) is dangerous because it does not allow to delete the critical pair $s \leftrightarrow u$ from $\mathcal{E}$ if it can be derived from two rewrite ambiguities $s \leftarrow t_1 \rightarrow u$ and $s \leftarrow t_2 \rightarrow u$ with $t_1 \succ t_2$ and the composite criterion applies only to the first ambiguity. In general combining arbitrary critical pair subsumption inferences and critical pair criteria is not complete. In that respect a promise based completion is a more secure formalism.

6.4 Equations in Compatible Systems

For compatible rewrite systems we have seen that equations can be used independently of the superposition object they have been derived from. This can be simulated as follows. For each equation $s \leftrightarrow t$ we add a new (pseudo) element $\bot(s,t)$ to $\mathcal{D}$ that is smaller than the least upper bound of s and t w.r.t. $\succ$ and for which $\bot(s,t) \succ s$ and $\bot(s,t) \succ t$. Then each promise $s \overset{u}{\longleftrightarrow} t$ is subsumed by the promise $s \overset{\bot(s,t)}{\longleftrightarrow} t$ and thus can be written simply as equation $s \leftrightarrow t$. Defining subsumption on promises of the form $s \overset{\bot(s,t)}{\longleftrightarrow} t$ yields the classical definition of subsumption. Yet the classical definition of subsumption is not compatible w.r.t. all critical pair criteria.

6.5 Weakly Compatible Systems

From the construction of $\gg_s$, we see that semi-compatibility of $\rightarrow_{\mathcal{R}}$ is actually a too strong requirement for our completeness proofs. All we need is that for an elementary reduction proof $\langle a \rightarrow_{\mathcal{R}} b \rangle$ there is a proof P for $C[a] = C[b]$ that is smaller than any rewrite ambiguity $\langle C[a] \leftarrow_{\mathcal{R}} c \rightarrow_{\mathcal{R}} C[b] \rangle$.

Definition 26. A reduction ambiguity $b \leftarrow_{\mathcal{R}} a \rightarrow_{\mathcal{R}} c$ or a pair (b,c) is *properly subconnected* w.r.t. $\succ$ if there are $d_1, \ldots, d_n \in \mathcal{D}$ with $b \succ d_i$ or $c \succ d_i$ for $1 \leq i \leq n$ and $b \leftrightarrow_{\mathcal{R}} d_1 \leftrightarrow_{\mathcal{R}} \ldots \leftrightarrow_{\mathcal{R}} d_n \leftrightarrow_{\mathcal{R}} c$.

Certainly proper subconnectedness of $b \leftarrow_{\mathcal{R}} a \rightarrow_{\mathcal{R}} c$ implies the subconnectedness of any rewrite ambiguity $b \leftarrow_{\mathcal{R}} d \rightarrow_{\mathcal{R}} c$.

Definition 27. A reduction relation $\rightarrow_{\mathcal{R}} \subseteq \mathcal{D} \times \mathcal{D}$ is *weakly compatible* w.r.t. a context class $\mathcal{C}$ if for all $a, b \in \mathcal{D}$ and all $C \in \mathcal{C}$, $a \rightarrow_{\mathcal{R}} b$ implies $(C[a], C[b])$ is properly subconnected.

Theorem 28. *Let $\rightarrow_{\mathcal{R}}$ be a weakly-compatible reduction relation included in an admissible ordering $\succ$. If the $a \overset{b}{\leftrightsquigarrow} c$ w.r.t. $\succ$ then $C[a] \overset{C[b]}{\leftrightsquigarrow} C[c]$ for all contexts that are redex preserving for b.* $\square$

Theorem 28 can be used to replace all applications of Theorem 14 and hence Theorems 16, 22, 23, 24 and 25 also hold if $\to_\mathcal{R}$ is weakly compatible.

7 Conclusion

We have presented a uniform framework that allows to investigate the completeness of completion procedures. Our framework can be used w. r. t. semi-compatible reduction relations and leads to a clear characterization of complete redundancy criteria. The latter point is important for implementing correct efficient completion procedures because a naive combination of redundancy criteria easily leads to incomplete procedures. In particular, our approach supplies us with a proof ordering to describe and investigate Buchberger's algorithm and its critical pair criteria. Understanding semi-compatible reduction relations is also important for understanding term completion modulo infinite congruences defined by a canonical term rewriting system. Normalized completion [14] and symmetrization based completion [3, 13, 7, 8] are candidates for these problems.

References

1. L. Bachmair and N. Dershowitz. Critical pair criteria for completion. *Journal of Symbolic Computation*, 6:1–18, 1988.
2. L. Bachmair and N. Dershowitz. Equational inference, canonical proofs, and proof orderings. *Journal of the ACM*, 41(2):236–276, 1994.
3. B. Benninghofen, S. Kemmerich, and M. M. Richter. *Systems of Reductions*. Springer-Verlag, 1987.
4. G. M. Bergman. The diamond lemma for ring theory. *Advances in Mathematics*, 29:178–218, 1978.
5. B. Buchberger. *Ein Algorithmus zum Auffinden der Basiselemente des Restklassenringes nach einem nulldimensionalen Polynomideal*. PhD thesis, Universität Innsbruck, 1965.
6. R. Bündgen. Abstract completion for non-compatible rewrite relations. Technical Report 96–13, Wilhelm-Schickard-Institut, Universität Tübingen, D-72076 Tübingen, 1996.
7. R. Bündgen. Buchberger's algorithm: the term rewriter's point of view. *Theoretical Computer Science*, 1996 (to appear).
8. R. Bündgen. Symmetrization based completion. In *Proceeding on Conference on Symbolic Rewriting Techniques*, 1996. (submitted, presented at SRT'95, Monte Verita, CH, May 1995).
9. N. Dershowitz and Z. Manna. Proving termination with multiset orderings. *Commun. ACM*, 22(8):465–476, 1979.
10. H. Ganzinger. Completion with history-dependent complexities for generated equations. In Sannellla and Tarlecki, editors, *Recent Trends in Data Type Specification*, 1987.
11. D. E. Knuth and P. B. Bendix. Simple word problems in universal algebra. In Leech, editor, *Computational Problems in Abstract Algebra*. Pergamon Press, 1970.
12. W. Küchlin. Inductive completion by ground proof transformation. In Aït-Kaci and Nivat, editors, *Rewriting Techniques*, volume 2 of *Resolution of Equations in Algebraic Structures*. Academic Press, 1989.
13. P. Le Chenadec. *Canonical Forms in Finitely Presented Algebras*. Pitman, London, 1986.
14. C. Marché. Normalised rewriting and normalised completion. In *Ninth Anual Symposium on Logic in Computer Science*. IEEE Computer Society Press, 1994.

15. M. H. A. Newman. On theories with a combinatorial definition of "equivalence". *Annals of Mathematics*, 43(2):223–243, 1942.

16. L. Pottier. *Algorithmes de complétion et généralison en logique du premier ordre*. PhD thesis, Université Nice — Sophia Antipolis, Parc Valrose, Fac. Sciences, F-06034 Nice, 1989.

17. F. Winkler and B. Buchberger. A criterion for eliminating unnecessary reductions in the Knuth-Bendix algorithm. In *Proc. Colloquium on Algebra, Combinatorics and Logic in Computer Science*. J. Bolyai Math. Soc., J. Bolyai Math. Soc. and North-Holland, 1985.

PATCH Graphs: An Efficient Data Structure for Completion of Finitely Presented Groups

Christopher Lynch[1] and Polina Strogova[2]

[1] INRIA Lorraine - CRIN
[2] INRIA Lorraine - CRIN and INRIA Rocquencourt
Technopôle de Nancy-Brabois, B.P. 101, 54602 Villers-lès-Nancy Cedex, France
phone: (+33) 83 59 30 00, fax: (+33) 83 27 83 19, e-mail: lynch, strogova@loria.fr

Abstract. Based on a new data structure called PATCH Graph, an efficient completion procedure for finitely presented groups is described. A PATCH Graph represents rules and their symmetrized forms as cycles in a Cayley graph structure. Completion is easily performed directly on the graph, and structure sharing is enforced. The structure of the graph allows us to avoid certain redundant inferences. The PATCH Graph data structure and inference rules complement other extensions of Knuth-Bendix completion for finitely presented groups.

1 Introduction

Important problems in computer algebra have been solved by applying the Knuth-Bendix completion procedure [KB70]. In particular it is useful for solving the word problem in a given finite presentation of an algebra. We will focus on computational group theory, where completion has been used to solve other problems, such that finding a confluent presentation [Gil79], testing whether a finitely presented group is a finite group [Gil79], verifying nilpotency [Sim87], and determining whether two finitely presented groups are isomorphic [HR89].

Our interest in Knuth-Bendix completion arises from the shortest path routing problem in Cayley Graphs. Cayley Graphs, based on group structure, represent an attractive model for interconnection networks. They are symmetric, dense, and have small diameters. Some classes of Cayley Graphs, such as hypercube, shuffle-exchange and de Bruijn networks for example, are parallel architectures with an effective routing. For all these classes of networks, existing routing algorithms take advantage of concrete form of group presentation relations. In [FKS94] we described a general routing algorithm, which enables us to find a shortest path between two given nodes of any Cayley Graph. To get a shortest path, we normalize a path, resulting from Todd-Coxeter coset enumeration, with respect to a canonical system describing the graph. Applying the standard Knuth-Bendix algorithm we were able to complete rewriting systems for groups of order up to 10^4 using RRL [KZ88], Herky [Zha92] and Otter [McC94] automated theorem provers. For larger systems, we need a more efficient completion algorithm specialized for group completion.

Several improvements of the Knuth-Bendix completion procedure have been developed for completion of finitely presented groups. Some of these algorithms extend the Knuth-Bendix completion procedure, to create an algorithm that halts for presentations where Knuth-Bendix completion does not. The disadvantage of these algorithms is that they only work for certain classes of groups. In [MNO91], the confluence property is weakened to only require that a string equivalent to the identity can be rewritten to the identity. An algorithm is given that works for monadic groups. In [EHR91], automata theory techniques are used to construct an automata representing an infinite rewrite system for automatic groups.

In this paper, instead of extending the Knuth-Bendix procedure, we choose to make it more efficient. Another paper applying this philosophy is [Mar96], which focuses on finding good orderings. Le Chenadec [LC86] remarks that the completion procedure founded on group reduction relation runs faster than the standard completion procedure. According to group reduction, a string is firstly reduced by a ground relation of group presentation and then simplified by group axioms. A similar relation, called normalized rewriting, is proposed by C. Marché [Mar93]. Also, Le Chenadec shows how to avoid inferences with the group axioms. He introduces the symmetrization procedure, specifying for groups the techniques of [Büc79]. Our procedure builds on Le Chenadec's ideas. We design special data structures so that the inferences can be performed efficiently. These data structures also use automatic structure sharing to store the data efficiently, and allow us to avoid performing certain redundant inferences. Our philosophy is to express the rewriting on the object level, as in [Lyn95] for paramodulation, and in [LS95] for term completion. We think our approach could be combined with the previously mentioned algorithms. The data structure we propose in this paper is called a *PATCH Graph*. A PATCH Graph represents in a very natural and compact way a string rewriting system for a group. The idea is to store strings as paths, and rewriting rules and equations as cycles (this interpretation is inspired by Cayley Graphs), and perform completion inference rules on cycles. Actually, an equation $a \cdot b = c$ can be seen as a cycle formed by concatenation of directed edges $e_1 = (v_1, v_2), e_2 = (v_2, v_3)$ and $e_3 = (v_1, v_3)$ labelled respectively a, b and c. The first advantage of this representation is that the same cycle can be seen as concatenation of edge e_1, edge e_3, and edge e_2, where e_2 is taken backwards, i.e. as (v_3, v_2). Reading out the edge labels, we get $a = c \cdot b^{-1}$. Equation $a = c \cdot b^{-1}$ is one of the symmetrized forms of equation $a \cdot b = c$. So, we can represent all the $O(n)$ symmetrized forms by one cycle of length n. Thereby, performing one inference on two cycles of respective lengths m and n, we calculate at one go $O(n \cdot m)$ critical pairs or simplifications for the rules represented by these cycles. This is the second advantage. Finally, from the technical point of view, operating on strings rather than terms avoids additional computations due to the associativity group axiom.

We formulate inference rules **Symmetrize**, **Patch** and **Eliminate** for group completion, based on PATCH Graphs. **Symmetrize** inference rule expresses the symmetrization procedure. **Patch** inference rule gives the law of creation of new cycles. **Eliminate** inference rule encodes cancellation and simplification with respect to group axioms.

A new cycle C in a PATCH Graph is generated from two cycle-ancestors C_1 and C_2, having common paths π_1 and π_2 (i.e. the paths representing the same string). Cycle C is as a combination of path $C_1 \backslash \pi_1$, which complements path π_1 to cycle C_1, and of path $C_2 \backslash \pi_2$, which complements path π_2 to cycle C_2. **Patch** inference rule is performed if and only if paths π_1 and π_2 are maximal non trivial common paths. Paths π_1 and π_2 are maximal, if they are not strict subpaths of other common paths for C_1

and C_2. Actually, rules represented by a cycle C, created for maximal common paths π_1 and π_2, simplify all the rules and equations, represented by a cycle C', generated for non-maximal common subpaths of paths π_1 and π_2. Paths π_1 and π_2 are trivial, if $C_1 = C_2$ and $\pi_1 = \pi_2$. A new cycle for trivial common paths π_1 and π_2 represents a critical pair between two symmetrized forms of the same rule. Such a critical pair can be simplified by group axioms to a trivial equation. Consequently, using our inference rules avoids generating some redundant equations and rules, whereas they are required in standard completion and Le Chenadec's completion procedure. These points enable us to expect a considerable improvement of the completion procedure for groups. We have started an implementation of the algorithm.

The paper is organized in five sections. Important notations and preliminary results are given in Section 2. After introducing in Section 3 the syntax and the semantics of PATCH Graphs, we formulate in Section 4 the inference rules for the completion procedure based on PATCH Graphs, and state the system of inference rules to be sound and complete. The improvements of the group completion procedure by using PATCH Graphs are discussed in Section 5.

2 Definitions and Preliminary Results

We recall in this section some definitions and results relative to string rewriting and group completion, according to the formalism of [BO93] and [LC86], and some notions of graph theory.

Let G be a finite alphabet of constants called *letters*. Consider a binary concatenation operation on G, denoted $\cdot$ in infix notation. Set G^* of all finite concatenations of letters, called *strings*, is the free monoid on G, with the empty string denoted ε as neutral element. For any string s, string t is a *prefix* of s if there exists $u \in G^*$ such that $s = t \cdot u$. String v is a *suffix* of s if there exists $w \in G^*$ such that $s = w \cdot v$. String t is a *cyclic permutation* of s if there exist $u, v \in G^*$ such that $s = u \cdot v$ and $t = v \cdot u$. String t is a *substring* of s if $s = u \cdot t \cdot v$ for some $u, v \in G^*$. For each string s, the *length* of s is the number of letters in it.

An *equation* on G^* is a pair $(l, r) \in G^* \times G^*$, denoted $l = r$, a *rewriting rule* on G^* is an ordered pair $(l, r) \in G^* \times G^*$, denoted $l \to r$. If (l, r) is a rule or an equation, then l is called its *left-hand side (lhs)*, and r its *right-hand side (rhs)*. A *string rewriting system* (or *rewriting system* for short) $\mathcal{R}$ on G^* is a set of rewriting rules on G^*.

Let $\mathcal{R}$ be a rewriting system. Relation $\to_{\mathcal{R}}$ such that for all $t, u \in G^*$ $t \to_{\mathcal{R}} u$ iff there exists a rule $l \to r \in \mathcal{R}$, and strings $p, s \in G^*$, such that $t = p \cdot l \cdot s$, and $u = p \cdot r \cdot s$, is called *single-step reduction relation* induced by $\mathcal{R}$, and we say that rule $l \to r$ *simplifies* string t. Rule $l \to r$ *simplifies* a rule or an equation $(t, u) \in G^*$ if $l \to r$ simplifies t or u. Let $\xrightarrow{*}_{\mathcal{R}}$ ($\xleftrightarrow{*}_{\mathcal{R}}$) denote the reflexive transitive (resp. the reflexive transitive symmetric) closure of $\to_{\mathcal{R}}$. Relation $\xrightarrow{*}_{\mathcal{R}}$ is called *reduction relation*, and $\xleftrightarrow{*}_{\mathcal{R}}$ is called *congruence relation*, induced by $\mathcal{R}$.

Relation $\to_{\mathcal{R}}$ is called *noetherian* if there is no infinite sequence $s_1 \to_{\mathcal{R}} s_2 \to_{\mathcal{R}} \cdots$ of single-step reductions. System $\mathcal{R}$ is *noetherian* if relation $\to_{\mathcal{R}}$ is noetherian, and $\mathcal{R}$ is *Church-Rosser* if for all $s, t \in G^*, s \xleftrightarrow{*}_{\mathcal{R}} t$ implies $s \xrightarrow{*}_{\mathcal{R}} u$ and $u \xleftarrow{*}_{\mathcal{R}} t$ for some $u \in G^*$. A noetherian Church-Rosser system $\mathcal{R}$ is called *canonical*.

An irreflexive antisymmetric transitive binary relation on G^*, denoted $>$ in infix notation, is called a *strict partial ordering*. Strict partial ordering $>$ is
- *well-founded* if there is no infinite decreasing sequence $s_1 > s_2 > \ldots$ in G^* ;
- a *reduction ordering* if $>$ is well-founded and if $s > t$ implies $u \cdot s \cdot v > u \cdot t \cdot v$ for all $u, v \in G^*$;
- *compatible* with a rewriting system $\mathcal{R}$ if $l > r$ for each rule $l \to r$ in $\mathcal{R}$.

Length-lexicographical ordering $>_{ll}$ is an example of a well-founded ordering. Let $g_1 > \ldots > g_n$ be a precedence on letters. Then for all strings s and t, $s >_{ll} t$ iff either $length(s) > length(t)$, or $length(s) = length(t)$ and there are strings $u, v, w \in G^*$ such that $s = u \cdot g_i \cdot v$ and $t = u \cdot g_j \cdot w$ for $g_i > g_j$.

Let $R_1 : u \to v$, $R_2 : l \to r$ be two rewriting rules in $\mathcal{R}$. A pair $(v \cdot s, p \cdot r)$ such that $u = p \cdot c$, $l = c \cdot s$ for some $c, p, s \in G^*$, is called *critical pair* between R_1 and R_2[3]. Critical pair $(v \cdot s, p \cdot r)$ is said to be *joinable*, if there is a $w \in G^*$ such that $v \cdot s \xrightarrow{*}_{\mathcal{R}} w$ and $w \xleftarrow{*}_{\mathcal{R}} p \cdot r$.

A completion procedure for a rewriting system $\mathcal{R}$ tries to construct a canonical rewriting system $\mathcal{R}'$, such that the congruence relations $\xleftrightarrow{*}_{\mathcal{R}}$ and $\xleftrightarrow{*}_{\mathcal{R}'}$ are the same. The standard Knuth-Bendix procedure [KB70] takes a reduction ordering $>$, compatible with rewriting system $\mathcal{R}$, and adds to the system all non joinable critical pairs oriented into rules with respect to ordering $>$. If the completion procedure terminates with success, then the resulting rewriting system is canonical.

Suppose that $G = G^+ \cup G^-$, where $G^+ = \{g_1, \ldots, g_n\}$ and $G^- = \{g_1^{-1}, \ldots g_n^{-1}\}$. Consider function $^{-1} : G^* \to G^*$, called *inverse function*, defined as follows:
- For any g_i in G^+, $(g_i)^{-1} = g_i^{-1}, 1 \le i \le n$.
- For any g_i^{-1} in G^-, $(g_i^{-1})^{-1} = g_i, 1 \le i \le n$.
- For the empty string, $\varepsilon^{-1} = \varepsilon$.
- For all strings s, t in G^*, $(s \cdot t)^{-1} = (t)^{-1} \cdot (s)^{-1}$.

Consider canonical rewriting system $\mathcal{A}$ of *group axioms*.

$$
\mathcal{A} = \begin{cases}
(x \cdot y) \cdot z \to x \cdot (y \cdot z) \\
x \cdot \varepsilon \to x, \ \varepsilon \cdot x \to x \\
x \cdot x^{-1} \to \varepsilon, \ x^{-1} \cdot x \to \varepsilon \\
(x \cdot y)^{-1} \to y^{-1} \cdot x^{-1} \\
x \cdot (x^{-1} \cdot y) \to y, \ x^{-1} \cdot (x \cdot y) \to y \\
(x^{-1})^{-1} \to x, \ \varepsilon^{-1} \to \varepsilon
\end{cases}
$$

Rewriting system $\mathcal{A}$ is a *term rewriting system*, where binary function $\cdot$ and unary function $^{-1}$ are arbitrary, and x, y and z are variables. Let us take as $\cdot$ binary concatenation operation on strings, and as $^{-1}$ inverse function defined above. Let $\mathcal{R}$ be a string rewriting system. Quotient $\mathcal{G}$ of monoid G^* by congruence relation $\xleftrightarrow{*}_{\{\mathcal{R},\mathcal{A}\}}$ induced by rewriting system $\{\mathcal{R}, \mathcal{A}\}$, is called a *group*. Elements of set G^+ are said to be *generators* of group $\mathcal{G}$, and the elements of set G^- are said to be *inverses of generators*. Pair $(G^+, \mathcal{R})$ is called a *presentation* of group $\mathcal{G}$. Group $\mathcal{G}$ is *finitely presented* if both set G^+ and rewriting system $\mathcal{R}$ are finite. Since concatenation operation

---

[3] This notion differs from the standard notion of critical pair, given in [BO93]. We specially distinguish between the case of simplification by a rewriting rule and the case of critical pair between rewriting rules, which are both included into the critical pair case in the standard definition

· is associative, associativity axiom $(x \cdot y) \cdot z = x \cdot (y \cdot z)$ is built in string structure. Therefore we omit parenthesis in group axioms. A string s is said to be in $\mathcal{A}$-*canonical form* if s is $\mathcal{A}$-irreducible. If string s is in $\mathcal{A}$-canonical form and $s \neq \varepsilon$, then s contains no consecutive inverse generators and empty substring ε.

Let $(G^+, \mathcal{R})$ be a presentation of group $\mathcal{G}$. Rewriting relation $\rightarrow_\mathcal{G}$ such that $s \rightarrow_\mathcal{G} t$ whenever there is a rule $R : l \rightarrow r$ in $\mathcal{R}$ and there are strings u, v such that $s = u \cdot l \cdot v$, $u \cdot r \cdot v \xrightarrow{*}_\mathcal{A} t$ and t is in $\mathcal{A}$-canonical form, is called the *group reduction*. The completion procedure implementing group reduction (introduced in [LC86]), runs faster than standard completion procedure. Also, since string rewriting for groups avoids using associativity axiom, computing critical pairs between ground rules (i.e. rules without variables) and group axioms becomes easy.

Theorem 1 [LC86]. *The completion procedure associates to each rule* $R : a_1 \cdot a_2 \cdot \ldots \cdot a_l \rightarrow b_1 \cdot b_2 \cdot \ldots \cdot b_m$, $a_i, b_j \in G$ *following critical pairs with group axioms, called normal pairs:* $(a_1 \cdot \ldots \cdot a_{l-1}, b_1 \cdot \ldots \cdot b_m \cdot a_l^{-1})$, $(a_2 \cdot \ldots \cdot a_l, a_1^{-1} \cdot b_1 \cdot \ldots \cdot b_m)$ *and* $(a_l^{-1} \cdot \ldots \cdot a_1^{-1}, b_m^{-1} \cdot \ldots \cdot b_1^{-1})$.

Precompiling normal pairs of each rewriting rule R and adding them directly to the rewriting system increases the efficiency of the completion procedure. Let $>$ be a reduction ordering. Computing normal pairs of rewriting rule R, normal pairs of oriented normal pairs of R, normal pairs of oriented normal pairs of normal pairs of R etc., we get the set of symmetrized forms of rewriting rule R.

Definition 2. Let $R : (a_1 \cdot \ldots \cdot a_l, b_1 \cdot \ldots \cdot b_m)$ be a rewriting rule or an equation. Let $Perm(R)$ be the largest set of rules $l \rightarrow r$ such that $l > r$ and $l \cdot r^{-1}$ is a cyclic permutation of $a_1 \cdot \ldots \cdot a_l \cdot b_m^{-1} \cdot \ldots \cdot b_1^{-1}$ or of $b_1 \cdot \ldots \cdot b_m \cdot a_l^{-1} \cdot \ldots \cdot a_1^{-1}$. For each letter $g \in \{a_i, a_i^{-1}\}_{i=1}^l \cup \{b_j, b_j^{-1}\}_{j=1}^m$ (g is taken as many times as it occurs in R) consider set $Perm(g) = \{l \rightarrow r \in Perm(g)$ such that l begins by $g\}$. Denote by $R(g)$ the rule in $Perm(g)$ with shortest l. The set $\{R(g)\}_g$ is called the *set of symmetrized forms of* R (with respect to $>$) and is denoted by $Sym(R)$.

We remark that if rule $R' \in Sym(R)$ then $Sym(R') = Sym(R)$.

A *graph* is defined by a set of *vertices* V and a set of *edges* E. An edge e is associated with a pair of vertices (v, v'). Different edges can be associated with the same pair of vertices. We will write $e = (v, v')$, when it is not important, which edge we take among all the edges associated with (v, v'). An edge $e = (v, v')$ is said to be *adjacent to vertices* v and v'. An edge e is said to be *directed* if the pair (v, v') is ordered; in this case v is called *initial vertex*, and v' *final vertex* of e. We also say that e *goes from* v *to* v' for *standard direction*. The same edge e is said to be taken in *opposite direction* if e is considered as going from v' to v. Edge e in standard direction is denoted by e^+, and edge e in opposite direction is denoted by e^-. By default, notation e means e^+.

A *path* π (denoted $\pi = e_1^{\alpha_1}.e_2^{\alpha_2}. \ldots .e_k^{\alpha_k}$) is the concatenation of a finite sequence of edges $e_1^{\alpha_1}, e_2^{\alpha_2}, \ldots, e_k^{\alpha_k}$, such that $\alpha_i \in \{+, -\}$, and $e_1^{\alpha_1} = (v_1, v_2)$, $e_2^{\alpha_2} = (v_2, v_3), \ldots, e_k^{\alpha_k} = (v_k, v_{k+1})$. The *empty path*, denoted by ϵ, is the empty sequence of edges. Vertex v_1 is called *initial vertex* or *beginning* of π, and v_{k+1} is called *final vertex* or *end* of π. Edge e is said to be *adjacent to a path* π if e is adjacent to the beginning or to the end of π. We say that π *goes from* v_1 *to* v_{k+1} for *standard direction* and from v_{k+1} to v_1 in *opposite direction*. We denote by π^+ path π in standard direction, and by

π^- path π in opposite direction. By default, π stands for π^+. Note that the edges in π^+ and π^- are the same, but their order and their directions in π^+ and π^- are opposite.

Let path $\pi_1 = e_1. \ldots .e_k$ goes from vertex v_1 to vertex v_2, and path $\pi_2 = e_{k+1}. \ldots .e_l$ goes from vertex v_2 to vertex v_3. Then path $\pi_1.\pi_2 = e_1. \ldots .e_k.e_{k+1}. \ldots .e_l$, called *concatenation* of paths π_1 and π_2 goes from v_1 to v_3.

Let $\pi = e_1. \ldots .e_k$ and $e_1 = (v_1, v_2), \ldots, e_k = (v_k, v_{k+1})$. We say vertices $v_1, \ldots, v_{k+1}$ *belong to the path* π and denote it $v_1, \ldots, v_{k+1} \in \pi$. A path $\pi' = e_i.e_{i+1}. \ldots .e_j$ for some $1 \leq i \leq j \leq k$ is a *subpath* of π. We denote it by $\pi' \subseteq \pi$. The empty path is not considered as a subpath. Let $v_{k+1} = v_1$ and π is not empty. Then the set $[\pi] = \{\pi_1, \ldots, \pi_{2k}\}$ of all paths, such that the sequences of their edges are cyclic permutations of the edges of π^+ or of π^-, is called a *cycle*. A path π' is a *subpath of a cycle* $[\pi]$, if $\pi' \subseteq \pi_i$ for some $i \in \{1, \ldots, k\}$. It is denoted by $\pi' \subseteq C$. Thereafter we will denote a cycle C by a path in $[\pi]$, writing C: $e_1^{\alpha_1}. \ldots .e_k^{\alpha_k}$. A pair of paths (π_1, π_2) is called a *cut* of a cycle C, if C: $\pi_1^+.\pi_2^-$. If (π_1, π_2) is a cut of C, then we write $\pi_1 = C \backslash \pi_2$ and $\pi_2 = C \backslash \pi_1$.

3 Patch Graphs

Here we explain intuitively, how a finite string rewriting system $\mathcal{R}$ on G^* for a group can be represented by a graph and how it can be completed. Section 3.2 introduces a formal definition of a PATCH Graph, and Sect.3.3 gives its semantics.

3.1 Informal Description of Graph Representation of a String Rewriting System for a Group and of its Completion

Let $\mathcal{R}$ be a rewriting system on G^*. Consider strings which are lhs and rhs of rules of $\mathcal{R}$. To a letter g or $g^{-1} \in G$ we associate a directed edge e labelled by g. A string $s \in G^*$ is associated with the path, composed of edges, corresponding to generators and inverses of generators in s. The edges are oriented depending on their "sign". For example, we make correspond to string $g_1 \cdot g_2^{-1}$ path $\pi = e_1^+.e_2^-$, where $e_1 = (v_1, v_2), e_2 = (v_3, v_2)$, and e_1 is labelled by g_1, and e_2 is labelled by g_2. A rewriting rule $l \to r$ corresponds to a cycle C and a cut (π_1, π_2) of C, such that the π_1 corresponds to l and π_2 corresponds to r. So paths π_1 and π_2 have common beginning and common end. A rewriting system $\mathcal{R}$ is represented as the union of cycles corresponding to its rules.

For example, system $\mathcal{R} = \{R_1 : a \cdot b \cdot c \to d, R_2 : b \to e\}$, is represented in Fig.1. Cycle C_1 represents rule R_1 and cycle C_2 represents rule R_2. The reason for storing rewriting rules as cycles is that one cycle can represent the set of all symmetrized forms of a rule, sharing edges. Let $>_{ll}$ be length-lexicographical ordering based on precedence $a > a^{-1} > b > b^{-1} > c > c^{-1} > d > d^{-1} > e > e^{-1}$. The set of all symmetrized forms of R_1 (with respect to $>_{ll}$) is $\{R_{(1,1)} : a \cdot b \to d \cdot c^{-1}, R_{(1,2)} : a^{-1} \cdot d \to b \cdot c, R_{(1,3)} : b^{-1} \cdot a^{-1} \to c \cdot d^{-1}, R_{(1,4)} : c^{-1} \cdot b^{-1} \to d^{-1} \cdot a\}$. Putting on C_1 vertices the following markers: $\{(\mathbf{B}, 1, 1, e_2, e_1), (\mathbf{E}, 1, 3, e_2, e_1)\}$ on v_1, $\{(\mathbf{B}, 1, 2, e_2, e_3), (\mathbf{E}, 1, 4, e_3, e_2)\}$ on v_2, $\{(\mathbf{B}, 1, 3, e_3, e_4), (\mathbf{E}, 1, 1, e_3, e_4)\}$ on v_3 and $\{(\mathbf{B}, 1, 4, e_4, e_1), (\mathbf{E}, 1, 2, e_1, e_4)\}$ on v_4 (see Fig.2), we represent by C_1 all four symetrized forms $R_{(1,i)}$. Take for example marker $(\mathbf{B}, 1, 2, e_2, e_3)$ of v_2. Number 1 in the label stands for the rule number one R_1 in system $\mathcal{R}$. Number 2 denotes symetrized form number 2 of R_1, i.e. rule $R_{(1,2)}$. $\mathbf{B}$ It indicates the beginning of paths $e_2^-.e_1$ and

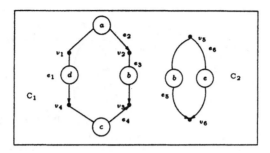

Fig. 1. Graph representation of $\mathcal{R} = \{R_1 : a \cdot b \cdot c \to d,\ R_2 : b \to e\}$

$e_3.e_4$ representing $a^{-1} \cdot d$ and $b \cdot c$ respectively. Edge e_2 is first in the path corresponding to the lhs and e_3 is first in the path corresponding to the rhs of $R_{(1,2)}$ (indicating first edges of lhs and rhs is sufficient to store the orientation). In a similar way, marker $(E, 1, 2, e_1, e_4)$ of v_4 stands for the end of paths $e_2^-.e_1$ and $e_3.e_4$, representing lhs and rhs of rule $R_{(1,2)}$ respectively. Edge e_1 is last in path $e_2^-.e_1$, and edge e_4 is last in path $e_3.e_4$. Taking separately each pair of markers $(B, 1, i, *, *), (E, 1, i, *, *)^4$, $1 \leq i \leq 4$, we obtain a representation of $R_{(1,i)}$. Similarly, to represent the set of all symmetrized forms of R_2, which is $\{R_{(2,1)} : b \to e, R_{(2,2)} : b^{-1} \to e^{-1}\}$, we put markers $\{(B, 2, 1, e_5, e_6), (E, 2, 2, e_5, e_6)\}$ on v_5, and $\{(B, 2, 2, e_5, e_6), (E, 2, 1, e_5, e_6)\}$ on v_6. Precompiling symmetrized forms and putting corresponding markers, we perform the *symmetrization procedure*, i.e. the completion procedure, such that only critical pairs with group axioms are computed. To compute other critical pairs and simplifications, we find common paths between cycles.

Definition 3. Let C_1, C_2 be cycles. Every pair π_1, π_2 of non-empty paths, such that $\pi_1 \subseteq C_1, \pi_2 \subseteq C_2$, and there is a string s, such that π_1 and π_2 both represent s, are called *common paths* for C_1 and C_2.

Adding new edges e_7 and e_8, labelled ε, going from v_2 to v_5 and from v_6 to v_3 respectively, we obtain new cycle $C_3 : e_2^+.e_7^+.e_6^+.e_8^+.e_4^+.e_1^-$ (in bold in Fig.2). Cycle C_3 represents the equation $a \cdot e \cdot c \cdot d^{-1} = \varepsilon$ (corresponding to the cut $(e_2^+.e_7^+.e_6^+.e_8^+.e_4^+.e_1^-, \epsilon)$. The set of all symmetrized forms of this equation with respect to $>_{ll}$ is $\{R_{(3,1)} : a \cdot e \to d \cdot c^{-1}, R_{(3,2)} : a^{-1} \cdot d \to e \cdot c, R_{(3,3)} : c \cdot d^{-1} \to e^{-1} \cdot a^{-1}, R_{(3,4)} : c^{-1} \cdot e^{-1} \to d^{-1} \cdot a\}$. We add following markers: $\{(B, 3, 1, e_2, e_1), (E, 3, 3, e_1, e_2)\}$ to v_1, $\{(B, 3, 2, e_2, e_7)\}$ to v_2, $\{(E, 3, 4, e_6, e_7)\}$ to v_5, $\{(E, 3, 1, e_6, e_8)\}$ to v_6, $\{(B, 3, 3, e_4, e_8)\}$ to v_3 and $\{(B, 3, 4, e_4, e_1), (E, 3, 2, e_1, e_4)\}$ to v_4.

Since edge e_5 represents lhs of both rules $R_{(2,1)}$ and $R_{(2,2)}$, rules represented by C_1 having e_3 as a subpath of path corresponding to lhs or to rhs, are simplified either by $R_{(2,1)}$, or by $R_{(2,2)}$. These are rules $R_{(1,1)}, \ldots, R_{(1,4)}$. We can delete them from the system. We remove markers corresponding to $R_{(1,1)}, \ldots, R_{(1,4)}$, but we keep all edges, because edges e_1, e_2, e_4 are shared by C_1 and C_3. The resulting graph, presented in Fig.2, corresponds to canonical rewriting system

---

[4] Here and below we denote by asterisks non-significant components

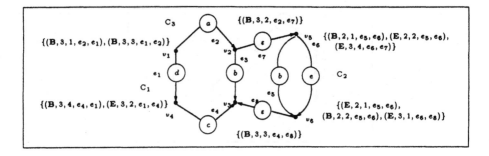

Fig. 2. Graph representation of the canonical rewriting system

$$\left\{ \begin{array}{c} R_{(2,1)}: \ b \to e, R_{(2,2)}: \ b^{-1} \to e^{-1}, \\ R_{(3,1)}: \ a \cdot e \to d \cdot c^{-1}, R_{(3,2)}: \ a^{-1} \cdot d \to e \cdot c, \\ R_{(3,3)}: \ c \cdot d^{-1} \to e^{-1} \cdot a^{-1}, R_{(3,4)}: \ c^{-1} \cdot e^{-1} \to d^{-1} \cdot a \end{array} \right\}.$$

3.2 Syntax

Before giving the formal definition of a PATCH Graph, let us introduce the notion of conjugate common paths.

Definition 4. Let π_1 and π_2 be paths. They are called *conjugate common paths* if there exist paths π_3, π_5 (possibly empty) and π_4 such that $\pi_1^+ = \pi_3^+ . \pi_4^+ . \pi_5^-$, π_2^+ and π_4^+ are common paths, and paths π_3^+ and π_5^+ represent the same string.

We need the notion of *conjugate* common paths, because the notion of common paths is not sufficient to define a PATCH Graph. We introduce in Sect. 4 inference rules **Patch**, **Symmetrize** and **Eliminate** for completion of rewriting systems in PATCH Graph representation. The reader will see that **Patch** and **Symmetrize** transform a PATCH Graph into another PATCH Graph, even if common paths are used instead of conjugate common paths, but **Eliminate** does not.

Definition 5 PATCH Graph. Let $\Gamma = (V, E, \lambda, \eta)$ be a finite labelled graph, where V is a set of vertices, E is a set of edges, λ is the labelling function on edges, and η is the labelling function on vertices. Γ is called a *PATCH Graph relative to a set of generators* G^+ iff

- $\Gamma = \bigcup_{n=1}^{m} C_n$, where for each $n \in \{1, \ldots, m\}$ C_n is a cycle number n in Γ.
  C_n is a *cycle number* n in Γ iff it is a minimal cycle, such that either
  - for any edge e of C_n, $\lambda(e) = (g_i, n)$ for some $g_i \in G^+$ and $n \in \mathbf{N}$. Such a cycle is called an *initial cycle*; or
  - there exist edges e_1 and e_2 in C_n,
    there exist cycles C_{n_1} and C_{n_2}, and
    there are cuts (π_1, π_3) of C_{n_1} and (π_2, π_4) of C_{n_2}, such that:

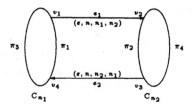

Fig. 3. Cycle in a PATCH Graph

* paths π_1 and π_2 are conjugate common paths, and
* $\lambda(e_1) = (\varepsilon, n, n_1, n_2)$ and $\lambda(e_2) = (\varepsilon, n, n_2, n_1)$,
  $e_1 = (v_1, v_2)$ and $e_2 = (v_3, v_4)$, where
  v_1 and v_4 are respectively initial and final vertices of π_1 and π_3, and
  v_2 and v_3 are respectively initial and final vertices of π_2 and π_4 (Figure 3).
In this case cycle number n is $C_n : e_1^+.\pi_4^+.e_2^+.\pi_3^-$. Pair (e_1, e_2) is called a *bridge* between cycles C_{n_1} and C_{n_2}, relative to conjugate common paths π_1 and π_2. Cycles C_{n_1} and C_{n_2} are said to be *ancestors* of cycle C_n. We also say that bridge (e_1, e_2) *defines* cycle C_n.

- For any vertex v label $\eta(v)$ is a (possibly empty) set of *beginning* and *end markers*, such that if $\eta(v)$ contains a marker $(\mathbf{B}, n, m, e_3, e_4)$ or $(\mathbf{E}, n, m, e_3, e_4)$ for some $m \in \mathbf{N}$, then $v \in C_n$ and edges $e_3, e_4 \in C_n$ are edges adjacent to v. For all n, there is exactly one $\mathbf{B}$-marker and one $\mathbf{E}$-marker with the same m.

In this notation, edge labels of the graph given by Figure 2 are: $\lambda(e_1) = (d, 1)$, $\lambda(e_2) = (a, 1)$, $\lambda(e_3) = (b, 1)$, $\lambda(e_4) = (c, 1)$, $\lambda(e_5) = (b, 2)$, $\lambda(e_6) = (e, 2)$, $\lambda(e_7) = (\varepsilon, 3, 1, 2)$ and $\lambda(e_8) = (\varepsilon, 3, 2, 1)$.

Constructing an initial PATCH graph Γ, representing rewriting system $\mathcal{R}$ we have to complete, we first cyclically reduce rewriting rules of $\mathcal{R}$ and then create for each reduced rule an initial cycle. Each new cycle receives a fresh number n[5], as in our example in Fig. 1.

3.3 Semantics

Now we give the semantics of PATCH Graphs. Function Δ defines inductively what a PATCH Graph, relative to a set of generators G^+, represents.

Definition 6. Let $\Gamma = (V, E, \lambda, \eta)$ be a PATCH Graph relative to a set G^+. Function Δ is the *semantic function* of Γ such that

1. For any edge $e \in E$ such that $\lambda(e) = (g, n)$ for some $g \in G^+$ and $n \in \mathbf{N}$, $\Delta(e^+) = g$, and $\Delta(e^-) = g^{-1}$.
2. For any edge $e \in E$ such that $\lambda(e) = (\varepsilon, n, n_1, n_2)$ for some $n, n_1, n_2 \in \mathbf{N}$, $\Delta(e^+) = \Delta(e^-) = \varepsilon$.

---

[5] $n = N + 1$, where N is the number of cycles in current graph

3. For empty path ϵ $\Delta(\epsilon) = \varepsilon$.
4. For each non-empty path π and an edge e
   - $\Delta(\pi.e) = \Delta(\pi)$ if $\Delta(e) = \varepsilon$ and $\Delta(\pi.e) = \Delta(\pi) \cdot \Delta(e)$ else,
   - $\Delta(e.\pi) = \Delta(\pi)$ if $\Delta(e) = \varepsilon$ and $\Delta(e.\pi) = \Delta(e) \cdot \Delta(\pi)$ else.
5. For each cycle $C_n : \pi$ in Γ, such that for any vertex $v \in C_n$, neither of markers $(\mathbf{B}, n, *, *, *)$ and $(\mathbf{E}, n, *, *, *)$ belongs to label $\eta(v)$, $\Delta(C_n)$ is the equation $\Delta(\pi) = \varepsilon$.
6. For each cycle C_n in Γ, such that there is a vertex $v \in C_n$ such that label $\eta(v)$ contain a marker $(\mathbf{B}, n, *, *, *)$ or $(\mathbf{E}, n, *, *, *)$, let (π_1, π_2) be a cut of C_n such that for some i
   - label $\eta(v_1)$ of initial vertex v_1 of paths π_1 and π_2 contains beginning marker $(\mathbf{B}, n, i, e_1, e_2)$, where e_1 (e_2) is the first edge of π_1 (resp. π_2), and
   - label $\eta(v_2)$ of final vertex v_2 of paths π_1 and π_2 contains end marker $(\mathbf{E}, n, i, e_3, e_4)$, where e_3 (e_4) is the last edge of π_1 (resp. π_2).

   Then $\Delta(C_n, (\pi_1, \pi_2))$ is the rule $R_{(n,i)} : \Delta(\pi_1) \rightarrow \Delta(\pi_2)$.
7. For each cycle $C_n \in \Gamma$, and set $\{(\pi_1, \pi_1'), \ldots, (\pi_k, \pi_k')\}$ of all cuts of C_n, such that for any $i \in \{1, \ldots, k\}$ cut (π_i, π_i') satisfies the conditions of item 6, $\Delta(C_n, \{(\pi_1, \pi_1'), \ldots, (\pi_k, \pi_k')\}) = \{R_{(n,i)}\}_{i=1}^{k}$.
8. Let cycles $C_1, \ldots, C_m$ are all cycles in graph Γ and for each $n \in \{1, \ldots, m\}$ set $\{(\pi_{k_1}, \pi_{k_1}'), \ldots, (\pi_{k_n}, \pi_{k_n}')\}$ of cuts of cycle C_n satisfies the conditions of item 7. Then $\Delta(\Gamma) = \{\Delta(C_n, \{(\pi_{k_1}, \pi_{k_1}'), \ldots, (\pi_{k_n}, \pi_{k_n}')\})\}_{n=1}^{m}$.

4 Inference Rules

In this section we give the inference rules of a new completion procedure for string rewriting systems for groups represented by PATCH Graphs, and state their soundness and completeness. We have three inference rules: **Patch, Symmetrize** and **Eliminate**. All of them are formulated as transformations of cycles in a PATCH Graph. **Patch** is the central rule, giving the law of creation of new cycles. It generalizes the critical pair and the simplification computations. **Symmetrize** expresses the symmetrization procedure, relative to a reduction ordering $>$ on G^*. It is applied to each new cycle, created by a **Patch** inference application. **Eliminate** defines the transformation of cycles encoding cancellation and simplification with respect to group axioms $x \cdot x^{-1} \rightarrow \varepsilon$ and $x^{-1} \cdot x \rightarrow \varepsilon$. **Eliminate** is applied to some of new cycles, created by a **Patch** inference rule application.

4.1 Patch Inference Rule

Figure 4 shows cycles C_i and C_j having maximal non-trivial common paths $\pi_1 \subseteq C_i$ and $\pi_2 \subseteq C_j$. Let us denote by π_3 path $C_i \backslash \pi_1$, and by π_4 path $C_j \backslash \pi_2$.

Definition 7. Let $\pi_1 \subseteq C_i$ and $\pi_2 \subseteq C_j$ be common paths. They are said to be *maximal common paths*, if there are no common paths $\pi_3 \subseteq C_i$ and $\pi_4 \subseteq C_j$ such that:
- $\pi_3 = \pi_5.\pi_1.\pi_6$ and $\pi_4 = \pi_7.\pi_2.\pi_8$,
- $\Delta(\pi_5) = \Delta(\pi_7)$ and $\Delta(\pi_6) = \Delta(\pi_8)$,
- π_5 or π_6 (and, consequently, π_7 or π_8) is not empty.

Paths π_1 and π_2 are called *trivial* if $C_i = C_j$ and $\pi_1 = \pi_2$.

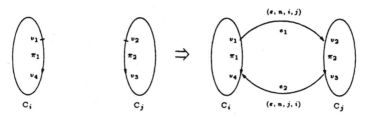

Fig. 4. Patch inference rule

If one of the following conditions is satisfied:

1. $\Delta(\pi_1) > \Delta(\pi_3)$ and $\Delta(C_i, (\pi_1, \pi_3))$ is a rule $R_{(i,l)}$: $\Delta(\pi_1) \to \Delta(\pi_3)$, and $\Delta(\pi_2)$ is a substring of the lhs or the rhs of a rule $R_{(j,m)}$,
2. $\Delta(\pi_1) \not> \Delta(\pi_3)$ or there are no rules $R_{(i,l)}$ and $R_{(j,m)}$ as in case 1, but there exist a rule $R_{(i,p)}$, and a rule $R_{(j,q)}$, such that $\Delta(\pi_1)$ is a suffix of the lhs of $R_{(i,p)}$ and $\Delta(\pi_2)$ is a prefix of the lhs of $R_{(j,q)}$,

then inference **Patch** creates a new cycle C_n, adding a new bridge (e_3, e_4), where $e_3 = (v_1, v_2), e_4 = (v_3, v_4), \lambda(e_1) = (\varepsilon, n, i, j), \lambda(e_2) = (\varepsilon, n, j, i)$. Cycles C_i and C_j are ancestors of C_n. If first condition is satisfied, beginning and end markers defining all the rules $R_{(j,m)}$, such that $\Delta(\pi_2)$ is a substring of the lhs or the rhs of $R_{(j,m)}$, are removed. **Patch** inference rule is applied only once to any pair of maximal common paths for C_i and C_j. We remark that it is easy to verify the conditions of **Patch** inference rule application, analyzing beginning and end markers of C_i and C_j. Actually, first condition is satisfied when $(\mathbf{B}, i, l, e_5, *) \in \eta(v_1)$ and $(\mathbf{E}, i, l, e_6, *) \in \eta(v_4)$, where e_5 and e_6 are respectively first and last edges of π_1, and beginning and end markers $(\mathbf{B}, j, m, *, *)$ and $(\mathbf{E}, j, m, *, *)$, defining $R_{(j,m)}$, are labels of vertices belonging to π_4. Second condition is satisfied when $(\mathbf{E}, i, l, e_6, *) \in \eta(v_4)$ and $(\mathbf{B}, i, l, *, *)$ appears in a label of a vertex belonging to π_3, and $(\mathbf{B}, j, m, e_7, *) \in \eta(v_2)$ (where e_7 is first edge of π_2) and $(\mathbf{E}, j, m, *, *)$ appears in a label

  Patch inference rule computes a critical pair or a simplification for each pair of rules represented by cycles C_i and C_j respectively. Actually, first case corresponds to a simplification of $R_{(j,m)}$ by $R_{(i,l)}$, and second case corresponds to a critical pair between $R_{(i,p)}$ and $R_{(j,q)}$.

  Beginning and end markers along cycle C_j, defining rules $R_{(j,m)}$ which can be simplified, are removed. We can not remove paths representing simplified strings, because these paths are shared between C_j and C_n. If symmetric conditions hold for C_i, then by symmetric transformation, beginning and end markers along cycle C_i, defining rules $R_{(i,l)}$, which can be simplified, are removed. We remark that when $\Delta(\pi_1) > \Delta(\pi_3)$ and $\Delta(\pi_2) > \Delta(\pi_4)$ (actually this case corresponds to a critical pair on the top), we never r emove beginning and end markers in both C_i and C_j, but only the markers of C_i, when $\Delta(\pi_3) > \Delta(\pi_4)$, or only the markers of C_j, when $\Delta(\pi_4) > \Delta(\pi_3)$. Then we keep smaller rewriting rules in the system represented by a PATCH graph.

  Since **Patch** is performed only for maximal non-trivial common paths, some cycles, representing redundant equations, are not produced.

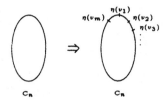

Fig. 5. Symmetrize inference rule

Lemma 8. *Let $\pi_1, \pi_3 \subseteq C_i$ and $\pi_2, \pi_4 \subseteq C_j$. Let π_1, π_2 and π_3, π_4 be common paths such that $\pi_3 = \pi_5 . \pi_1 . \pi_6$, $\pi_4 = \pi_7 . \pi_2 . \pi_8$, and $\Delta(\pi_5) = \Delta(\pi_7)$, $\Delta(\pi_6) = \Delta(\pi_8)$. Let π_5 or π_6 be non-empty. Denote by C_k (resp. C_l) the new cycle, generated applying **Patch** inference rule to common paths π_1 and π_2 (resp. to common paths π_3 and π_4). Then for any cut (π_9, π_{10}) of cycle C_k, there is some cut (π_{11}, π_{12}) of cycle C_l, such that rewriting rule $R_2 = \Delta(C_l, (\pi_{11}, \pi_{12}))$ belongs to set $Sym(\Delta(C_l))$, and rule $R_1 = \Delta(C_k, (\pi_9, \pi_{10}))$ can be simplified to a trivial equation by rewriting rule R_2 and group axioms $\mathcal{A}$.*

Lemma 9. *Inference **Patch**, performed for trivial common paths π_1 and π_2, generates a new cycle C such that for every cut (π_3, π_4) of C, rewriting rule $R = \Delta(C, (\pi_3, \pi_4))$ can be simplified by group axioms to a trivial equation.*

4.2 Symmetrize Inference Rule

This inference rule encodes the symmetrization procedure for rewriting rules represented by a cycle of a PATCH Graph. Translated in terms of PATCH Graphs, the definition of symmetrized forms looks like follows.

Definition 10. Let C be a cycle in a PATCH Graph. Let $\{(\pi_1, \pi_1'), \ldots, (\pi_k, \pi_k')\}$ be the largest set of cuts of C such that for all $i, j \in \{1, \ldots, k\}, i \neq j$, the conditions $\pi_i \not\subseteq \pi_j$, $\Delta(\pi_i) > \Delta(\pi_i')$ and $\Delta(\pi_j) > \Delta(\pi_j')$ hold. The set of rules $\Delta(C, \{(\pi_1, \pi_1'), \ldots, (\pi_k, \pi_k')\})$ is called the *set of symmetrized forms* of $\Delta(C)$.

For each new cycle C_n, we put along C_n beginning and end markers $(\mathbf{B}, n, j, *, *)$, $(\mathbf{E}, n, k, *, *)$, corresponding to the set of all symmetrized forms of $\Delta(C_n)$.

4.3 Eliminate Inference Rule

Inference rule **Eliminate** encodes the cancellation and the simplification by group axioms $x \cdot x^{-1} \to \varepsilon$ and $x^{-1} \cdot x \to \varepsilon$.

Let C_n be a non-initial cycle, and let (e_1', e_2') be the bridge between cycles C_i and C_j, defining cycle C_n. Let $\lambda(e_1') = (\varepsilon, n, i, j)$ and $\lambda(e_2') = (\varepsilon, n, j, i)$. Let $\pi = e_1^{\alpha_1} . e_2^{\alpha_2} . \ldots . e_{k-1}^{\alpha_{k-1}} . e_k^{\alpha_k}$, $\alpha_i \in \{+, -\}, 1 \leq i \leq k$, be a subpath of C_n satisfying the following properties:

(1) e_1 and e_k are labelled by the same $g \in G^+$, and $e_2, \ldots, e_{k-1}$ by ε,
(2) $\alpha_k = -(\alpha_1)$, i.e. either $\alpha_1 = +$ and $\alpha_k = -$, or $\alpha_1 = -$ and $\alpha_k = +$.

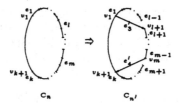

Fig. 6. Eliminate inference rule

We perform **Eliminate** on each new cycle C_n as it is produced by application of **Patch** inference to cycles C_i and C_j. Then we suppose that cycles-ancestors of C_n have no subpath satisfying properties (1) and (2). It is easy to see that since **Patch** inference is applied only to maximal common paths, both of edges e'_1 and e'_2 belong to π. Let l and m be numbers such that $1 < l < m < k$ and $e'_1 = e_l, e'_2 = e_m$. Inference rule **Eliminate** consists in removing cycle C_n from the graph, and adding a new cycle $C_{n'}$, where n' is a fresh number. Precisely, we delete bridge (e'_1, e'_2) between cycles C_i and C_j, defining C_n, and add a new bridge (e'_3, e'_4) between C_i and C_j, defining $C_{n'}$. Let $e_i^{\alpha_i} = (v_i, v_{i+1}), 1 \le i \le k$. Then $e'_3 = (v_1, v_{l+1})$. Initial vertex of edge e'_3 is the same as initial (when $\alpha_1 = +$) or final (when $\alpha_1 = -$) vertex of edge e_1, and final vertex of edge e'_3 is the same as final vertex of edge e'_1. Edge $e'_4 = (v_m, v_{k+1})$ goes from vertex v_m, which is initial vertex of edge e'_2, to vertex v_{k+1}, which is initial (when $\alpha_1 = +$) or final (when $\alpha_1 = -$) vertex of edge e_k. Labels of new edges are: $\lambda(e'_3) = (\varepsilon, n', i, j)$ and $\lambda(e'_4) = (\varepsilon, n', j, i)$. New cycle is $C_{n'} : (C_n \backslash \pi)^+ . (e'_4)^- . e_{l+1}^- . \ldots . e_{m-1}^- . (e'_3)^-$.

We remark that if $\pi_1 \subseteq C_i$ and $\pi_2 \subseteq C_j$ are common paths, relative to bridge (e'_1, e'_2), then new bridge (e'_3, e'_4) is relative to common paths $e_1^{\alpha_1} . \ldots . e_{l-1}^{\alpha_{l-1}} . \pi_1 . e_{m+1}^{\alpha_{m+1}} \ldots . e_k^{\alpha_k} \subseteq C_i$ and π_2, conjugate by string g (or g^{-1}, according to the values of α_1 and α_k). Figure 6 illustrates inference rule **Eliminate** when $\alpha_1 = +$ and $\alpha_k = -$. Case $\alpha_1 = -$ and $\alpha_k = +$ is similar.

4.4 Completeness and Soundness of the Inference System

Our inference rules preserve PATCH Graph structure. In other words, each of inference rules **Patch**, **Symmetrize** and **Eliminate**, applied to a PATCH Graph relative to a set of generators G, produces another PATCH Graph relative to the set G.

Theorem 11. *Inference rules* **Patch**, **Symmetrize** *and* **Eliminate** *for finitely presented group completion are sound, i.e. they do not change the equational theory defined by rewriting rules and equations represented by a PATCH Graph Γ and group axioms.*

Let $\mathcal{R}$ be a string rewriting system. Let $>$ be a reduction ordering, compatible with system $\{\mathcal{R}, \mathcal{A}\}$. Let Γ be a PATCH Graph representing $\mathcal{R}$. The *Completion Procedure of PATCH Graph Γ* transforms graph Γ, applying as much as possible inference rules **Patch**, **Eliminate** and **Symmetrize**. New cycle are generated by **Patch** inference, possibly reduced by **Eliminate** inference, and their vertices are labelled by **Symmetrize** inference. Let Γ_0 be initial graph Γ, and for any natural j, graph $\Gamma_{j+1} = Inf(\Gamma_j)$ be the graph generated by application of an inference rule $Inf \in \{\textbf{Patch}, \textbf{Eliminate}, \textbf{Symmetrize}\}$ to graph Γ_j. Graph $\Gamma^\infty = \bigcup_{i \geq 0} \bigcap_{j > i} \Gamma_j$ is called *persisting graph*. Graph Γ^∞ is saturated with respect to inference rules **Patch**, **Eliminate** and **Symmetrize**, i.e. no inferences can be applied to Γ^∞. Denote by $\mathcal{R}^\infty$ rewriting system $\Delta(\Gamma^\infty)$, represented by persisting graph.

Theorem 12. *The system of inference rules* **Patch**, **Symmetrize** *and* **Eliminate** *for finitely presented group completion is complete, i.e. rewriting system* $\{\mathcal{R}^\infty, \mathcal{A}\}$ *is canonical.*

5 Concluding Remarks

We have developed PATCH Graphs, a new data structure, representing rewriting systems for groups in a compact way. All the symmetrized forms of a rule or an equation are represented by a cycle in a PATCH Graph. Then each transformation performed on a cycle corresponds to a transformation of a class of rewriting rules by the standard completion procedure or the group completion procedure by Le Chenadec [LC86].

We have formulated inference rules **Symmetrize**, **Patch** and **Eliminate** for completion of finitely presented groups using PATCH Graphs. The inference rules can be efficiently performed on cycles of a PATCH Graph. **Symmetrize** inference rule gives the symmetrization procedure for groups. **Eliminate** inference rule expresses the simplification by group axioms and the cancellation of ground rules. **Patch** inference rule generalizes the critical pair and the simplification computations. Actually, inference rule **Patch** translates **Deduce**, **Simplify**, **Compose** and **Collapse** inference rules of the standard completion procedure [DJ90] into the formalism of PATCH Graphs. Each application of **Patch** inference to cycles C_1 and C_2 in a PATCH Graph corresponds to several applications of **Deduce**, **Simplify**, **Compose** and **Collapse** inference rules to symmetrized forms of rewriting rules represented by cycles C_1 and C_2.

Our inference rules avoid generating cycles giving some redundant rewriting rules. A cycle C_3 created by **Patch** inference rule for any pair of maximal non trivial common paths π_1 and π_2 represent rewriting rules which simplify equations or rewriting rules, defined by a cycle C_4, created by **Patch** inference rule for some non maximal common subpaths of paths π_1 and π_2. Every equation or rewriting rule, defined by a cycle C_5, generated by **Patch** inference rule applied to any pair of trivial common paths can be simplified to a trivial equation by group axioms. In the standard algorithm as well as in the group completion algorithm by Le Chenadec, equations and rewriting rules represented by such cycles C_4 and C_5 are introduced into the rewriting system, and the corresponding simplifications are done when they are discovered.

We believe that our work will be useful in solving open problems in the field of finitely presented groups. The simplicity, compactness, and avoidance of redundancies will help to focus the Knuth-Bendix completion procedure. Our PATCH Graph data structure complements other improvements of the Knuth-Bendix completion procedure, and we believe the combination of ideas is what will be most useful in solving the open problems.

Acknowledgements. We thank Dominique Fortin and Claude Kirchner for fruitful discussions on PATCH Graphs.

References

[BO93]   R. V. Book and F. Otto. *String-Rewriting Sytsems*. Springer-Verlag, 1993.

[Büc79]   Hans Bücken. Reduction-systems and small cancellation theory. In *Proceedings of 4th Workshop on Automated Deduction*, pages 53–59, 1979.

[DJ90]   N. Dershowitz and J.-P. Jouannaud. *Handbook of Theoretical Computer Science*, volume B, chapter 6: Rewrite Systems, pages 244–320. Elsevier Science Publishers B. V. (North-Holland), 1990. Also as: Research report 478, LRI.

[EHR91]   D. B. A. Epstein, D. F. Holt, and S. Rees. The use of knuth-bendix methods to solve the word problem in authomatic groups. *Journal of Symbolic Computation*, 12:397–414, 1991.

[FKS94]   D. Fortin, C. Kirchner, and P. Strogova. Routing in Regular Networks Using Rewriting. In J. Slaney, editor, *Proceedings of the CADE international workshop on automated reasoning in algebra (ARIA)*, pages 5–8, June 1994.

[Gil79]   R. H. Gilman. Presentations of groups and monoids. *J. of Algebra*, 57:544–554, 1979.

[HR89]   D. Holt and S. Rees. Testing for isomorphism between finitely presented groups. In *Proceedings of the Conference on Groups and Combinatorics*, Cambridge, 1989. Cambridge University Press.

[KB70]   Donald E. Knuth and P. B. Bendix. Simple word problems in universal algebras. In J. Leech, editor, *Computational Problems in Abstract Algebra*, pages 263–297. Pergamon Press, Oxford, 1970.

[KZ88]   D. Kapur and H. Zhang. RRL: A rewrite rule laboratory. In *Proceedings 9th International Conference on Automated Deduction, Argonne (Ill., USA)*, volume 310 of *Lecture Notes in Computer Science*, pages 768–769. Springer-Verlag, 1988.

[LC86]   Ph. Le Chenadec. *Canonical Forms in Finitely Presented Algebras*. John Wiley & Sons, 1986.

[LS95]   Christopher Lynch and Polina Strogova. Sour graphs for efficient completion. Technical Report 95-R-343, CRIN, 1995.

[Lyn95]   Christopher Lynch. Paramodulation without duplication. In Dexter Kozen, editor, *Proceedings of LICS'95*, San Diego, June 1995.

[Mar93]   C. Marché. *Réécriture modulo une théorie présentée par un système convergent et décidabilité du problème du mot dans certaines classes de théories équationnelles*. Thèse de Doctorat d'Université, Université de Paris-Sud, Orsay (France), October 1993.

[Mar96]   U. Martin. Theorem proving with group presentations: examples and questions. In *Proceedings of CADE-13*, 1996. To appear.

[McC94]   W. W. McCune. Otter 3.0: Reference manual and guide. Technical Report 6, Argonne National Laboratory, 1994.

[MNO91]   K. Madlener, P. Narendran, and F. Otto. A specialized completion procedure for monadic string-rewriting systems presenting groups. In *Proceedings 18th ICALP Conference, Madrid (Spain)*, volume 516 of *Lecture Notes in Computer Science*, pages 279–290. Springer-Verlag, 1991.

[Sim87]   C. C. Sims. Verifying nilpotence. *Journal of Symbolic Computation*, 3:231–247, 1987.

[Zha92]   H. Zhang. Herky: High-performance rewriting techniques in rrl. In D. Kapur, editor, *Proceedings of 1992 International Conference of Automated Deduction*, volume 607 of *Lecture Notes in Artificial Intelligence*, pages 696–700. Springer-Verlag, 1992.

Measuring the Likely Effectiveness of Strategies*

Brian J. Dupée **

School of Mathematical Sciences   University of Bath
Claverton Down   Bath BA2 7AY   United Kingdom

Abstract. Where we have measurable attributes and multiple possible strategies within an expert system there are situations where the theory of belief functions requires refinement and explanation. Given a set of attributes for which mapping exists to a topology of strategies, I will show how we refine Dempster-Shafer theory to interpret both combinations of strategies based on qualitative measures and combinations of possibly conflicting quantitative measures. This is then applied in an expert system for selecting appropriate numerical routines for the solution of a range of mathematical problems.

1 Introduction

In an expert system we are faced with evaluating potential strategies for obtaining solutions to a set of mathematical problems. This entails measurement of various attributes of each problem for use as *evidence*. We then use *belief functions* to give values to the effectiveness of each strategy.

Sometimes we use strategies which are themselves multiple strategies i.e. we wish to split the problem into a number of pieces and wish to give appropriate values to them for comparison with other multiple and singleton strategies. This process forces a structure on the underlying singleton values.

Other strategies may have attributes which conflict in some way. We therefore have to use functions describing the compatibility of each attribute *in combination* with each possible strategy.

Bayesian theory is restricted to singleton hypotheses and so cannot on its own be applied. A more respectable solution is applying Dempster-Shafer theory (Gordon & Shortliffe 1983) whereby we can assign measures to any number of subsets of the set of strategies. However, such a set of subsets does not normally include multiple copies of the same strategy.

Gordon & Shortliffe (1985), Pearl (1986) and Shafer & Logan (1987) show us how we can construct belief spaces for evidential reasoning using Dempster-Shafer theory on a hierarchy of hypotheses. This is used as a basis for considering hierarchies of the complete belief space and some of the effects this necessitates on the system.

---
* The project "More Intelligent Delivery of Numerical Analysis to a Wider Audience" is funded by the UK Govt. Joint Information Systems Committee under their New Technologies Initiative (NTI–24)
** Email: bjd@maths.bath.ac.uk

2   Multiple Strategies

Definition 1. Let Θ be a **Measurable Discrete Topological Space** of the set of subsets of a set of methods S. This is the **Dempster-Shafer frame of discernment**. So the set of hypotheses is Θ. It is not discounted that two or more members of S represent the same method.

Example 1. Let the set of methods $S = \{A, B, C\}$. Also, let A and B represent the *same* method and C represent *two copies* of method A. We can construct Θ as all subsets of S. This set Θ contains the subsets $\{A, B\}$ and $\{C\}$; each, ostensibly, representing the same strategy i.e. using method A twice. We will have to reconcile these and judge their respective values within our proposed model.

What we propose to do is to work theoretically on multiple levels i.e. treat singleton strategies differently to multiple strategies. To do this, we consider S as the set of methods *without copies*. An element of Θ which represents a multiple method strategy, instead of using the singleton elements within Θ, *creates a number of copies of Θ*. In the terminology of Shafer (1976), this is a *refinement* of Θ such that an element of Θ is recursively refined as being one or more copies of Θ.

The extension of the Dempster-Shafer frame of reference to include multiple topologies has an effect on the inference architecture, since we are now having to work on a number of different levels. This forces us to maintain levels of belief in Θ even though we can completely assign values to all singleton elements of Θ. The effect can be seen in an example.

Example 2. Let S be the set of strategies $\{M_1, M_2, \ldots M_n\}$ where M_1 is a strategy representing a number of any of the methods $M_2 \ldots M_n$. The *basic probability assignments (b.p.a.)* for the singleton methods of Θ are calculated in the usual way, but the *b.p.a.* of M_1 is calculated from one or more *copies* of Θ, say $\Theta_1, \ldots \Theta_m$, by combining the largest *b.p.a.s* of the $\Theta_1, \ldots \Theta_m$.

This therefore splits the original problem into two or more pieces and a *b.p.a.* obtained for each of these separate subproblems.

The effect of this is to cause difficulties with consistency of the system. This can be solved using a further normalisation. Now that we have the foundations of a hierarchical system, we can investigate some of the features and peculiarities the system provides.

Example 3. Let S be as above and M_2 represent a singleton method which splits the problem and uses a particular method on each sub-problem.

Furthermore, let the remaining $M_3 \ldots M_n$ be subdivided into two groups i.e. *specific* and *general*. A *specific* method is one that is specific to a single type of problem. A *general* method is one which can be addressed to many different types of problem (with, perhaps, a differing degree of reliability).

Let us consider the $b.p.a.$ for each of M_1 and M_2, i.e. we have a problem which can be solved using these two strategies. If the individual methods used both by M_1 and M_2 are *general* methods, the *cost* of using M_1 is likely to be higher so the system should give a higher $b.p.a.$ to M_2. However, if *one* or *both* of the methods selected for use by M_1 is *specific*, and *specific* methods are always preferred, a higher $b.p.a.$ should be allocated to M_1.

Using **Dempster's Rule of Combination**, this would only be possible if the $b.p.a.$ of *specific* methods (having the highest $b.p.a.$ in the sub-topologies above) is greater than 0.5 and all *general* methods have a best $b.p.a.$ of less than 0.5.

Since there can be no difference between the pattern of the topology Θ and the copies of Θ used to calculate the $b.p.a.$ of M_1, this structure must be applied throughout. Furthermore, the maximum $b.p.a.$ for M_2 must be finely judged as being slightly greater than the maximum $b.p.a.$ for the combination of the singleton *general* methods it represents.

Whilst I have used only a single multiple strategy extending the topologies, it is possible, using the same reasoning, to expand the example to have more than one strategy with this feature.

3   Conflicting Evidence

Where we have conflicting evidence, Dempster-Shafer theory can be applied to calculate the plausibility of each singleton method. In combination with possible multiple methods, the normalisation process is more complex since we must always maintain belief in Θ to allow for the extended topologies.

However, the implementation we have used deserves a little explanation. In keeping with the system recommended by Lucks & Gladwell (1992) we have introduced four types of functions:

measurement functions quantifying the degree of presence of features in an input elementary problem;

intensity functions conversion of the measurement of features onto a standard scale;

compatibility functions describing relationships between the degree of presence of features and the behaviour of the inner workings of the methods;

aggregation functions describing the overall behaviour of a method based on the aggregate effect of the individual problem features.

The compatibility functions are the input to our D-S model which is implemented within the aggregate functions. We can therefore economise on code by using dynamic table lookup for values obtained for the intensity functions. The behaviour of individual methods under the influence of various features is an area that takes as its basis the *judgement* of Numerical Analysis "experts" whether that be from documentation or alternative sources. However, its assessment of the suitability or otherwise of a particular method to a particular problem is reflected in a single normalised value facilitating the direct comparison of the suitability of a number of possible methods or strategies.

Example 4. Let $S = \{A, B, C, D, E\}$ be a set of strategies for solving a particular problem and let Ω be a set of attributes of the problem which gives us an idea of the evidence of effectiveness of each of the members of S. Let strategy A be such that it splits the problem in two and each of the separate sections of the problem solved using some member of S. This is to say that the frame of discernment Θ has a recursive refinement to A. Furthermore, let B be a strategy which implements strategy C twice, and D and E be favoured strategies i.e. if evidence shows that they can be used, they will be preferred over C.

Let us say that evidence points to either A or B being effective, but we have, at the moment no clues as to which is the better. The evidence for B can be calculated fairly easily, but the evidence for A is more difficult. So we split the problem into two and re-enter each of the sub-problems into the system. The attributes for each of the sub-problems will be different, and so will give different measures of belief in each of the strategies of S.

Now, if each of the recommendations of the sub-problems is to implement strategy C, this should not be preferable to using strategy B in the first place, so the *joint measure of belief* for A should be less than that for B i.e. for consistency, $\mathrm{Bel}(B) > 0.5 > \mathrm{Bel}(A)$.

If *one or both* of the sub-problems maintains that D or E is preferred over strategy C, the joint measure of belief for A must be greater than that for strategy B i.e. $\mathrm{Bel}(A) > 0.5 > \mathrm{Bel}(B)$. Assuming that we are using the Dempster rule of combination, if C is preferred for a sub-problem, its highest value of belief must be less than 0.5, therefore forcing the joint measure of belief of two separate implementations of strategy C to be as required.

4 Application

Due to the diverse nature of mathematical functions which are considered *integrable*, it has become necessary to use a number of different methods or strategies for their evaluation. A numerical library, such as the NAG Fortran Library, can have as many as 25 different routines (NAG 1993), each of use in evaluating a subset of integrals. In building an expert system to choose and implement such routines (Dupée & Davenport 1995), we have to create an inference mechanism for making such a choice, either from the list of numerical routines or from specially created combined symbolic/numeric strategies.

We can think of these methods as falling into three subgroups – those that implement a general strategy which can be applied to a large subset of integrals; those that implement a specific strategy for applying to particular subsets of integrals i.e. of those functions which reveal a particular attribute or set of attributes; those that implement a number of strategies e.g. those that split the function into two and perform different strategies on each part. However, it is of extreme benefit to maintain a consistent interface to each of these methods and each must be considered equally.

So we have, for example, a routine 'd01ajf' which implements a strategy which can be addressed to a wide variety of different classes of integrals, but which may fail to work, or give an inaccurate result, under certain difficult conditions. We have a routine 'd01apf' which is ideal for use in situations where the values of the integral at the end points of integration are undefined or uncalculable, but which are of little or no benefit in other cases. We also have routines such as 'd01amf' and 'd01transform' which, should one or both of the end points of integration be infinite, will split the function and transform each part onto a finite region before implementing one or more of the other routines. The routine 'd01amf' performs this internally (hard-wired into the Fortran code) using a general routine for the implementation and 'd01transform' does the splitting and transformation externally (in the Axiom interface to the NAG routines) and can then implement any of the other routines depending on the attributes it finds.

So specific routines such as 'd01apf' give a positive $b.p.a.$ if the specific attributes or combination of attributes is present; 'd01ajf' and other general routines give a positive $b.p.a.$ *unless* certain difficult conditions prevail; 'd01amf' gives a positive $b.p.a.$ if the integral is finite or semi-infinite (since the details of transformation and implementation is hidden, no other analysis can take place) whilst 'd01transform' performs full analysis of the different parts of the integral and calculates its $b.p.a.$ from the $b.p.a.$s of the individual routines which should be considered appropriate for each separate part. Therefore, if specific routines are appropriate, this strategy should be applied in preference to 'd01amf'.

5   Concluding Remarks

The principles above have been used and implemented in an expert system ANNA, created with Axiom version 2.0 and its links to the NAG Fortran Library. ANNA is nearing completion and the results will be made available for evaluation by UK Higher Education Institutions by October 1996 and incorporated within Axiom for release at a future date. Incomplete code for testing is available from the author.

References

Buchanen, B.G., Shortliffe, E.H. [eds]: *Rule-Based Expert Systems: The MYCIN Experiments of the Stanford Heuristic Programming Project.* Reading, Mass.: Addison Wesley. 1984.

Dupée, B.J., Davenport, J.H.: Using Computer Algebra to Choose and Apply Numerical Routines. *AXIS*, **2(3)**, 31–41. 1995.

Gordon, J., Shortliffe, E.H.: The Dempster-Shafer Theory of Evidence and its Relevance to Expert Systems. *In:* Buchanen & Shortliffe (1984). 1983.

Gordon, J., Shortliffe, E.H.: A Method of Managing Evidential Reasoning in a Hierarchical Hypothesis Space. *Artificial Intelligence*, **26(3)**, 323–357. 1985.

Houstis, E. N., Rice, J. R., Vichnevetsky, R. [eds]: *Expert Systems for Scientific Computing*. Amsterdam: North-Holland. Proceedings of the Second IMACS International Conference on Expert Systems for Numerical Computing, Purdue University, April 1992.

Kruse, R., Schwecke, E., Heinsohn, J.: *Uncertainty and Vagueness in Knowlege Based Systems: Numerical Methods*. Berlin: Springer Verlag. 1991.

Lucks, M., Gladwell, I.: Automated Selection of Mathematical Software. *ACM Transactions on Mathematical Software*, **18(1)**, 11–34. Also published in Houstis et al. (1992, pp 421–459) as 'A Functional Representation for Software Selection Expertise'. 1992.

NAG.: *Fortran Library Manual – Mark 16*. NAG Ltd, Oxford, UK. NAG Publication Code NP2478. 1993.

Pearl, J.: On Evidential Reasoning in a Hierarchy of Hypotheses. *Artificial Intelligence*, **28(1)**, 9–15. 1986

Pearl, J.: *Probabilistic Reasoning in Intelligent Systems: Networks of Plausible Inference*. San Mateo, Calif.: Morgan Kaufmann. 1988.

Shafer, G.: *A Mathematical Theory of Evidence*. Princeton, NJ: Princeton University Press. 1976.

Shafer, G., Logan, R.: Implementing Dempster's Rule for Hierarchical Evidence. *Artificial Intelligence*, **33**, 271–298. 1987.

A New Approach on Solving 3-Satisfiability

Robert Rodošek

IC-Parc, Imperial College,
London SW7 2AZ, England

Abstract. In this paper we describe and analyse an algorithm for solving the 3-satisfiability problem. If clauses are regarded as constraints of Constraint Satisfaction Problems, then every clause presents a constraint with a special property, namely *subquadrangle*. We show that the algorithm on subquadrangles guarantees a solution in time less than $O(1.476^n)$, which improves the current well-known 3-satisfiability algorithms. Tests have shown the number of steps to be significantly smaller also in the average compared with the other algorithms.

1 Introduction

The NP-completeness of the 3-Satisfiability problem (3-SAT) is a strong argument that there do not exist algorithms, which solve these problems in polynomial time [5]. There has been done a lot of work studying the complexity of algorithms to solve 3-SAT (see, e.g., [6], [10], [15]). Papers on worst case analysis of algorithms represent the bounds relative to different parameters of NP-hard problems.

Consider the number m of clauses in the 3-SAT problem, Beigel and Eppstein [1] represent an algorithm which takes time $O(1.381^m)$. Their algorithm is based on the idea to only partially solve the problem by restricting each 3-clause to a 2-clause. It is tested in polynomial time if the partial solution can be extended to a complete satisfiability (e.g. as a 2-SAT instance). Consider the number n of variables in the 3-SAT problem, the worst case analysis we are aware of are by Monien and Speckenmeyer [10], and Schiermeyer [12]. Monien and Speckenmeyer give an algorithm using autark truth assignments test and it takes time $O(1.619^n)$. Otherwise, Schiermeyer represents an algorithm which is based on procedures for 21 different "patterns" (i.e., specific combinations of clauses over some variables) and it takes time $O(1.579^n)$.

Unlike all of these algorithms we use the structure of constraints to choose a variable. To guarantee that an instantiation of k, $k \leq n$, variables does not cause an instantiation of other k variables, only specific structures of constraints are allowed. Using binary decisions like an instantiation of a variable or the equality between two variables, our algorithm solves the 3-SAT problem in $O(1.476^n)$ time. As the bound depends on variables, the result is incomparable with bound $O(1.381^m)$ derived by Beigel and Eppstein. Consider the relationship between parameters m and n (see [1]), our bound represents an improvement for 3-SAT problems in which the number of clauses is greater than $1.206n$.

The paper is organised as follows. In Section 2 we define the notion of subquadrangles and some basic operations on them. Section 3 presents a reduction of a given 3-SAT formula using subquadrangles. A complete algorithm for solving the 3-SAT problem is described and analysed in Section 4. Some empirical results on the comparison between the proposed algorithm and the other 3-SAT solution techniques are given. Section 5 concludes the paper.

2   Clauses and Subquadrangles

To analyse our algorithm on subquadrangles we introduce the definition of satisfiability formulae of the 3-SAT problem and constraints of the CSP.

Let $V = \{v_1, ..., v_n\}$ be a set of *variables* and let $L = \{l_1, \overline{l_1}, ..., l_n, \overline{l_n}\}$ be the corresponding set of *literals*. A disjunction $c = (l_1 \vee ... \vee l_k)$ of literals is called a *k-clause*, or simply clause. As usual we demand that a literal occurs at most once in a clause and a clause does not contain both, a literal l and its complement $\overline{l}$. A conjunction $c_1 \wedge ... \wedge c_m$ of clauses is called a formula in CNF form. A solution to the 3-SAT problem is an assignment of logical values *true* and *false* to variables satisfying this formula. By $\square$ we denote the empty clause, which is unsatisfiable by definition and $\emptyset$ the empty formula, which is a tautology by definition.

A *constraint* C is usually defined in the CSP notation as a set $\{t_1, ..., t_k\}$ of tuples where each tuple $t = (d_1, ..., d_k)$ presents a combination of values for the same k-subset[1] of a finite set of variables (cf. [4]). We say that this subset of variables is the *scope* of constraint C. Solving the CSP with a set $\mathcal{C}$ of constraints implies searching for a tuple of values for all variables $\bigcup_{C \in \mathcal{C}} scope(C)$ satisfying these constraints.

Consider the variables of the 3-SAT problem as the variables of the CSP problem, every clause c can be regarded as a constraint, denoted by $C(c)$. For example, clause $l_1 \vee l_2$ represents the constraint

$$C(l_1 \vee l_2) = \{(false, true), (true, false), (true, true)\}.$$

Every solution of the 3-SAT problem is also a solution of the CSP problem, and vice versa. We show that clauses represent special constraints, namely *subquadrangles*. First we need the definition of the projection of constraints onto subsets of variables. The *projection* of a tuple t with a value d_i for each variable v_i of V onto set $W \subset V$ is a tuple $t[W]$ with the same values for variables of W. Similarly, for any constraint C, the projection C onto W is the constraint

$$C{\downarrow}_W := \{t[W] \mid t \in C\}.$$

We define a *subquadrangle* as a constraint that constrains no proper subset of its scope. It means that if M denotes a proper subset of the scope, the projection onto M is a *quadrangle*, i.e., a Cartesian product of sets of logical values.

Definition 1 (subquadrangle). A constraint C is a *subquadrangle* if and only if $C{\downarrow}_M$ is a quadrangle for every $M, \emptyset \subset M \subset scope(C)$.

---

[1] A *k-set* is a set of k elements.

Notice that to recognise a subquadrangle, it is enough to test only "the last but one" projections of a given constraint. A constraint C is a subquadrangle if and only if for any subset M of $scope(C)$, $|M| = |scope(C)| - 1$, it holds that $C{\downarrow}_M$ is a quadrangle. The information the subquadrangles tell us is of particular interest. If we are given a constraint on three variables we could not have been given the same information by knowing the three derived pairwise constraints between the variables. In fact ternary constraints give us much more "expressive power" [7]. If a ternary constraint is a subquadrangle, then the derived binary constraints give us no information!

Proposition 2. *Every clause is a subquadrangle.*

Proof of proposition. If a literal of a given clause c is instantiated by *true*, then every other literal of c can be instantiated by *true* or *false*. Constraint $C(c)$ satisfies the property of subquadrangles, and the result follows. □

Let us give an example of clauses where even the conjunction of them is a subquadrangle.

Example 1. Consider clauses

$$c_1 = v_1 \vee v_2 \vee v_3$$
$$c_2 = v_1 \vee \overline{v_2} \vee \overline{v_3}$$
$$c_3 = \overline{v_1} \vee \overline{v_2} \vee v_3$$
$$c_4 = \overline{v_1} \vee v_2 \vee \overline{v_3},$$

constraints $C(c_1)$, ..., $C(c_4)$ and even $C(c_1 \wedge c_2 \wedge c_3 \wedge c_4)$ are subquadrangles. The last subquadrangle follows from the fact that any combination of values is allowed for any non-empty proper subset of $\{v_1, v_2, v_3\}$.

To solve the 3-SAT problem we consider only constraints over at most three variables. The join[2] of clauses over three variables is called a *3-SAT constraint*. Using the result on subquadrangles [11], every 3-SAT constraint can be expressed by subquadrangles with different scopes. A graphical representation of the scopes of such subquadrangles is given in Fig. 1. Cycles denote variables of clauses and polygons denote the scopes.

3 A Reduction of Constraints

In this paper two types of binary decisions are considered:

Definition 3 (two binary decisions). The *binary decisions* on variables are:

1. *Instantiate $v :=$ true or $v :=$ false.* (DEC1)
2. *Decide $v_1 = v_2$ or $v_1 \neq v_2$.* (DEC2)

---

[2] The *join* of constraints C and C' with scopes W and W', respectively, is the constraint over variables $W \cup W'$ given by $C \bowtie C' := \{l \mid l[W] \in C \text{ and } l[W'] \in C'\}$.

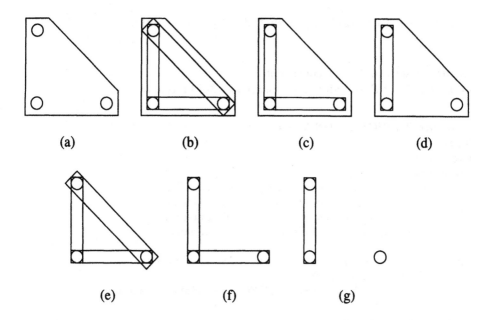

Fig. 1. Scopes of subquadrangles to 3-SAT constraints.

Let us give two examples of such binary decisions. Consider the 3-SAT constraint in Example 1 the instantiation $v_1 := true$ causes $v_2 = v_3$ and the instantiation $v_1 := false$ causes $v_2 = \overline{v_3}$. The problem is reduced for two variables. Another example, 2-clauses $v_1 \vee v_2$ and $\overline{v_1} \vee v_3$ with complemented literals to variable v cause that every instantiation of v leads to the instantiation of v_2 or v_3. In the following, two variables which are 'equal' or 'not equal' are called *direct dependent* variables.

The main idea of our search strategy is to reduce the problem for two variables using one of binary decisions (DEC1) and (DEC2), and to repeat the same strategy on the reduced problem. If the problem can be solved with this strategy, then we needed at most $2^{n/2}$ steps (i.e., 1.415^n steps), and hence, less than the desired 1.476^n steps.

3.1 Binary Decisions on Each 3-SAT Constraint

To follow the main idea of our strategy we reduce and simplify each 3-SAT constraint. We say that a 3-SAT constraint can be *reduced* if a binary decision causes a reduction of the problem for two variables. A 3-SAT constraint can be *simplified* if it is also represented by the join of less 3-clauses and possibly more 2-clauses. We show that there are only four different types of 3-SAT constraints using the following algorithm:

Algorithm 4 (a reduction on each 3-SAT constraint).

Input: A set C of 3-SAT constraints.
Output: A set C' of reduced and simplified 3-SAT constraints.

```
procedure REDUCTION_ONE( in C; out C');
begin
flag := true;
while  flag  do
if       every instantiation of v₁ reduces a constraint C ∈ C with
```

procedure REDUCTION_ONE(**in** C; **out** C');
begin
flag := true;
while flag **do**
if   every instantiation of v_1 reduces a constraint $C \in C$ with
            $scope(C) = \{v_1, v_2, v_3\}$ to allow only one value for v_2 or to
            represent a direct dependency between v_2 and v_3,
            then   apply binary decision $(DEC1)$ on variable v_1

elseif every direct dependency between v_1 and v_2 reduces
            C to allow only one value for v_3,
            then   apply binary decision $(DEC2)$ on variables v_1 and v_2

else   flag := false, and
            minimise the number of 3-clauses and maximise
            the number of 2-clauses to represent constraint C
endif
endwhile
end procedure

Every binary decision causes an instantiation of at least two variables. We show that the proposed reduction in Algorithm 4 leads to particular constraints. A 3-SAT constraint is denoted by C_{n_1, n_2} if it can be represented by the join of n_1 3-clauses and n_2 2-clauses.

Lemma 5. *By performing Algorithm 4, every 3-SAT constraint over variables v_1, v_2, v_3 is one of the constraints:*

$$C_{2,0} = C(l_1 \vee l_2 \vee l_3) \bowtie C(\overline{l_1} \vee \overline{l_2} \vee \overline{l_3})$$
$$C_{1,1} = C(l_1 \vee l_2 \vee l_3) \bowtie C(\overline{l_1} \vee \overline{l_2})$$
$$C_{1,0} = C(l_1 \vee l_2 \vee l_3)$$
$$C_{0,2} = C(l_1 \vee l_2) \bowtie C(l_1 \vee l_3)$$
$$C_{0,1} = C(l_1 \vee l_2).$$

Proof in Appendix A.   □

Since $C_{2,0}$ is a subquadrangle it follows that possible scopes of subquadrangles to 3-SAT constraints are reduced from seven sets in Fig. 1 to four sets in Fig. 2.

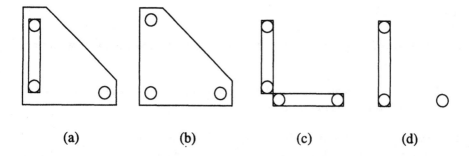

(a)   (b)   (c)   (d)

Fig. 2. Scopes of subquadrangles to 3-SAT constraints by Algorithm 4.

3.2 Binary Decisions on a Set of 3-SAT Constraints

The reduced and simplified 3-SAT constraints in the previous subsection allow only particular relationships between the constraints. Using the following algorithm, additional binary decisions can be performed:

Algorithm 6 (a reduction on a set of 3-SAT constraints).

Input: A set C of 3-SAT constraints.
Output: A set C' of reduced and simplified 3-SAT constraints.

```
procedure REDUCTION_ALL( in C; out C');
begin
flag := true;
while flag do
1.  if a literal v does not belong to any subquadrangle,
        then v := false
2.  elseif there exists a constraint C(v₁ ∨ v₂) but not C(v̄₁ ∨ v̄₂ ∨ v),
            then replace every literal v̄₁ with v₂
3.  elseif there exist constraints C(v₁ ∨ v₂) and exactly one C(v̄₁ ∨ v̄₂ ∨ v),
            then apply binary decision (DEC1) on v
    else flag := false
    endif
endwhile
end procedure
```

Consider Step 1 and Step 2, there exists a value to a variable which does not reduce any other constraint on the other variables. By Step 3, binary decision (DEC1) splits the problem into two subproblems relative to an instantiation of a binary variable. It follows that the algorithm guarantees a solution if a solution of the given problem exists.

Lemma 7. *Binary decision (DEC1) in Algorithm 6 causes a reduction of the problem for at least two variables.*

Proof in Appendix A.   □

The proposed reductions of constraints in this section allow problems with particular properties:

Corollary 8. *By Algorithm 4 and Algorithm 6, the 3-SAT problem is reduced to a problem with the following properties:*

1. *every 3-SAT constraint is $C_{2,0}$, $C_{1,1}$, $C_{1,0}$, $C_{0,2}$ or $C_{0,1}$*
2. *for every $C(l_1 \vee l_2)$, there are $C(\overline{l_1} \vee \overline{l_2} \vee l_3)$ and $C(\overline{l_1} \vee \overline{l_2} \vee l_4)$ such that l_3 and l_4 are literals of different variables.*

Proof in Appendix A.   □

4   A Complete Algorithm

In the previous section we have shown a part of the solution search where each binary decision guarantees a reduction for at least two variables. In this section we present a complete search algorithm for the 3-SAT problem.

Algorithm 9 (satisfiability of the 3-SAT problem).

Input: A set C of 3-SAT constraints.
Output: True, if C is satisfiable, and false, otherwise.

```
function SEARCH( in C): boolean;
begin
1. call REDUCTION_ONE(C, C');
2. return SEARCH_REST(C');
end function
```

```
function SEARCH_REST( in C'): boolean;
begin
1. call REDUCTION_ALL(C', C'');
2. if there exists a C(v₁ ∨ v₂) ∈ C'',
        then apply binary decision (DEC2) on v₁ and v₂
        else apply a binary decision on a variable such that
             two new 2-clauses are derived
   endif
3. if □ ∈ C'', then apply backtracking to the previous binary decisions endif
4. if  C'' = ∅, then return true else return SEARCH_REST(C'') endif
6. return false
end function
```

Let us show the correctness of Algorithm 9. By using the reduction of constraints in the previous section, set C'' contains only constraints represented in Corollary 8. If there exists a constraint $C(v \lor v')$, then binary decision (DEC2) on variables v and v' is performed. If there is no 2-clause, then we apply a binary decision according to Step 2 of SEARCH_REST. Thus, by the recursive call of SEARCH_REST, all combinations of values to the variables can be checked. If there exists a solution, then the algorithm returns value $true$, otherwise it returns value $false$.

To compute the upper bound complexity of the algorithm, we use the information how much a given problem is reduced by Step 2 of SEARCH_REST:

Proposition 10. *Let C be a set of $C_{1,0}$ and $C_{2,0}$ constraints after applying Algorithm 4 and Algorithm 6. There exists a binary decision to reduce C to a set with at least two constraints $C_{0,1}$ or to reduce the problem for two variables.*

Proof in Appendix A. □

Proposition 11. *Let C be a set of constraints containing constraints $C(v_1 \lor v_2)$, $C(\bar{v}_1 \lor \bar{v}_2 \lor v_3)$ and $C(\bar{v}_1 \lor \bar{v}_2 \lor v_4)$ after applying Algorithm 4 and Algorithm 6. There exists a binary decision such that for one value the problem is reduced for four variables and the number of constraints representing a 2-clause becomes not smaller.*

Proof in Appendix A. □

By using both propositions, the desired complexity result follows:

Theorem 12. *Algorithm 9 solves the 3-SAT problem in less than $O(1.476^n)$ time.*

Proof of theorem. Consider the proof of Lemma 5, every binary decision in RE-DUCTION_ONE procedure reduces the problem for at least two variables. This leads to the complexity bound $O(1.415^n)$ which is less than the desired $O(1.476^n)$. Since REDUCTION_ONE procedure does not necessarily solve the problem, it is enough to show the desired complexity result for SEARCH_REST procedure.

Let n be the number of variables of the problem and let k be the number of constraints representing 2-clauses of that problem. There are three different properties of the sets of constraints:

- there exists $C(v_1 \lor v_2)$ but not different $C(\bar{v}_1 \lor \bar{v}_2 \lor v_3)$ and $C(\bar{v}_1 \lor \bar{v}_2 \lor v_4)$: binary decision (DEC1) in Step 3 of REDUCTION_ALL guarantees a reduction to a problem containing at most $n-2$ variables and k 2-clauses for one value and $(k-1)$ 2-clauses for another value of the binary decision (see the proof of Lemma 7).
- there exist $C(v_1 \lor v_2)$ and different $C(\bar{v}_1 \lor \bar{v}_2 \lor v_3)$, $C(\bar{v}_1 \lor \bar{v}_2 \lor v_4)$: a binary decision in Step 2 of SEARCH_REST guarantees a reduction to a problem with at most $n-4$ variables and k 2-clauses for one value and a reduction to a problem with at most $n-1$ variables and $(k-1)$ 2-clauses for another value of the decision (see Proposition 11).

– there does not exist $C(v_1 \lor v_2)$: a binary decision in Step 2 of SEARCH_REST guarantees a reduction to a problem with at most $n - 1$ variables and two 2-clauses (see Proposition 10).

The number of decisions which are needed to solve the 3-SAT problem can be represented by recursive functions. We denote $F_k(n)$ the number of decisions needed to solve the problem with n variables and k 2-clauses. According to the mentioned three binary decisions in SEARCH_REST, the following recursive functions are derived:

$$F_0(n) = 2 * F_2(n - 1)$$
$$F_2(n) = F_1(n - 2) + F_2(n - 2)$$
$$F_2(n) = F_1(n - 1) + F_2(n - 4)$$
$$F_1(n) = F_0(n - 2) + F_1(n - 2)$$
$$F_1(n) = F_0(n - 1) + F_1(n - 4)$$

Using the recursive techniques [2] to extract the explicit formula for $F_0(n)$, it follows that the upper bound complexity is $F_0(n) = 1.4756^n$. Since for every binary decision the algorithm needs $O(m)$ steps, the whole complexity is $O(m * 1.4756^n)$ which is less than $O(1.476^n)$. $\qquad\qquad\square$

4.1 Empirical Results

Empirical tests have shown that the proposed algorithm is faster in the average compared with algorithms based on the resolution techniques for problems with a relative small number of solutions. Moreover, by empirical results on several satisfiability problems in [9] (also the pigeon-hole problem) the proposed algorithm gives better results than the other well-known satisfiability algorithms (see in [9]):

1. *Davis/Putnam/Loveland-algorithm (DPL):* the version of this algorithm was implemented by prof. John Hooker, Pittsburgh, USA. The well-known algorithm is based on unit resolution.
2. *Jeroslow/Wang-algorithm (Jer-Wa):* the version of this algorithm was implemented by prof. John Hooker. The algorithm is a backtracking method which uses a complex branching scheme in which, at each node of a DPL search tree, a search heuristic tries to find a satisfying solution.
3. *Branch-and-cut-algorithm as an integer linear program (IP):* the version of this algorithm was implemented by prof. John Hooker. The algorithm combines the branch-and-bound method and the cutting-plane technique of integer programming.
4. *SAT:* this algorithm was implemented by Cosimo Spera (Computational results for solving large general satisfiability problems, TR-222, University of Siena, 1990). The algorithm is based on the equivalence of a conjunctive satisfiability problem with an appropriate linear complementarity problem. A solution to the linear complementarity problem and therefore a solution

to the satisfiability problem is found by finding the optimal solutions to the linear program, while varying parametrically a scalar parameter.

5. *CHIP:* it is an equation solving method for boolean algebra use symbolic boolean terms to represent functions (see specialised unification algorithms for boolean algebra: (i) Siekmann: Universal unification, (ii) Martin-Nipkow: Boolean Unification, (iii) Buchberger's algorithm for Groebner bases).

6. *CHIPD:* this method of solving is to interleave labeling of the variables with values from the domain $\{0, 1\}$ with local constraint propagation.

7. *SUBQ:* our algorithm.

The results of the algorithms 1-6 on the pigeon-hole problem described in Table 1 were derived by Franz Josef Radermacher, University of Passau in the cooperation with ECRC. All timings in the table are in CPU seconds on a SUN-SPARC/20. Symbol '#' means that the search was stopped after 10 minutes without solving the problem. The results show that the problems are solved faster by using our algorithm. Moreover, we started our algorithm on another

Table 1. Empirical results of different algorithms

Name	Variables	Clauses	DPL	Jer-Wa	IP	SAT	CHIP	CHIPD	SUBQ
hole6	42	133	10,98	8,81	58,76	162,32	88,70	68,66	10,23
hole7	56	204	181,77	75,48	710,53	447,85	1875,48	1042,24	43,12
hole8	72	297	2.112,61	642,53	#	1518,06	#	#	206,91
hole9	90	415	#	5.512,02	#	5.770,87	#	#	812,03
hole10	110	561	#	#	#	#	#	#	2.768,94

set of 3-satisfiability benchmarks which are suggested in [14]. These problems were recursively generated by addition of binary equalities. Surprisingly, our algorithm solved the problems without backtracking while the other algorithms needed much more search effort. For all benchmarks in [14] our algorithm does not need more time than the satisfiability algorithms which are mentioned in this section.

5 Conclusions

The proposed algorithm for the 3-SAT problem is based on subquadrangles which allow simply to identify k variables such that their instantiation causes an instantiation of other k variables. We have shown that the algorithm solves the 3-SAT problem in $O(1.476^n)$ steps which improves the known 3-satisfiability algorithms. The empirical tests show that the algorithm gives also good results in the average compared with the well-known complete satisfiability algorithms.

References

1. Beigel, R., Eppstein, D.: 3-Coloring in Time $O(1.3446^n)$: A No-MIS Algorithm. In Technical Report ECCC TR95-33 (1995)
2. Biggs, N. L.: Discrete Mathematics. Oxford Science Publications, New York (1994)
3. Davis, M., Putnam, H.: A Computing Procedure for Quantification Theory. Journal of the Association for Computing Machinery **7(3)** (1960) 201–215
4. Fikes, R. E.: REF-ARF: A System for Solving Problems Stated as Procedures. Artificial Intelligence **1** (1970) 27–120
5. Garey, M. R., Johnson, D. S.: Computers and Intractability. W. H. Freeman and Company, San Francisco (1979)
6. Haken, A.: The Intractability of Resolution. Theoretical Computer Science **39** (1985) 297–308
7. Jeavons, P. G.: The Expressive Power of Constraint Networks. In Technical Report CSD-TR-640, Royal Holloway and Bedford New College, University of London (1990)
8. Lawler, E. L.: A Note on the Complexity of the Chromatic Number Problem. Inf. Proc. Lett. **5** (1976) 66–67
9. Mayer, J., Mitterreiter, I., Radermacher, F. J.: Running Time Experiments on some Algorithms for Solving Propositional Satisfiability Problems. In Technical Report, Forschungsinstitut für anwendungsorientierte Wissensverarbeitung, Ulm (1994)
10. Monien, B., Speckenmeyer, E.: Solving Satisfiability in less than 2^n Steps. Discrete Appl. Math. **10** (1985) 287–295
11. Rodošek, R.: Combining Heuristics for Constraint Satisfaction Problems. Proceedings of the CP'95 Workshop on Studying and Solving Really Hard Problems, Cassis (1995) 147–156
12. Schiermeyer, I.: Solving 3-Satisfiability in less than 1.579^n Steps. 6th Workshop Computer Science Logic, Spring-Verlag (1993) 379–394
13. Schiermeyer, I.: Deciding 3-Colourability in less than $O(1.415^n)$ Steps. 19th Int. Workshop Graph-Theoretic Concepts in Computer Science, Spring-Verlag (1994) 177–182
14. Trümper, K., Radermacher, F. J.: Analyse der Leistungsfähigkeit eines neuen Systems zur Auswertung aussagenlogischer Probleme. In Technical Report FAW-TR-90003 (1990)
15. Urquhart, A.: Hard Examples for Resolution. JACM **34** (1987) 209–219

A   Appendix: Proofs

Lemma 5 *By performing Algorithm 4, every 3-SAT constraint over variables v_1, v_2, v_3 is one of the constraints:*

$$C_{2,0} = C(l_1 \vee l_2 \vee l_3) \bowtie C(\overline{l_1} \vee \overline{l_2} \vee \overline{l_3})$$
$$C_{1,1} = C(l_1 \vee l_2 \vee l_3) \bowtie C(\overline{l_1} \vee \overline{l_2})$$
$$C_{1,0} = C(l_1 \vee l_2 \vee l_3)$$
$$C_{0,2} = C(l_1 \vee l_2) \bowtie C(l_1 \vee l_3)$$
$$C_{0,1} = C(l_1 \vee l_2).$$

Proof of lemma. Every 3-SAT constraint over variables v_1, v_2, v_3 is a non-empty proper subset of $\{true, false\}^3$. We show how to transform and to reduce 3-SAT constraints by preserving the proposed upper-bound complexity, i.e., an instantiation of a variable causes an instantiation of another variable.

We do not consider 1-clauses since the instantiation of their literals follows directly. Since the constraints are defined over three variables number n_2 of 2-clauses can not be greater than three. Moreover, there are no two 2-clauses on the same two variables. Either they present 1-clause or they are direct dependent.

We investigate each 3-SAT constraint relative to parameters n_1 and n_2:

- **case $n_1 > 4$:** Every 3-clause is an extension of $l_1 \vee l_2, l_1 \vee \overline{l_2}, \overline{l_1} \vee l_2, \overline{l_1} \vee \overline{l_2}$ by a literal to the third variable. If there are more than four 3-clauses, then two of them are reduced to one 2-clause. ($\rightarrow$ not accepted)
- **case $n_1 = 4$:** There is only one $C_{4,0}$ and it is equal to $\{(l_1, l_2, l_3), (l_1, \overline{l_2}, \overline{l_3}), (\overline{l_1}, \overline{l_2}, l_3), (\overline{l_1}, l_2, \overline{l_3})\}$.

  $n_2 = 3$: Since every 2-clause reduces $C_{4,0}$ by one tuple, three 2-clauses reduce this constraint to a set with only one tuple, and hence, variables v_1, v_2 and v_3 are instantiated. ($\rightarrow$ not accepted)

  $n_2 = 2$: Two 2-clauses reduce $C_{4,0}$ for two elements, and hence, one variable is instantiated. ($\rightarrow$ not accepted)

  $n_2 = 1$: Without loss of generality, assume 2-clause $l_1 \vee l_2$. Constraint $C_{4,0}$ is reduced to $C_{1,3}$, represented by clauses $l_1 \vee l_2 \vee l_3$, $\overline{l_1} \vee \overline{l_2}$, $\overline{l_1} \vee l_3$, $\overline{l_2} \vee l_3$. ($\rightarrow$ not accepted)

  $n_2 = 0$: An instantiation of variable v_1 causes direct dependent v_2 and v_3. ($\rightarrow$ not accepted)
- **case $n_1 = 3$:** Constraint $C_{3,0}$ represents 3-clauses $l_1 \vee l_2 \vee l_3$, $l_1 \vee \overline{l_2} \vee \overline{l_3}$, $\overline{l_1} \vee l_2 \vee \overline{l_3}$. It contains 5 tuples.

  $n_2 = 3$: Three 2-clauses cause that $C_{3,3}$ contains at most two tuples. If $C_{3,3}$ contains one tuple, then all three variables are instantiated, otherwise $C_{3,3}$ is equal to $C_{2,0}$. ($\rightarrow$ not accepted)

  $n_2 = 2$: There is the following case (the other cases follow on the same way):
    1. by adding 2-clauses $l_1 \vee l_3$ and $\overline{l_2} \vee \overline{l_3}$ to $C_{3,0}$, the constraint $C_{3,2}$ is $C_{1,2}$, represented by $l_1 \vee \overline{l_2} \vee \overline{l_3}$, $l_1 \vee l_3$, $\overline{l_1} \vee \overline{l_3}$ ($\rightarrow$ not accepted)

  $n_2 = 1$: There are the following different cases (the other cases follow on the same way):
    1. by adding 2-clause $l_1 \vee l_3$ to $C_{3,0}$, the constraint $C_{3,1}$ is $C_{2,1}$, represented by $l_1 \vee \overline{l_2} \vee \overline{l_3}$, $l_1 \vee \overline{l_2} \vee \overline{l_3}$, $l_1 \vee l_3$ ($\rightarrow$ not accepted)
    2. by adding 2-clause $l_1 \vee \overline{l_3}$ to $C_{3,0}$, the constraint $C_{3,1}$ is $C_{0,3}$, represented by $l_1 \vee l_2$, $l_1 \vee \overline{l_3}$, $l_2 \vee \overline{l_3}$ ($\rightarrow$ not accepted)
    3. by adding 2-clause $\overline{l_1} \vee \overline{l_3}$ to $C_{3,0}$, the constraint $C_{3,1}$ is $C_{1,2}$, represented by $l_1 \vee l_2 \vee l_3$, $\overline{l_1} \vee \overline{l_3}$, $\overline{l_2} \vee \overline{l_3}$. ($\rightarrow$ not accepted)

  $n_2 = 0$: By binary decision $v_1 = v_2$, variables v_1 and v_2 are instantiated by *true*. By $v_1 \neq v_2$ it follows instantiation $v_3 = false$. ($\rightarrow$ not accepted)
- **case $n_1 = 2$:** There are two constraints: $C_{2,0}^1$ represents 3-clauses $l_1 \vee l_2 \vee l_3$, $l_1 \vee \overline{l_2} \vee \overline{l_3}$, and $C_{2,0}^2$ represents 3-clauses $l_1 \vee l_2 \vee l_3$, $\overline{l_1} \vee \overline{l_2} \vee \overline{l_3}$.

$n_2 = 3$: By adding three 2-clauses to $C_{2,0}^1$ or $C_{2,0}^2$, the constraint contains at most two tuples, and hence, an instantiation of variable v_1 causes direct dependent v_2 and v_3. ($\rightarrow$ not accepted)

$n_2 = 2$: There are the following different cases (the other cases follow on the same way):

1. by adding 2-clauses $l_1 \vee l_3$ and $l_2 \vee l_3$ to $C_{2,0}^2$, the constraint $C_{2,2}$ is $C_{1,2}$, represented by $\overline{l_1} \vee \overline{l_2} \vee \overline{l_3}$, $l_1 \vee l_3$, $l_2 \vee l_3$ ($\rightarrow$ not accepted)
2. by adding 2-clauses $l_1 \vee l_3$ and $l_2 \vee \overline{l_3}$ to $C_{2,0}^1$, the constraint $C_{2,2}$ is $C_{1,2}$, represented by $l_1 \vee l_2 \vee l_3$, $l_2 \vee l_3$, $\overline{l_2} \vee \overline{l_3}$ ($\rightarrow$ not accepted)
3. by adding 2-clauses $l_1 \vee l_3$ and $\overline{l_2} \vee l_3$ to $C_{2,0}^2$, the constraint $C_{2,2}$ is $C_{1,2}$, represented by $\overline{l_1} \vee \overline{l_2} \vee l_3$, $l_1 \vee l_3$, $\overline{l_2} \vee l_3$ ($\rightarrow$ not accepted)
4. by adding 2-clauses $l_1 \vee l_3$ and $\overline{l_2} \vee \overline{l_3}$ to $C_{2,0}^2$, the constraint $C_{2,2}$ is $C_{0,2}$, represented by $l_1 \vee l_3$, $\overline{l_2} \vee \overline{l_3}$ ($\rightarrow$ not accepted)
5. by adding 2-clause $l_1 \vee \overline{l_3}$ and $\overline{l_2} \vee \overline{l_3}$ to $C_{2,0}^1$, the constraint $C_{2,2}$ is $C_{1,2}$, represented by $l_1 \vee \overline{l_2} \vee \overline{l_3}$, $l_1 \vee \overline{l_2}$, $\overline{l_1} \vee l_2$. ($\rightarrow$ not accepted)

$n_2 = 1$: There are the following different cases (the other cases follow on the same way):

1. by adding 2-clause $l_1 \vee l_2$ to $C_{2,0}^1$, the constraint $C_{2,1}$ is $C_{1,1}$, represented by $\overline{l_1} \vee \overline{l_2} \vee l_3$, $l_1 \vee l_2$ ($\rightarrow$ not accepted)
2. by adding 2-clause $l_1 \vee \overline{l_2}$ to $C_{2,0}^1$, the constraint $C_{2,1}$ is $C_{0,3}$, represented by $l_1 \vee \overline{l_2}$, $l_1 \vee l_3$, $\overline{l_2} \vee l_3$ ($\rightarrow$ not accepted)
3. by adding 2-clause $l_1 \vee l_3$ to $C_{2,0}^1$, the constraint $C_{2,1}$ is $C_{0,2}$, represented by $l_1 \vee l_3$, $\overline{l_2} \vee l_3$ ($\rightarrow$ not accepted)
4. by adding 2-clause $\overline{l_2} \vee \overline{l_3}$ to $C_{2,0}^1$, the constraint $C_{2,1}$ is $C_{1,2}$, represented by $l_1 \vee l_2 \vee l_3$, $l_1 \vee l_2$, $\overline{l_2} \vee \overline{l_3}$ ($\rightarrow$ not accepted)
5. by adding 2-clause $l_2 \vee l_3$ to $C_{2,0}^2$, the constraint $C_{2,1}$ is $C_{1,1}$, represented by $\overline{l_1} \vee \overline{l_2} \vee \overline{l_3}$, $l_2 \vee l_3$ ($\rightarrow$ not accepted)
6. by adding 2-clause $l_2 \vee \overline{l_3}$ to $C_{2,0}^2$, the constraint $C_{2,1}$ is $C_{0,3}$, represented by $l_1 \vee l_2$, $l_2 \vee \overline{l_3}$, $\overline{l_1} \vee \overline{l_3}$. ($\rightarrow$ not accepted)

$n_2 = 0$: By binary decision $v_1 = v_2$, variables v_1 and v_2 are instantiated by *true*. By $v_1 \neq v_2$ it follows instantiations $v_1 = true$, $v_2 = false$. ($\rightarrow$ $C_{2,1}$ not accepted, $C_{2,0}$ accepted)

- **case $n_1 = 1$** :

$n_2 = 3$: There are the following different cases (the other cases follow on the same way):

1. By adding a 3-clause, e.g., $l_1 \vee \overline{l_2} \vee \overline{l_3}$, to $C_{0,3}^1$ (see the definition of $C_{0,3}^i$, $i = 1..4$, below in case $n_1 = 0$, $n_2 = 3$), the constraint $C_{1,3}$ contains only one tuple, and hence, all variables are instantiated. ($\rightarrow$ not accepted)
2. Variable v_2 is instantiated by adding 3-clause $l_1 \vee \overline{l_2} \vee \overline{l_3}$ to $C_{0,3}^2$. ($\rightarrow$ not accepted)
3. There are two different $C_{1,3}$ by adding 3-clause $l_1 \vee \overline{l_2} \vee \overline{l_3}$ to $C_{0,3}^3$ such that the variables are not yet instantiated. In both cases $C_{1,3}$ is $C_{0,3}$, represented by 2-clauses $l_1 \vee \overline{l_3}$, $\overline{l_1} \vee l_3$, $l_2 \vee l_3$, or 2-clauses $l_1 \vee l_2$, $\overline{l_1} \vee \overline{l_2}$, $l_2 \vee l_3$. ($\rightarrow$ not accepted)

4. By adding 3-clause $\overline{l_1} \vee \overline{l_2} \vee \overline{l_3}$ to $C_{0,3}^4$, the constraint $C_{1,3}$ has the property that any instantiation of v_1 causes direct dependent v_2 and v_3. ($\to$ not accepted)

$n_2 = 2$: There are the following different cases (the other cases follow on the same way):

1. by adding 2-clauses $l_1 \vee \overline{l_3}$ and $l_2 \vee \overline{l_3}$ to $C_{1,0}$, the constraint $C_{1,2}$ is $C_{0,3}$, represented by $l_1 \vee l_2$, $l_1 \vee \overline{l_3}$ and $l_2 \vee \overline{l_3}$ ($\to$ not accepted)
2. by adding 2-clauses $\overline{l_1} \vee \overline{l_2}$ and $\overline{l_2} \vee \overline{l_3}$ to $C_{1,0}$, the constraint $C_{1,2}$ has the property that any instantiation of v_1 causes direct dependent v_2 and v_3. ($\to$ not accepted)

$n_2 = 1$: There are the following different cases (the other cases follow on the same way):

1. by adding 2-clause $\overline{l_2} \vee l_3$ to $C_{1,0}$, the constraint $C_{1,1}$ is $C_{0,2}$, represented by $l_1 \vee l_3$ and $\overline{l_2} \vee l_3$ ($\to$ not accepted)
2. by adding 2-clause $\overline{l_1} \vee \overline{l_2}$ to $C_{1,0}$, the constraint $C_{1,1}$ can not be further reduced. ($\to$ accepted)

- **case $n_1 = 0$:**

$n_2 = 3$: Every $C_{0,3}$ is represented by one of four different sets of clauses:

1. $l_1 \vee l_2$, $\overline{l_2} \vee l_3$, $\overline{l_1} \vee \overline{l_3}$ (denoted by $C_{0,3}^1$)
2. $l_1 \vee l_2$, $l_2 \vee l_3$, $\overline{l_1} \vee \overline{l_3}$ (denoted by $C_{0,3}^2$)
3. $l_1 \vee l_2$, $l_2 \vee l_3$, $\overline{l_1} \vee l_3$ (denoted by $C_{0,3}^3$)
4. $l_1 \vee l_2$, $l_2 \vee l_3$, $l_1 \vee l_3$ (denoted by $C_{0,3}^4$).

Consider $C_{0,3}^k$, $k = 1..3$, there exist different literals over the same variable. An instantiation of such a variable causes the instantiation of another variable. It is not the case for $C_{0,3}^4$. By binary decision $v_1 = v_2$, variables v_1 and v_2 are instantiated by *true*. By $v_1 \neq v_2$ it follows instantiation $v_3 = true$. ($\to$ not accepted)

$n_2 = 2$: There are two $C_{0,2}$ constraints. First, the join of 2-clauses $l_1 \vee l_2$ and $\overline{l_1} \vee l_3$ where an instantiation of variable v_1 causes the instantiation of v_2 or v_3, and hence, it is not accepted. Second, the join of $l_1 \vee l_2$ and $l_2 \vee l_3$ which can not be further reduced. ($\to$ accepted)

$n_2 = 1$: The constraint can not be further reduced. ($\to$ accepted) $\quad\square$

Lemma 7 *Binary decision (DEC1) in Algorithm 6 causes a reduction of the problem for at least two variables.*

Proof of lemma. Let us consider steps 1,2 and 3 of Algorithm 6:

1. Step 1 reduces the problem for one variable if a literal v does not belong to any subquadrangle. A solution of the original problem is preserved.
2. Step 2 replaces every literal $\overline{v}$ with v_1, and by Step 1, $v := true$ preserves a solution of the original problem. The problem is reduced for two variables.
3. Step 3 leads to two reduced problems. First, instantiation $v := true$ leads to Step 2 for $C(v_1 \vee v_2)$, and the problem is reduced for two variables. To prevent the property of Step 1, there exists another constraint containing $\overline{v}$, and this constraint is also reduced! (It is reduced to a constraint representing

a 2-clause or a 1-clause.) Second, instantiation $v := false$ causes $v_1 = \bar{v}_2$, and the problem is also reduced for two variables.

Thus, the reductions in Step 1 and Step 2 guarantee that binary decision (DEC1) in Step 3 reduces the problem for at least two variables, and we are done. $\square$

Corollary 8 *By Algorithm 4 and Algorithm 6, the 3-SAT problem is reduced to a problem with the following properties:*

1. *every 3-SAT constraint is $C_{2,0}$, $C_{1,1}$, $C_{1,0}$, $C_{0,2}$ or $C_{0,1}$*
2. *for every $C(l_1 \vee l_2)$, there are $C(\bar{l}_1 \vee \bar{l}_2 \vee l_3)$ and $C(\bar{l}_1 \vee \bar{l}_2 \vee l_4)$ such that l_3 and l_4 are literals of different variables.*

Proof of corollary. The first property represents the result of Lemma 5. The second property we derive from the steps of Algorithm 6. Let us consider a 3-SAT constraint $C_{1,1}$ with constraint $C(v_1 \vee v_2)$. Using the "if"-conditions in Algorithm 6, it follows that if there is no $C(\bar{l}_1 \vee \bar{l}_2 \vee l_3)$ and there is also not only one $C(\bar{l}_1 \vee \bar{l}_2 \vee l_3)$, then there are two such constraints, i.e., $C(\bar{l}_1 \vee \bar{l}_2 \vee l_3)$ and $C(\bar{l}_1 \vee \bar{l}_2 \vee l_4)$ where l_3 and l_4 are literals of different variables. $\square$

Proposition 10 *Let C be a set of $C_{1,0}$ and $C_{2,0}$ constraints after applying Algorithm 4 and Algorithm 6. There exists a binary decision to reduce C to a set with at least two constraints $C_{0,1}$ or to reduce the problem for two variables.*

Proof of proposition. Let P denote the property that there exist constraints $C(v \vee v_1 \vee v_2)$ and $C(\bar{v} \vee v_3 \vee v_4)$ but no another constraint containing v_1 and v_2, and no another constraint containing v_3 and v_4. If the problem has property P, then binary decision (DEC1) on variable v leads to a reduction for at least two variables. Instantiation $v := true$ leads to Step 2 of Algorithm 6 for $C(v_1 \vee v_2)$, and instantiation $v := false$ leads to Step 2 of Algorithm 6 for $C(v_3 \vee v_4)$.

It is enough to show the proposed result for problems without property P. Let $C(v_1 \vee v_2 \vee v_3)$ be a 3-clause constraint $C_{1,0}$. The same result can be also shown for $C_{2,0}$. To prevent the property of Step 1 of REDUCTION_ALL on variable v_1, there exists a constraint $C(\bar{v}_1 \vee v_4 \vee v_5)$. To prevent property P w.r.t. variable v_1, there exists a constraint $C(\bar{v}_4 \vee \bar{v}_5 \vee v_6)$. To prevent property P w.r.t. variable v_5, there exists a constraint C' containing variable v_4. It follows that variable v_4 appears in at least three constraints. Without loss of generality, constraint C' can be $C(\bar{v}_1 \vee \bar{v}_4 \vee v_7)$. Finally, to prevent property P w.r.t. variable v_4, there exists a constraint $C(v_1 \vee \bar{v}_7 \vee v_8)$ or $C(\bar{v}_1 \vee \bar{v}_5 \vee v_9)$. In the first case, literal v_1 appears in two 3-clauses and literal $\bar{v}_1$ appears in two 3-clauses, and the result follows by using binary decision (DEC1) on variable v_1. In the second case, binary decision (DEC2) on variables v_4 and v_5 is used. If $v_4 = v_5$, then constraints $C(v_1 \vee v_5)$ and $C(v_5 \vee v_6)$ representing two 2-clauses are derived. If $v_4 \neq v_5$, then Step 1 is used on variable v_1, and the problem is reduced for 2 variables. $\square$

Proposition 11 *Let C be a set of constraints containing constraints $C(v_1 \vee v_2)$, $C(\bar{v}_1 \vee \bar{v}_2 \vee v_3)$ and $C(\bar{v}_1 \vee \bar{v}_2 \vee v_4)$ after applying Algorithm 4 and Algorithm 6.*

There exists a binary decision such that for one value the problem is reduced for four variables and the number of constraints representing a 2-clause becomes not smaller.

Proof of proposition. Binary decision $v_1 = v_2$ causes $v_1 = v_2 = true$ and $v_3 = v_4 = false$, and the problem is reduced for four variables. If these instantiations do not reduce any constraint representing a 3-clause to a constraint representing a 2-clause, then $v_1 = v_2$ does not represent a decision! To solve the whole problem it is enough to solve the reduced problem. It follows that the number of constraints representing a 2-clause does not become smaller. □

Geometry Machines: From AI to SMC

Dongming Wang

Laboratoire LEIBNIZ – Institut IMAG
46, avenue Félix Viallet, 38031 Grenoble Cedex, France

Abstract. The existing techniques and software tools for automated geometry theorem proving (GTP) are examined and reviewed. The underlying ideas of various approaches are explained with a set of selected examples. Comments and analyses are provided to illustrate the encouraging success of GTP which interrelates AI and SMC. We also present some technological applications of GTP and discuss its challenges ahead.

1 Introduction

We recall *Geometry Machine*, the first GTP system developed by H. L. Gelernter on modern computer in the late 1950's. This historical machine, popularized for years in the AI (artificial intelligence) community, represents one of the early AI successes. Now the term *theorem prover* is often used instead of *machine* and refers to theorem proving systems and sometimes methods as well. In this paper we give an overview on how geometry machines have moved ahead from the past AI to the present SMC (symbolic mathematical computation) community. It is not meant that AI is no more a field to accommodate the machines, rather we want to demonstrate the interrelation between AI and SMC which has made and will continue making the machines more and more powerful. A set of selected examples is presented to illustrate the basic ideas underlying the various approaches that have offered convincing indication of the capability and fruitfulness of AI and SMC techniques in solving intelligence-required problems. Comments and remarks are provided for understanding the notable success of GTP. It is hoped that our analysis on the instance of geometry will be helpful and instructive for accomplishing automated theorem proving on other mathematical subjects.

2 Synthetical and Logical Approaches

2.1 Early Developments in AI

The geometry machine realized by Gelernter [58] on an IBM 704 proved its first theorem mechanically in early 1959. It was extended and improved to prove many other theorems; further work in this direction was reported in [50, 59, 60, 61, 101] and other relevant papers [1, 2] (see [61] for the prover BTP/PTP developed by I. Goldstein in MICRO-PLANNER).

To see the underlying idea of Gelernter's machine, we consider a class of geometric theorems with a set $\mathfrak{A}$ of axioms (and already established lemmas and theorems). For any concrete theorem $\mathcal{T}$, let its hypothesis be given as a set of geometric statements $H_1, H_2, \ldots, H_s$ and its conclusion given as a single geometric statement C. Proving $\mathcal{T}$ amounts to deciding whether the formula

$$H_1 \wedge H_2 \wedge \cdots \wedge H_s \implies C$$

is valid with respect to (w.r.t.) $\mathfrak{A}$.

To reach the *goal* C, one proceeds to construct labeled trees by generating *subgoals* from C with application of possible axioms of $\mathfrak{A}$. As $\mathfrak{A}$ is finite and only finitely many letters are involved in a given theorem, only finitely many subgoals can be generated at each level. One can generate *sub-subgoals* from the already generated subgoals in the same way. Up to any fixed level, a finite number of labeled trees will be generated. If it happens that all the statements associated with the leaves of a labeled tree are contained in the set of hypothesis-statements and $\mathfrak{A}$, then the theorem is proved and this tree is called a *proof tree*. The following example illustrates the idea explained.

Example 1 [59, 61]. If the segment joining the midpoints of the diagonals of a trapezoid is extended to intersect a side of the trapezoid, it bisects that side.[1]

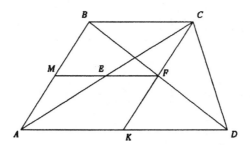

Let midp(A, B) stand for the midpoint of the segment AB. Then the hypothesis and conclusion of the theorem are:

$$\mathcal{H}: \quad AD \| BC, \ E = \mathsf{midp}(A, C), \ F = \mathsf{midp}(B, D),$$
$$\mathcal{C}: \quad M = \mathsf{midp}(A, B).$$

To prove the theorem, one need construct an auxiliary line CFK (which can be done heuristically [59, 61]). A proof tree can be generated as shown below (cf. [61]), where backward chaining is made by applying the corresponding facts and axioms indicated in the ovals. All the statements in the dashed boxes (leaves of the tree) are hypotheses and axioms, so the theorem is proved to be true.

Note that before a proof tree is obtained in the way described above, a large number of labeled trees may be generated. So practically, various heuristics

---

[1] The theorem may be false or meaningless when BC coincides with AD or AB is parallel to CD. Such degenerate cases can be ruled out automatically when algebraic methods are applied (see Example 2 and relevant discussions in [27, 41, 74, 86, 87, 126, 132, 142, 152]).

have to be and were developed to guide the search in order to generate a smaller number of labeled trees.

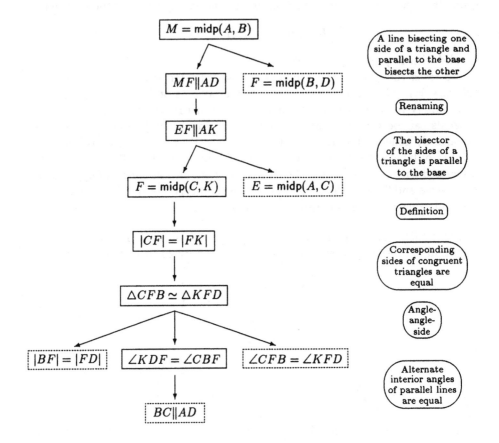

2.2 Late and Recent Developments

Nevertheless, the axiomatic approaches along with the line of Gelernter were hardly possible to reach the desire of proving and discovering hard theorems. Although various strategies and heuristics were subsequently adopted and implemented (see the comprehensive study of H. Coelho and L. M. Pereira [43] and the system GEOM developed by them in Prolog [42]), the problem of search space amongst the others still remains and makes the method highly inefficient.

In [104] A. Quaife applied the general-purpose system OTTER to theorem proving in Tarski's geometry. The procedures he described were able to prove a set of interesting theorems from axioms. Some of the theorems concern the betweenness relation. Much work has been done recently by P. Balbiani, L. Fariñas del Cerro and their collaborators [4, 5, 6, 7, 8]. They have studied extensively the application of logical deduction techniques, in particular term rewriting and diagrammatic reasoning, for GTP.

3 The Success of Algebraic Methods

3.1 Historical Remarks

The idea of mechanizing the proof of geometric theorems may be traced back to R. Descartes and G. W. Leibniz. One of the earliest theoretical contributions to the mechanization of geometry, according to W.-t. Wu, is due to D. Hilbert who gave a method for proving pure point of intersection theorems in his book "Grundlagen der Geometrie" [65]. However, Hilbert's method had never been recognized before Wu pointed out its merit [139, 142]. This method was extended late by the author to a wider application domain [117, 120]. On the other hand, in 1930 A. Tarski discovered a general decision method for the theory of elementary algebra and geometry. His method, published in 1948, provides a remarkable positive result on the mechanizability problem for a large domain. This method as well as some others subsequently proposed has been considered of high complexity and not yet been implemented on any computer up to the present. It is G. E. Collins who invented in 1975 another method based on cylindrical algebraic decomposition (CAD) which deals with the same decision problem as Tarski's and is practically more efficient.

In 1969, E. Cerutti and P. Davis applied a symbolic manipulation system, called FORMAC, to prove theorems in elementary geometry, following the analytic method of Descartes. They proved the well-known theorem of Pappus, pointed out the way of attacking Pascal's theorem and obtained a couple of new theorems by inspecting the machine printout. Also contained in [18] are interesting discussions on the aspects of machine proofs. This work is much in the spirit of the algebraic approaches to GTP to be discussed in the following section. Unfortunately, the techniques of Cerutti and Davis were not developed further to a general GTP method.

3.2 Wu's Method

A breakthrough of GTP is marked by the work of Wu [136]. His method has stimulated the lasting activities on the subject. Wu observed that most of the fundamental geometric relations can be expressed by means of polynomial equations via algebraization and coordinatization. Thus, proving most of the theorems in elementary geometries can be reduced to manipulating the corresponding algebraic relations, for which he developed a powerful method based on J. F. Ritt's concept of characteristic sets (CS [107]). Wu implemented his method as a system, called *China Prover*, in Fortran and used it to prove and discover several non-trivial theorems including the Morley theorem, Steiner theorem and Pascal-Conic theorem in elementary geometry [141, 142] as well as theorems about curve pairs of Bertrand type in differential [148, 153] geometry. Wu's work has become widely known outside China since 1984 when the book [9] edited by W. W. Bledsoe and D. W. Loveland was published. This book reprints Wu's 1978 paper and contains a notable paper by S.-C. Chou [19] in which he described a prover based on Wu's method that had already proved 130 theorems by that time.

An elementary version of Wu's method is simple: using the techniques of analytic geometry, one expresses the hypothesis and conclusion of a geometric theorem $\mathcal{T}$ as polynomial equations and inequations of the form

$$\mathcal{H}: \quad h_1 = 0, \ldots, h_s = 0, \quad d_1 \neq 0, \ldots, d_t \neq 0,$$
$$\mathcal{C}: \quad c = 0.$$

Proving $\mathcal{T}$ is equivalent to deciding whether the formula

$$(\forall x_1 \cdots \forall x_n)\, [h_1 = 0 \wedge \cdots \wedge h_s = 0 \wedge d_1 \neq 0 \wedge \cdots \wedge d_t \neq 0 \Longrightarrow c = 0] \quad (1)$$

is valid, where h_i, d_j and c are polynomials in $x_1, \ldots, x_n$ with coefficients in a geometry-associated field $\mathbf{K}$. For this, one computes a characteristic set $\mathbb{C}$ of $\{h_1, \ldots, h_s\}$ and the pseudo-remainder r of c w.r.t. $\mathbb{C}$. If r is identically equal to 0, then $\mathcal{T}$ is proved to be true under the subsidiary condition $J \neq 0$, where J is the product of initials of the polynomials in $\mathbb{C}$ [141, 142].

Example 2 (Secant Theorem). Let AB and CD be any two secants of a circle, intersecting at T. Then $|TA||TB| = |TC||TD|$.

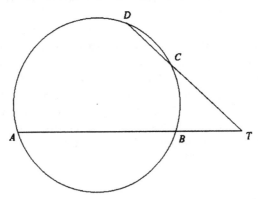

Without loss of generality, we take the coordinates of the points as $A(-u_1, 0)$, $B(u_1, 0), C(u_2, u_3), D(u_4, y_1), T(y_2, 0)$. Then the hypothesis of the theorem consists of

$$\mathcal{H}: \begin{cases} h_1 = u_1(u_3 y_1^2 - u_3^2 y_1 - u_2^2 y_1 + u_1^2 y_1 + u_3 u_4^2 - u_1^2 u_3) = 0, \\ \hspace{5cm} \text{\% } A, B, C, D \text{ are co-circular} \\ h_2 = y_1 y_2 - u_3 y_2 - u_2 y_1 + u_3 u_4 = 0. \hspace{0.8cm} \text{\% } AB \text{ intersects } CD \text{ at } T \end{cases}$$

Let collinear$(A, B, C, \ldots)$ state that the points $A, B, C, \ldots$ are collinear. We may add the following condition to $\mathcal{H}$ to exclude the obvious degenerate case:

$$d = u_1 u_3 \neq 0. \hspace{4cm} \text{\% not collinear}(A, B, C)$$

The conclusion of the theorem is

$$\mathcal{C}: \quad c = (y_2 + u_1)^2 (y_2 - u_1)^2 - \left[(y_2 - u_2)^2 + u_3^2\right]\left[(y_2 - u_4)^2 + y_1^2\right] = 0.$$
$$\text{\% } |TA|^2|TB|^2 = |TC|^2|TD|^2$$

Here $\{h_1, h_2\}$ is already a characteristic set of itself w.r.t. the variable ordering

$y_1 \prec y_2$ and the product of the initials of h_1 and h_2 is $u_1 u_3 (y_1 - u_3)$. Clearly, $u_1 u_3 \neq 0$ as we have assumed. Under the subsidiary condition $y_1 \neq u_3$, pseudo-divide c successively by h_2 w.r.t. y_2 and by h_1 w.r.t. y_1. One should find that the final remainder is 0, so the theorem is proved to be true. It is easy to see that $y_1 = u_3$ corresponds to the degenerate case in which AB and CD are parallel and thus do not have any intersection point.

The above simple procedure works effectively for confirming a large number of theorems. However, if r happens to be non-zero, one cannot tell immediately whether the theorem is false or not. In this case, one has to examine the reducibility of $\mathbb{C}$ and to perform decomposition further. $\mathbb{C}$ is reducible often when some geometric ambiguities such as bisection of angles and contact of circles are involved in the theorem. Moreover, if one wishes to know whether the theorem is true when $J = 0$ (which in general corresponds to some degenerate cases of the theorem), further decomposition need be carried out as well [26, 81, 124, 141, 142]. For a complete version of Wu's method, much deep mathematical theory and methods have to be employed, for which we refer to [142].

3.3 Gröbner Bases and CAD for GTP

Awareness of Wu's method motivated researchers to apply B. Buchberger's well-known method of Gröbner bases (GB [12]) to prove the same type of theorems that Wu's method addresses. In 1986 several papers [38, 72, 73, 88, 89] on the application of GB to GTP were published. See also [16, 39, 69, 131, 133]. Roughly, there are two ways of applying GB for GTP: one as proposed by B. Kutzler, S. Stifter and Chou is to compute first a Gröbner basis $\mathbb{G}$ of the set $\mathbb{H}$ of hypothesis-polynomials and then the normal form h of the conclusion-polynomial c modulo $\mathbb{G}$. The theorem is true if $h = 0$. The other proposed by D. Kapur is a refutational approach by computing a Gröbner basis $\mathbb{G}^*$ of $\mathbb{H} \cup \{zc - 1\}$. The theorem is true if $1 \in \mathbb{G}^*$. In practice, one has to take care of some details in dealing with non-degeneracy conditions.

Example 3 (Secant Theorem Revisited). Recall the theorem considered in Example 2. A Gröbner basis of $\{h_1, h_2\}$ w.r.t. the purely lexicographical term ordering determined by $y_1 \prec y_2$ is

$$\mathbb{G} = \left\{ h_1 / u_1, (u_4 + u_2) y_2 - u_3 y_1 - u_2 u_4 - u_1^2 \right\}.$$

The normal form of c modulo $\mathbb{G}$ is 0, so the theorem is true under some subsidiary/non-degeneracy conditions.

Alternatively, one can compute a Gröbner basis of $\{h_1, h_2, zc - 1\}$ w.r.t. the purely lexical term ordering determined by $y_1 \prec y_2 \prec z$ and should find that it contains 1. Thus the theorem is proved to be true under some conditions as well. The conditions may be expressed as an inequation in the parametric variables $u_1, \ldots, u_4$.

Researchers developed different provers including the GEO-Prover by Chou [19, 20, 22] (in Pascal, Macsyma and Lisp), ALGE-Prover/ALGE-Prover II/GEO-

METER by H.-P. Ko, Kapur and their co-workers [47, 80, 84] (in Macsyma and on a Symbolics Lisp machine) and GPP by Kutzler, Stifter and their co-workers [85, 90] (in SAC-2 and Scratchpad II), based on Wu's and Buchberger's method. Extensive experiments with comparisons on/between the two methods were made for a large number of theorems around that time [22, 83, 84, 91, 109]. Chou's prover is one of the most remarkable and has been used to prove more than 500 theorems in elementary (differential) geometries [22, 23]. A late version of the prover is also available for Macintoshes [24].

Meanwhile, Wu together with his students continued improving his method and applying it to theorem proving and discovering [53, 130, 140, 144, 146, 149]. With the limitation of computing facilities and access to high level software, members of Wu's group wrote their own Fortran code, e.g., the systems CPS/PS/DPMS1/DPMS2 by X.-S. Gao and D. Wang [129], for polynomial manipulation and GTP. More than hundred theorems were proved [129], with several "new" theorems discovered [119, 143, 148, 159].

Wu's method also inspired the invention of another method — proving by examples [67, 68, 118] and was extended and applied to prove theorems in elementary geometries besides plane Euclidean geometry [37, 54, 117] and in differential geometry [137, 138, 147]. Procedures based on GB were also developed to prove theorems in differential geometry [16].

Methods based on CS and GB mainly deal with geometric theorems whose algebraic formulations only involve polynomial equations and inequations. When inequalities occur, they may be inapplicable. The general quantifier elimination method based on CAD introduced by Collins [44] provides a decision procedure for the theory of elementary algebra and geometry. Collins' method has been subsequently improved and is applicable for a variety of decision problems that may be expressed as formulae of the prenex form

$$(Q_1 x_1 \cdots Q_n x_n) \Phi(x_1, \ldots, x_n).$$

In the above expression, $Q_i x_i$'s are existential ($\exists$) and universal ($\forall$) quantifiers and Φ a standard quantifier-free formula in $x_1, \ldots, x_n$, composed from standard atomic formulae of the forms $F = 0$ and $F > 0$, where F is any non-zero polynomial with rational coefficients in $x_1, \ldots, x_n$. The method was applied to GTP by D. S. Arnon [3].

Example 4 (Pedoe's Inequality [121, 150]). Let ABC and $A_1 B_1 C_1$ be two triangles in the same plane with sides a, b, c; a_1, b_1, c_1 and areas R, R_1 respectively. Then

$$a_1^2(b^2 + c^2 - a^2) + b_1^2(c^2 + a^2 - b^2) + c_1^2(a^2 + b^2 - c^2) \geq 16 R_1 R. \qquad (2)$$

The equality holds only if the two triangles are similar.

To prove the inequality by CAD, we assume, without loss of generality, that the two triangles have a common vertex, say $B = B_1$, and the side BC coincides with $B_1 C_1$. Choose the coordinates of the points as $A(x, y), A_1(x_1, y_1), B = B_1(0, 0), C(a, 0)$ and $C_1(a_1, 0)$.

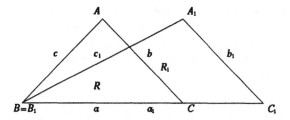

We may further assume that $a > 0, a_1 > 0, y > 0$ and $y_1 > 0$. Then

$$b^2 = (x - a)^2 + y^2, \quad b_1^2 = (x_1 - a_1)^2 + y_1^2,$$
$$c^2 = x^2 + y^2, \quad c_1^2 = x_1^2 + y_1^2,$$
$$R = \frac{1}{2}ay, \quad R_1 = \frac{1}{2}a_1 y_1.$$

Substituting them into (2), the inequality to be proved becomes

$$a^2 x_1^2 + a^2 y_1^2 - 2aa_1 xx_1 + a_1^2 x^2 + a_1^2 y^2 \geq 2aa_1 yy_1. \tag{3}$$

By eliminating the quantifiers of the following formula (with x, x_1 considered as parameters)

$$(\forall a, a_1, y, y_1)[a > 0 \wedge a_1 > 0 \wedge y > 0 \wedge y_1 > 0 \Longrightarrow (3)],$$

one finds that the formula is true for any x, x_1. Clearly there are a, a_1, x, x_1, y, y_1 such that $a > 0, a_1 > 0, y > 0, y_1 > 0$. Hence, the inequality is proved to be true universally (i.e., without non-degeneracy condition). Similarly, one can prove that the equality holds only if the two triangles are similar. It may also be proved by using Wu's method in combination with the technique of Lagrangian multiplier [150].

3.4 Further Developments

Theoretical and practical improvements for the methods of CS, GB and CAD are made constantly through the research activities on the methods' subjects, for which the reader is referred to [14, 121] and references therein for more information. Below we briefly mention some improvements and recent developments directed to GTP.

Wu's method has been further studied and improved [26, 53, 78, 81, 82, 150, 154, 155, 156, 157], implemented [56, 66, 128], extended and applied to prove theorems in space geometry [116], geometry over finite fields [97] and differential geometry [29, 30, 96, 153], and combined with Collins' method for proving theorems involving inequalities [31, 98].

The dimension method of Carrà Ferro and Gallo has been extended in [17]. See also the prover GEO developed by M. A. Alberti, B. Lammoglia and M. Torelli in Mathematica [66].

The method of proving by examples developed by M.-K. Deng, L. Yang, J.-Z. Zhang and their co-workers [49, 164, 168, 171] follows the idea of the single-instance method of J. Hong. It uses multi-instances (m examples) with parallel numerical verification incorporated, where m is calculated from the degrees of the hypothesis-polynomials.

Other methods based on elimination techniques include those proposed by M. Kalkbrener [70], Wang [126, 127], Yang, Zhang and X.-R. Hou [165, 166, 167] and Kapur, T. Saxena and Yang [77]. Kalkbrener's method is based on a generalized Euclidean algorithm which computes unmixed-dimensional decompositions of algebraic varieties with no need of polynomial factorization. The underlying idea of this method also appears in the method presented in [166, 167]. Resultant computation is used for the methods suggested in [77, 108, 165, 166, 167]. See [62] for a group of mechanically proved theorems in real geometry and [105] for GTP over both the complexes and the reals.

In what follows we discuss some examples for the author's method which is described in detail in [126] (for elementary geometry) and [127] (for differential geometry). In this method as well as our implementation of Wu's method in GEOTHER [128], we make use of polynomial factorization over algebraic extension fields. This is appropriate in particular when the reducibility problem is of concern [124, 149].

For Thébault-Taylor's theorem, the first mechanized proof taking about 44 CPU hours on a Symbolics 3600 was given by Chou in 1986 [22, pp. 66–69], and a simplified proof taking about 6 CPU hours on a Dual was soon given by Wu [149]. Another mechanized proof was given by Yang and Zhang [165] via resultant computation. Their proof took 1042 and 268 CPU seconds on a SUN386i and a CONVEX C210 respectively. With Yang-Zhang's formulation, the set of hypothesis-polynomials may be decomposed by our method into four irreducible triangular sets. It can be verified that the pseudo-remainder of the conclusion-polynomial is 0 w.r.t. one triangular set, but not 0 w.r.t. the others. Therefore, the theorem is true for one component, and false for all the others. The proof took about 60 CPU seconds in Maple 4.3 on an Apollo DN10000 [126].

The previous proofs of Morley's theorem have all followed the tricky formulation of Wu [22, 141]. By using our method with algebraic factorization, we were able to compute an irreducible zero decomposition and to prove the theorem in less than hundred seconds without using Wu's trick [126]. The example considered below also illustrates how algebraic factorization can be helpful for GTP. The theorem is of special interest mainly because it is true over the reals but not the complexes.

Example 5 (MacLane's 8_3 Configuration [45, 86, 98]). Given eight points $A, B,$ C, D, E, F, G, H such that the following eight triples are collinear: $ABD, BCE,$ $CDF, DEG, EFH, FGA, GHB, HAC$. Then the eight points lie on the same line.

Take the coordinates of the points as $A(0, 0), B(1, 0), D(u_1, 0), C(x_1, y_1),$ $E(x_2, y_2), F(x_3, y_3), G(x_4, y_4), H(x_5, y_5)$. The hypothesis and conclusion of the theorem may be expressed as follows

$$\mathcal{H}: \begin{cases} h_1 = y_1 x_2 - y_2 x_1 + y_2 - y_1 = 0, & \text{\% collinear}(B, C, E) \\ h_2 = y_1 x_3 - y_3 x_1 + u_1(y_3 - y_1) = 0, & \text{\% collinear}(C, D, F) \\ h_3 = y_2 x_4 - y_4 x_2 + u_1(y_4 - y_2) = 0, & \text{\% collinear}(D, E, G) \\ h_4 = (y_2 - y_3)x_5 + (y_5 - y_2)x_3 + (y_3 - y_5)x_2 = 0, & \text{\% collinear}(E, F, H) \\ h_5 = y_3 x_4 - y_4 x_3 = 0, & \text{\% collinear}(F, G, A) \\ h_6 = y_4 x_5 - y_5 x_4 + y_5 - y_4 = 0, & \text{\% collinear}(G, H, B) \\ h_7 = y_1 x_5 - y_5 x_1 = 0, & \text{\% collinear}(H, A, C) \end{cases}$$

$$\mathcal{C}: \qquad c = y_1 = y_2 = y_3 = y_4 = y_5 = 0. \quad \text{\% collinear}(A, B, C, D, E, F, G, H)$$

Computing an irreducible zero decomposition of $\{h_1, \ldots, h_7\}$ w.r.t. the variable ordering $u_1 \prec y_1 \prec \cdots \prec y_5 \prec x_1 \prec \cdots \prec x_5$, we get 20 irreducible triangular sets $\mathbb{T}_1, \ldots, \mathbb{T}_{20}$ (in 27 CPU seconds on a SUN SparcServer 690/51). One of them is $[y_1, y_2, y_3, y_4, y_5]$, for which the theorem is clearly true. With a careful analysis, one can figure out that 17 of the triangular sets correspond to some degenerate cases[2] of the theorem such as two points coincide. For the remaining two non-degenerate triangular sets, the first polynomials are quadratic and all the others are linear w.r.t. their leading variables. The theorem is not true over the complex number field for these two components. To see whether it is true over the reals, we consider one of the components: the first polynomial is

$$c_1 = u_1^2 y_2^2 - u_1 y_2^2 + y_2^2 + u_1 y_1 y_2 - 2 y_1 y_2 + y_1^2.$$

It is easy to see that c_1 can be factorized into two linear factors (cf. [117, 124]):

$$c_1 = \frac{1}{4(u_1^2 - u_1 + 1)} \left(2 u_1^2 y_2 - 2 u_1 y_2 + 2 y_2 + u_1 y_1 - 2 y_1 + u_1 y_1 \alpha\right)$$
$$\left(2 u_1^2 y_2 - 2 u_1 y_2 + 2 y_2 + u_1 y_1 - 2 y_1 - u_1 y_1 \alpha\right),$$

where $\alpha = \sqrt{-3}$. So c_1 has real zeros if and only if the imaginary part $u_1 y_1$ of the two linear factors are zero. A further analysis shows that $u_1 y_1 = 0$ correspond to some degenerate cases. Therefore, the theorem is not true over the reals for this component. The same discussion holds for the other component. So the theorem is true for one component and false in the other degenerate cases over the reals.

4 Coordinate-Free Techniques

A drawback of the coordinate approaches is that the constructed proofs of geometric theorems are neither readable nor geometrically interpretable. In order that the attraction of geometry is not drowned by complicated algebraic expressions, one way is to create friendly environments for manipulating and proving theorems, leaving out the algebraic proof steps as black-box or representing them

---

[2] One of them is not exactly a degenerate case but a specialized one with $u_1^2 - u_1 + 1 = 0$.

with good structure. The other is to express geometric statements in terms of other geometric entities such as distances, vectors, brackets, angles, areas and volumes, rather than point coordinates. This is feasible for some classes of geometric theorems, and in such cases the algebraic manipulations performed among the geometric entities are usually simple. This is the so-called *coordinate-free* approach. Work in this direction started in [46, 51, 64, 134] and has been highlighted recently [32, 33, 34, 35, 36, 93, 94, 106, 111, 170]. The methods suggested are efficient and capable of producing proofs composed of expressions in geometrically interpretable quantities.

4.1 Bracket Algebra

A method based on bracket algebra is proposed in [106] for proving theorems in projective geometry. To get an impression about how the method proceeds, let $A_1, \ldots, A_n$ be n points in an $(n-1)$-dimensional projective space. We define a bracket

$$[A_1 \cdots A_n] = \det(\boldsymbol{x}_1, \ldots, \boldsymbol{x}_n),$$

where $\boldsymbol{x}_i = (x_{i1}, \ldots, x_{in})^\tau$ is the n-tuple of coordinates of the point A_i for each i and det denotes the determinant of the matrix. Then the incidence relations in projective geometry can be translated into algebraic equations in terms of brackets. Proving a theorem is reduced to manipulating the bracket equations (note that coordinates here are used only for the definition of brackets and will not be involved in the algebraic manipulation). A system called CINDERELLA has been developed by H. Crapo and J. Richter-Gebert to deal with configurations in projective, Euclidean and spheric geometry [66]. This system contains a prover based on the bi-quadratic final polynomial method that generates proofs in terms of syzygies of determinant (bracket) expressions [106]. A number of incidence theorems have been proved by this prover.

Example 6 (Pappus' Theorem [18, 106, 119]). Let A, B, C and A_1, B_1, C_1 be two sets of points respectively on two distinct lines. Then the points P, Q, R of intersection are collinear.

From the hypothesis and conclusion of the theorem the following bi-quadratic equations among the brackets can be produced:

$$\mathcal{H}: \begin{cases} [ABQ][RA_1B_1] = -[ABA_1][QRB_1], & \% \ AB, QA_1, RB_1 \text{ are concurrent} \\ [AQR][BA_1B_1] = -[ABQ][RA_1B_1], & \% \ AQ, A_1B_1, BR \text{ are concurrent} \\ [ABP][PQB_1] = [APQ][BPB_1], & \% \text{ collinear}(A, P, B_1) \\ [ABA_1][BPB_1] = [ABP][BA_1B_1], & \% \text{ collinear}(B, P, A_1) \end{cases}$$

$$\Downarrow$$

$$\mathcal{C}: \quad [AQR][PQB_1] = [APQ][QRB_1]. \qquad \% \text{ collinear}(P, Q, R)$$

Multiplying the two sides of the hypothesis-equations and canceling the brackets

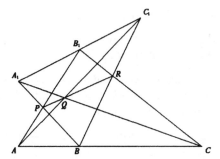

that occur on both sides yield the conclusion-equation, and thus the theorem is proved.

4.2 Area Method

The key idea of the area method [169] is to represent geometric relations in terms of areas rather than coordinates. The generation of proofs is accomplished by eliminating points from the conclusion-relation successively according to some well-designed rules and basic propositions. Extensive work on implementing the method has been done by Chou, Gao and Zhang [33, 35, 36, 170]. They have developed a system named EUCLID [35, 66] based on this method and the vector approach [34]. EUCLID can produce *short and readable* proofs[3] for many constructive statements in plane as well as solid geometry [32, 35].

Let $\overline{AB}$ denote the length of the oriented segment AB, Δ_{ABC} the signed area of the oriented triangle ABC and define

$$\Delta_{A_1 \cdots A_n} = \sum_{i=4}^{n} \Delta_{A_1 A_{i-1} A_i}, \quad n \geq 3.$$

The following example shows how the area method works.

Example 7 (Pappus' Theorem Revisited [170]). Refer to the preceding example. From the hypothesis of the theorem and some fundamental geometric propositions, one can establish the following relations among the areas:

$$\Delta_{QBC_1} = \frac{\Delta_{QBC_1}}{\Delta_{ABC_1}} \cdot \Delta_{ABC_1} = \frac{\overline{QC_1}}{\overline{AC_1}} \cdot \Delta_{ABC_1} = \frac{\Delta_{A_1CC_1} \cdot \Delta_{ABC_1}}{\Delta_{ACC_1A_1}}, \qquad (4)$$

$$\Delta_{QCB_1} = \frac{\Delta_{QCB_1}}{\Delta_{A_1CB_1}} \cdot \Delta_{A_1CB_1} = \frac{\overline{QC}}{\overline{A_1C}} \cdot \Delta_{A_1CB_1} = \frac{\Delta_{ACC_1} \cdot \Delta_{A_1CB_1}}{\Delta_{ACC_1A_1}}, \qquad (5)$$

$$\Delta_{PBC_1} = \frac{\Delta_{PQC_1}}{\Delta_{A_1BC_1}} \cdot \Delta_{A_1BC_1} = \frac{\overline{PB}}{\overline{A_1B}} \cdot \Delta_{A_1BC_1} = \frac{\Delta_{ABB_1} \cdot \Delta_{A_1BC_1}}{\Delta_{ABB_1A_1}}, \qquad (6)$$

$$\Delta_{PCB_1} = \frac{\Delta_{PCB_1}}{\Delta_{ACB_1}} \cdot \Delta_{ACB_1} = \frac{\overline{PB_1}}{\overline{AB_1}} \cdot \Delta_{ACB_1} = \frac{\Delta_{A_1BB_1} \cdot \Delta_{ACB_1}}{\Delta_{ABB_1A_1}}. \qquad (7)$$

---

[3] The length and readability of proofs depend on how many and which kind of lemmas/propositions have been built into the prover and thus do not have great significance in the sense that a theorem may be proved in two steps by introducing two lemmas.

The conclusion of the theorem corresponds to

$$\mathcal{C}: \quad \frac{\Delta_{PBC_1}}{\Delta_{QBC_1}} \cdot \frac{\Delta_{QCB_1}}{\Delta_{PCB_1}} = \frac{\overline{PO}}{\overline{QO}} \cdot \frac{\overline{QO'}}{\overline{PO'}} = 1, \qquad \text{\% collinear}(P, Q, R)$$

where O and O' are the intersection points of PQ with BC_1 and with B_1C, respectively. Using the formulas (4)–(7) to eliminate the points Q and P, one obtains a proof of the theorem with the following calculation

$$\frac{\Delta_{PBC_1}}{\Delta_{QBC_1}} \cdot \frac{\Delta_{QCB_1}}{\Delta_{PCB_1}} = \frac{\Delta_{ABB_1}}{\Delta_{A_1CC_1}} \cdot \frac{\Delta_{A_1BC_1}}{\Delta_{ABC_1}} \cdot \frac{\Delta_{ACC_1}}{\Delta_{A_1BB_1}} \cdot \frac{\Delta_{A_1CB_1}}{\Delta_{ACB_1}}$$

$$= \frac{\overline{AB}}{\overline{AC}} \cdot \frac{\overline{A_1C_1}}{\overline{A_1B_1}} \cdot \frac{\overline{AC}}{\overline{AB}} \cdot \frac{\overline{A_1B_1}}{\overline{A_1C_1}} = 1.$$

The above methods employ different notations (brackets vs. areas) and different techniques for algebraic manipulation, but they have obvious relevance. This is simply because areas can be defined by means of determinants.

4.3 Clifford Algebra

The application of Clifford algebra (representation and reduction) to GTP has been investigated by H. Li and M. Cheng [93, 94]. Theorems have been proved by combining Clifford algebraic technique and Wu's method. Li also points out that the area method can be unified by means of Clifford algebras. See [34, 111, 134] for some related work. The author has noticed that for Clifford algebras (which are non-commutative algebras of solvable type) non-commutative Gröbner bases can be computed [71]. Thus, a systematic approach for GTP based on non-commutative GB may be developed. However, there are a few technical difficulties involved, for which we shall discuss elsewhere.

Let $\mathbf{v}_1, \ldots, \mathbf{v}_m$ be m elements of a vector space $\mathbf{V}$. Consider the free associative algebra $K\langle \mathbf{v}_1, \ldots, \mathbf{v}_m \rangle$ over K generated by $\{x_1, \ldots, x_n\}$ with commutation system

$$\mathbb{Q} = \{\mathbf{v}_j\mathbf{v}_i + \mathbf{v}_i\mathbf{v}_j \mid 1 \le i < j \le n\}.$$

Let $\mathcal{Q}$ denote the two-side ideal generated by $\mathbb{Q}$ in $K\langle \mathbf{v}_1, \ldots, \mathbf{v}_m \rangle$; then

$$\mathbf{R} = K\{\mathbf{v}_1, \ldots, \mathbf{v}_m\} = K\langle \mathbf{v}_1, \ldots, \mathbf{v}_m \rangle / \mathcal{Q}$$

is a solvable polynomial ring. Let $\mathcal{I}$ be the two-side ideal in $\mathbf{R}$ generated by $\mathbb{P} = \{\mathbf{v}_1^2, \ldots, \mathbf{v}_m^2\}$. Then $\mathbb{P}$ is a two-side Gröbner basis in $\mathbf{R}$ and the quotient $\mathbf{R}/\mathcal{I}$ is Grassmann algebra (i.e., the Clifford algebra associated to the null quadratic form). The ideal membership in $\mathbf{R}/\mathcal{I}$ can be tested as follows [71]: given the generating set $\mathbb{F}$ of a two-side (left, right) ideal $\mathcal{J} \subset \mathbf{R}/\mathcal{I}$, construct a (left, right) Gröbner basis $\mathbb{G}$ of $\mathcal{I} + \mathcal{J}$ from $\mathbb{P} \cup \mathbb{F}$ and compute a normal form h of any given polynomial f in $\mathbf{R}/\mathcal{I}$ modulo $\mathbb{G}$. Then $f \in \mathcal{J}$ if and only if $h = 0$. The decision procedure of ideal membership provides therefore a simple way for GTP. The following example illustrates this point.

Example 8 (Gauss Line [117, 120]). For any four lines $A_1A_2A_3$, $A_1B_2B_3$, $B_1A_2B_3$ and $B_1B_2A_3$, the midpoints M_1, M_2, M_3 of A_1B_1, A_2B_2, A_3B_3 are collinear.

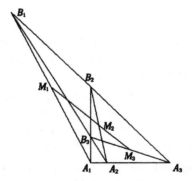

The hypothesis and conclusion of the theorem may be expressed as

$$\mathcal{H}: \begin{cases} h_1 = (A_1 - A_2) \wedge (A_1 - A_3) = 0, & \text{\% collinear}(A_1, A_2, A_3) \\ h_2 = (A_1 - B_2) \wedge (A_1 - B_3) = 0, & \text{\% collinear}(A_1, B_2, B_3) \\ h_3 = (B_1 - A_2) \wedge (B_1 - B_3) = 0, & \text{\% collinear}(B_1, A_2, B_3) \\ h_4 = (B_1 - B_2) \wedge (B_1 - A_3) = 0, & \text{\% collinear}(B_1, B_2, A_3) \\ h_5 = 2M_1 - (A_1 + B_1) = 0, & \text{\% } M_1 = \text{midp}(A_1, B_1) \\ h_6 = 2M_2 - (A_2 + B_2) = 0, & \text{\% } M_2 = \text{midp}(A_2, B_2) \\ h_7 = 2M_3 - (A_3 + B_3) = 0. & \text{\% } M_3 = \text{midp}(A_3, B_3) \end{cases}$$

$$\mathcal{C}: \qquad c = (M_1 - M_2) \wedge (M_1 - M_3) = 0, \qquad \text{\% collinear}(M_1, M_2, M_3)$$

where $\wedge$ denotes the outer product. A two-side (non-commutative) Gröbner basis of $\{h_1, \ldots, h_7\} \cup \{A_1 \wedge A_1, \ldots, M_3 \wedge M_3\}$ w.r.t. the lexical term ordering determined by $A_1 \prec A_2 \prec A_3 \prec B_1 \prec B_2 \prec B_3 \prec M_1 \prec M_2 \prec M_3$ is

$$\mathbb{G} = \begin{cases} A_2 \wedge A_3 - A_1 \wedge A_3 + A_1 \wedge A_2, \\ B_1 \wedge B_2 - A_3 \wedge B_2 + A_3 \wedge B_1, \\ B_1 \wedge B_3 - A_2 \wedge B_3 + A_2 \wedge B_1, \\ B_2 \wedge B_3 - A_1 \wedge B_3 + A_1 \wedge B_2, \\ M_1 - \frac{1}{2}B_1 - \frac{1}{2}A_1, \\ M_2 - \frac{1}{2}B_2 - \frac{1}{2}A_2, \\ M_3 - \frac{1}{2}B_3 - \frac{1}{2}A_3 \end{cases} \cup \{A_1 \wedge A_1, \ldots, B_3 \wedge B_3\}.$$

The normal form of c modulo $\mathbb{G}$ is 0, so the theorem is proved to be true.

Note that if the normal form is non-zero, one does not know if the theorem is true or not. In other words, the procedure explained above is not complete. Its applicability can be extended by generating more hypothesis-relations using Clifford algebra operations (preprocessing). Further research in this direction is in progress.

5 Applications

The methods and tools of GTP have applications in various domains of science, engineering and industry. We list some of the applications with pointers for access to the existing literature:

Geometric computations such as automated derivation of unknown relations and locus equations [21, 25, 28, 57, 121, 125, 145, 153], conditions and detection of singularities [13, 121, 125], decomposition of algebraic varieties [40, 121, 122, 125, 142] and arrangement problems [3, 10, 156];

Geometric modeling and CAGD [76] such as parametrization and implicitization of curves and surfaces [13, 55, 95, 121, 125], construction with constraints [11] and surface fitting [160];

Computer vision such as perspective viewing [75, 100, 112], fusing multiple geometric sensor outputs [48] and global stereo vision [161];

Robot inverse and direct kinematics [13, 99, 121] and motion planning [10, 100, 103, 121, 135];

Mechanics such as automated derivation and discovery of motion equations and physical laws [28, 99, 127, 151, 153, 158];

Geometry education [2, 35, 63, 92, 159].

In what follows are provided two simple examples from geometric modeling and robot motion planning for illustration.

Example 9 (Biarcs [121, 125]). Given two points A and C of two different circular arcs which have given tangent directions at A and C, determine the locus of an intermediate point B at which the two circular arcs join together with a common tangent.

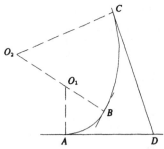

Choose the coordinates of the points as $A(0,0), D(u_1,0), C(u_2,u_3), B(X,Y)$, $O_1(0,x_1), O_2(x_2,x_3)$. The geometric conditions may be expressed as

$$\mathcal{H}: \begin{cases} h_1 = (u_2 - u_1)(x_2 - u_2) + u_3(x_3 - u_3) = 0, & \text{\% } CO_2 \perp DC \\ h_2 = X^2 + (x_1 - Y)^2 - x_1^2 = 0, & \text{\% } |O_1A| = |O_1B| \\ h_3 = (x_2-u_2)^2+(x_3-u_3)^2-(x_2-X)^2-(x_3-Y)^2=0, & \text{\% } |O_2C| = |O_2B| \\ h_4 = X(x_3 - x_1) - x_2(Y - x_1) = 0. & \text{\% collinear}(B,O_1,O_2) \end{cases}$$

From the above hypothesis-relations one can derive automatically, by using Wu's method, the GB method or the author's, the "new" relation $r_1 r_2 = 0$, where

$$r_1 = (X - \frac{u_1 - a}{2})^2 + (Y - \frac{u_1 u_2 + a u_2 - u_1^2 + a^2}{2u_3})^2 - \frac{a(2u_1 u_2 - u_1^2 + a^2)}{2(-u_2 + u_1 + a)},$$

$$r_2 = (X - \frac{u_1 + a}{2})^2 + (Y - \frac{u_1 u_2 - a u_2 - u_1^2 + a^2}{2u_3})^2 - \frac{a(2u_1 u_2 - u_1^2 + a^2)}{2(u_2 - u_1 + a)},$$

$$a = \sqrt{u_3^2 + (u_1 - u_2)^2} = |CD|.$$

Therefore, the locus of B has two components for any fixed u_1, u_2, u_3. $r_1 = 0$ and $r_2 = 0$ represent two circles passing through A and C, of which one corresponds to the biarc of convex shape and the other to the biarc of S-shape.

Example 10 ("Piano Movers" Problem [121]). Determine whether an infinitesimally thin "ladder" of length 3 can traverse a right angle corner in a corridor with width 1.

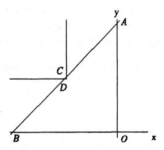

Mathematically, the corridor is described by

$$\{(x, y)| \ x \le 0, 0 \le y \le 1\} \cup \{(x, y)| \ y \ge 0, -1 \le x \le 0\}.$$

By simple reasoning, one sees that the ladder cannot pass the corridor if and only if it intersects all the four walls $\{x = 0, y \ge 0\}, \{y = 0, x < 0\}, \{x = -1, y \ge 1\}$ and $\{y = 1, x < -1\}$.

If the ladder does intersect the four walls, say at $A(0, a), B(b, 0), C(-1, c)$ and $D(d, 1)$, then the constraints among the coordinates of these points are given as

$$\begin{cases} a \ge 0, b < 0, c \ge 1, d < -1, & \text{\% } A, B, C, D \text{ are resp. on 4 walls} \\ a^2 + b^2 \le 9, & \text{\% } |AB| \le 3 = \text{ length of the ladder} \\ d - (1 - a)(d - b) = 0, & \text{\% collinear}(A, D, B) \\ c - (1 + b)(c - a) = 0. & \text{\% collinear}(A, C, B) \end{cases}$$

Therefore, the problem is formulated to determine the truth of

$$(\exists a, b, c, d)[a \ge 0 \wedge b < 0 \wedge c \ge 1 \wedge d < -1 \wedge a^2 + b^2 \le 9$$
$$\wedge d - (1 - a)(d - b) = 0 \wedge c - (1 + b)(c - a) = 0].$$

Applying the CAD method to this formula, one gets a positive answer. This means that the ladder may intersect simultaneously all the four walls and thus cannot traverse the corridor. The quantifier elimination for the above formula took about one CPU second on a DECstation 5000/200 by using H. Hong's QEPCAD package [66].

Going one step further, we may suppose that the length r of the ladder is not given and ask of what length the ladder can traverse the corridor. By formulating the problem and applying the CAD method, one can get an equivalent quantifier-free formula $r^2 > 8 \wedge r > 0$, which says that the ladder can traverse the corridor if its length is less than or equal to $2\sqrt{2}$ (see [121]).

6 Comments, Future Research and Prospects

6.1 Why GTP Successful?

Geometry is attractive and comprehensive in mathematics. It is rich both in theory and for applications. Proving fascinating geometric theorems is difficult and requires high human intelligence and ingenuity. Mechanizing this process has been the dream of geometers and other scientists for hundreds of years and is a concrete challenge that stimulated its early investigations in AI.

Heuristics are in place for GTP. One can imagine how far a proof of an interesting theorem has to go while starting from the very initial axioms without using any previously proved theorems. GTP began with a detailed exploration of heuristics and made use of the domain-specific knowledge that has been accumulated previously. This is a key to GTP's early success in AI. In fact, heuristics are used for GTP also in coordinate-free and coordinate-based approaches, such as automatically assigning coordinates, generating geometric non-degeneracy conditions and drawing diagrams (e.g., in GEOTHER).

Mathematical tools are well employed for GTP. The subsequent advances of GTP have largely benefited from non-trivial application and extension of mathematical theory and tools that have been deeply developed for centuries. The power of algebraic approaches comes from the algebraization of geometry, to which many theorems and results have already been proved. In other words, one is now allowed to bypass the long way of establishing algebraic relations from geometric axioms and to employ the full power and properties of algebraic systems. Thus, the focus of GTP may be turned into developing and applying effective algebraic algorithms.

GTP interrelates AI and SMC. It goes from heuristics (in AI) to algorithms (in SMC). The exploration of GTP within the two disciplines demonstrates each's advantages and disadvantages and calls for cooperation and combination of techniques from each other. This leads GTP to step forward: seeking for coordinate-free techniques, traditional proofs, database support, mixing algebraic and synthetical approaches, and introducing hybrid methods. It is our belief that the success of GTP has been the result of developing and applying powerful AI heuristics and SMC algorithms.

6.2 What to Do?

Integration. Whenever proving a geometric theorem using whatever approach, one uses, intentionally or unwittingly, more or less, principles and techniques from other disciplines (e.g., logical deduction, algebraic calculation, numerical verification and graphic drawing). Moreover, the axiomatic approaches of GTP may be supported, for instance, by diagrams [43, 60] and algebraic methods [121]. It is possible to translate algebraic proofs into logical ones [52]. The synthetic and algebraic approaches can be combined to generate desirable proofs [35]. Knowledge-bases may be used to aid GTP [63, 162].

The capability of GTP will be considerably enhanced when the most advanced techniques from different approaches are incorporated. We are convinced that one of the most challenging research directions for GTP now is an integrated study of logical, combinatorical, algebraical, numerical and graphical algorithms with heuristics, knowledge bases and reasoning mechanisms for geometry research, education and application.

System Construction. It is now of time to initiate research projects on the design and implementation of integrated systems for computer geometry (designing, analyzing, implementing and applying geometric algorithms). The theoretical foundation of such systems should be based on the outcome of the integrated study mentioned before. The potential systems can be rather sophisticated (involving various fundamental algorithms and making use of different software tools). The design issues include the internal data structures, strong data types of geometric objects and communication with other systems. The object-oriented paradigm with abstract specification of geometric structures and objects may be considered for such systems. In our group, the integration of GEOTHER [128] with the inference laboratory ATINF [15] has been under investigation (with initial experiments towards the implementation of a computer geometry system).

Application. Although significant applications of GTP have been found as presented in the preceding section, there is an increasing demand on extending GTP's application domains and power. In addition to pure geometry and other subjects of mathematics, GTP should be applied to problems from physics, chemistry and, in particular, geometry engineering such as geometric modeling, computer-added design/manufacturing, computer vision, computer graphics, air traffic control and robotics.

Higher Geometry. The existing algorithms should be extended and improved, new and advanced algorithms be developed to deal with reasoning problems in real geometry [3, 62, 98], finite geometry [97, 114], non-Euclidean geometry [32, 37, 54, 142, 145], non-commutative geometry [102], solid geometry [36, 117], algebraic geometry [122, 142, 147], differential geometry [16, 17, 29, 30, 96, 127, 137, 138, 147, 148, 153] and topology (the references are cited for some of the existing work). In these geometries, many un-attackable problems remain for further research.

Geometric Algebra. It is Wu's vision that powerful approaches for GTP have to go with three steps: algebraization of geometry, manipulation of algebraic relations, interpretation/analysis of proof steps and subsidiary conditions. There are different ways for each step. As for the first step — representing geometric notions and relations, one has the choice among different levels from geometric (Clifford, Grassmann) algebra to coordinate algebra (while bracket algebra is an example at the intermediate level of the hierarchy).

Geometric algebra at the highest level provides an expressive and unified language for its subordinate algebras. It can be taken as a basis of developing systematic approaches for coordinate-free techniques. Designing and implementing efficient algorithms for non-commutative geometric algebras should be a promising direction of research. The outcome will be useful not only for GTP, but also for the development of geometric reasoning systems, in which geometric algebra can be taken as a specification language.

6.3   Does GTP Have a Future?

Yes, GTP has a future; but one must take care of which direction it is about to go. Despite its past success, current GTP research activities may deserve questions from several aspects. For example, what is GTP's use? A number of people have been impressed by our prover GEOTHER after seeing a demo, but none really wants to use it. We are not satisfied with spending time and effort to produce something for joy, but there is little chance for the time being to bring GTP systems into academic institutions for popular use, nor to mention the possibility of getting the software into industrial organizations and commercial markets. The author dare claim that GTP will lose its impact in our community if no more use can be found in the future.

Proving hundreds of elementary geometric theorems, or even with several dozens discovered, would not contribute much to the development of the classical area of geometry, nor to the advance of modern geometry engineering. Notwithstanding the significance of such work on testing the practical value of existing methods, introducing novel approaches and exploiting their full power in scientific and high-technological applications are essential and crucial.

Geometry education is one of GTP's potential applications, but this is much related to educational policy and curriculum. In which way students should be guided for theorem proving on computer is still a non-trivial question that remains to be settled by educators.

We believe that to keep GTP well in progress, most of the topics listed in the preceding subsection, among others, need be thoroughly investigated. The focus of GTP research should be shifted to more advanced geometries than plane Euclidean geometry. It is our hope that GTP as a subject comprising techniques from AI, SMC and other areas of mathematics and computer science will play a truly significant role in foundational sciences with technological and industrial applications in the next decade.

References

Note. The author has seen several recent papers at an early stage. Some of them will appear in the Journal of Automated Reasoning (JAR) and a volume to be published in connection with the workshop on Automated Deduction in Geometry (Toulouse, September 27–29, 1996). The following abbreviations are used in the list of references: LNCS (Lecture Notes in Computer Science, Springer, Berlin-Heidelberg), JSC (Journal of Symbolic Computation), SSMS (Systems Science and Mathematical Sciences). Additional references can be found in [99, 123].

1. Anderson, J. R.: Tuning of search of the problem space for geometry proofs. In: Proc. IJCAI '81 (Vancouver, August 24–28, 1981), pp. 165–170.
2. Anderson, J. R., Boyle, C. F., Yost, G.: The geometry tutor. In: Proc. IJCAI '85 (Los Angeles, August 18–23, 1985), pp. 1–7.
3. Arnon, D. S.: Geometric reasoning with logic and algebra. Artif. Intell. **37**: 37–60 (1988).
4. Balbiani, P.: Equation solving in projective planes and planar ternary rings. In: LNCS **850**, pp. 95–113 (1994).
5. Balbiani, P.: Equation solving in geometrical theories. In: LNCS **968**, pp. 31–55 (1995).
6. Balbiani, P., Dugat, V., Fariñas del Cerro, L., Lopez, A.: Eléments de géométrie mécanique. Hermès, Paris (1994).
7. Balbiani, P., Fariñas del Cerro, L.: Affine geometry of collinearity and conditional term rewriting. In: LNCS **909**, pp. 196–213 (1995).
8. Balbiani, P., Lopez, A.: Simplification des figures de la géométrie affine plane d'incidence. In: 9e congrès reconnaissance des formes et intell. artif. (Paris, January 11-14, 1994), pp. 341–351.
9. Bledsoe, W. W., Loveland, D. W. (eds.): Automated theorem proving: After 25 years. Comptemp. Math. **29**, Amer. Math. Soc., Providence (1984).
10. Boissonnat, J.-D., Laumond, J.-P. (eds.): Geometry and robotics. Springer, Berlin-Heidelberg (1988).
11. Brüderlin, B.: Automatizing geometric proofs and constructions. In: LNCS **333**, pp. 232–252 (1988).
12. Buchberger, B.: Gröbner bases: An algorithmic method in polynomial ideal theory. In: Multidimensional systems theory (Bose, N. K., ed.), Reidel, Dordrecht-Boston, pp. 184–232 (1985).
13. Buchberger, B.: Applications of Gröbner bases in non-linear computational geometry. In: Mathematical aspects of scientific software (Rice, J. R., ed.), Springer, New York, pp. 59–87 (1987).
14. Buchberger, B., Collins, G. E., Kutzler, B.: Algebraic methods for geometric reasoning. Ann. Rev. Comput. Sci. **3**: 85–119 (1988).
15. Caferra, R., Herment, M.: A generic graphic framework for combining inference tools and editing proofs and formulae. JSC **19**: 217–243 (1995).
16. Carrà Ferro, G., Gallo, G.: A procedure to prove statements in differential geometry. JAR **6**: 203–209 (1990).
17. Carrà Ferro, G.: An extension of a procedure to prove statements in differential geometry. JAR **12**: 351–358 (1994).
18. Cerutti, E., Davis, P. J.: Formac meets Pappus: Some observations on elementary analytic geometry by computer. Amer. Math. Monthly **76**: 895–905 (1969).

19. Chou, S.-C.: Proving elementary geometry theorems using Wu's algorithm. In [9], pp. 243–286 (1984).
20. Chou, S.-C.: GEO-prover — A geometry theorem prover developed at UT. In: LNCS **230**, pp. 679–680 (1986).
21. Chou, S.-C.: A method for the mechanical derivation of formulas in elementary geometry. JAR **3**: 291–299 (1987).
22. Chou, S.-C.: Mechanical geometry theorem proving. Reidel, Dordrecht-Boston (1988).
23. Chou, S.-C.: Automated reasoning in geometries using the characteristic set method and Gröbner basis method. In: Proc. ISSAC '90 (Tokyo, August 20–24, 1990), pp. 255–260.
24. Chou, S.-C.: A geometry theorem prover for Macintoshes. In: LNCS **607**, pp. 687–690 (1992).
25. Chou, S.-C., Gao, X.-S.: Mechanical formula derivation in elementary geometries. In: Proc. ISSAC '90 (Tokyo, August 20–24, 1990), pp. 265–270.
26. Chou, S.-C., Gao, X.-S.: Ritt-Wu's decomposition algorithm and geometry theorem proving. In: LNCS **449**, pp. 207–220 (1990).
27. Chou, S.-C., Gao, X.-S.: Proving geometry statements of constructive type. In: LNCS **607**, pp. 20–34 (1992).
28. Chou, S.-C., Gao, X.-S.: Automated reasoning in differential geometry and mechanics using characteristic method III. In: Automated reasoning (Shi, Z., ed.), Elsevier, North-Holland, pp. 1–12 (1992).
29. Chou, S.-C., Gao, X.-S.: Automated reasoning in differential geometry and mechanics using the characteristic set method — Parts I, II. JAR **10**: 161–189 (1993).
30. Chou, S.-C., Gao, X.-S.: Automated reasoning in differential geometry and mechanics using the characteristic method IV. SSMS **6**: 186–192 (1993).
31. Chou, S.-C., Gao, X.-S., Arnon, D. S.: On the mechanical proof of geometry theorems involving inequalities. In: Issues in robotics and nonlinear geometry (Hoffmann, C., ed.), JAI Press, Greenwich, pp. 139–181 (1992).
32. Chou, S.-C., Gao, X.-S., Yang, L., Zhang, J.-Z.: Automated production of readable proofs for theorems in non-Euclidean geometries. Tech. Rep. TR-WSU-94-9, Wichita State Univ., USA (1994).
33. Chou, S.-C., Gao, X.-S., Zhang, J.-Z.: Automated production of traditional proofs for constructive geometry theorems. In: Proc. 8th IEEE Symp. LICS (Montreal, June 19–23, 1993), pp. 48–56.
34. Chou, S.-C., Gao, X.-S., Zhang, J.-Z.: Automated geometry theorem proving by vector calculation. In: Proc. ISSAC '93 (Kiev, July 6–8, 1993), pp. 284–291.
35. Chou, S.-C., Gao, X.-S., Zhang, J.-Z.: Machine proofs in geometry. World Scientific, Singapore (1994).
36. Chou, S.-C., Gao, X.-S., Zhang, J.-Z.: Automated production of traditional proofs in solid geometry. JAR **14**: 257–291 (1995).
37. Chou, S.-C., Ko, H.-P.: On mechanical theorem proving in Minkowskian plane geometry. In: Proc. IEEE Symp. LICS (Cambridge, June 16–18, 1986), pp. 187–192.
38. Chou, S.-C., Schelter, W. F.: Proving geometry theorems with rewrite rules. JAR **2**: 253–273 (1986).
39. Chou, S.-C., Schelter, W. F., Yang, J.-G.: Characteristic sets and Gröbner bases in geometry theorem proving. In: Resolution of equations in algebraic structures (Ait-Kaaci, H., Nivat, M., eds.), Academic Press, San Diego, pp. 33–92 (1989).

40. Chou, S.-C., Schelter, W. F., Yang, J.-G.: An algorithm for constructing Gröbner bases from characteristic sets and its application to geometry. Algorithmica 5: 147–154 (1990).

41. Chou, S.-C., Yang, J. G.: On the algebraic formulation of certain geometry statements and mechanical geometry theorem proving. Algorithmica 4: 237–262 (1989).

42. Coelho, H., Pereira, L. M.: GEOM: A Prolog geometry theorem prover. Laboratório Nacional de Engenhaaria Civil Memória no. 525, Ministerio de Habitacao e Obrass Publicas, Portugal (1979).

43. Coelho, H., Pereira, L. M.: Automated reasoning in geometry theorem proving with Prolog. JAR 2: 329–390 (1986).

44. Collins, G. E.: Quantifier elimination for real closed fields by cylindrical algebraic decomposition. In: LNCS 33, pp. 134–165 (1975).

45. Conti, P., Traverso, C.: A case of automatic theorem proving in Euclidean geometry: The Maclane 8_3 theorem. In: LNCS 948, pp. 183–193 (1995).

46. Crapo, H. (ed.): Computer-aided geometric reasoning. Proc. INRIA Workshop (INRIA Sophia-Antipolis, June 22–26, 1987), INRIA Rocquencourt, France.

47. Cyrluk, D. A., Harris, R. M., Kapur, D.: GEOMETER: A theorem prover for algebraic geometry. In: LNCS 310, pp. 770–771 (1988).

48. Deguchi, K.: An algebraic framework for fusing geometric constraints of vision and range sensor data. In: Proc. IEEE Int. Conf. MFI '94 (Las Vegas, October 2–5, 1994), pp. 329–336.

49. Deng, M.-K.: The parallel numerical method of proving the constructive geometric theorem. Chinese Sci. Bull. 34: 1066–1070 (1989).

50. Elcock, E. W.: Representation of knowledge in a geometry machine. Machine Intell. 8: 11–29 (1977).

51. Fearnley-Sander, D.: The idea of a diagram. In: Resolution of equations in algebraic structures (Aït-Kaaci, H., Nivat, M., eds.), Academic Press, San Diego, pp. 27–150 (1989).

52. Fevre, S.: A hybrid method for proving theorems in elementary geometry. In: Proc. ASCM '95 (Beijing, August 18–20, 1995), pp. 113–123.

53. Gao, X.-S.: Transcendental functions and mechanical theorem proving in elementary geometries. JAR 6: 403–417 (1990).

54. Gao, X.-S.: Transformation theorems among Cayley-Klein geometries. SSMS 5: 263–273 (1992).

55. Gao, X.-S., Chou, S.-C.: Computations with parametric equations. In: Proc. ISSAC '91 (Bonn, July 15–17, 1991), pp. 122–127.

56. Gao, X.-S., Li, Y.-L., Lin, D.-D., Lü, X.-S.: A geometric theorem prover based on Wu's method. In: Proc. IWMM '92 (Beijing, July 16–18, 1992), pp. 201–205.

57. Gao, X.-S., Wang, D.-K.: On the automatic derivation of a set of geometric formulae. J. Geom. 53: 79–88 (1995).

58. Gelernter, H.: Realization of a geometry theorem proving machine. In: Proc. Int. Conf. Info. Process. (Paris, June 15–20, 1959), pp. 273–282.

59. Gelernter, H., Hansen, J. R., Loveland, D. W.: Empirical explorations of the geometry-theorem proving machine. In: Proc. Western Joint Comput. Conf. (San Francisco, May 3–5, 1960), pp. 143–147.

60. Gilmore, P. C.: An examination of the geometry theorem machine. Artif. Intell. 1: 171–187 (1970).

61. Goldstein, I.: Elementary geometry theorem proving. MIT AI Memo no. 28, MIT, USA (1973).

62. Guergueb, A., Mainguené, J., Roy, M.-F.: Examples of automatic theorem proving in real geometry. In: Proc. ISSAC '94 (Oxford, July 20–22, 1994), pp. 20–24.

63. Hadzikadic, M., Lichtenberger, F., Yun, D. Y. Y.: An application of knowledge-base technology in education: A geometry theorem prover. In: Proc. SYMSAC '86 (Waterloo, July 21–23, 1986), pp. 141–147.

64. Havel, T. F.: Some examples of the use of distances as coordinates for Euclidean geometry. JSC 11: 579–593 (1991).

65. Hilbert, D.: Grundlagen der Geometrie. Teubner, Stuttgart (1899).

66. Hong, H., Wang, D., Winkler, F. (eds.): Algebraic approaches to geometric reasoning. Special issue of Ann. Math. Artif. Intell. 13(1,2), Baltzer, Basel (1995).

67. Hong, J.: Can we prove geometry theorems by computing an example? Sci. Sinica 29: 824–834 (1986).

68. Hong, J.: Proving by example and gap theorems. In: Proc. 27th Ann. Symp. Foundations Comput. Sci. (Toronto, October 27–29, 1986), pp. 107–116.

69. Hussain, M. A., Drew, M. A., Noble, B.: Using a computer for automatic proving of geometric theorems. Comput. Mech. Eng. 5: 56–69 (1986).

70. Kalkbrener, M.: A generalized Euclidean algorithm for geometry theorem proving. In [66]: 73–95 (1995).

71. Kandri-Rody, A., Weispfenning, V.: Non-commutative Gröbner bases in algebras of solvable type. JSC 9: 1–26 (1990).

72. Kapur, D.: Geometry theorem proving using Hilbert's Nullstellensatz. In: Proc. SYMSAC '86 (Waterloo, July 21–23, 1986), pp. 202–208.

73. Kapur, D.: Using Gröbner bases to reason about geometry problems. JSC 2: 399–408 (1986).

74. Kapur, D.: A refutational approach to geometry theorem proving. Artif. Intell. 37: 61–93 (1988).

75. Kapur, D., Mundy, J. L.: Wu's method and its application to perspective viewing. Artif. Intell. 37: 15–26 (1988).

76. Kapur, D., Mundy, J. L. (eds.): Geometric reasoning. Special issue of Artif. Intell. 37, The MIT Press, Cambridge (1989).

77. Kapur, D., Saxena, T., Yang, L.: Algebraic and geometric reasoning using Dixon resultants. In: Proc. ISSAC '94 (Oxford, July 20–22, 1994), pp. 99–107.

78. Kapur, D., Wan, H. K.: Refutational proofs of geometry theorems via characteristic set computation. In: Proc. ISSAC '90 (Tokyo, August 20–24, 1990), pp. 277–284.

79. Kelanic, T. J.: Theorem-proving with EUCLID. Creative Comput. 4/4: 60–63 (1978).

80. Ko, H.-P.: ALGE-prover II: A new edition of ALGE-prover. Tech. Rep. 86CRD-081, General Electric Co., Schenectady, USA (1986).

81. Ko, H.-P.: Geometry theorem proving by decomposition of quasi-algebraic sets: An application of the Ritt-Wu principle. Artif. Intell. 37: 95–122 (1988).

82. Ko, H.-P., Chou, S.-C.: A decision method for certain algebraic geometry problems. Rocky Mountain J. Math. 19: 709–724 (1989).

83. Ko, H.-P., Hussain, M. A.: A study of Wu's method — A method to prove certain theorems in elementary geometry. Congr. Numer. 48: 225–242 (1985).

84. Ko, H.-P., Hussain, M. A.: ALGE-prover: An algebraic geometry theorem proving software. Tech. Rep. 85CRD139, General Electric Co., Schenectady, USA (1985).

85. Kusche, K., Kutzler, B., Stifter, S.: Implementation of a geometry theorem proving package in Scratchpad II. In: LNCS 387, pp. 246–257 (1989).

86. Kutzler, B.: Algebraic approaches to automated geometry theorem proving. Ph.D thesis, RISC-LINZ, Johannes Kepler Univ., Austria (1988).

87. Kutzler, B.: Careful algebraic translations of geometry theorems. In: Proc. ISSAC '89 (Portland, July 17–19, 1989), pp. 254–263.

88. Kutzler, B., Stifter, S.: Automated geometry theorem proving using Buchberger's algorithm. In: Proc. SYMSAC '86 (Waterloo, July 21–23,, 1986), pp. 209–214.

89. Kutzler, B., Stifter, S.: On the application of Buchberger's algorithm to automated geometry theorem proving. JSC 2: 389–397 (1986).

90. Kutzler, B., Stifter, S.: A geometry theorem prover based on Buchberger's algorithm. In: LNCS 230, pp. 693–694 (1986).

91. Kutzler, B., Stifter, S.: Collection of computerized proofs of geometry theorems. Tech. Rep. 86-12, RISC-LINZ, Johannes Kepler Univ., Austria (1986).

92. Laborde, J.-M. (ed.): Intelligent learning environments — The case of geometry. Springer, Berlin-New York (1996).

93. Li, H.: Clifford algebra and area method. In [99] no. 14: 37–69 (1996).

94. Li, H., Cheng, M.: Proving theorems in elementary geometry with Clifford algebraic method. Preprint, MMRC, Academia Sinica, China (1995).

95. Li, Z.: Automatic implicitization of parametric objects. In [99] no. 4: 54–62 (1989).

96. Li, Z.: Mechanical theorem proving in the local theory of surfaces. In [66]: 25–46 (1995).

97. Lin, D., Liu, Z.: Some results on theorem proving in geometry over finite fields. In: Proc. ISSAC '93 (Kiev, July 6–8, 1993), pp. 292–300.

98. McPhee, N. F., Chou, S.-C., Gao, X.-S.: Mechanically proving geometry theorems using a combination of Wu's method and Collins' method. In: LNCS 814, pp. 401–415 (1994).

99. MMRC (ed.): Mathematics-Mechanization Research Preprints, nos. 1–14. Academia Sinica, China (1987–1996).

100. Mundy, J. L.: Reasoning about 3-D space with algebraic deduction. In: Robotics research: The third international symposium, The MIT Press, Cambridge-London, pp. 117–124 (1986).

101. Nevins, A. J.: Plane geometry theorem proving using forward chaining. Artif. Intell. 6: 1–23 (1975).

102. Pfalzgraf, J.: A category of geometric spaces: Some computational aspects. In [66]: 173–193 (1995).

103. Pfalzgraf, J., Stokkermans, K., Wang, D.: The robotics benchmark. In: Proc. 12-Month MEDLAR Workshop (Weinberg Castle, November 4–7, 1990), DOC, Imperial College, Univ. of London, England.

104. Quaife, A.: Automated development of Tarski's geometry. JAR 5: 97–118 (1989).

105. Rege, A.: A complete and practical algorithm for geometric theorem proving. In: Proc. 11th Ann. Symp. Comput. Geom. (Vancouver, June 5–7, 1995), pp. 277–286.

106. Richter-Gebert, J.: Mechanical theorem proving in projective geometry. In [66]: 139–172 (1995).

107. Ritt, J. F.: Differential algebra. Amer. Math. Soc., New York (1950).

108. Shi, H.: On the resultant formula for mechanical theorem proving. In [99] no. 4: 77–86 (1989).

109. Smietanski, F.: Systèmes de réécriture sur des idéaux de polynômes, géométrie et calcul formel. RAPPORT de DEA, E.N.S., Université de Jussieu Paris 7, France (1986/87).

110. Starkey, J. D.: EUCLID: A program which conjectures, proves and evaluates theorems in elementary geometry. Order no. 75-2780, Univ. Microfilms (1975).
111. Stifter, S.: Geometry theorem proving in vector spaces by means of Gröbner bases. In: Proc. ISSAC '93 (Kiev, July 6-8, 1993), pp. 301-310.
112. Swain, M. J., Mundy, J. L.: Experiments in using a theorem prover to prove and develop geometrical theorems in computer vision. In: Proc. IEEE Int. Conf. Robotics Automat. (San Francisco, April 7-10, 1986), pp. 280-285.
113. Tarski, A.: A decision method for elementary algebra and geometry. The RAND Corporation, Santa Monica (1948).
114. Ueberberg, J.: Interactive theorem proving and computer algebra. In: LNCS **958**, pp. 1-9 (1995).
115. Ullmen, S.: A model-driven geometry theorem prover. AI Lab Memo no. 321, MIT, Cambridge, USA (1975).
116. Wang, D.-K.: Mechanical solution of a group of space geometry problems. In: Proc. IWMM '92 (Beijing, July 16-18, 1992), pp. 236-243.
117. Wang, D.: Mechanical approach for polynomial set and its related fields. Ph.D thesis, Academia Sinica, China (1987) [in Chinese].
118. Wang, D.: Proving-by-examples method and inclusion of varieties. Kexue Tongbao **33**: 1121-1123 (1988).
119. Wang, D.: A new theorem discovered by computer prover. J. Geom. **36**: 173-182 (1989).
120. Wang, D.: On Wu's method for proving constructive geometric theorems. In: Proc. IJCAI '89 (Detroit, August 20-25, 1989), pp. 419-424.
121. Wang, D.: Reasoning about geometric problems using algebraic methods. In: Medlar 24-month deliverables, DOC, Imperial College, Univ. of London, England (1991).
122. Wang, D.: Irreducible decomposition of algebraic varieties via characteristic sets and Gröbner bases. Comput. Aided Geom. Design **9**: 471-484 (1992).
123. Wang, D.: Geometry theorem proving with existing technology. In: Medlar II Report PPR1, DOC, Imperial College, Univ. of London, England (1993). Also in: Proc. 1st Asian Tech. Conf. Math. (Singapore, December 18-21, 1995), 561-570.
124. Wang, D.: Algebraic factoring and geometry theorem proving. In: LNCS **814**, pp. 386-400 (1994).
125. Wang, D.: Reasoning about geometric problems using an elimination method. In: Automated practical reasoning: Algebraic approaches (Pfalzgraf, J., Wang, D., eds.), Springer, Wien-New York, pp. 147-185 (1995).
126. Wang, D.: Elimination procedures for mechanical theorem proving in geometry. In [66]: 1-24 (1995).
127. Wang, D.: A method for proving theorems in differential geometry and mechanics. J. Univ. Comput. Sci. **1**: 658-673 (1995).
128. Wang, D.: GEOTHER: A geometry theorem prover. In: Proc. CADE-13 (New Brunswick, July 30 - August 3, 1996), to appear.
129. Wang, D., Gao, X.-S.: Geometry theorems proved mechanically using Wu's method — Part on Euclidean geometry. In [99] no. 2: 75-106 (1987).
130. Wang, D., Hu, S.: A mechanical proving system for constructible theorems in elementary geometry. J. SSMS **7**: 163-172 (1987) [in Chinese].
131. Winkler, F.: A geometrical decision algorithm based on the Gröbner bases algorithm. In: LNCS **358**, pp. 356-363 (1988).
132. Winkler, F.: Gröbner bases in geometry theorem proving and simplest degeneracy conditions. Math. Pannonica **1**: 15-32 (1990).

133. Winkler, F.: Automated theorem proving in nonlinear geometry. In: Issues in robotics and nonlinear geometry (Hoffmann, C., ed.), JAI Press, Greenwich, pp. 183–197 (1992).

134. Wong, R.: Construction heuristics for geometry and a vector algebra representation of geometry. Tech. Memo. 28, Project MAC, MIT, Cambridge, USA (1972).

135. Wu, T.-J.: On a collision problem. In [99] no. 7: 96–104 (1991).

136. Wu, W.-t.: On the decision problem and the mechanization of theorem-proving in elementary geometry. Sci. Sinica 21: 159–172 (1978). Also in [9], pp. 213–234 (1984).

137. Wu, W.-t.: On the mechanization of theorem-proving in elementary differential geometry. Sci. Sinica Special Issue on Math. (I): 94–102 (1979) [in Chinese].

138. Wu, W.-t.: Mechanical theorem proving in elementary geometry and elementary differential geometry. In: Proc. 1980 Beijing Symp. Diff. Geom. Diff. Eqs. (Beijing, August 18 – September 21, 1980), pp. 1073–1092.

139. Wu, W.-t.: Toward mechanization of geometry — Some comments on Hilbert's "Grundlagen der Geometrie". Acta Math. Scientia 2: 125–138 (1982).

140. Wu, W.-t.: Some remarks on mechanical theorem-proving in elementary geometry. Acta Math. Scientia 3: 357–360 (1983).

141. Wu, W.-t.: Basic principles of mechanical theorem proving in elementary geometries. J. SSMS 4: 207–235 (1984). Also in JAR 2: 221–252 (1986).

142. Wu, W.-t.: Basic principles of mechanical theorem proving in geometries (part on elementary geometries). Science Press, Beijing (1984) [in Chinese]. English Translation, Springer, Wien-New York (1994).

143. Wu, W.-t.: Some recent advances in mechanical theorem-proving of geometries. In [9], pp. 235–241 (1984).

144. Wu, W.-t.: A mechanization method of geometry I. Chinese Quart. J. Math. 1: 1–14 (1986).

145. Wu, W.-t.: A mechanization method of geometry and its applications I. J. SSMS 6: 204–216 (1986).

146. Wu, W.-t.: A report on mechanical theorem proving and mechanical theorem discovering in geometries. Adv. Sci. China Math. 1: 175–198 (1986).

147. Wu, W.-t.: A constructive theory of differential algebraic geometry. In: Lecture Notes in Math. 1255, pp. 173–189 (1987).

148. Wu, W.-t.: A mechanization method of geometry and its applications II. Kexue Tongbao 32: 585–588 (1987).

149. Wu, W.-t.: On reducibility problem in mechanical theorem proving of elementary geometries. Chinese Quart. J. Math. 2: 1–20 (1987).

150. Wu, W.-t.: A mechanization method of geometry and its applications III. SSMS 1: 1–17 (1988).

151. Wu, W.-t.: A mechanization method of geometry and its applications IV. SSMS 2: 97–109 (1989).

152. Wu, W.-t.: Equations-solving and theorems-proving — Zero-set formulation and ideal formulation. In: Proc. Asian Math. Conf. (Hong Kong, August 14–18, 1990), pp. 1–10.

153. Wu, W.-t.: Mechanical theorem proving of differential geometries and some of its applications in mechanics. JAR 7: 171–191 (1991).

154. Wu, W.-t.: A report on mechanical geometry theorem proving. Progr. Natur. Sci. 2: 1-17 (1992).

155. Wu, W.-t.: A mechanization method of equations solving and theorem proving. In: Issues in robotics and nonlinear geometry (Hoffmann, C., ed.), JAI Press, Greenwich, pp. 103–138 (1992).
156. Wu, W.-t.: On problems involving inequalities. In [99] no. 7: 1–13 (1992).
157. Wu, W.-t.: On a finiteness theorem about problems involving inequalities. SSMS 7: 193–200 (1994).
158. Wu, W.-t.: Central configurations in planet motions and vortex motions. In [99] no. 13: 1–14 (1995).
159. Wu, W.-t., Lü, X.-L.: Triangles with equal bisectors. People's Edu. Press, Beijing (1985) [in Chinese].
160. Wu, W.-t., Wang, D.-K.: The algebraic surface fitting problem in CAGD. Math. Practice Theory no. 3: 26–31 (1994) [in Chinese].
161. Xu, C., Shi, Q., Cheng, M.: A global stereo vision method based on Wu-solver. In: Proc. GMICV '95 (Xi'an, April 27–29, 1995), pp. 198–205.
162. Xu, L., Chen, J.: AUTOBASE: A system which automatically establishes the geometry knowledge base. In: Proc. COMPINT: Computer aided technologies (Montreal, September 8–12, 1985), pp. 708–714.
163. Xu, L., Chen, J., Yang, L.: Solving plane geometry problem by learning. In: Proc. IEEE 1st Int. Conf. Comput. Appl. (Beijing, June 20–22, 1984), pp. 862–869.
164. Yang, L.: A new method of automated theorem proving. In: The mathematical revolution inspired by computing (Johnson, J., Loomes, M., eds.), Oxford Univ. Press, New York, pp. 115–126 (1991).
165. Yang, L., Zhang, J.-Z.: Searching dependency between algebraic equations: An algorithm applied to automated reasoning. In: Artificial intelligence in mathematics (Johnson, J., McKee, S., Vella, A., eds.), Oxford Univ. Press, Oxford, pp. 147–156 (1994).
166. Yang, L., Zhang, J.-Z., Hou, X.-R.: A criterion for dependency of algebraic equations, with applications to automated theorem proving. Sci. China Ser. A 37: 547–554 (1994).
167. Yang, L., Zhang, J.-Z., Hou, X.-R.: An efficient decomposition algorithm for geometry theorem proving without factorization. In: Proc. ASCM '95 (Beijing, August 18–20, 1995), pp. 33–41.
168. Yang, L., Zhang, J.-Z., Li, C.-Z.: A prover for parallel numerical verification of a class of constructive geometry theorems. In: Proc. IWMM '92 (Beijing, July 16–18, 1992), pp. 244–250.
169. Zhang, J.-Z.: How to solve geometry problems using areas. Shanghai Edu. Publ., Shanghai (1982) [in Chinese].
170. Zhang, J.-Z., Chou, S.-C., Gao, X.-S.: Automated production of traditional proofs for theorems in Euclidean geometry I. In [66]: 109–137 (1995).
171. Zhang, J.-Z., Yang, L., Deng, M.-K.: The parallel numerical method of mechanical theorem proving. Theoret. Comput. Sci. 74: 253–271 (1990).

Interactive Theorem Proving and Finite Projective Planes

Johannes Ueberberg

1 Introduction

Interactive Theorem Proving, ITP for short, is a concept to use current computer algebra systems to support mathematicians in proving theorems. Roughly speaking, the programs of ITP are intended to be some sort of "intelligent paper and pencil". They are conceived as a dialog between the user and the system, where the user directs the proof and the system applies the basic techniques. The idea of ITP is not to develop a general Theorem Prover but to formalize and to implement proof techniques that turned out to be useful in specific mathematical areas. ITP grew out of a more general project – called Symbolic Incidence Geometry – which is concerned with the systematic use of computers in incidence geometry. For a detailed description of Symbolic Incidence Geometry the interested reader is referred to [2]. At present concepts for ITP are only developed for finite geometries. In the present paper we shall discuss ITP for the analysis of point sets in finite projective planes. More precisely we shall describe the design of a system supporting geometers in investigating the following problem.

Problem. Let $0 \leq m_0 < m_1 < \ldots < m_r \leq q + 1$ be non-negative integers, and let S be a set of points in a finite projective plane of order q such that any line of P has exactly $m_0, m_1, \ldots,$ or m_r points in common with S. Determine the combinatorial properties of S, for example the cardinality of S, the number of tangent lines through a point of S or a point outside of S.

In UEBERBERG [9] we presented a concept for ITP in order to investigate finite linear spaces. We can use many of the concepts developed in [9] which I do not want to repeat here. On the other hand my aim was to produce a self-contained readable text. So I decided to mention those aspects which are already contained in [9] very briefly with exact references to [9] and to explain the new concepts in detail.

In Section 2 we shall give a short survey about point sets in finite projective planes. The design of the system is described in Sections 3 – 6. The problem of implementing the geometric properties of a projective plane and of an abstract point set S is discussed in Section 3. Section 4 is devoted to the administration of the combinatorial properties (the cardinality of S, the number of tangent lines, etc.) The idea of ITP is to apply basic proof techniques automatically. In Proposition 2.1 some of these proof techniques for the study of point sets in finite projective planes are listed. The algorithms for their automatic application are the content of Section 5. In Section 6 we briefly mention the concepts developed in [9] which can be used for the present work without or with only a few changes. Finally, we illustrate how to use a system for ITP and finite projective planes for studying sets of class $\{1, 3\}$ (in Section 7) and ovals (in Section 8).

There is a wide literature about Geometry Theorem Proving. An excellent survey is provided by WANG [10].

2 Finite Projective Planes

A **projective plane** is a geometry of points and lines satisfying the following axioms.

(P1) Any two points are incident with exactly one line.

(P2) Any two lines intersect in a unique point.

(P3) Any line is incident with at least three points.

A projective plane is called **finite**, if it has finitely many points and lines. For any finite projective plane P there exists an integer q such that P contains $q^2 + q + 1$ points and $q^2 + q + 1$ lines, any line is incident with $q + 1$ points and any point is incident with $q + 1$ lines. The number q is called the **order** of P.

The picture shows a projective plane of order 2, where the "circle" symbolizes one of the seven lines of P.

Let P be a finite projective plane of order q, and let S be a set of points of P. For the study of S it is particularly useful to consider the intersections of S with the lines of P. More precisely, one is interested in the number of points, a line can share with S. Many of the most important definitions in the study of finite projective planes are based on these numbers. Here are some examples.

A $\{v, n\}$**-arc** of P is a set S of v points such that any line of P has at most n points in common with S. A v**-arc** is a $\{v, 2\}$-arc. A **maximal arc** is a $\{v, n\}$-arc S such that any line of P intersects S in either 0 or n points. A **blocking set** is a set B of points such that any line has at least one and at most q points in common with B. Let $0 \le m_0 < m_1 < \ldots < m_r \le q + 1$. A point set S is called of class $\{m_0, \ldots, m_r\}$ if any line of P intersects S either in $m_0, m_1, \ldots,$ or in m_r points. For example, a maximal arc is a set of class $\{0, n\}$ for some integer n.

Let S be a set of points in a finite projective plane P of order q. For every $i \in \{0, 1, 2, \ldots, q + 1\}$ we denote by t_i the number of lines of P intersecting S in exactly i points. The number t_i is called the ith **character** of S. If S is a set of class $\{0, 1, n\}$, then $t_i = 0$ for all $i \ne 0, 1, n$. If x is a point of P, then we denote by $r_i(x)$ the number of lines through x intersecting S in exactly i points. Obviously, if $t_i = 0$ for some i, then $r_i(x) = 0$ for all points x. The following proposition is a main tool in the study of point sets of finite projective planes.

2.1 Proposition. *Let P be a finite projective plane of order q, and let S be a set of v points of P.*

(i) $\sum\limits_{i=0}^{q+1} t_i = q^2 + q + 1.$

(ii) $\sum\limits_{i=1}^{q+1} it_i = v(q+1)$.

(iii) $\sum\limits_{i=2}^{q+1} i(i-1)t_i = v(v-1)$.

(iv) $\sum\limits_{i=0}^{q+1} r_i(x) = q+1$ *for all points* x *of* P.

(v) $1 + \sum\limits_{i=2}^{q+1} (i-1)r_i(x) = v$ *for all* $x \in S$.

(vi) $\sum\limits_{i=1}^{q+1} ir_i(x) = v$ *for all* $x \notin S$.

(vii) $t_i = \frac{1}{i} \sum\limits_{x \in S} r_i(x)$ *for* $i \neq 0$.

(viii) Let g *be a* i-*secant. Then* $t_i = -q + \sum\limits_{x \in g} r_i(x)$.

(ix) Let g *be a* j-*secant with* $j \neq i$. *Then* $t_i = \sum\limits_{x \in g} r_i(x)$. $\square$

For further information about finite projective planes the interested reader is referred to DEMBOWSKI [3], HUGHES, PIPER [4], HIRSCHFELD [5], BEUTELSPACHER [1], THAS [6] or Section 2 of UEBERBERG [7] which is a survey on $\{v, n\}$-arcs.

3 Partitions

For the design of ITP in the context of finite projective planes we have to describe how the system administrates the geometric and combinatorial information and how the combinatorial techniques stated in Proposition 2.1 are applied. In this section we shall deal with the representation of the geometric information. Our general setting for this paper is that we want to study a set S of points of a certain class $\mathcal{C}$ in a finite projective plane P of order q. ITP is conceived as a dialog between the user and the system. Typical inputs of the user are

- Let p be a point of S.
- Let g be a line through p.

3.1 Definition. a) The points and lines indicated by the user are called **singular**.

b) If l is a non-singular line joining two singular points, then l is called a **special line**.

c) If p is a non-singular intersection point of two singular lines, then p is called a **special point**.

If the user's input is

- Let p be a point of S,

then p is a singular point. The formal description of this input is described below.

3.2 Definition. Let S be a set of points in a projective plane P.

a) If p is a singular point, then $\{p\}$ is called a **singular point component**.

b) Let G be a set of singular lines such that no two of them intersect in a singular point. Let $Sp(G)$ be the set of all points incident with any line of G and non-incident with any singular line not contained in G. Then $Sp_S(G) := Sp(G) \cap S$ and $Sp_{SC}(G) := Sp(G) \cap S^C$ are called the **special point components defined by** G^1.

c) Let g be a singular line, and let $L(g)$ be the set of non-singular and non-special points on g. Then $L_S(g) := L(g) \cap S$ and $L_{SC}(g) := L(g) \cap S^C$ are called the **linear point components defined by** g.

d) Let F be the set of all non-singular points that are not incident with any singular line. Then $F_S := F \cap S$ and $F_{SC} := F \cap S^C$ are called the **free point components**.

e) The attributes singular, special, linear and free of a component are called the **type** of it.

f) The set of all point components is called the **point partition**.[2]

Let us consider the following example.

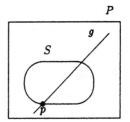

Suppose that the user has made the following two inputs.
- Let p be a point of S.
- Let g be a line through p.

Then the corresponding point partition consists of the following five components.

$$\{p\}, \{x \in S \mid x \neq p, x \in g\}, \{x \in S^C \mid x \in g\}, \{x \in S \mid x \notin g\}, \{x \in S^C \mid x \notin g\}.$$

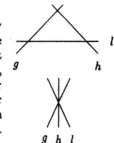

We observe that a singular point component consists of exactly one point and a special point component consists of at most one point. Given three singular lines g, h, l there are four special point components, namely the special components defined by $\{g, h\}$, $\{g, l\}$, $\{h, l\}$, and $\{g, h, l\}$, respectively. Either the first three or the fourth component are non-empty. By considering these four special point components the cases that g, h, l have a common point or do not have a common point are treated simultaneously.

The point components and the point partitions are the tools for representing the projective plane P, the point set S and the inputs of the user of the form

---

[1] S^C denotes the complement of S in P.

[2] The definition of the point components and of the point partition are context-sensitive. The point components in the context of finite linear spaces (2.3.1 in [9]) are different from the point components in the context of finite projective planes.

• Let p be a point of S.

The point partition depends on the inputs of the user. In other words, after any input of the user the point partition has to be recomputed. Since any point component C is completely characterized by its type, the name of its singular point (if C is singular), the singular lines incident with the points of C and the information whether $C \subseteq S$ or $C \subseteq S^C$, it can be represented by a record or a list containing these four data. The point partition can be represented as a list of its point components. A representation of the partition of the above example using lists could look as follows:

$\{\texttt{singular}, p, \{g\}, S\}, \{\texttt{linear}, \emptyset, \{g\}, S\}, \{\texttt{linear}, \emptyset, \{g\}, S^C\},$
$\{\texttt{free}, \emptyset, \emptyset, S\}, \{\texttt{free}, \emptyset, \emptyset, S^C\}.$

In a similar way the line set of P is partitioned.

3.3 Definition. Let S be a set of points in a projective plane P.

a) If g is a singular line, then $\{g\}$ is called a **singular line component**.

b) Let R be a set of at least two singular points such that no two of them are joined by a singular line. Let $Sp(R)$ be the set of all lines incident with all points of R and non-incident with any singular point not contained in R. Then $Sp(R)$ is called the **special line component defined by** R.

c) Let p be a singular point, and let $P(p)$ be the set of all non-singular and non-special lines through p. Then $P(p)$ is called the **pencil component defined by** p.

d) Let F be the set of all non-singular lines that are not incident with any singular point. Then F is called the **free line component**.

e) The attributes singular, special, pencil and free of a line component are called the **type** of it.

f) The set of all line components is called the **line partition** of P.

Note that the statements of Definition 3.3 are almost the dual of the statements of Definition 3.2. The asymmetry in some cases comes from the fact that S is a point set and not a line set.

A data structure for the representation of a line component D is a list or a record containing its type, the name of the singular line (if D is singular) and the singular points incident with the lines of D.

With the above definitions we are able to describe the possible user inputs in order to introduce new singular points and lines. Let $\mathcal{P} = \{C_1, \ldots, C_r\}$ be the actual point partition. Then the user can introduce a singular point with respect to any non-singular component $C \in \mathcal{P}$. The introduction of a new singular point $p \in C$ is called to **focus on** C.

In the user-system dialog the system prints out the non-singular point components and the user can select one of them to focus on it. Let us consider an example. The initial point partition $\mathcal{P}_0$ (before any input of the user) consists of the two free point components $C_1 := \{x \in P \mid x \in S\}$ and $C_2 := \{x \in P \mid x \notin S\}$. The user has the possibility either to focus on C_1 or on C_2. In other words the following two inputs are allowed.

• Let p be a point of S.

- Let p be a point outside of S.

If the user focusses on C_1, then the system computes the new point partition $\mathcal{P}_1$ with components $\{p\}$, $\{x \in P \mid x \in S, x \neq p\}$ and $\{x \in P \mid x \notin S\}$ which can be represented by $\{\text{singular}, p, \emptyset, S\}$, $\{\text{free}, \emptyset, \emptyset, S\}$, $\{\text{free}, \emptyset, \emptyset, S^C\}$. The name of the singular element (in our example p) is chosen by the user.

The algorithms for the automatic computation of the point and line partition are not very complicated and similar to the algorithms developed for the treatment of finite linear spaces. We refer to Chapter 3 of [9]. The number of components grows with the number of singular elements. Therefore one might want to delete ("forget") some singular elements. The deletion of a singular element is called **retraction** (of the corresponding singular component). For details see Section 3.2 of [9].

4 The Knowledge Base

The **knowledge base** is a data base, where the system administrates the combinatorial information. The system has to be able to deal with the following parameters: the cardinality of S, the characters of S, the cardinality $|g \cap S|$ for a given line g.

4.1 Definition. Let S be a set of points in a finite projective plane P. A **free parameter** is one of the following notations.

   a) The cardinality of S is denoted by $\bar{v}$.

   b) The ith character of S is denoted by $\bar{t}_i$.

   c) The cardinality of a component C is denoted by $\bar{m}(C)$.

   d) The cardinality of the set $l \cap S$ for some line l is denoted by $\bar{s}(l)$.

   e) The number of i-secants through a point x is denoted by $\bar{r}_i(x)$.

   f) The free parameters $\bar{v}$ and $\bar{t}_i$ are called **global free parameters**, the parameters $\bar{m}$, $\bar{s}$ and $\bar{r}_i$ are called **local free parameters**.

The free parameters are used as key words (more formally: as key attributes) to store the combinatorial information.

4.2 Definition. Bounded parameters are the variables introduced by the user and the order of the projective plane.

We are mainly interested in investigating an abstract projective plane of order q (and not in investigating projective planes of a concrete order, say 5).[3] The order q is an example of a bounded parameter. Often it is useful to introduce new variables, for example the variable v for the cardinality of the set S (to distinguish from $\bar{v}$ which is a key word to store the information about the cardinality of S). Then v is a bounded parameter. The free parameters are used to store the combinatorial information, whereas the bounded parameters are used for computations.

Our aim is to analyze point sets S admitting only a few characters. At the beginning of the user-system dialog the user is asked to specify the set $\mathcal{C} \subseteq$

---

[3] Although this is not excluded.

$\{0, 1, \ldots, q+1\}$ such that $t_i = 0$ for all $i \notin C$ (in other words S is of class C). Possible inputs of the user are (finite) sets of non-negative integers and bounded parameters like $C = \{0, 1, 2\}$, $C = \{0, 1, n\}$ or $C = \{0, 1, 2, q+1\}$. Throughout this paper we shall suppose that the set C has been specified by the user.

4.3 Definition. A fact is a 5-tuple $(Z, C, j, \square, E)$ satisfying the following properties.

(i) Z is a free parameter.

(ii) If Z is a global parameter, then $C = \emptyset$. If $Z = \bar{m}$, then C is a point or a line component. If $Z = \bar{s}$, then C is a line component. If $Z = \bar{r}_i$, then C is a point component.

(iii) If $Z = \bar{v}$, $Z = \bar{m}$ or $Z = \bar{s}$, then $j = -1$. If $Z = \bar{t}_i$ or $Z = \bar{r}_i$, then j is some element of C.

(iv) $\square$ is one of the relations $=, \neq, \leq, \geq$, or $\in$.

(v) If $\square$ is one of the relations $=, \neq, \leq, \geq$, then E is a polynomial expression whose variables are bounded parameters. If $\square$ is the relation $\in$, then E is a set of expressions whose variables are bounded parameters.

b) A fact is called **global** or **local** according to whether the free parameter Z is global or local.

4.4 Definition. The **knowledge base** is a data base, where the facts are stored.

We explain the meaning of the facts by two examples. Suppose that the user wants to introduce the variable v for the cardinality of S. The free parameter for the cardinality is $\bar{v}$. Hence the user wants to introduce the fact $\bar{v} = v$, or, more formally, the fact $(\bar{v}, \emptyset, -1, =, v)$. As a second example we consider the component C of the points not in S. We want to express that any point of C is incident with either one or two tangent lines of S. The corresponding fact is $F = (\bar{r}_i, C, 1, \in, \{1, 2\})$. The component C is the point component the local parameter $\bar{r}_i$ refers to. The entry $j = 1$ means that we are interested in the parameter $\bar{r}_1(x)$ with $x \in C$. In other words, the fact $F = (\bar{r}_i, C, 1, \in, \{1, 2\})$ signifies the relation

$$r_1(x) \in \{1, 2\} \text{ for all } x \in C.$$

A fact can be represented as a list or a record containing the five data described above.

Sometimes we also need concrete estimates about a bounded parameter, like the inequality $q \geq 4$ or we have to deal with relations between several bounded parameters like the equation $n^2 = q$. These relations are called **estimates of a bounded parameter** and **relations of bounded parameters**, respectively. They can be represented in a way similar to the representation of the facts. They are also contained in the knowledge base.

5 Application of Combinatorial Techniques

In this section we shall explain how the system applies the equations of Proposition 2.1. The equations of 2.1 fall into two different families; in the equation

(i) to (vi) the summation is over the set $\{0, 1, \ldots, q+1\}$, whereas in (vii) to (ix) the summation is over a set of points in P.

At some places these two families of equations require different treatments. However, the evaluation algorithms are so similar that we can treat the two cases simultaneously. We shall concentrate on the equation (ii) (exemplary for (i) to (vi)) and (vii) (exemplary for (vii) to (ix)).

Step 1. Reduction. The sum $\sum_{i=1}^{q+1} i t_i$ has to be transformed into a sum with a concrete number of terms. At the beginning of the user-system dialog the user had to specify the set $\mathcal{C} = \{i_1, \ldots, i_m\}$ such that S is a set of class $\mathcal{C}$. Hence the sum $\sum_{i=1}^{q+1} i t_i$ changes into the sum $i_1 t_{i_1} + \cdots + i_m t_{i_m}$.

So the first step of the system in order to apply equation (ii) is to produce the equation

$$i_1 \bar{t}_{i_1} + \cdots + i_m \bar{t}_{i_m} = \bar{v}(q + 1).$$

This equation is represented using the facilities of the underlying computer algebra system. For $\mathcal{C} = \{0, 1, 2\}$ or $\mathcal{C} = \{1, n, q + 1\}$, we get the equations $\bar{t}_1 + 2\bar{t}_2 = \bar{v}(q + 1)$ or $\bar{t}_1 + n\bar{t}_n + (q + 1)\bar{t}_{q+1} = \bar{v}(q + 1)$, respectively.

The sum $\frac{1}{i} \sum_{x \in S} r_i(x)$ of equation (vii) has also to be transformed into a sum with a concrete number of terms. To do so, let $\mathcal{P}$ be the actual point partition, and let $\mathcal{P}_S = \{C_1, \ldots, C_r\}$ be the set of point components such that $S = C_1 \cup \ldots \cup C_r$. The set $\mathcal{P}_S$ can be computed by the system. We get

$$\frac{1}{i} \sum_{x \in C_1} \bar{r}_i(x) + \cdots + \frac{1}{i} \sum_{x \in C_r} \bar{r}_i(x) = \bar{t}_i.$$

This equation can also be represented using the underlying computer algebra system. A sum of the form $\sum_{x \in C_1} \bar{r}_i(x)$ can be represented by a list containing the parameter $\bar{r}_i$ and the component C_1.

Step 2. Substitution. In this step of the algorithm the free parameters appearing in the equations (i) to (ix) have to be substituted by bounded parameters such that the resulting relation contains either one or no free parameters. If the final result contains one free parameter Z, then we get a fact about Z. If it does not contain any free parameters, we obtain a relation of bounded parameters.

We first treat equation (ii). The reduction resulted in the following simplification

$$\sum_{i=1}^{q+1} i\bar{t}_i = \bar{v}(q + 1) \rightarrow i_1 \bar{t}_{i_1} + \cdots + i_m \bar{t}_{i_m} = \bar{v}(q + 1).$$

The system prints out all facts about the free parameters involved in the above equation. Let us consider the following example. Suppose that $\mathcal{C} = \{1, n, q+1\}$. Then the above equation is of the form

$$\bar{t}_1 + n\bar{t}_n + (q + 1)\bar{t}_{q+1} = \bar{v}(q + 1).$$

The involved free parameters are $\bar{t}_1, \bar{t}_n, \bar{t}_{q+1}$ and $\bar{v}$. Then the system prints out all facts about these parameters, that is, all facts $F = (Z, \emptyset, j, \Box, E)$ with $Z = \bar{t}_i$, $j = 1$, $j = n$ or $j = q + 1$ and with $Z = \bar{v}$ and $j = -1$.

The user can select (at most) one fact with respect to any free parameter with the consequence that the system substitutes the free parameters by the expressions of the correspondig facts. For example, if the user selects the facts

$$\bar{t}_1 = q + 1 - n, \bar{t}_n \geq q^2, \text{ and } \bar{v} = nq + 1,$$

then the system simplifies the above equation to

$$q + 1 - n + nq^2 + (q + 1)\bar{t}_{q+1} \geq (nq + 1)(q + 1).$$

Let $p(Z_1, \ldots, Z_n)$ be a polynomial expression whose variables are the free parameters $Z_1, \ldots, Z_n$ and whose coefficients are bounded parameters. Let M be a subset of $\{1, \ldots, n\}$. For $j \in M$ let F_j be a fact about Z_j. The substitution of the free parameters $Z_1, \ldots, Z_n$ by the expressions of the facts $F_j, j \in M$ is called **the evaluation of** $p(Z_1, \ldots, Z_n)$ **with respect to** $F_j, j \in M$. The algorithms for the automatic evaluation of an expression $p(Z_1, \ldots, Z_n)$ have to respect many technical details which are explained in Section 5.1 of [9]. The result of the evaluation of $p(Z_1, \ldots, Z_n)$ with respect to the facts $F_j, j \in M$ is again a polynomial expression $q(X_1, \ldots, X_t)$, whose variables $X_1, \ldots, X_t$ are free parameters and whose coefficients are bounded parameters.

Next we shall treat the more complicated application of equation (vii). By the reduction we got the simplification

$$\frac{1}{i} \sum_{x \in S} \bar{r}_i(x) = \bar{t}_i \rightarrow \frac{1}{i} \sum_{x \in C_1} \bar{r}_i(x) + \cdots + \sum_{x \in C_r} \bar{r}_i(x) = \bar{t}_i.$$

The system prints out the facts about the free parameters $\bar{t}_i$ and $\bar{r}_i$ with respect to the components $C_1, \ldots, C_r$. Again the idea is that the free parameters are replaced by bounded parameters. If $\bar{r}_i(x) \geq a$ for all $x \in C_1$, then the system has to realize the substitution $\sum_{x \in C_1} \bar{r}_i(x) \geq \bar{m}(C_1)a$.

For each non-singular and non-special component $C \in \{C_1, \ldots, C_r\}$ the user has to select one fact about $\bar{r}_i$ with respect to C. (If we allowed the user not to select a fact about such a component, then the substitution would be as follows:

$$\sum_{x \in C} \bar{r}_i(x) \rightarrow \bar{m}(C)\bar{r}_i(x),$$

where the parameter $\bar{r}_i(x)$ still depends on the point $x \in C$. For singular and special point components, the problem does not arise, since these components consist of at most one point.) For the other free parameters the user can select at most one fact. Then the system substitutes the free parameters by the expressions of the selected facts and replaces the sums $\sum_{x \in C}$ by the free parameter for the cardinality of C, that is, $\bar{m}(C)$. As an example we suppose that the user has selected r facts of the form

$$\bar{r}_i(x) \geq a_l \text{ for all } x \in C_l, l = 1, \ldots, r,$$

where $a_1, \ldots, a_r$ are bounded parameters (or expressions, whose variables are bounded parameters). Then the equation

$$\frac{1}{i} \sum_{x \in C_1} \bar{r}_i(x) + \cdots + \frac{1}{i} \sum_{x \in C_r} \bar{r}_i(x) = \bar{t}_i$$

is transformed to the inequality $\frac{1}{i}\bar{m}(C_1)a_1 + \cdots + \frac{1}{i}\bar{m}(C_r)a_r \leq \bar{t}_i$.

More generally, the equation

$$\frac{1}{i} \sum_{x \in C_1} \bar{r}_i(x) + \cdots + \frac{1}{i} \sum_{x \in C_r} \bar{r}_i(x) = \bar{t}_i$$

is transformed to a relation of the form $\frac{1}{i}\bar{m}(C_1)a_1 + \cdots + \frac{1}{i}\bar{m}(C_r)a_r \;\square\; c$, where $\square$ is one of the relations $\{=, \leq, \geq, \in\}$ and c is either a polynomial expression or a list of polynomial expressions according to whether $\square$ is one of the relations $=, \leq, \geq$ or $\in$. The algorithms for this transformation are described in Section 6.3 of [9].

At this point we encounter the difficulty that we lack explicit information about the cardinalities $\bar{m}(C_1), \ldots, \bar{m}(C_r)$, but we often have implicit information. To illustrate this let us consider the following example. Suppose that the user has introduced one singular line g with the effect that there are two components $C_1 := \{x \in S \mid x \in g\}$ and $C_2 := \{x \in S \mid x \notin g\}$ such that $S = C_1 \cup C_2$. Furthermore suppose that we know that $\bar{r}_1(x) = 1$ for all $x \in S$ (that is, every point of S is incident with exactly one tangent line) and that $\bar{v} = q+1$. In order to apply equation (vii) for $i = 1$, the system proceeds as follows

$$\sum_{x \in S} \bar{r}_1(x) = \bar{t}_1$$

$$\rightarrow \sum_{x \in C_1} \bar{r}_1(x) + \sum_{x \in C_2} \bar{r}_1(x) = \bar{t}_1$$

$$\rightarrow \bar{m}(C_1) + \bar{m}(C_2) = \bar{t}_1.$$

We do not know the cardinalities $\bar{m}(C_1)$ and $\bar{m}(C_2)$ but we know that $\bar{v} = q+1$, that is, that $\bar{m}(C_1)+\bar{m}(C_2) = q+1$ from which we deduce that $\bar{t}_1 = q+1$. In order to use the implicit information about the cardinalities of the components we introduce the systems of linear constraints.

5.1 Definition. Let S be a point set of a finite projective plane P of order q, and let $\mathcal{F} = \{C_1, \ldots, C_r\}$ be the components of the actual point partition such that $S = C_1 \cup \ldots \cup C_r$.

(i) Let $X_1, \ldots, X_l$ be the components X of $\mathcal{F}$ such that the knowledge base contains an equation about the cardinality of X, say

$$\bar{m}(X_1) = a_1, \ldots, \bar{m}(X_l) = a_l,$$

where $a_1, \ldots, a_l$ are expressions of bounded parameters.[4]

(ii) Let g be a singular line, and let $D_1, \ldots, D_s$ be the components $D \in \mathcal{F}$ such that $D_1 \cup \ldots \cup D_s = g \cap S$. Then

$$\bar{m}(D_1) + \cdots + \bar{m}(D_s) = \bar{s}(g).$$

(iii) We have $\bar{m}(C_1) + \cdots + \bar{m}(C_r) = \bar{v}$.

The system of equations

$$\bar{m}(X_1) = a_1, \ldots, \bar{m}(X_l) = a_l,$$
$$\bar{m}(D_1) + \cdots + \bar{m}(D_s) = \bar{s}(g),$$
$$\bar{m}(C_1) + \cdots + \bar{m}(C_r) = \bar{v}$$

is called the **system of linear constraints with respect to** S.

In equations (viii) and (ix) the sum is taken over all points of a line g. We commit that equations (viii) and (ix) can only be applied if g is singular. The corresponding system of linear constraints is as follows.

5.2 Definition. Let P be a finite projective plane of order q, and let g be a singular line. Let $\mathcal{F} = \{C_1, \ldots, C_r\}$ be the components of the actual point partition such that $S = C_1 \cup \ldots \cup C_r$.

(i) Let $X_1, \ldots, X_l$ be the components X of $\mathcal{F}$ such that the knowledge base contains an equation about the cardinality of X, say

$$\bar{m}(X_1) = a_1, \ldots, \bar{m}(X_l) = a_l,$$

where $a_1, \ldots, a_l$ are expressions of bounded parameters.

(ii) Let $D_1, \ldots, D_s$ be the components $D \in \mathcal{F}$ such that $D_1 \cup \ldots \cup D_s = g \cap S$. Then

$$\bar{m}(D_1) + \cdots + \bar{m}(D_s) = \bar{s}(g).$$

(iii) Let h be a singular line intersecting g in a non-singular point, let $E_1, \ldots, E_t$ be the special point components $D \in \mathcal{F}$ consisting of an intersection point of g and h. Then

$$\bar{m}(E_1) + \cdots + \bar{m}(E_t) = 1.$$

(iv) We have $\bar{m}(C_1) + \cdots + \bar{m}(C_r) = q + 1$.

The system of equations

$$\bar{m}(X_1) = a_1, \ldots, \bar{m}(X_l) = a_l,$$
$$\bar{m}(D_1) + \cdots + \bar{m}(D_s) = \bar{s}(g),$$
$$\bar{m}(E_1) + \cdots + \bar{m}(E_t) = 1,$$
$$\bar{m}(C_1) + \cdots + \bar{m}(C_r) = q + 1$$

is called the **system of linear constraints with respect to** g.

---

[4] For example, if C is a singular component, then $\bar{m}(C) = 1$.

The systems of linear constraints are automatically built by the system. Once the linear constraints are formulated the corresponding system of linear equations is solved by the underlying computer algebra system with respect to the variables $\bar{m}(C_1), \ldots, \bar{m}(C_r)$, and the solution is used to simplify the relation

$$\frac{1}{i}\bar{m}(C_1)a_1 + \cdots + \frac{1}{i}\bar{m}(C_r)a_r \,\square\, c.$$

By this procedure we replace as many free parameters $\bar{m}(C_i)$ as possible. The resulting relation is of the form

$$p(Z_1, \ldots, Z_s) \,\square\, 0,$$

where $\square$ is one of the relations $=, \leq, \geq$.[5] $Z_1, \ldots, Z_s$ are free parameters and $p(Z_1, \ldots, Z_s)$ is a polynomial expression, whose coefficients are bounded parameters.

As a next step the system prints out the facts about the remaining free parameters $Z_1, \ldots, Z_s$, and the user can select (at most) one fact about each free parameter. As before the system evaluates the expression $p(Z_1, \ldots, Z_s)$ with respect to the selected facts. The result is a polynomial expression $q(X_1, \ldots, X_t)$ with free parameters $X_1, \ldots, X_t$.

Step 3. Final Conclusions. Applying the steps Reduction and Substitution, the original equations

$$\sum_{i=1}^{q+1} i\bar{t}_i = \bar{v}(q+1) \text{ and } \frac{1}{i}\sum_{x \in S} \bar{r}_i(x) = \bar{t}_i$$

have been transformed to a relation of the form $p(Z_1, \ldots, Z_r) \,\square\, 0$, where $\square$ is one of the relations $=, \leq, \geq$. $Z_1, \ldots, Z_r$ are free parameters and $p(Z_1, \ldots, Z_r)$ is a polynomial expression whose coefficients are bounded parameters. It follows from the structure of the equations (i) to (ix) that the partial functions $Z_k \mapsto p(Z_1, \ldots, Z_r)$ are linear or quadratic for all $k = 1, \ldots, r$.

If $r \geq 2$, that is, if the expression $p(Z_1, \ldots, Z_r)$ contains at least two free parameters, then the evaluation algorithm terminates without any result.

If $r = 1$, then $p(Z_1, \ldots, Z_r) = p(Z_1)$ is a linear or quadratic polynomial in Z_1. Therefore the relation $p(Z_1) \,\square\, 0$ can be transformed to a relation of the form $Z_1 \,\square\, c$, where c is a polynomial expression, whose variables are bounded parameters. The relation $Z_1 \,\square\, c$ is a fact and is inserted into the knowledge base.

If $r = 0$, then the expression $p(Z_1, \ldots, Z_r)$ does not contain any free parameter, so the relation $p(Z_1, \ldots, Z_r) \,\square\, 0$ reduces to $c \,\square\, 0$, where c is a polynomial expression, whose variables are bounded parameters. If $c \neq 0$, then $c \,\square\, 0$ is a relation of bounded parameters which is inserted into the knowledge base. Particularly interesting is the case, where $c = 0$. Then we have the relation $0 \,\square\, 0$. If $\square$ equals $=$, then the algorithm terminates without any result. The situation is completely different, if $\square$ equals $\leq$ or $\geq$. To illustrate this phenomenon let

---

[5] For sake of simplicity we omit the case that $\square$ equals $\in$.

us consider the following example. Suppose that S is a point set in a projective plane of order q admitting $q + 1$ tangent lines such that there is at least one tangent through every point of S and at least two tangent lines through every point outside of S. Let g be a tangent line of S, and let $p := g \cap S$. Then, by (viii), we have

$$q + 1 = t_1 = -q + \sum_{x \in g} r_1(x)$$

$$= -q + r_1(p) + \sum_{x \in g \setminus \{p\}} r_1(p)$$

$$\geq -q + 1 + q \cdot 2 = q + 1.$$

It follows that the above inequalities are in fact equations. Hence $r_1(p) = 1$ and $r_1(x) = 2$ for all $x \in g \setminus \{p\}$. The evaluation algorithm for this example proceeds as follows. By reduction (Step 1) the equation

$$-q + \sum_{x \in g} \bar{r}_1(x) = \bar{t}_1$$

is simplified as follows (let $C_1 := g \cap S$ and $C_2 := g \cap S^C$)

$$-q + \sum_{x \in C_1} \bar{r}_1(x) + \sum_{x \in C_2} \bar{r}_1(x) = \bar{t}_1.$$

The substitution algorithm uses the facts $\bar{r}_1(x) \geq 1$ for all $x \in C_1$ and $\bar{r}_1(x) \geq 2$ for all $x \in C_2$ and $\bar{t}_1 = q + 1^6$ in order to simplify the above equation to

$$-q + \bar{m}(C_1) \cdot 1 + \bar{m}(C_2) \cdot 2 \leq q + 1.$$

The system of linear constraints with respect to g is

$$\bar{m}(C_1) = \bar{s}(g), \ \bar{m}(C_1) + \bar{m}(C_2) = q + 1.$$

Solving this system of linear constraints the above inequality is simplified to

$$-q + \bar{s}(g) + (q + 1 - \bar{s}(g)) \cdot 2 \leq q + 1.$$

Finally the fact $\bar{s}(g) = 1$ (g a tangent line) is used for the final simplification

$$q + 1 \leq q + 1.$$

We come back to the description of Step 3 (final conclusions) of the evaluation algorithm. Suppose that the substitution (Step 2) used the facts $F_1, \ldots, F_s$ with $F_l = (Z_l, C_l, j_l, \Box_l, E_l)$ for $l = 1, \ldots, s$ and obtained the relation $0 \Box 0$, where $\Box$ equals either $\leq$ or $\geq$. Then in the knowledge base the system replaces the facts $F_1, \ldots, F_s$ by the facts $F'_1, \ldots, F'_s$ with $F'_l := (Z_l, C_l, j_l, =, E_l)$ for $l = 1, \ldots, s$.

---

[6] This means that the system prints all facts about $\bar{t}_1$ and $\bar{r}_1$ with respect to C_1 and C_2 and that the user selects the three facts mentioned above.

6 Miscellanea

In Sections 3, 4 and 5 we left out of consideration those aspects of the design of a system for ITP and finite projective planes that are more or less the same as for ITP and finite linear spaces. In this section we shall briefly mention these aspects and refer to the corresponding sections in [9].

Partition Graphs. Suppose that we know that $\bar{r}_1(x) = 1$ for all points $x \in S$. Focussing on the component $C := S$ the component C splits into the two components $C_1 := \{p\}$ (new singular point) and $C_2 := \{x \in S \mid x \neq p\}$. The system has to transfer the information about C to C_1 and C_2, that is, to insert the facts $\bar{r}_1(x) = 1$ for all $x \in C_1$ and $\bar{r}_1(x) = 1$ for all $x \in C_2$, into the knowledge base. One function of the partition graph is to facilitate this transfer of information. The second function is related to the shifting principle (see below). See 2.6 of [9].

Weight and Cardinality of a component. Focussing on a component C means to say
- Let x be a (singular) element of C.

In particular, C has to be non-empty. A component C is said to be of **weight** 1 or -1, if C is non-empty or empty, respectively. If it is not clear whether C is empty or not, then C is said to be of **weight** 0 (see 2.5 and 3.3 of [9]). For the computation of the weight of a component and for the application of equations (vii) - (ix) it is important to compute the cardinality of a component (see 5.2 of [9]).

Knowledge Base. Focussing on a component or retracting a singular component has consequences for the knowledge base (see 7.4 of [9]). It is useful to keep the knowledge base as small as possible. If the knowledge base contains the fact $F : \bar{v} \leq q + 1$ and if the fact $F' : \bar{v} = q + 1$ is inserted, then F has to be *replaced* by F' (see 7.1 – 7.3 of [9]).

Shifting Principle. Suppose that the user focussed on a point $p \in S$ and obtained the result $\bar{r}_1(p) = n$ by applying one of the equations of Proposition 2.1. Then p is a representative element of S with the consequence that it follows $\bar{r}_1(x) = n$ for all points $x \in S$. The "shift" of information from the singular component $\{p\}$ to the original component S is called the shifting principle (see 8.1 of [9]).

Proof Tree. In general, a proof is divided into several steps, and often there are different cases that have to be treated separately. The proof tree is used for the administration of the structure of a proof (see 8.2 of [9]).

Contradictions. The organization of a proof by contradiction is explained in 8.2 of [9].

User-System Dialog. A complete list of the possible user inputs and the reaction of the system is described in 8.3 of [9]. This list has to be adapted from the context of finite linear spaces to point sets in finite projective planes.

7 Case Study I. Arcs of Class $\{1,3\}$

In our survey article on Symbolic Incidence Geometry (BEUTELSPACHER, UE-BERBERG [2]) we included a rough sketch about the treatment of characters in finite projective planes (Section 5.2.2). As an example we considered sets of class $\{1,3\}$ in a finite projective plane. In this section we shall return to this example for a detailed discussion.

Problem. Classify all arcs of class $\{1,3\}$ in a finite projective plane of order q.

We shall describe one possible user-system dialog in order to solve this problem. Most inputs of the user are preceeded by a short explanation motivating this input. After the input of the user we discuss the reaction of the system. We start the user-system dialog with a description of the initial state of the system.

System. The point partition consists of the two components S and S^C. The line partition consists of one component containing all lines of P.
Motivation. As a first step the user has to specify the set C such that S is of class C.
User's Input. Let $C = \{1,3\}$.
System. The knowledge base is initialized with the free parameters $\bar{v}, \bar{t}_1, \bar{t}_3, \bar{m}, \bar{r}_1, \bar{r}_3$.

User's Input. The user specifies the bounded parameters t_1 and t_3 for the free parameters $\bar{t}_1$ and $\bar{t}_3$. This is the formal translation of the input
- Let t_1 and t_3 be the number of tangent lines and of 3-secants, respectively.
System. The system inserts the facts $\bar{t}_1 = t_1$ and $\bar{t}_3 = t_3$ into the knowledge base.
User's Input. Apply equation (i).
System. *Step 1. Reduction.* Equation (i) is simplified to $\bar{t}_1 + \bar{t}_3 = q^2 + q + 1$.
   Step 2. Substitution. The facts $\bar{t}_1 = t_1$ and $\bar{t}_3 = t_3$ are printed. Suppose that the user selects the fact $\bar{t}_1 = t_1$. Then the system obtains $t_1 + \bar{t}_3 = q^2 + q + 1$. Hence $\bar{t}_3 = q^2 + q + 1 - t_1$. This fact is inserted into the knowledge base.
User's Input. Apply equation (ii).
System. *Step 1. Reduction.* Equation (ii) is simplified to $\bar{t}_1 + 3\bar{t}_3 = \bar{v}(q+1)$.
   Step 2. Substitution. The facts $\bar{t}_1 = t_1$, $\bar{t}_3 = t_3$ and $\bar{t}_3 = q^2 + q + 1 - t_1$ are printed. Suppose that the user selects the facts $\bar{t}_1 = t_1$ and $\bar{t}_3 = q^2 + q + 1 - t_1$. Then the system obtains the equation $t_1 + 3(q^2 + q + 1 - t_1) = \bar{v}(q+1)$. Hence $\bar{v} = \frac{1}{q+1}\left(3(q^2 + q + 1) - 2t_1\right)$. This fact is inserted into the knowledge base.
User's Input. Apply equation (iii).
System. *Step 1. Reduction.* Equation (iii) is simplified to $6\bar{t}_3 = \bar{v}(\bar{v} - 1)$.
   Step 2. Substitution. Suppose that the user selects the facts $\bar{t}_3 = q^2 + q + 1 - t_1$ and $\bar{v} = \frac{1}{q+1}\left(3(q^2 + q + 1) - 2t_1\right)$. Then

$$6(q^2+q+1-t_1) = \frac{1}{q+1}\left(3(q^2 + q + 1) - 2t_1\right)\left(\frac{1}{q+1}\left(3(q^2 + q + 1) - 2t_1\right) - 1\right).$$

This relation of bounded parameters is inserted into the knowledge base.

Motivation. The relations of bounded parameters provide additional informations about the set S. They are not analysed by the system. The user can analyse them by hand or use the facilities of a computer algebra system. We suppose that the user solves the above equation with respect to the variable t_1. Then the two solutions are

$$t_1^{(1)} = \frac{1}{4}\left(3q^2 - q + 2 + (q+1)\sqrt{-3q^2 + 12q + 4}\right)$$

$$t_1^{(2)} = \frac{1}{4}\left(3q^2 - q + 2 - (q+1)\sqrt{-3q^2 + 12q + 4}\right)$$

Since t_1 is a positive integer, it follows that $-3q^2 + 12q + 4$ is a square. In particular $12q + 4 \geq 3q^2$, hence $q \leq 4$. For $q = 2, 3, 4$ we have $-3q^2 + 12q + 4 = 16, 13, 4$, respectively. Since 13 is a non-square, the only possible values for q are 2 and 4. These results are inserted into the knowledge base.

User's Input.

$$\text{Let } \bar{t}_1 \in \left\{\frac{1}{4}\left(3q^2 - q + 2 + (q+1)\sqrt{-3q^2 + 12q + 4}\right),\right.$$

$$\left.\frac{1}{4}\left(3q^2 - q + 2 - (q+1)\sqrt{-3q^2 + 12q + 4}\right)\right\}.$$

A case by case study yields the following possible parameter constellations.

	Case 1	Case 2	Case 3	Case 4
q	2	2	4	4
$\bar{t}_1$	0	6	9	14
$\bar{t}_3$	7	1	12	7
$\bar{v}$	7	3	9	7

They all occur. Case 1 is the whole point set in a projective plane of order 2. Case 2 is the set of points of a line in a projective plane of order 2. Case 3 is an affine plane of order 3 embedded in a projective plane of order 4. Finally, Case 4 is a projective plane of order 2 embedded in a projective plane of order 4.

8 Case Study II. Ovals in finite projective planes

An irreducible conic $\mathcal{O}$ of a finite desarguesian projective plane P of order q has the following two important combinatorial properties. Any line of P intersects $\mathcal{O}$ in at most two points, and $\mathcal{O}$ has exactly $q + 1$ points. In other words $\mathcal{O}$ is a $(q + 1)$-arc. Generalizing irreducible conics we say that an **oval** of a finite projective plane of order q is a $(q + 1)$-arc. In this section we shall study the combinatorial properties of ovals. The section is organized in the same way as the preceeding section.

User's Input. Let $C := \{0, 1, 2\}$.

System. The knowledge base is initialized with the free parameters $\bar{v}$, $\bar{t}_0$, $\bar{t}_1$, $\bar{t}_2$, $\bar{m}$, $\bar{r}_0$, $\bar{r}_1$, $\bar{r}_2$.

User's Input. Let $\bar{v} = q + 1$.

System. The fact $\bar{v} = q + 1$ is inserted into the knowledge base.

User's Input. Let p be a point of S.

System. The system focusses on the free point component containing the points of S and computes the new partitions.

User's Input. Apply equation (v).

System. *Step 1. Reduction.* Equation (v) is simplified to $1 + \bar{r}_2(p) = \bar{v}$.

Step 2. Substitution. Suppose that the user selects the fact $\bar{v} = q + 1$. It follows $1 + \bar{r}_2(p) = q + 1$, hence $\bar{r}_2(p) = q$. Applying the shifting principle the system obtains $\bar{r}_2(x) = q$ for all $x \in S$. This fact is inserted into the knowledge base.

User's Input. Apply equation (iv).

System. *Step 1. Reduction.* Equation (iv) is simplified to $\bar{r}_1(p) + \bar{r}_2(p) = q + 1$.

Step 2. Substitution. Suppose that the user selects the fact $\bar{r}_2(p) = q$. Then $\bar{r}_1(p) = 1$. By the shifting principle it follows that $\bar{r}_1(x) = 1$ for all $x \in S$. This fact is inserted into the knowledge base.

User's Input. Retract p. Let z be a point outside of S.

System. The resulting partitions are computed.

User's Input. Apply equation (vi).

System. *Step 1. Reduction.* Equation (vi) is simplified to $\bar{r}_1(z) + 2\bar{r}_2(z) = \bar{v}$.

Step 2. Substitution. Suppose that the user selects the fact $\bar{v} = q + 1$. Then the system obtains $\bar{r}_1(z) + 2\bar{r}_2(z) = q + 1$.

Motivation. The system is not able to use this equation. We suppose that the user observes that the (additional) condition q even implies that $\bar{r}_1(z) \geq 1$ and decides to investigate this case. At present a condition like q even cannot be administrated by the system. So we suppose that the user inserts the fact $\bar{r}_1(z) \geq 1$ into the knowledge base.

User's Input. Insert the fact $\bar{r}_1(z) \geq 1$.

System. By the shifting principle the system inserts the fact $\bar{r}_1(x) \geq 1$ for all $x \in S^C$.

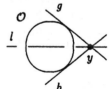

User's Input. Retract z. Let g and h be two tangent lines intersecting in a point $y \in S^C$. Let l be a non-tangent line through y.

System. The corresponding partitions are computed. Furthermore the fact $\bar{s}(l) \neq 1$ is inserted into the knowledge base.

User's Input. Apply equation (ix) with $i = 1$ with respect to l.

System. *Step 1. Reduction.* Let $C_1 := l \cap S$, $C_2 := \{x \in l \cap S^C \mid x \neq y\}$ and $C_3 := \{y\}$. Then equation (ix) is simplified to

$$\bar{t}_1 = \sum_{x \in C_1} \bar{r}_1(x) + \sum_{x \in C_2} \bar{r}_1(x) + \sum_{x \in C_3} \bar{r}_1(x).$$

Step 2. Substitution. Suppose that the user selects the facts $\bar{t}_1 = q + 1$, $\bar{r}_1(x) = 1$ for all $x \in C_1$, $\bar{r}_1(x) \geq 1$ for all $x \in C_2$ and $\bar{r}_1(y) \geq 2$. It follows

$$q + 1 \geq \bar{m}(C_1) \cdot 1 + \bar{m}(C_2) \cdot 1 + \bar{m}(C_3) \cdot 2.$$

The system of linear constraints is $\bar{m}(C_1)+\bar{m}(C_2)+\bar{m}(C_3) = q+1$ and $\bar{m}(C_3) = 1$ (C_3 is singular). Hence $q + 1 \geq q + 2$, a contradiction.

It follows that l is a tangent line. Thus any line through y is a tangent line. Because of $t_1 = q + 1$ all tangent lines of $\mathcal{O}$ meet in y. So the result obtained by the above user-system dialog is as follows.

If $\mathcal{O}$ is an oval of a finite projective plane of even order q, then there exists a point y outside of $\mathcal{O}$ such that the lines through y are exactly the tangent lines of $\mathcal{O}$.

This point y is called the **nucleus** of $\mathcal{O}$. For q odd one can show in a similar way that any point outside of an oval $\mathcal{O}$ is incident with either 0 or 2 tangent lines.

References

1. A. BEUTELSPACHER: Projective Planes, in F. BUEKENHOUT (ed.): *Handbook of Incidence Geometry*, Elsevier Amsterdam (1995), 107 – 137.
2. A. BEUTELSPACHER, J. UEBERBERG: Symbolic Incidence Geometry – A proposal for doing Geometry with a Computer, *SIGSAM Bull.* **27**, No. 2 (1993), 19 – 29 and No. 3 (1993), 9 – 24.
3. P. DEMBOWSKI: *Finite Geometries*, Springer Berlin, Heidelberg (1968).
4. D. HUGHES, F. PIPER: *Projective Planes*, Springer Berlin, Heidelberg (1973).
5. J. W. P. HIRSCHFELD: *Projective Geometries over Finite Fields*, Oxford University Press (1979).
6. J. A. THAS: Projective Geometry over a Finite Field, in F. BUEKENHOUT (ed.): *Handbook of Incidence Geometry*, Elsevier Amsterdam (1995), 295 – 349.
7. J. UEBERBERG: On regular $\{v, n\}$-arcs in Finite Projective Spaces, *J. Comb. Designs*, 1, No. 6 (1993), 395 – 409.
8. J. UEBERBERG: Interactive Theorem Proving and Computer Algebra, in J. CALMET, J. A. CAMPBELL (eds.): *Integrating Symbolic Mathematical Computing and Artificial Intelligence*, Springer Lecture Notes Comp. Science **958** (1995), 1 – 9.
9. J. UEBERBERG: *Interactive Theorem Proving in Symbolic Incidence Geometry*, submitted (1995).
10. D. WANG: *Geometry Theorem Proving with Existing Technology*, RISC-Linz Report Series No. 93-40.

Author's address:
Johannes Ueberberg, Mathematisches Institut
Arndtstrasse 2, D-35392 Giessen
e-mail: Johannes.Ueberberg@math.uni-giessen.de

Towards Modelling the Topology
of Homogeneous Manifolds
by Means of Symbolic Computation

Michael Joswig*

RISC-Linz
Johannes-Kepler-Universität Linz
A-4040 Linz
Austria

Abstract. We describe the implementation of a program which semi-decides whether a Lie group with certain specified properties can possibly act (continuously and transitively) on a given manifold or not. As a criterion the exactness of the induced homotopy sequence is used.

1 Introduction

The theory of Lie groups is among the central topics in modern mathematics. It represents a natural link between geometry, group theory and mathematical physics. Problems arising from physics, e.g. from the theory of special relativity, are transferable to the language of geometry. Often the geometrical questions can be answered by group-theoretical means in case the geometrical structure has enough symmetry, i.e. it admits sufficiently many automorphisms. In this respect, transitive actions are most important. Manifolds which admit a transitive action of some Lie group are called *homogeneous*.

There is a profound theory of Lie groups. It is important to note that the purely algebraic properties of a Lie group are intimately related to its topological properties. Many fundamental results in geometry and other fields of mathematics have been achieved via methods from Lie theory. Regularly the proofs depend on a detailed knowledge of the topology of homogeneous manifolds. Although only in rare cases their topological structure is known completely, it is often sufficient to have some algebraic invariants of these topological spaces, e.g. the cohomology ring or its homotopy groups.

In recent years quite a few attempts have been made to bring Lie theory to the computer, e.g. see van Leeuwen, Cohen and Lisser [18], Moody, Patera and Rand [26], Bödi and Joswig [1], not to mention the various packages devoted to the study of finite groups of Lie type. When it comes to solving problems arising from real Lie groups by means of a computer, frequently the derived

* The author gratefully acknowledges the financial support by the Austrian "Fonds zur Förderung der wissenschaftlichen Forschung"; Meitner post-doc grant, project M00295-MAT

linearized problems are studied instead. This means one tries to translate the initial problem about Lie groups into a (usually simpler one) about Lie algebras. However, the correspondence between Lie groups and Lie algebras is *not* 1–1. In fact, for a given Lie algebra $\mathfrak{g}$ there are usually several (not isomorphic) groups having $\mathfrak{g}$ as their associated Lie algebra. They are only *locally* isomorphic, which means that they are algebraically somehow similar but differ with respect to their global topological properties. Roughly speaking, during the linearization process the global topological information is lost.

Here we want to suggest an approach which is in some sense complementary. Instead of skipping the topological information we intend to focus on it.

We want to look for a computational method to solve problems of the following kind and related ones: Given a Lie group Γ and a homogeneous manifold M. Does there exist a (continuous) transitive action of Γ on M?

Without any doubt this immediately raises at least two questions. How should a non-discrete Lie group (being of uncountable cardinality as a set) be represented in the computer? How can one even expect that a question like this is decidable at all? Instead of trying to solve the problem in this generality we want to discuss a certain approximation of the problem which can be outlined as follows. A transitive action of Γ on M gives rise to an infinitely long exact sequence of their respective homotopy groups (see below). The group Γ and the manifold M are described by some of their algebraic and topological properties such as the dimension, information about compactness, maybe some homotopy groups which are known etc. It is explicitly admitted that only partial information will be given. Assume that a transitive action exists. Select finitely many among the infinitely many pieces of information obtained from the long exact homotopy sequence. Translate this into an arithmetical problem. Check for a contradiction. In case we actually arrive at a contradiction then this falsifies the assumption concerning the existence of a transitive action. However, if we do not obtain a contradiction, we might not gain any information.

To the unfamiliar reader it might be doubtful whether by the proposed strategy even a single interesting problem can be solved. Let us give just a few examples from topological geometry where mathematical proofs are at least partially based on this very reasoning: Salzmann et al. [31], proof of 62.9/62.10, first proof of 63.8, proof of 81.17, proof of 86.34, Joswig [20], proof of 3.2.

The paper is structured as follows. We start with two sections providing the background material needed from Lie theory and the theory of finitely generated abelian groups, respectively. Then we describe the algorithm. The last two sections cover an outline of the implementation as well as some examples.

I am indebted to Richard Bödi for many valuable discussions concerning the subject. I am grateful to Markus Stroppel for being unwilling to accept the occasional sloppiness that occurred in the previous version of this paper.

2 Lie groups and homogeneous manifolds

Let Γ be a Lie transformation group acting transitively on a manifold M. Choose a point $p \in M$, and let $\Delta = \Gamma_p$ be its stabilizer in Γ. Note that Δ is closed in Γ and is therefore a Lie group itself.

For information about Lie groups and homogeneous manifolds, e.g. see Bourbaki [8] [4] [5] [6], Warner [36], Helgason [14], Hilgert and Neeb [15], Gorbatsevich and Onishchik [13], Onishchik [27]. The books by Rotman [30] and Bredon [9] might serve as an introduction to algebraic topology. See also the appendix in Salzmann et al. [31] for a convenient collection of results (with few proofs) related to Lie groups and their topology.

For a topological space X and a point $x_0 \in X$ let $\pi_k(X, x_0)$ denote the k-th homotopy group of X with base point x_0. For $k \geq 2$ the group $\pi_k(X, x_0)$ is abelian. In case the topological space X is arcwise connected the isomorphism type of the homotopy groups does not depend on the chosen base point. We therefore write $\pi_k(X)$. For Lie groups connectedness implies arcwise connectedness. Thus the connected components of a Lie group are exactly the arcwise connected components. The connected component Γ^1 (of the unit element) of a Lie group Γ is a subgroup, the other components are cosets of Γ^1 and therefore pairwise homeomorphic. Analogously, coset spaces (of closed subgroups) and orbits (of continuous actions) of Lie groups have homeomorphic connected components. Being only interested in homotopy groups of Lie groups and their homogeneous spaces this means that we do not have to consider base points at all. Also, by Hilton's lemma, e.g. see Hilgert and Neeb [15], I.8.18, the fundamental group (i.e. first homotopy group) of a topological group is abelian. From the Mal'cev-Iwasawa decomposition theorem, see Iwasawa [19], Thm. 13, and also Hofmann and Terp [16], it follows that a connected Lie group is homotopic to any of its (mutually conjugate) maximal compact subgroups.

We have the *long exact homotopy sequence*, see Bredon [9], VII.6.7, Rotman [30], 11.51,

$$\ldots \xrightarrow{\delta_k} \pi_k(\Delta) \xrightarrow{\iota_k} \pi_k(\Gamma) \xrightarrow{\phi_k} \pi_k(M) \xrightarrow{\delta_{k-1}} \ldots \xrightarrow{\delta_1} \pi_1(\Delta) \xrightarrow{\iota_1} \pi_1(\Gamma) \xrightarrow{\phi_1} \pi_1(M).$$

In fact, it is possible to extend the exact sequence given above by $\pi_0(\Delta) \rightarrow \pi_0(\Gamma) \rightarrow \pi_0(M)$ where $\pi_0(X)$ denotes the set of arcwise connected components of the topological space X. In general, however, these sets do not carry a canonical group structure.

Theorem 1. *The homotopy groups of a homogeneous manifold are finitely generated abelian.*

We want to give a brief outline of a proof for this known result. Note that this theorem, in general, does not hold for arbitrary manifolds.

From the long exact homotopy sequence we infer that it is sufficient to show that the homotopy groups of an arbitrary Lie group are finitely generated. The Mal'cev-Iwasawa theorem implies that we only have to consider the case of a compact group. Assume first that the group is simply connected. A simply connected compact Lie group is semi-simple. The cohomology rings of the simple compact Lie groups are known, see Borel et al. [2]. In particular, they are finitely generated. Because of simple connectedness this implies that their homotopy groups are also finitely generated, see Spanier [33], §9.6, 19. The universal covering of a compact quasi-simple Lie group K is again a compact quasi-simple Lie group $\tilde{K}$. Their homotopy groups π_k coincide for $k \geq 2$. A Lie group is called *quasi-simple* if its Lie algebra is (non-abelian) simple, i.e. it is locally isomorphic to a simple group. Now $\pi_1(K)$ is a factor of the center of $\tilde{K}$, which is finite. Due to the product structure of a compact connected Lie group the only case left to consider is the case of the torus $\mathbb{T}$ for which $\pi_k(\mathbb{T}) = 0$, if $k \geq 2$, and $\pi_1(\mathbb{T}) \cong \mathbb{Z}$.

3 Exact sequences of finitely generated abelian groups

As it had already been pointed out we essentially want to utilize the long exact homotopy sequence in order to semi-decide whether a given Lie group can possibly act on a given manifold or not. We saw in the previous section that we always have to deal with finitely generated abelian groups with the possible exception that $\pi_1(M)$ might not be abelian if the stabilizer Δ is not connected.

Let $\mathbf{Ab_{fg}}$ denote the category of finitely generated abelian groups. A group $A \in \mathbf{Ab_{fg}}$ is isomorphic to $\mathbb{Z}^{f(A)} \times T(A)$ where $T(A)$ is finite, i.e $T(A)$ is isomorphic to a product of cyclic groups. The number $f(A)$ is called the *free rank* of A. We set $t(A) = |T(A)|$, the *torsion number* of A, and $\hat{r}(A) = $ order of a minimal generating system for $T(A)$, the *torsion rank* of A.

Now A can be represented as a sequence of natural numbers $(a_i)_{i=0}^{\hat{r}(A)}$ where $A \cong \mathbb{Z}^{a_0} \times C_{a_1} \times \ldots \times C_{a_{\hat{r}(A)}}$.

For information about finitely generated abelian groups see Zassenhaus [37], in particular §III.3 and §III.4.

Lemma 2. *Exact sequences of arbitrary length can be interpreted as sets of short exact sequences. Namely, the sequence*

$$A_0 \xrightarrow{\alpha_1} A_1 \xrightarrow{\alpha_2} \cdots \xrightarrow{\alpha_n} A_n$$

in $\mathbf{Ab_{fg}}$ *with groups* $A_0, \ldots, A_n$ *and homomorphisms* $\alpha_i : A_{i-1} \to A_i$ *such that* $A_0 = A_n = 0$ *is exact if and only if*

$$0 \to \operatorname{coker} \alpha_{i-1} \xrightarrow{\alpha_i} A_i \xrightarrow{\alpha_{i+1}} \operatorname{im} \alpha_{i+1} \to 0$$

is a short exact sequence, for $1 \leq i \leq n-1$, *where* $\operatorname{coker} \alpha_{i-1} = A_{i-1}/\operatorname{im} \alpha_{i-1} = A_{i-1}/\ker \alpha_i \cong \operatorname{im} \alpha_i$.

For basic information on sequences and their exactness e.g. see Bourbaki [7], I§1, no. 3.

The exactness of a short sequence $0 \to A \to B \to C \to 0$ is equivalent to a set of sentences in the theory of elementary arithmetic, see Joswig [21], 3.4

and its proof. If the torsion ranks of the groups A, B, C are bounded then this set of sentences is finite. Under the additional restriction that the variable sets for the free ranks is disjoint from the variable set of the torsion factors, the exactness problem for finitely generated abelian groups with bounded torsion ranks becomes decidable, [21], 3.8. This result heavily relies on Presburger's arithmetic, [29]. As this algorithm is very expensive, see Fischer and Rabin [12], and as, moreover, a bound on the torsion ranks might not always be obvious we decided to use a different approach.

In our system a group $A \in \mathbf{Ab_{fg}}$ is represented as the pair $(f(A), t(A))$. The exactness of the sequence $0 \to A \to B \to C \to 0$ then implies the following, see [21], 2.3.

(ES$_1$) $f(B) = f(A) + f(C)$,
(ES$_2$) $t(A) \mid t(B)$,
(ES$_3$) $t(B) \mid t(A) \cdot t(C)$,
(ES$_4$) $f(A) \neq 0 \vee t(B) = t(A) \cdot t(C)$.

These are the properties which are used here to model the exactness of sequences in $\mathbf{Ab_{fg}}$ within the language of elementary arithmetic.

Note that in our model non-isomorphic groups in $\mathbf{Ab_{fg}}$ might have the same representation. This simplification is done in order to severely cut down the overall complexity. For actual problems quite often one does not loose as much information as one might expect, however. This is due to the fact that many of the first few homotopy groups of a compact quasi-simple group are trivial or cyclic, e.g. see Mimura [24], §3.2.

4 An outline of the algorithm

4.1 Input

A triple (Δ, Γ, M) where Δ and Γ are Lie groups and M is a manifold.

Each of these objects is represented by a certain collection of its (algebraic and topological) properties. In general, these properties are far from characterizing the objects in question. Actually, a situation in which only partial information is supplied is the usual case.

4.2 Output

A sentence ϕ in the theory of elementary arithmetic such that ϕ is valid if there is a (continuous and transitive) action of Γ on M such that the stabilizer of a point is isomorphic to Δ.

The result gives certain restrictions on the structure of the homotopy groups of the manifolds (and groups) involved. They arise from the assumption that Γ acts continuously and transitively on M such that the stabilizer of a point is isomorphic to Δ. Typically, we aim for the most restrictive result possible, namely the rejection of the aforementioned assumption.

4.3 Procedure

We assume that there is such a transitive action. Then we try to derive consequences.

1. Consider the first $3k$ terms of the long exact homotopy sequence

$$\pi_k(\Delta) \xrightarrow{\iota_k} \pi_k(\Gamma) \xrightarrow{\phi_k} \pi_k(M) \xrightarrow{\delta_{k-1}} \ldots \xrightarrow{\delta_1} \pi_1(\Delta) \xrightarrow{\iota_1} \pi_1(\Gamma) \xrightarrow{\phi_1} \pi_1(M)$$

   for some $k \in \mathbb{N}$. There are several ways to choose k; we will not discuss this here. The reader might think of k as set to some global value (definitely less than 10 for practical purposes). If Δ is not connected then the group $\pi_1(M)$ might not be abelian. In this case we remove it from the sequence.
   We obtain a *finite* sequence in $\mathbf{Ab_{fg}}$ where, in general, only some of the groups are known. The information about the homomorphisms in the sequence is reduced to the isomorphism classes of their respective kernels and images. No information is provided about the actual mappings.

2. A group $A \in \mathbf{Ab_{fg}}$ is represented as $(f(A), t(A))$. Using Lemma 2 and (ES_1) – (ES_4) we translate the sequence into a sentence ϕ_0 in the theory of elementary arithmetic. If a homotopy group is not known then choose two new variables v_1, v_2 instead and represent the group by (v_1, v_2).
   In principal, this step already delivers what we are aiming at. We obtain a sentence (of a special syntactical structure) which is valid in the theory of elementary arithmetic if the input action exists. Quite typically though, this sentence is lengthy and unreadable.

3. Use some algorithm to compute the solution set for all the free variables introduced in the previous step. Actually, we are doing a little bit less by only simplifying ϕ_0 to some equivalent sentence ϕ by using rewriting techniques. This does not necessarily come up with the optimal answer but it works quite well in practice.
   This is the step where most of the work is done. Actually, the input for this step are two sentences from the first order theory of elementary arithmetic. In one of them addition is the only arithmetic operation which arises. In the other sentence multiplication is the only arithmetic operation. The variable sets of the two sentences are disjoint.

For the last step there is a certain freedom of choice what kind of algorithm one wants to use in order to simplify the formula obtained in step 2. Basically, one has to trade the (average) quality of the result versus the time necessary to compute it.

The first order theory of the natural numbers with addition can be decided by applying Presburger's algorithm [29]. Analogously, the first order theory of the natural numbers with multiplication is decidable due to Skolem [32]. Alternatively, Mostowski's algorithm [25], 2.3.6 and 5.31, together with Presburger's arithmetic could be applied. However, any solution of this problem is super-exponentially expensive, see Fischer and Rabin [12]. This is why a different approach was used.

5 About the implementation

We want to give an overview of the implementation, however coarse it might be. It has to be stressed that the author wants to see this program as a prototype. Therefore, more or less obvious optimizations are usually skipped for the sake of clarity of the mathematics behind. Also, the built-in algebraic topology is by now very limited. Mostly, simplified pieces of code are listed in order to clarify the structure. In particular, the implementation is highly modularized, whereas here we pretend that everything is defined within a single scope.

All the implementation has been done in Standard ML of New Jersey [34]. For a description of ML, e.g. see Paulson [28].

5.1 Data types

We want to represent Lie groups and manifolds as abstract objects with certain properties. Sometimes we explicitly want to allow that it is undefined whether a Lie group has a specific property or not. We model this by introducing a three-valued logic in a straightforward way. The data type is called **datur** in order to express that we do have **notSure** as a third truth value, thereby contradicting the famous law of Aristotelian logic.

```
datatype datur = truth of bool | notSure;

fun Not(truth(x)) = truth(not(x))
  | Not(_) = notSure;

fun And(truth(x),truth(y)) = truth(x andalso y)
  | And(_,_) = notSure;

fun Or(truth(x),truth(y)) = truth(x orelse y)
  | Or(_,_) = notSure;
```

As already mentioned in the introduction we explicitly admit the usage of objects which are only partially specified. We introduce the data type **intObject**, which extends the built-in data type **int** by adding the symbol **noValue** which can be used in situations where the actual value is not known or not defined. Occurrences of **noValue** are usually replaced by variables before the actual evaluation starts. Variables with the same names will be identified.

```
datatype intObject = num of int | noValue | var of string;
```

The data type **grp** is used to represent objects of the category $\mathbf{Ab_{fg}}$. A finitely generated abelian group is modelled as a pair (f, t) of integers, where f denotes the free rank and t the torsion number, respectively.

```
type grp = intObject*intObject;
```

We have constants to denote the trivial group and an unspecified group, respectively. There is also a function which returns the representation of a finite group of given order.

```
val trivialGrp = (num(0),num(1));
val unGrp = (noValue,noValue);
fun finGrp(n:int) = (num(0),num(n));
```

Arbitrary topological spaces are reduced to their list of homotopy groups. Note that in our model the first homotopy group of a space is not defined (return value unGrp) if this group is not abelian. The return value unGrp is also used for situations where the homotopy group is not known.

```
type topSpace = int -> grp;
```

We think of manifolds as of topological spaces (i.e. a sequence of homotopy groups) with a dimension and which can be compact or connected. It does not matter which definition of topological dimension is used, as all relevant notions (small/big inductive dimension, covering dimension) coincide for manifolds, e.g. see Engelking [11], 3.1.29. Our manifolds have names[2].

```
type manifold =
  {name : string,
   top : topSpace,
   dim : intObject,
   compact : datur,
   connected : datur};
```

Now (#top M)(k) gives us the k-th homotopy group of the manifold M.

Of course, Lie groups are implemented as special manifolds. Apart from their topological properties they can be abelian, soluble or quasi-simple. The direct product of Lie groups is again a Lie group; this is the product in the category of groups as well as the product in the category of topological spaces. In case a group can be written as a product, we prefer to keep the product structure rather than implementing a function directProd which would return a new record with the appropriate entries.

```
type LieGrpRec =
  {mf : manifold,
   abelian : datur,
   soluble : datur,
   quasisimple : datur};

datatype LieGrp = Rec of LieGrpRec
                | directProd of LieGrp*LieGrp;
```

---

[2] Usually AMS-TEX-syntax is used.

It is straightforward to write functions which determine whether a Lie group
(which might be given as a direct product) is abelian, soluble, semi-simple, compact or connected. Note that these functions are projections from our three-
valued logic to the classical logic. The truth value notSure is mapped to false,
i.e. if, say, compactness cannot be assured then this does *not* mean that the
group is not compact.

Real vector spaces, e.g., can be implemented like this.

```
fun RasManifold(n:int):manifold =
    {name = "\\RR^{" ^ intToStr(n) ^ "}",
     top = (fn(_) => (num(0),num(1))), dim = num(n),
     compact = truth(n=0), connected = truth(true)};

val R : int->grp = fn(n) =>
    Rec({mf = RasManifold(n),
     abelian = truth(true), soluble = truth(true),
     quasisimple = truth(false)});
```

Another example would be the compact form of exceptional type G_2. Only
the first six homotopy groups are implemented (for the second and third one see
the definition of the function Pi below).

```
val G2 =
    Rec({abelian = truth(false),
     soluble = truth(false),
     quasisimple = truth(true),
     mf = {name = "\\text{G}_{2(-14)}",
        top = fn(r) =>
            case r of
                1 => trivialGrp
              | 4 => trivialGrp
              | 5 => trivialGrp
              | 6 => finGrp(3)
              | _ => unGrp,
        dim = num(14),
        compact = truth(true),
        connected = truth(true)}});
```

Most of the classical Lie groups are implemented in a similar manner. The implementation concerning the homotopy groups, however, is essentially restricted
to what can be derived from Bott periodicity, e.g. see Bredon [9], VII.8.5. This
will be extended in the near future.

A (continuous and transitive) action of the group Γ on the manifold M with
the stabilizer Γ_p for some point $p \in M$ is represented as the triple (Γ_p, Γ, M).

```
type action = LieGrp*LieGrp*manifold;
```

5.2 Translating homogeneous spaces into elementary arithmetic

Unfortunately, due to limits of space we cannot describe this very heart of the algorithm in any detail. We will give a brief outline instead.

We have to mention the two functions Hot : action -> separated and solve : separated -> separated, where separated is a data type which allows to represent separated open formulas in the first order theory of the structure $\langle \mathbb{N}, +, \cdot \rangle$. A formula is called *separated* if it is actually in the subtheory of the structure $\langle \mathbb{N}, + \rangle \times \langle \mathbb{N}, \cdot \rangle$; i.e. in a separated formula addition and multiplication are separated.

The function Hot selects a finite part of the long exact homotopy sequence arising from the input action and translates it into the first order language of elementary arithmetic in the way it has been described in the previous section.

Hot calls the function Pi which returns the n-th homotopy group of the Lie group G if it is known. There is a certain general knowledge about topology of Lie groups coded in the function Pi. Firstly, it "knows" how to compute homotopy groups of products. Secondly it "knows" that the second homotopy group of a Lie group is trivial and that the third homotopy group of a quasi-simple compact Lie group is isomorphic to $\mathbb{Z}$, see Bott [3] and also Browder [10]. Only in the case when none of these built-in rules can be applied we rely on the sequence of homotopy groups stored.

We have the **forget** functor which maps a Lie group to the underlying manifold.

```
fun Pi(n,directProd(G,G')) = grpProd(Pi(n,G),Pi(n,G'))
  | Pi(n,G) =
      if (n=2) then
        trivialGrp
      else if (n=3) andalso compact(G)
                    andalso quasisimple(G) then
        (num(1),num(1))
      else
        (#top(forget(G)))(n);
```

The function solve keeps rewriting the input formula until it does not change any more. The termination of solve is guaranteed by means of a partial ordering on the set of formulas with respect to which the intermediate results of solve become simpler in each step.

5.3 Main program

The input to the main program is a set of actions, i.e. the input is of type action list. In order to keep the system flexible the main program consists of three steps where one step can take the output of the previous step as an input. The idea is that firstly each of the actions is analyzed individually. For each action the function SeparateEvaluation gives back a sentence in the theory of elementary arithmetic in disjunctive normal form.

```
fun SepEvalStep(accumulated,nil) = accumulated
  | SepEvalStep(accumulated,(H,G,M)::tail) =
        SepEvalStep(solve([Hot(H,G,M)])::accumulated,tail);

fun SeparateEvaluation(T:action list) = SepEvalStep(nil,T);
```

In the second step the function `CombinedEvaluation` extracts the *facts* from all the input actions, i.e. we are given a sentence which expresses necessary conditions without any case distinctions. This is useful in situations where a full combined evaluation is not feasible. And finally, in the third step, the conjunction of all sentences obtained from the separate evaluation is simplified again by basically just another call to `solve`. This is accomplished by the function `CombinedFullEvaluation`.

In the case where just a single action is analyzed the last step is trivial.

The ML style output is barely readable. Therefore, we have a TEX back end.

5.4 Future development

The overall "knowledge" in algebraic topology of the program should be increased, i.e. we have to code more homotopy groups explicitly.

So far we are only able to work with actual direct products of groups. From the view point of Lie theory, however, this is somehow inadequate. A generalization to quasi-direct products would be more natural. A Lie group Γ is called a *quasi-direct product* of Lie groups A and B if the Lie algebra of Γ is isomorphic to the direct product of the respective Lie algebras of A and B.

In order to be able to use the full strength of the method discussed this purely topological approach should be combined with packages dealing with Lie algebras and their representation theory. At least some information on the subgroup lattice of compact quasi-simple groups should be provided. This could be used to generate all groups of a given (small) dimension with given properties. This is a typical demand in situations as in the second example below.

6 Some examples

We present two examples, both arising in topological geometry. For the general theory of projective planes and generalized polygons see Hughes and Piper [17] and Van Maldeghem [35], respectively. For the topological situation see Salzmann et al. [31] and Kramer [22], respectively.

6.1 Projective planes

The point (and line) spaces of the classical compact connected projective planes (over $\mathbb{R}, \mathbb{C}, \mathbb{H}, \mathbb{O}$) are homogeneous manifolds. They admit a transitive action of the respective collineation group. In fact, Löwen [23] showed that these are the only examples with this property. In Salzmann et al. [31], 63.8, a variety of proofs are given. One step in the first proof given there requires to show that the compact group of type G_2 cannot act continuously and transitively on the point set P of a compact connected projective plane. Assuming the contrary, a dimension argument yields that the stabilizer of a point is isomorphic to $SO_4\mathbb{R}$. We infer that P is 8-dimensional. Therefore, we know that $\pi_2(P) = 0$ by [31], 52.14c.

We define the manifold P.

```
val P = {name = "P",
         top = fn(r) => if r=2 then trivialGrp else unGrp,
         dim = num(8),
         compact = truth(true),
         connected = truth(true)}:manifold;
```

The groups $SO_4\mathbb{R}$ and $G_{2(-14)}$ are predefined. We can immediately ask for analyzing the action.

```
SeparateEvaluation [ (SO(4),G2,P) ];
```

The TEX output looks like this.

– Evaluating the action of $G_{2(-14)}$ on P with stabilizer $SO_4\mathbb{R}$ ...
  (FALSE)

Thus we infer that $G_{2(-14)}$ does not admit a transitive action on any compact connected projective plane.

6.2 Generalized quadrangles

In Joswig [20] the compact connected quadrangles admitting a group of collineations which acts pentagon-transitively are classified. One step in the proof is to exclude the quadrangle related to the Lie group of exceptional type $E_{6(-14)}$. Assuming that this quadrangle does admit a pentagon-transitive action, we end up in a situation which can be described as follows.

We have a Lie group Σ of dimension 18 whose maximal compact subgroup is isomorphic to the unitary group $U_4\mathbb{C}$.

```
val Sigma = Rec({abelian = truth(false),
     soluble = truth(false),
     quasisimple = truth(false),
     mf = {name = "\\Sigma",
           top = #top(forget(U(4))),
           dim = num(18),
           compact = truth(false),
           connected = notSure}}):LieGrp;
```

We assume further that this group acts transitively on a 9-sphere with two points removed. This space is clearly homotopic to the 8-sphere.

```
val doublyPuncturedLine =
        {name = "L\\setminus\\{p,q\\}",
         top = #top(S(8)),
         dim = num(9),
         compact = truth(false),
         connected = truth(true)}:manifold;
```

Let Γ be the stabilizer of a point. Note that the function **Stabilizer** does not perform deep computations. It only sets up a record with the appropriate entries. A name for Γ is generated automatically. Considering an action of Λ on M the stabilizer will be denoted as $\Lambda_i(M)$ where i is an index which makes this group unique. Note that there might be several actions of Λ on M. In our example the name of Γ will be $\Sigma_1(L \setminus \{p, q\})$.

```
val Gamma = Stabilizer(Sigma,doublyPuncturedLine);
```

The last assumption is that Γ acts transitively on a 6-sphere with two points removed. Similarly to the situation above, this space is homotopic to the 5-sphere.

```
val doublyPuncturedPencil =
        {name = "\\cL_p\\setminus\\{L,M\\}",
         top = #top(S(5)),
         dim = num(6),
         compact = truth(false),
         connected = truth(true)}:manifold;
```

As $\dim \Gamma = 9$ the stabilizer of a point in Γ is three-dimensional. Let it be denoted by Δ. There are not many choices for Δ. In fact, Δ is either soluble or quasi-simple. In the latter case Δ is locally isomorphic to either $SO_3\mathbb{R}$ or $SL_2\mathbb{R}$. Each of these cases has to be checked individually. Let us consider the case where Δ is actually isomorphic to $SO_3\mathbb{R}$.

```
val Delta = SO(3);
```

We start with a separate evaluation of both actions.

```
SeparateEvaluation [ (Gamma,Sigma,doublyPuncturedLine),
                     (Delta,Gamma,doublyPuncturedPencil) ]
```

And this is what we get.

- Evaluating the action of Σ on $L \setminus \{p, q\}$ with stabilizer $\Sigma_1(L \setminus \{p, q\})$

  . . .
  $(1 = t(\pi_1(\Sigma_1(L \setminus \{p, q\})))) \wedge 0 = f(\pi_1(\Sigma_1(L \setminus \{p, q\}))) \wedge f(\pi_3(\Sigma_1(L \setminus \{p, q\}))) = 1 + \mathrm{var}_1) \vee (1 = f(\pi_1(\Sigma_1(L \setminus \{p, q\})))) = t(\pi_1(\Sigma_1(L \setminus \{p, q\}))) \wedge f(\pi_3(\Sigma_1(L \setminus \{p, q\}))) = 1 + \mathrm{var}_1)$

- Evaluating the action of $\Sigma_1(L \setminus \{p, q\})$ on $\mathcal{L}_p \setminus \{L, M\}$ with stabilizer $SO_3\mathbb{R}$ ...

$(1 = f(\pi_3(\Sigma_1(L \setminus \{p, q\}))) \wedge 0 = f(\pi_1(\Sigma_1(L \setminus \{p, q\}))) \wedge 2 = t(\pi_1(\Sigma_1(L \setminus \{p, q\})))) \vee (1 = f(\pi_3(\Sigma_1(L \setminus \{p, q\}))) = t(\pi_1(\Sigma_1(L \setminus \{p, q\}))) \wedge 0 = f(\pi_1(\Sigma_1(L \setminus \{p, q\})))) \vee (1 = t(\pi_3(\Sigma_1(L \setminus \{p, q\}))) \wedge 0 = f(\pi_1(\Sigma_1(L \setminus \{p, q\}))) = f(\pi_3(\Sigma_1(L \setminus \{p, q\}))) \wedge 2 = t(\pi_1(\Sigma_1(L \setminus \{p, q\})))) \vee (1 = t(\pi_1(\Sigma_1(L \setminus \{p, q\}))) = t(\pi_3(\Sigma_1(L \setminus \{p, q\}))) \wedge 0 = f(\pi_1(\Sigma_1(L \setminus \{p, q\}))) = f(\pi_3(\Sigma_1(L \setminus \{p, q\})))) \vee (0 = f(\pi_1(\Sigma_1(L \setminus \{p, q\}))) = f(\pi_3(\Sigma_1(L \setminus \{p, q\}))) \wedge 2 = t(\pi_1(\Sigma_1(L \setminus \{p, q\}))) \wedge \text{var}_{27} \neq 0) \vee (1 = t(\pi_1(\Sigma_1(L \setminus \{p, q\}))) \wedge \text{var}_{27} \neq 0 \wedge 0 = f(\pi_1(\Sigma_1(L \setminus \{p, q\}))) = f(\pi_3(\Sigma_1(L \setminus \{p, q\})))))$

We can now extract the facts.

CombinedEvaluation it;

- Combined evaluation ...
  - Combining facts ...
    $(f(\pi_3(\Sigma_1(L \setminus \{p, q\})))) = 1 + \text{var}_1)$

We obtain that $f(\pi_3(\Gamma)) \geq 1$. And now we are asking for maximum information.

CombinedFullEvaluation it;

This took more than 90% of the total time.

  - Combining all partial results ...
    (FALSE)

We infer that Δ is not isomorphic to $SO_3\mathbb{R}$. The other cases can be treated in a similar way.

References

1. R. Bödi and M. Joswig. Tables for an effective enumeration of real representations of quasi-simple Lie groups. *Sem. Sophus Lie*, 3(2):239–253, 1993.
2. A. Borel et al. *Seminar on transformation groups* Ann. of Math. Stud. 46, Princeton Univ. Press, 1960.
3. R. Bott. An application of Morse theory to the topology of Lie groups. *Bull. Soc. Math. France*, 84:353–411, 1956.
4. N. Bourbaki. *Groupes et algèbres de Lie, Chap. 4, 5, 6*. Hermann, Paris, 1968.
5. N. Bourbaki. *Groupes et algèbres de Lie, Chap. 7, 8*. Hermann, Paris, 1975.
6. N. Bourbaki. *Groupes et algèbres de Lie, Chap. 9, Groupes de Lie réels compactes*. Masson, Paris, 1982.
7. N. Bourbaki. *Commutative Algebra, Chap. 1–7*. Springer, Heidelberg–Berlin–New York, 1989.

272

8. N. Bourbaki. *Lie groups and Lie algebras, Chap. 1-3*. Springer, Heidelberg–Berlin–New York, 1989.
9. G.E. Bredon. *Topology and geometry*. Springer, Heidelberg–Berlin–New York, 1993.
10. W. Browder. Torsion in h-spaces. *Ann. Math. (2)*, 74:24–51, 1961.
11. R. Engelking. *Dimension theory*. North-Holland, Amsterdam, 1978.
12. M.J. Fischer and M.O. Rabin. Super exponential complexity of Presburger's arithmetic. *SIAM-AMS Proceedings*, 7:27–41, 1974.
13. V.V. Gorbatsevich and A.L. Onishchik. Transformation Lie groups. *Itogi Nauki i Tekhniki: Sovremennye Problemy Mat.: Fundamental'nye Napravleniya*, 20:5–101, 1988.
14. S. Helgason. *Differential geometry, Lie groups and symmetric spaces*. Academic Press, New York, 1978.
15. J. Hilgert and K.-H. Neeb. *Lie-Gruppen und Lie-Algebren*. Vieweg, Braunschweig, 1991.
16. K.H. Hofmann and Ch. Terp. Compact subgroups of Lie groups and locally compact groups. *Proc. Amer. Math. Soc.*, 120:623–634, 1994.
17. D.R. Hughes and F.C. Piper. *Projective planes*. Springer, Heidelberg–Berlin–New York, 1973.
18. M.A.A. van Leeuwen, A.M. Cohen, and B. Lisser. *LiE — A package for Lie group computations*. Computer Algebra Nederland, Amsterdam, 1992.
19. K. Iwasawa. On some types of topological groups. *Ann. Math. (2)*, 50:507–558, 1949.
20. M. Joswig. Generalized polygons with highly transitive collineation groups. *Geom. Ded.*, 58:91–100, 1995.
21. M. Joswig. Deciding the exactness of sequences of finitely generated abelian groups. Preprint.
22. L. Kramer. *Compact polygons*. Phd thesis, Univ. Tübingen, 1994.
23. R. Löwen. Homogeneous compact projective planes. *J. Reine Angew. Math.*, 321:217–220, 1981.
24. M. Mimura. Homotopy theory of Lie groups. Chap. 9 of I.M. James (ed.): Handbook of algebraic topology, Amsterdam, 1995.
25. A. Mostowski. On direct products of theories. J. Symbolic Logic, 17:1–20, 1952.
26. R.V. Moody, J. Patera, and D.W. Rand <Rand@ERE.UMontreal.CA>. simpLie — Macintosh software for simple Lie algebras. Internet WWW page, at URL: <http://www.crm.umontreal.ca/~rand/simpLie.html>, (version of June 13, 1995).
27. A.L. Onishchik. *Topology of transitive transformation groups*. Barth, Leipzig–Berlin–Heidelberg, 1994.
28. L.C. Paulson. *ML for the working programmer*. Cambridge Univ. Press, 1991.
29. M. Presburger. Über die Vollständigkeit eines gewissen Systems der Arithmetik ganzer Zahlen, in welchem die Addition als einzige Operation hervortritt. pages 92–101. Comptes rendus du 1er Congrès des Mathématiciens des Pays Slaves, Warszawa, 1929.
30. J.J. Rotman. *An introduction to algebraic topology*. Springer, Heidelberg–Berlin–New York, 1988.
31. H. Salzmann, D. Betten, T. Grundhöfer, H. Hähl, R. Löwen, and M. Stroppel. *Compact projective planes*. De Gruyter, Berlin, 1995.
32. T. Skolem. Über gewisse Satzfunktionen in der Arithmetik. Skrifter utgit av Videnskapsselskapet i Kristiania, I. klasse, no. 7, 1930.

33. E.H. Spanier. *Algebraic topology*. McGraw-Hill, New York, 1966.
34. Standard ML of New Jersey. Copyright by AT&T Bell Laboratories. Version 109, 1996.
35. H. Van Maldeghem. *Generalized polygons - a geometric approach*. To appear.
36. F.W. Warner. *Foundations of differentiable manifolds and Lie groups*. Springer, Heidelberg–Berlin–New York, 1983. Corrected reprint of the 1971 edition.
37. H. Zassenhaus. *The theory of groups*. Chelsea, New York, 1949.

Solving Geometrical Constraint Systems Using CLP Based on Linear Constraint Solver

Denis Bouhineau

Laboratoire IMAG - LSR
BP 53 X, 38041 Grenoble Cedex 9
FRANCE

Abstract. Euclidean geometrical configurations obtained with ruler, square and compass may be described as arithmetic constraint systems over rational numbers and consequently belong to the domain of CLP(R). Unfortunately, CLP based on linear constraint solvers which are efficient and can deal with geometrical constraints such as parallelism, perpendicularity, belonging to a line i.e. pseudo-linear constraints, cannot handle quadratic constraints introduced when using circles.

Two problems arise with quadratic constraints : the first problem is how to solve mixed constraint systems i.e. linear constraints combined with quadratic constraints; the second problem is how to represent the real numbers involved in the resolution of mixed constraints, so that correctness and completeness of linear constraint solvers are preserved.

In this paper we present a naive algorithm for mixed constraints based on a cooperation with a linear constraint solver. We define a representation for the real numbers, i.e. constructible numbers, occuring in Euclidean geometry. This representation preserves correctness and completeness of above algorithms. A survey over 512 theorems of Euclidean geometry shows that from both theoretical and experimental points of view, this representation is appropriate. This work is intended to be used to verify geometrical properties in Intelligent Tutoring System for geometry.

Keywords: quadratic algebraic extension, representation of rational, real, algebraic, and constructible numbers, cooperation between solvers for mixed constraints.

1 Introduction

Constraint Logic Programming (CLP) is based on extensions of Prolog incorporating constraint solving algorithms tailored to specific domains such as trees, booleans, finite domains or real numbers. For example, in the domain of arithmetic constraints over rational numbers, Linear Programming algorithms have been incorporated to CLP languages such as Prolog III, CHIP, and CLP(R) see [Col90, Din88, Jaf92] respectively. The expressive power and flexibility gained by combining Linear Programming, Logic Programming, and Constraint Programming allows one to express systems of pseudo-linear[1] constraints that can be solved using linear constraint solvers. This is particularly true in the case of

---

[1] equations which become linear after solving some other equations

geometrical configuration defined by points and lines, often used in geometry courses. Moreover in practical issues in the domain of Intelligent Tutoring Systems in geometry, coordinates of objects defining geometrical configurations are directly obtained from the screen and correspond to rational numbers. When an exact representation of rational numbers is used by the solver, solutions given by the linear constraints solver are exact rational numbers not approximate. In this case, as a consequence that linear constraint systems solver over rational numbers algorithms are correct and complete, overconstraints resulting from geometrical overspecifications can be verified.

Unfortunately, Euclidean geometry is not based on lines and points, ruler and square only. Euclidean geometry is based on ruler, square **and** compass. An other point, rational numbers are not sufficient, algebraic numbers are needed to express some geometrical constructions. For example, a equilateral triangle necessarily has one coordinate in an algebraic extension of Q with $\sqrt{3}$. In consequence of the introduction of the compass, geometrical configurations deal quadratic constraints (second degree equations) which cannot be solved using linear programming algorithms over rational numbers. On the other hand, the numbers introduced to express coordinates of all constructible configurations occuring in Euclidean geometry do not belong to the numerical domain of rational numbers where the linear algorithms are complete and correct. Two main problems arise with quadratic constraints and real numbers of geometry :

1. The first problem, namely how to solve set of mixed constraints i.e. linear constraints combined with quadratic constraints, may be solved for certain geometrical configurations. For the particular geometrical configurations called "constructive" in [Chou92], we present an algorithm which is complete. See [Pes95] for complementary approach.
2. The second problem which will be essentially addressed in this paper is how to represent real numbers, called constructible numbers (a special class of algebraic numbers, cf. [Leb92]), involved in the resolution of mixed constraints, and in Euclidean geometry.

We propose in this paper one exact representation for constructible numbers compatible with the linear constraint solvers used in CLP. On the practical side, this representation aims at avoiding numerical errors due to the lack of precision with representations of real numbers with floating point arithmetic. On the theoretical side, this representation preserves the correctness and completeness of the linear constraint solvers over rational numbers. These are two essential perspectives in the application domain of Teaching Geometry in which we are concerned, cf. [All93, Lab95].

The paper is organized as follows. Section 2 sketches the resolution of geometrical constraints with the help of the linear solver of Prolog III. Section 3 introduces a basic algorithm for solving mixed constraints. Sections 4 and 5 are concerned with the representation of constructible numbers. Section 6 provides examples of solution and includes details about the complexity of geometrical configurations arising in practice : a survey of about 512 geometrical configurations from [Chou88] is exposed. A short conclusion ends the paper.

2 Geometric Configuration and Linear Constraint Solvers

Let's take one configuration from the 170 theorems without circles among the 512 theorems given by Chou[2] in [Chou88]. For example, let us consider theorem 82 page 143 :

Theorem 82 In triangle ABC, let F be the midpoint of the side BC, D and E the feet of the altitudes on AB and AC respectively, FG is perpendicular to DE at G. Show that G is the midpoint of DE.

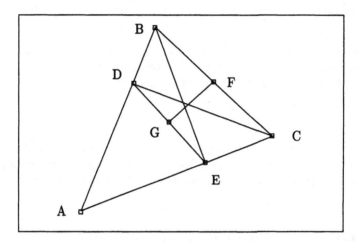

Fig. 1. Configuration 82

The configuration observed in theorem 82 can be specified in Prolog III in the following way :

```
Collinear(<A_x,A_y>,<B_x,B_y>,<C_x,C_y>) ->
    { (A_x-B_x)*(B_y-C_y)-(A_y-B_y)*(B_x-C_x)=0 };
Perpendicular(<A_x,A_y>,<B_x,B_y>,<C_x,C_y>,<D_x,D_y>) ->
    { (A_x-B_x)*(C_x-D_x)+(A_y-B_y)*(C_y-D_y)=0 };
Midpoint(<A_x,A_y>,<B_x,B_y>,<C_x,C_y>) ->
    { 2*B_x-A_x-C_x=0, 2*B_y-A_y-C_y=0 };

Configuration82(A,B,C,D,E,F,G) ->
```

---

[2] Numerous examples can be also found among theorems in Projective Geometry.

```
        Midpoint(B,F,C)
        Perpendicular(D,C,A,B)
        Collinear(D,A,B)
        Perpendicular(E,B,A,C)
        Collinear(E,A,C)
        Perpendicular(G,F,E,D)
        Collinear(G,E,D);


    Theorem82(A,B,C)  ->
        Configuration82(A,B,C,D,E,F,G)
        Midpoint(D,G,E);
```

The representation of the configuration using Prolog predicates is straightforward, and elegant. Within the same formalism, the theorem can be expressed as well. The solution of the linear constraint system gives complete configuration.

```
> Configuration82(<3,0>,<6,8>,<12,5>,D,E,F,G);
  {D = <420/73,536/73>, E = <921/106,335/106>, F = <9,13/2>,
   G = <111753/15476,81271/15476>}
```

Since the linear constraint solver of Prolog III computes exact solution of linear system over rational numbers, the overconstraint Midpoint(D,G,E) resulting from geometry in the theorem 82 is numerically verified :

```
> Theorem82(<3,0>,<6,8>,<12,5>);
    {}
```

The method informally introduced in this section provides sound results when the constraints introduced are pseudo-linear, see [Col93]. For the class of constructive configuration, the linear constraint solver approach is complete and correct.

3 Solution of Mixed Constraint Systems

This section is devoted to the solution of mixed constraint systems obtained when geometrical properties about circles are introduced. In [Chou88], 342 theorems out of 512 correspond to configurations with circles. For example let us consider theorem 110 on page 156 :

Theorem 110 Let D, E be two points on sides AC and BC of a triangle ABC such that AD=BE, F the intersection of DE and AB. Show that FD.AC=EF.BC.

The configuration in theorem 110 can be specified in Prolog III as follows :

```
Collinear(<A_x,A_y>,<B_x,B_y>,<C_x,C_y>) ->
    { (A_x-B_x)*(B_y-C_y)-(A_y-B_y)*(B_x-C_x)=0 };

EquiDist(<A_x,A_y>,<B_x,B_y>,<C_x,C_y>,<D_x,D_y>) ->
    { (A_x-B_x)^2+(A_y-B_y)^2-(C_x-D_x)^2-(C_y-D_y)^2=0 };
```

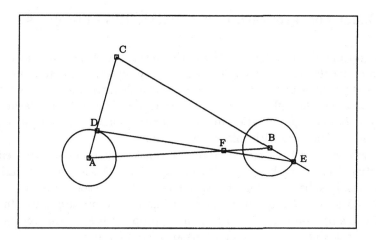

Fig. 2. Configuration 110

```
Configuration110(A,B,C,D,E,F) ->
    Collinear(A,C,D)
    EquiDist(E,B,A,D)
    Collinear(E,C,B)
    Collinear(F,E,D)
    Collinear(F,A,B);
```

The constraint system obtained here cannot be solved directly with linear constraint solvers. So we used the interaction between the linear constraint solver of Prolog III and a quadratic constraint solver. Each linear geometric constraint is given to the linear constraint solver. Pseudo-linear constraints are "frozen" until they become linear and then given to the linear constraint solver (as it is done automatically in Prolog III). Quadratic constraints are handled by a special program whose purpose is to transform them symbolically by interacting with the linear solver.

The quadratic constraint solver operates as follows : Let (E) $aX^2 + bX + cXY + dY^2 + eY + f = 0$ be a symbolic constraint

1. if it is found that $Y = mX + p$, then (E) is transformed into (E') $a'X^2 + b'X + c' = 0$, and equation (E') is solved in a classical way, i.e. $X = (-b' \pm \sqrt{b'^2 - 4a'c'})/(2a')$.
2. else if (E) becomes linear this result is passed on to the linear constraints solver.
3. else the process is dis-activated until a new activation.

The naive algorithm presented here is complete for geometric configuration of constructive type. Assuming that the square root computed in step 1. is correct, the algorithm is correct.

4 Exact representation of constructible numbers

The algorithm given in the previous section introduces true real numbers during square root calculations. This does not involve an approximate representation with floating point arithmetic. An exact representation of square root numbers can be obtained through a symbolic representation. This section is devoted to the definition of one representation for constructible numbers and to the definition of the arithmetic used with this representation.

In subsection 4.1 a representation for constructible numbers is defined; in subsection 4.2 and 4.3 known algorithms are given for the four basic arithmetic operations. In subsection 4.4 a trivial square root operation is presented. In subsection 4.5 an original definition of nullity and positivity are given. In subsection 4.6 examples are given showing the appropriateness of this representation for exact calculations Some limitations are also point out.

4.1 Representation of constructible numbers

Suppose we have a sequence of square root calculation, then numbers belong to the following quadratic extension of Q :

- $k_0 = Q$,
- $k_1 = k_0[\sqrt{\alpha_0}]$ where $\alpha_0 \in k_0, \alpha_0 \geq 0$, α_0 is the first square root introduced during the calculations.
- $k_n = k_{n-1}[\sqrt{\alpha_{n-1}}]$ where $\alpha_{n-1} \in k_{n-1}, \alpha_{n-1} \geq 0$, α_{n-1} is the last square root introduced during the calculations .

Let $A \in k_n$, we write $A : (a_1, a_2)$ where $a_1, a_2 \in k_{n-1}$ when $A = a_1 + a_2 * \sqrt{\alpha_{n-1}}$.

For example, in $Q[\sqrt{2}][\sqrt{1 + \sqrt{2}}]$, the real number $e_0 : ((5, 2), (3, 1))$ represents the value : $e_0 = 5 + 2\sqrt{2} + (3 + 1\sqrt{2})\sqrt{1 + \sqrt{2}}$

In the following paragraphs, rational numbers will be represented as simple numerical values; constructive numbers, like $e_0 : ((5, 2), (3, 1))$, will be represented as a Prolog binary trees, like $[[5, 2], [3, 1]]$. The current field used for calculation, Q will be represented as $[]$, and sequence of quadratic extension $Q[\sqrt{2}][\sqrt{1 + 1\sqrt{2}}]$ is represented by a Prolog list of constructible numbers, here $[[1, 1]|[2]]$.

4.2 Addition and Subtraction

Given $A, B \in k_n$, with $A : (a_1, a_2)_{k_n} = a_1 + a_2\sqrt{\alpha_{n-1}}$ and $B : (b_1, b_2)_{k_n} = b_1 + b_2\sqrt{\alpha_{n-1}}$. Then $(A + B) : (a_1 + b_1, a_2 + b_2)$ since $A + B = a_1 + b_1 + (a_2 + b_2)\sqrt{\alpha_{n-1}}$. Let us distinguish additions between elements of k_n and additions between elements of k_{n-1}, we have : $A +^n B : (a_1 +^{n-1} b_1, a_2 +^{n-1} b_2)$. The operation $+^0$ denote the usual addition in Q.

We could define $(A - B) : (a_1 - b_1, a_2 - b_2)$, but this definition is superfluous : unification between $(A - B)$ and $(a_1 - b_1, a_2 - b_2)$ is equivalent to the unification of A with $(A - B) + B$. This definition stands with CLP, thanks to logic programming with linear constraints, as the constraints involved for the unification of A with $(A - B) + B$ are all linear constraints.

The predicate ng_plus(E,A,B,C), true if C is equal to A+B in the extension E and predicate ng_minus(E,A,B,C) true if C is equal to A-B are described by :

```
ng_plus([],A,B,A+B).
ng_plus([E|L],[A1,A2],[B1,B2],[C1,C2]) :-
    ng_plus(L,A1,B1,C1),
    ng_plus(L,A2,B2,C2).

ng_minus(L,A,B,C) :-
    ng_plus(L,B,C,A).
```

4.3 Multiplication and Division

Given $A, B \in k_n$ where $A : (a_1, a_2)_{k_n} = a_1 + a_2\sqrt{\alpha_{n-1}}$ and $B : (b_1, b_2)_{k_n} = b_1 + b_2\sqrt{\alpha_{n-1}}$. Then $(A*B) : (a_1*b_1 + a_2*b_2*\alpha_{n-1}, a_1*b_2 + a_2*b_1)$. Distinguishing operations in k_n and operations in k_{n-1}, we get : $A *^n B : (a_1 *^{n-1} b_1 +^{n-1} a_2 *^{n-1} b_2 *^{n-1} \alpha_{n-1}, a_1 *^{n-1} b_2 +^{n-1} a_2 *^{n-1} b_1)$, where the operation $*^0$ denotes usual multiplication in Q.

We may define $(A/B) = A * B^{-1}$ where $B^{-1} : (b_1/(b_1^2 - b_2^2\alpha_{n-1}), -b_2/(b_1^2 - b_2^2\alpha_{n-1}))$, but this definition is superfluous too. Unification between (A/B) and $A*B^{-1}$ is equivalent to the unification between A with $(A/B)*B$. This definition stands with CLP since the system of constraints, on the coordinates of A, B and (A/B), equivalent to the unification $A = (A/B) * B$ is a linear system.

The predicate ng_mult(E,A,B,C), true if C is equal to A*B in the extension E and predicate ng_div(E,A,B,C) true if C is equal to A/B can be described by :

```
ng_mult(<>,A,B,A*B).
ng_mult(<E>.L,<A1,A2>,<B1,B2>,<C1,C2>) :-
    ng_mult(L,A1,B1,T1),
    ng_mult(L,A1,B2,T2),
    ng_mult(L,A2,B1,T3),
    ng_mult(L,A2,B2,T4),
    ng_mult(L,E,T4,T5),
    ng_plus(L,T1,T5,C1),
    ng_plus(L,T2,T3,C2).

ng_div(L,A,B,C) :-
    ng_mult(L,B,C,A).
```

4.4 Square root

Given $A \in k_n$. We define $\sqrt{A}$ by $(0,1)_{k_{n+1}}$ where $k_{n+1} = k_n[\sqrt{A}]$.

The predicate ng_sqrt(L,A,L',S), true if S in L' is equal to $\sqrt{A}$ when A is in the extension L, can be described by :

```
ng_sqrt(L,A,[A|L],[Z,0]) :-
   zero(L,Z)
   one(L,0).
```

4.5 Positivity and Nullity

Given $A \in k_n$ with $A : (a_1, a_2)_{k_n} = a_1 + a_2\sqrt{\alpha_{n-1}}$. Number A is positive in the following cases :

- if a_1 and a_2 are positive because $(a_1 \geq 0, a_2 \geq 0 \rightarrow a_1 + a_2\sqrt{\alpha_{n-1}} \geq 0)$
- if a_1 is positive, a_2 is negative and $a_1^2 - a_2^2\alpha_{n-1}$ is positive, because $(a_1 \geq 0, a_2 \leq 0) \rightarrow (a_1 + a_2\sqrt{\alpha_{n-1}} \geq 0 \iff a_1^2 \geq a_2^2\alpha_{n-1})$
- if a_2 is positive, a_1 is negative and $a_2^2\alpha_{n-1} - a_1^2$ is positive.

4.6 Examples

The representation and arithmetic given in section 4.1, through,4.5 may be used to set mixed constraint system over constructible numbers. The linear constraint solver and the quadratic solver are obviously correct over constructible number in this representation, that is a first important step compared to the linear constraint solvers over real represented as floating point number which is not correct nor complete. But the algorithms are not complete as multiple representations of one number are possible. For example the following system fails : $3 = sqrt(9)$

We consider in this section two examples on symbolic manipulation involving constructible numbers. The first example shows the advantages of the symbolic representation, the second example shows some limitations which can be eliminated.

First Example Consider the series :

$$\begin{cases} U_0 = 1, U_1 = 2, U_2 = \sqrt{2} \\ U_n = U_{n-1} * U_{n-2} + U_{n-3} \end{cases}$$

This series is defined in $Q[\sqrt{2}]$

The predicate series(U,N), true if U is the element N of the series is given by :

```
series(U,N) :-
  ng_series(E,N,U,V,W).

ng_series([[2]],3,[0,1],[2,0],[1,0]).
```

```
ng_series(E,N,U,V,W) :-
  ng_series(E,N-1,V,W,X),
  ng_mult(E,V,W,Z),
  ng_plus(E,X,Z,U).
```

Calculation of the U_n series, with our representation gives :

U_0 : $(1,0)$
U_1 : $(2,0)$
U_2 : $(0,1)$
U_3 : $(1,2)$
U_4 : $(6,1)$
U_5 : $(10,14)$
U_6 : $(89,96)$
U_7 : $(3584,2207)$
U_8 : $(742730,540501)$
U_9 : $(5047715823,3576360790)$
U_{10} : $(7615143139931954,5384565899606230)$
U_{11} : $(769533792308900022173294272,544142578273187201946 01451)$
U_{12} : $(11720053122434177925779227135614805189 62771$
$,828732903874311492000197638627684580540604)$

Significant growth of the coefficients is observed. This growth is common in symbolic calculation. The calculations have been executed in less than 0.1 second with a Mac II SI and Prolog III.

Let us compare the basic manipulations of algebraic numbers with those of a symbolic processor like Mathematica or Maple; the calculation of the U_n series using Mathematica yields :

$$U_3 = 2\sqrt{2}+1$$
$$U_4 = \sqrt{2}(2\sqrt{2}+1)+2$$
$$U_5 = (\sqrt{2}(2\sqrt{2}+1)+2)(2\sqrt{2}+1)+\sqrt{2}$$
$$U_6 = ((\sqrt{2}(2\sqrt{2}+1)+2)(2\sqrt{2}+1)+\sqrt{2})(\sqrt{2}(2\sqrt{2}+1)+2)+2\sqrt{2}+1$$

Some limitations become apparent. They depend on the strategy used for calculation. When simplification are privileged, computing time becomes important, when efficiency is desirable memory consumption is neglected. In both cases, the calculation of the U_n series fails to compute U_{20} on a Mac II SI in less than 30 minutes.

Second example How is represented $A = \sqrt{2} + \sqrt{3}$ and $B = \sqrt{5 + 2\sqrt{6}}$ taken from [Bor85] ? We calculate successively :

$$R_0 : (0,1) \text{ in } Q[\sqrt{2}]$$
$$= \sqrt{2}$$
$$R_1 : ((0,0),(1,0)) \text{ in } Q[\sqrt{2}][\sqrt{3}]$$
$$= \sqrt{3}$$
$$R_2 : (((0,0),(0,0)),((1,0),(0,0))) \text{ in } Q[\sqrt{2}][\sqrt{3}][\sqrt{6}]$$
$$= \sqrt{6}$$
$$R_3 : (((5,0),(0,0)),((2,0),(0,0))) \text{ in } Q[\sqrt{2}][\sqrt{3}][\sqrt{6}]$$
$$= 5 + 2\sqrt{6}$$

And then, $R_5 : ((((0,0),(0,0)),((0,0),(0,0))),(((1,0),(0,0)),((0,0),(0,0))))$
$$\text{in } Q[\sqrt{2}][\sqrt{3}][\sqrt{6}][\sqrt{5 + 2\sqrt{6}}]$$
$$= B \quad = \sqrt{5 + 2\sqrt{6}}$$

and, $R_6 : ((((0,1),(1,0)),((0,0),(0,0))),(((0,0),(0,0)),((0,0),(0,0))))$
$$\text{in } Q[\sqrt{2}][\sqrt{3}][\sqrt{6}][\sqrt{5 + 2\sqrt{6}}]$$
$$= A \quad = \sqrt{2} + \sqrt{3}$$

One can observe the following limitations :

- an exponential growth of the representation size with the number of square root extractions.
- a significant difficulty to prove equality between expressions ($A = B$ indeed !). As a consequence an additional algorithm would be necessary to verify overconstraints.

In fact, this section defined the constraints system (D,O,R), where D is the domain of fixed size binary trees representing constructible numbers, O is the set of operators $(+, -, *, /)$ on these trees, and R is the set of relational predicates $(=, \geq)$. The definition of this constraint systems is only partially correct because equality is not equivalent to unification. The next section considers the constraints system (D',O',R') where D' is the domain of fixed minimum size binary trees representing constructible numbers, O=O', and R'=$(\geq)$. In (D',O',R') trees representing real numbers are compacted, and as a consequence, unification is equivalent to =.

Note that the problem of equality of expression is difficult to solve. Mathematica and Maple have identical behavior with expressions A and B : both of them numerically evaluate $(A - B)$ to 0 with floating point arithmetic, but evaluate the boolean $(A - B == 0)$ to false.

5 Normal representation for constructible numbers

The representation proposed in section 4 may be definitely improved if quadratic extensions are only introduced when necessary. This is the principal and original part of this work.

Suppose we have a sequence of quadratic extensions of Q :

- $k_0 = Q$,
- $k_1 = k_0[\sqrt{\alpha_0}]$ where $\alpha_0 \in k_0$, $\sqrt{\alpha_0} \notin k_0$. We suppose that the introduction of $\sqrt{\alpha_0}$ has been necessary to express the square root of an element A_0 of k_0 which is not a square in k_0. Note that in this condition $(1, \sqrt{\alpha_0})$ constitutes a basis of the vector space k_1 over k_0.
- we set $k_n = k_{n-1}[\sqrt{\alpha_{n-1}}]$ where $\alpha_{n-1} \in k_{n-1}$, $\sqrt{\alpha_{n-1}} \notin k_{n-1}$. We suppose that the introduction of $\sqrt{\alpha_{n-1}}$ has been necessary to express the square root of an element A_{n-1} of k_{n-1} which is not a square in k_{n-1}. Note that in this condition $(1, \sqrt{\alpha_{n-1}})$ constitutes a basis of the vector space k_n over k_{n-1}.

Let $A \in k_n$. Since k_n is a vector space over k_{n-1} and that $(1, \sqrt{\alpha_{n-1}})$ is a basis of that vector space, then there are unique elements $a_1, a_2 \in k_{n-1}$ such that $A = a_1 + a_2 * \sqrt{\alpha_{n-1}}$. The representation of A is unique. It is a normal representation.

The first improvement in this new definition concerns the uniqueness of the representation. This establishes a correspondence between equality defined for constructible numbers and unification defined for Prolog binary trees and rational numbers representing constructible numbers. So, overconstraint systems on constructible numbers are verified without supplementary work as overconstraint systems on rational numbers and binary trees are. As a consequence, completeness and correctness of the linear constraint solver over constructible numbers are preserved in this normal representation. An other direct consequence of this normal representation is provided by the following predicate ng_null(E,N), true if value of N is zero in the sequence of extension E, defined by :

```
ng_null([],0) .
ng_null([E|L],[A,B]) :-
    ng_null(L,A),
    ng_null(L,B).
```

5.1 Square root

The normal definition of the quadratic extension of k_n needs to change the algorithm given in subsection 4.4 with the following :

Let $A \in k_n$, we want to calculate the square root of A, i.e. $\sqrt{A}$.

If A is a square in k_n, i.e. $A = a^2$ with $a \in k_n$, then $\sqrt{A} = a$ (we show below how a square root of A can be find in k_n).

If A is not a square in k_n, the calculation of $\sqrt{A}$ introduces a new quadratic extension : $k_{n+1} = k_n[\sqrt{\alpha_n}]$ where $\alpha_n = A$ and we get $\sqrt{A} : (0,1)_{k_{n+1}}$.

We have now to show how to compute a square root in k_n. This is the particularity of constructible number : explicit square root can be obtained, if one exists.

Let $A \in k_n$, is there $a \in k_n$ such that $a^2 = A$? Consider the following cases :

$n = 0, k_n = k_0 = Q$.

Then A is rational, i.e. $A = X/Y$ with X, Y integer and $gcd(X, Y) = 1$, $X > 0, Y > 0$.
Then a square root of A exists if and only if X and Y have a square root in N. Thus, if $X = x^2, Y = y^2$ then $A = a^2$ with $a = x/y$, (we always choose the positive root). Calculation of integer square root can be found in [Rol87].

$n > 0$.

Then $a = a_1 + a_2 * \sqrt{\alpha_{n-1}}$ with $a_1, a_2 \in k_{n-1}$.
A is a square in k_n if and only if there are $x, y \in k_{n-1}$ such that :

$$A = (x + y * \sqrt{\alpha_{n-1}})^2 \quad (0)$$

Rewriting this equation as an equation in vector space k_n over k_{n-1} we get :

$$\begin{cases} a_1 = x^2 + y^2 \alpha_{n-1} \\ a_2 = \quad 2xy \end{cases} (1)$$

We have to consider different cases in order to solve this non-linear system of equations.

$a_2 = 0$.

Then A is a square in k_n if and only if $x = 0$ or $y = 0$.
Assume $x = 0$ then A is a square in k_n if and only if A/α_{n-1} is a square in k_{n-1}.
Assume $y = 0$ then A is a square in k_n if and only if A is square in k_{n-1}.

$a_2 \neq 0$.

Then we have to solve the following equation, obtained from (1) by substitution of x :

$$y^4 \alpha_{n-1} - a_1 y^2 + (a_2)^2/4 = 0 \quad (2)$$

We note $Y = y^2$. This leads to the second degree equation :

$$Y^2 \alpha_{n-1} - a_1 Y + (a_2)^2/4 = 0 \quad (3)$$

whose discriminant is : $\Delta = (a_1)^2 - (a_2)^2 \alpha_{n-1}$.
Δ must be a square in k_{n-1}. [proof : Actually, (1) has solution for the variable y in k_{n-1}, is equivalent to (3) has solutions for the variable Y in k_{n-1}. As $\Delta = (\pm 2\alpha_{n-1}Y - a_1)^2$, with $\pm 2\alpha_{n-1}Y - a_1 \in k_{n-1}$, (1) has solution implies that Δ has roots in k_{n-1}. Conversely, if Δ has no root in k_{n-1} then (3) has no solution for Y in k_{n-1}; therefore (1) has no solution in k_n]
If, by recurrence, a root δ of Δ is found in k_{n-1}, then solutions for Y are :

$$Y_1 = (a_1 + \delta)/2\alpha_{n-1}, \quad and Y_2 = (a_1 - \delta)/2\alpha_{n-1}$$

Then square root y_1 et y_2, of Y_1 et Y_2 in k_{n-1} are calculated by recurrence. If square roots exist for Y_1 or Y_2 then y_1 and y_2 are solutions of (1) in k_{n-1}. If Y_1 and Y_2 are not squares in k_{n-1} then (1) has no solution and A is not a square in k_n.

5.2 Examples with normal representation

Uniqueness of the representation Let us consider A and B defined in 4.6 with the new representation model. We compute successively :

$$\begin{aligned}
R_0 &: & (0,1)_{Q[\sqrt{2}]} &= & \sqrt{2} \\
R_1 &: ((0,0),(1,0))_{Q[\sqrt{2}][\sqrt{3}]} &&= & \sqrt{3} \\
R_2 &: ((0,1),(1,0))_{Q[\sqrt{2}][\sqrt{3}]} &&= & \sqrt{2}+\sqrt{3} \\
R_3 &: ((0,0),(0,1))_{Q[\sqrt{2}][\sqrt{3}]} &&= & \sqrt{6} \\
R_4 &: ((5,0),(0,2))_{Q[\sqrt{2}][\sqrt{3}]} &&= & 5+2\sqrt{6} \\
R_5 &: ((0,1),(1,0))_{Q[\sqrt{2}][\sqrt{3}]} &&= B = & \sqrt{5+2\sqrt{6}} \\
R_6 &: ((0,1),(1,0))_{Q[\sqrt{2}][\sqrt{3}]} &&= A = & \sqrt{2}+\sqrt{3}
\end{aligned}$$

First we observe a diminution in the size of the representation. Secondly, $A = B$ is now quite trivial to check. The overconstraint $A = B$ is reduced without any extra algorithm.

Survey of 512 theorems in Euclidean geometry Two evaluations conclude this article. First we have tested whether Euclidean configurations are really hard or not. If they need too many quadratic extensions the exponential growth of the representation of the constructible numbers will be a difficult task to cope with. The second evaluation will test time needed for calculation with constructible numbers in normal representation.

A survey over the 512 theorems of Euclidean geometry proved in [Chou88], shows that :

- at least 479 theorems concern geometrical configurations where all coordinates may be rational numbers,
- at most 29 theorems concern geometrical configurations where coordinates need one quadratic extension,
- only 4 theorems need two quadratic extensions to represent coordinates of all the elements.

Therefore, most of the geometrical configuration do not require quadratic extension, just rational numbers. Almost all square root calculations done in geometrical configuration, do not introduce quadratic extension, but remain rational numbers. But in few cases, an exact representation, with quadratic extension is required, for example with equilateral triangle (1 quadratic extensions), or regular pentagon (2 quadratic extensions).

From a performance point of view, we have expressed and solved 30 configurations found in [Chou88] grouped in 3 sets. The first group contains the 10 configurations where Chou's demonstrator spends the least time to prove the corresponding theorem. The second group corresponds to the 10 times between the 250 and the 260 time for Chou's demonstrator. And last group corresponds to the hardest theorems.

We obtain the following tables, where "Ex." stands for the number given in [Chou88] for the problem, "Tms" is the time needed for the calculation, and "Ext." is the number of quadratic extensions involved.

Group 1.			Group 2.			Group 3.		
Ex.	Tms	Ext.	Ex.	Tms	Ext.	Ex.	Tms	Ext.
173	3 ms	0	260	5 ms	0	440	24 ms	0
86	3 ms	0	90	6 ms	0	80	28 ms	0
141	6 ms	0	110	350 ms	1	393	330 ms	1
140	5 ms	0	339	11 ms	0	17	23 ms	0
193	7 ms	0	116	12 ms	0	144	30 ms	0
92	6 ms	0	336	10 ms	0	347	31 ms	0
121	4 ms	0	434	13 ms	0	401	28 ms	0
181	7 ms	0	82	6 ms	0	396	35 ms	0
31	6 ms	0	330	13 ms	0	453	18 ms	1
37	6 ms	0	506	15 ms	0	316	40 ms	0

In group 1. the times to compute the configuration and to check numerically the theorem are 100 times faster than Chou's prover times. In the second group times are 1000 times faster except for 110 which is only 10 times faster. In the last group, speed-up are even more important. For 316, Chou's prover needs 20058s.

Comparisons between numerical evaluations and proving times may seem a bit tricky. But it is worth to recall that "Given a geometric proposition, we can easily present a concrete numerical example such that in order to determine whether the proposition is generally true, one need only to try this example" from [Hon86]. Within the perspective of an Euclidean prover based on numerical evaluations, the proposed comparison is worth considering.

From both theoretical and experimental point of view, the representation seems appropriate.

6 Conclusion and further works

We have proposed a normal representation for constructible number occuring in Euclidean geometry. This normal representation is compatible with linear constraint solver and preserves correctness and completeness over constructible numbers. The normal representation is also compatible with the quadratic constraint solver presented briefly in section 3. We have shown on 512 examples that the complexity of the representation is not too high in general cases. We have shown on 30 runtime examples that times consuming are closer to floating point representation times, than to symbolic representation prover times. In consequence, this approach seems appropriate. It is already used in one Intelligent Tutoring System for geometry, see [All93].

Our next goal is to introduce this exact arithmetic over constructible number in a Euclidean Prover based on numerical evaluations, cf. [Hon86, Dav77,

Chou92]. In future work we also plan to define a notion of constructible numbers and a normal representation over finite fields, cf. [Nau85]

Open problem The main result of this paper relies on the possibility to find explicit square root in algebraic extension of Q with square roots. Can this be extended to root of arbitrary degree ?

References

[All93] Allen, R., Idt, J., Trilling, L., *Constrained based automatic construction and manipulation of geometric figures*, Proceedings of the 13th IJCAI Conference, Chambery, Morgan Kaufmann Publishers, Los Altos, 1993.

[Bor85] Borodin, A., Fagin, R., Hopcroft, J.E., Tompa, M., *Decreasing the nested depth of expression involving square roots*, Journal of Symbolic Computation, no. 1 page 169-188, 1985.

[Chou92] S.C. Chou and X.S. Gao, *Proving Geometry Statement of Constructive Type*, CADE, LNCS 607, D. Kapur Eds, 1992.

[Chou88] S.C. Chou, *Mechanical Geometry Theorem Proving*, D. Reidel Publishing Company, 1988.

[Col90] A. Colmerauer, *An Introduction to Prolog III*, Commun. ACM, 28(4):412-418,1990.

[Col93] A. Colmerauer, *Naive Solving of Non-linear Constraints*, Constraint Logic Programming : Selected Research, F. Benhamou and A. Colmerauer eds, The MIT Press, page 89-112, 1993.

[Dav77] Davis, P.J., *Proof, Completeness, Transcendental, and Sampling*. Journal of the ACM, Vol. 24, no. 2, pages 298-310, 1977.

[Din88] M. Dincbas, P. Van Hentenryck, H. Simonis, A. Aggoun, T. Graf, and F. Berthier. *The Constraint Logic Programming Language CHIP*, In Proceedings of the Internationnal Conference on 5 Fifth Generation Computer Systems, Tokyo, Japan, December 1988.

[Jaf92] J. Jaffar, S. Michaylov, P-J. Stuckey, , and R. Yap, *The CLP(R) Language and System*. *ACM Trans. on Programming Languages and Systems*, 14(3):339-395,1992.

[Hon86] J. Hong, *Proving by Example and Gap Theorems* 27th An. Symp. on Foundations of Computer Science, Toronto, Ontario, Canada, IEEE press, p107-116, Oct 1986.

[Lab95] J.M. Laborde, *Des connaissances abstraites aux réalités artificielles, le concept de micromonde Cabri*, IVèmes journées EIAO de Cachan, ed Eyrolles, 1995.

[Lan92] Landau, S., *Simplification of nested radicals*, SIAM Journal of Computing Vol 21, no.1 page 85-110, February 1992.

[Leb92] Lebesgue, H., *Leçons sur les constructions géométriques*, réédition, Gauthier-Villard, Paris, 1989.

[Nau85] Naudin, P., Quitté, C. *Algorithmique algébrique* Collection Logique mathématiques informatique, édition Masson, pages 312–324, 1985.

[Pes95] Pesant, G., *Une approche géométrique aux contraintes arithmétiques quadratiques en programmation logique avec contraintes* Thèse de l'Universite de Montréal, 1995.

[Rol87] Rolfe, T., *On a fast integer square root algorithm*. ACM SIGNUM Newsletter, Vol. 22, no. 4, pages 6–11, 1987.

Towards a Sheaf Semantics for Cooperating Agents Scenarios

Viorica Sofronie*

RISC-Linz, Johannes Kepler University, A-4040 Linz, Austria
e-mail Viorica.Sofronie@risc.uni-linz.ac.at

Abstract. The ultimate goal of our work is to show how sheaf theory can be used for studying cooperating robotics scenarios. In this paper we propose a formal definition for systems and define a category of systems. The main idea of the paper is that relationships between systems can be expressed by a suitable Grothendieck topology on the category of systems. We show that states and (parallel) actions can be expressed by sheaves and use this in order to study the behavior of systems in time.

1 Introduction

Originally, the theory of sheaves was conceived as a tool in topology and algebraic geometry. The influence of sheaf theory has spread in many areas of mathematics: besides the fields where its origins are, such as analysis, topology and algebraic geometry, it is now widely used in algebra (representations of algebras by continuous sections; global subdirect products, see e.g. [KC79]), as well as logic (cf. [FS79] and related work – for details see also [MLM92]). Sheaf theory was developed in mathematics because of the necessity of studying the relationship between "local" and "global" phenomena.

In what follows we will briefly explain why we think that sheaf theory can be a useful framework for modeling cooperating agents scenarios.

When modeling states, actions or behavior it is often necessary to make a link between "local" properties (characteristic for given subsystems) and "global" properties (relevant for the whole system). The goal of our study is an analysis of subsystem interaction, taking into account the contribution of subsystems to the description (states, actions, behavior) of the whole system. The alternance "local - global" that occurs in this case suggests that a sheaf-theoretical approach to the study of systems of cooperating agents (and in the study of concurrency in general) would be natural. Moreover, the tools of sheaf theory, and of topos theory in general, in particular geometric logic, should explain why some properties of systems are preserved when restricting to subsystems, and why there are cases when properties of subsystems are not transferred to the system obtained by their interconnection. This is part of ongoing work (cf. [Sof96]).

There already exist a number of approaches to concurrency based on "fiberings" and sheaf theory. Among them, [MP86], [Pfa91], [Gog92], [Lil93], [Mal94].

---

* Partially supported by the Austrian Science Foundation under ESPRIT BRP 6471 "MEDLAR II"

In [MP86] the authors aim at developing a structural theory of concurrency in which the locality of interaction between subsystems is described with the mathematical tools of the theory of sheaves. They show that the behavior of a given family of interconnected systems can be modeled by so-called behavior monoids (which turn to form sheaves of monoids). In [Gog92], a sheaf semantics aiming at modeling the behavior of concurrent interacting objects is presented. The approach is based on an earlier paper [Gog75]. These ideas have been applied to Petri Nets by Lilius [Lil93]. The ideas from [Gog92] have been further developed by Malcolm in [Mal94], where a formalization of object classes and systems of objects is given, in order to study basic properties of ways in which systems of objects may be interconnected. He expresses the hope that by using a more general notion of sheaf as a functor on a category with a Grothendieck topology, an adjunction between system specifications and sheaves of objects can be obtained.

In a series of papers, J. Pfalzgraf develops the idea of "logical fiberings", with the goal of developing a (non-classical) "fibered logical calculus", by means of which one could construct logical controllers for multi-tasking scenarios in a formal way (for details see e.g. [Pfa91], [Pfa95]). In [Pfa95], he points out that the notion of a fibering is closely related to indexed systems (indexed categories) and to sheaves (cf. their relation with topos theory and the geometric aspects). These methods and concepts arose in the process of modeling cooperating agents scenarios (see [Pfa91]) and have been illustrated, in the frame of the MEDLAR project, on concrete examples (see for example [Pfa93],[PS95], [PSS95], [PSS96]).

Finally, we would like to point out another application of sheaf theory in computer science in a paper by Srinivas (cf. [Sri93]). Although the topic of [Sri93] (pattern matching) has no link with our approach, the idea of using Grothendieck topologies occurred to us after reading that paper.

In this paper we make first steps towards a theoretical approach, based on sheaf theory, for describing systems of cooperating agents. The basic idea of our formalism is that even relatively simple agents, such as a robot that provides an assembly bench with pieces, are in fact complex systems composed of interacting subsystems, like joints and wrists, locomotion modules, etc. Therefore, instead of "individual" agents, a category of "systems" is considered, where the morphisms describe the relation "is a subsystem of". We show that under certain (non-restrictive) hypotheses a Grothendieck topology (describing a "covering" relation between systems) can be defined on this category, and that states as well as admissible parallel actions can be modeled by sheaves. We use these results for studying the behavior in time of the systems, adapting one of the ideas presented in [Gog92]. The difference between our approach and the approach of Monteiro and Pereira in [MP86] lies in the fact that here we do not model the behavior of interconnected systems by sheaves of monoids. These aspects are analyzed in [Sof96].

The paper is structured as follows:

In section 2, as a motivation for our theoretical study, we illustrate the problems that appear by means of a simple example, adapted from [Pfa93]. This example leads to a formal definition of a system, given in section 3.

In section 3 we give a formal definition of a system and introduce a category of systems on which a Grothendieck topology is defined. We show that states and parallel actions can be defined as sheaves over this site.

In section 4 we focus on modeling the behavior of a system. We start with the formalism developed by Goguen in [Gog92], and slightly modify Goguen's definition for the behavior of a system, explicitly indicating for every moment of time not only the state of the system but also the action performed at the given moment, and obtain a contravariant functor from the category of systems to the category of sheaves over time. We also show that a contravariant functor from $\mathcal{T}$ to the category of sheaves over Sys can be defined.

The goal of the present paper is to clarify some of the basic notions required in the study of cooperating agents. At this level we do not yet aim at formalizing scheduling, especially when a synchronization of the actions is needed. In further work we will take into account the time necessary for the execution of the actions, and also precedence relations between actions; moreover we will consider space-dependency of formulas. We hope that our approach will help in clarifying some of the problems stated in [PS95] and [Pfa95].

2 An Example

We begin with a simple example (adapted from [Pfa93]) as a motivation for our theoretical study, for the definitions that will be given, and for the assumptions that will be made.

Let R_0, R_1, R_2, R_3 be four robots performing the following task:

- R_0 receives a work piece a and a work piece b and performs an assembly task. The work piece r obtained from assembling a and b is placed on the assembly bench.
- R_1 furnishes pieces of type a. He checks whether there are pieces of type a left in stock, and whether a piece of type a or an r resulting from assembling a and b is placed on the assembly bench of R_0. If there are pieces of type a in stock, and if no a or r are placed on the table, R_1 brings a piece of type a to R_0.
- R_2 furnishes pieces of type b. He checks whether there are pieces of type b left in stock, and whether a b or an r is placed on the table. If there are pieces of type b in stock, and no b or r is on the table, R_2 brings a piece of type b to R_0.
- After R_0 has assembled a and b, R_3 receives the result r and transports it to the stock.

Let S be the system resulting from the interaction of these robots. We can assume that the system can be "described" by the interconnected subsystems S_0, S_1, S_2, and S_3, which correspond to the robots R_0, R_1, R_2 and R_3.

2.1 States

The states of the system S can be expressed using the *control variables* described in the table below. The set of control variables relevant for system S_0 is $X_0 = \{p_a, p_b, p_r\}$, the one for S_1 is $X_1 = \{s_a, p_a, p_r\}$, for S_2, $X_2 = \{s_b, p_b, p_r\}$, and for S_3, $X_3 = \{p_r\}$. We will assume that the subsystems $S_0, \ldots, S_3$ communicate via common control variables. In Figure 1 we show how the control variables are shared, and how they can be used for communication with "external" systems (e.g. Stock-a, Stock-b).

Variable	Description	System
$s_a =$ in-stock-a	"there is at least one piece of type a in stock"	S_1
$s_b =$ in-stock-b	"there is at least one piece of type b in stock"	S_2
$p_a =$ on-table-a	"a piece of type a is on the assembly bench"	S_0, S_1
$p_b =$ on-table-b	"a piece of type b is on the assembly bench"	S_0, S_2
$p_r =$ on-table-res	"the result r is on the assembly bench"	S_0, S_1, S_2, S_3

A *state* of the system S is a possible assignment of truth values to the relevant control variables. We might additionally assume that only some of these assignments are admissible, imposing some *constraints* on the values that can be taken by the control variables. This turns out to be especially useful when the control variables are not independent.

Let us assume for example that in the given system it is not allowed to have a result piece and a piece of type a or b on the working bench at the same time, but it is allowed to have a piece of type a and one of type b. That means that p_a and p_r cannot both be true, and p_b and p_r cannot both be true. This can be expressed by a set of identities on the boolean algebra freely generated by the control variables of the system, in this case $Id = \{p_a \wedge p_r = 0, p_b \wedge p_r = 0\}$.

The agents R_0 and R_1 can "communicate" using the control variables common to these systems, namely the set $X_{01} = \{p_a, p_r\}$. Analogously, the agents R_0 and R_2 can "communicate" using the set $X_{02} = \{p_b, p_r\}$, and R_1 and R_2 using the set $X_{12} = \{p_r\}(= X_3)$. We can therefore assume that the structure of the given systems of cooperating agents as an interconnection of subsystems determines a topology on the set of control variables. Also the set of constraints (if they do not link variables in different subsystems) can be "distributed" over the subsystems in the same way.

Consider the basis $\mathcal{B} = \{X_0, X_1, X_2, X_3, X_{01}, X_{02}\}$, consisting of the sets of control variables corresponding respectively to the subsystems consisting of the robots $R_0, \ldots, R_3$, as well as to their "subsystems" by means of which the communication is done. The corresponding restrictions of the set of constraints are $\{Id_0, Id_1, Id_2, Id_3, Id_{01}, Id_{02}\}$, where $Id_0 = Id$, $Id_1 = \{p_a \wedge p_r = 0\}$, $Id_2 =$

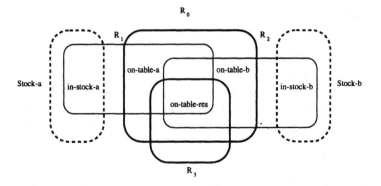

Fig. 1. Control Variables

$\{p_b \wedge p_r = 0\}$, and $Id_{01} = Id_{02} = Id_3 = \emptyset$, corresponding to the subsystems mentioned above.

The set of states of the system will be the set of those assignments of truth values to the control variables that satisfy this set of identities. Similarly, the set of states for system S_i (corresponding to the agent R_i) is $St(S_i) = \{s_i : X_i \rightarrow \{0,1\} \mid s_i \models Id_i\}$. Thus $St(S_0) = \{s_0 : X_0 \rightarrow \{0,1\} \mid s_0(p_a)s_0(p_r) = 0, s_0(p_b)s_0(p_r) = 0\}$, $St(S_1) = \{s_1 : X_1 \rightarrow \{0,1\} \mid s_1(p_a)s_1(p_r) = 0\}$, $St(S_2) = \{s_2 : X_2 \rightarrow \{0,1\} \mid s_2(p_b)s_2(p_r) = 0\}$, and $St(S_3) = \{s_3 \mid s_3 : X_3 \rightarrow \{0,1\}\}$.

It is easy to see that for every family $(s_i)_{i=0,\ldots,3}$ with the property that s_i is a state for the system S_i (corresponding to R_i), and such that for every i, j, s_i and s_j coincide on the common control variables, there is exactly one state of the system, s, such that the restriction of s to the control variables P_i is s_i for every i. This means that the following gluing condition is satisfied:

- For every $\{s_B\}_{B \in \mathcal{B}}$, where $s_B : B \rightarrow \{0,1\}$ satisfies the equations in $Id_B = \{e \in Id \mid Var(e) \subseteq B\}$, such that for every $B_1, B_2 \in \mathcal{B}$, $s_{B_1|B_1 \cap B_2} = s_{B_2|B_1 \cap B_2}$, there exists a unique $s : P \rightarrow \{0,1\}$ that satisfies the set of constraints Id, such that for every $B \in \mathcal{B}$, $s_{|B} = s_B$.

Since there are typical properties of a sheaf visible, this leads to the idea that the link between local and global states could best be described by sheaves over a suitable topology on the set of control variables (or over a suitable Grothendieck topology on a category of systems) defined by the structure of the given system.

Remark: Note that the gluing property described above does not hold for every topology on the set X of control variables of the system. Consider for example the discrete topology on X. Then X can be covered by the family $\{\{s_a\}, \{s_b\}, \{p_a\}, \{p_b\}, \{p_r\}\}$. Then the following family of assignments of truth values to the control variables: $s_{\{s_a\}}(s_a) = 0$, $s_{\{s_b\}}(s_b) = 0$, $s_{\{p_a\}}(p_a) = 1$, $s_{\{p_b\}}(p_b) = 1$, $s_{\{p_r\}}(p_r) = 1$ agrees on common control variables (because the domains are disjoint), but no information about the constraints can be recovered, hence by "gluing" these mappings together one obtains a map $s : X \to \{0,1\}$ which does not satisfy the set Id of identities. This shows that an appropriate topology on X has to respect the way the constraints are shared between subsystems.

2.2 Actions

The system S is also characterized by a set of (atomic) actions. Below we will give the list of the atomic actions (with pre- and postconditions) and the agent that performs them.

Action	Description	Precond.	Postcond.	Agent/Interpretation
A	Assemble a piece of type a with one of type b	$p_a = 1$ $p_b = 1$ $p_r = 0$	$p_a = 0$ $p_b = 0$ $p_r = 1$	R_0: assemble
B_a	Bring a piece of type a	$p_a = 0$ $s_a = 1$ $p_r = 0$	$p_a = 1$ $s_a = 0$ $p_r = 0$	R_1: give-a R_0: receive-a
B_b	Bring a piece of type b	$p_b = 0$ $s_a = 1$ $p_r = 0$	$p_b = 1$ $s_a = 0$ $p_r = 0$	R_2: give-b R_0: receive-b
T_r	Store the result	$p_r = 1$	$p_r = 0$	R_3: receive-r R_0: give-r

We can assume for example that R_1 and R_2 can perform the actions of bringing a piece of type a respectively b in parallel but that R_0 is not allowed to execute in parallel the action of taking a piece of type a and of giving the result to R_3. We also can assume that other actions, as for example give-a (by R_1) and receive-a (by R_0) have to be executed in the same time. Therefore, they have been "identified" in the larger subsystem under the name B_a. Figure 2 shows how actions are shared between subsystems, and some relations between them (in what follows for the sake of simplicity we do not consider the actions get-a, get-b and transport). It is also natural to suppose that some *constraints* are given, expressing which of these actions can be performed in parallel, which cannot, which must be executed in the same time, and so on.

There are many possibilities to specify this kind of constraints. One solution is to consider parallel actions as subsets of the set of atomic actions A, or equivalently by maps $f : A \to \{0,1\}$, where $f(a) = 1$ means that a is executed and $f(a) = 0$ means that a is not executed. In this case the constraints can be described by imposing restrictions on the combinations of the values (0 or 1) that can be assigned to the atomic actions in an admissible parallel action. This approach is very similar to the one adopted in the description of states (section 2.1). We will assume that the constraints can be described by identities on the boolean algebra freely generated by the set of atomic actions A. (On

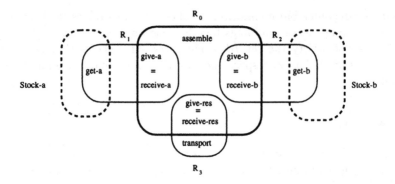

Fig. 2. Actions

our example, among the constraints are: **give-a** = **receive-a**, **give-b** = **receive-b**, **give-res** = **receive-res**, **give-res** ∧ **receive-a** = 0, **give-res** ∧ **receive-b** = 0, etc.).

There exist approaches for modeling concurrency where it is necessary to specify which actions can be performed in parallel and which not. One such approach is based on considering a "dependence" relation on the set of (atomic) actions, i.e. a reflexive and symmetric relation $D \subseteq A \times A$: the parallel execution is then only allowed for those actions which are "independent" w.r.t. D. This leads to the study of partial-commutative monoids (cf. [Die90]). The link with this approach is analyzed in [Sof96].

As when considering states, the structure of the system as an interconnection of subsystems induces a topology on the set of all actions. Correspondingly, the constraints distribute over the subsystems (it seems natural to make the assumption that all the constraints are made "inside" some specified subsystem). A basis $\mathcal{B}$ for this topology is obtained as explained in the study of states, taking the family of all actions corresponding to the subsystems $S_0, \ldots, S_3$, as well as to the "subsystems" by means of which the communication is done (i.e. finite intersections of those sets). In the case where there are constraints that link actions in different subsystems (that could be for example expressed in some special "scheduling systems"), the actions that correspond to these scheduling systems will also be considered elements in the basis. Also in this case a similar gluing property holds:

- For every family of parallel actions $\{f_B\}_{B \in \mathcal{B}}$, where $f_B : A_B \to \{0, 1\}$ satisfies the constraints $Id_B = \{e \in Id \mid Var(e) \subseteq B\}$, such that for every $B_1, B_2 \in \mathcal{B}$, $f_{B_1|B_1 \cap B_2} = f_{B_2|B_1 \cap B_2}$, there exists a unique $f : A \to \{0, 1\}$ that satisfies Id such that for every $B \in \mathcal{B}$, $f_{|B} = f_B$.

Note that the fact that the gluing property holds is strongly related to the specific form of the constraints in C_A (boolean equations in our case). If for example a parallel action $f : A \to \{0, 1\}$ is allowed if and only if $f^{-1}(1)$ is finite, then the infinite gluing property does not hold.

3 Systems

In what follows we will use basic notions of category theory. We will also assume known the definitions for Grothendieck topologies and sheaves with respect to a Grothendieck topology. For details see, for example, [MLM92].

Taking into account the considerations in section 2, we will assume that a system S can be described by:

- A set X of control variables of the system (for every control variable $x \in X$ a set V_x of possible values for x), and a set Γ of constraints, specifying which combinations of values for the variables are admissible (i.e. satisfy Γ). An admissible combination of values of the control variables will describe a state of the system S. The set of states of the system will be denoted $St(S)$,
- A set A of atomic actions (where for every $a \in A$, X_a denotes the minimal set of control variables a depends on and $Tr_a \subseteq St(S)_{|X_a} \times St(S)_{|X_a}$ a relation indicating how the values of these variables change when the action a is performed), together with a set C of constraints that shows which actions are incompatible and cannot be performed in parallel.

In order to formally express the "constraints" on the possible combinations of values for the control variables, we need a "language" in which these are expressed.

Definition 1. A *system* $S = (\Sigma, X, \Gamma, M, A, C)$ consists of:

(1) A language $\mathcal{L}_S$ consisting of
(1a) a signature $\Sigma = ($ Sort$, O, P)$ (Sort is the set of sorts, O the set of operation symbols, and P the set of predicate symbols),
(1b) a (many-sorted) set of variables $X = \{X_s\}_{s \in \text{Sort}}$,
(2) A set $\Gamma \subseteq \text{Fma}_\Sigma(X)$ (the set of constraints or axioms of S),
(3) A model M (structure of similarity type Σ),
(4) A set of actions A; for every $a \in A$ a set $X_a \subseteq X$ of variables a depends upon, and a transition relation Tr_a.
(5) A set C of constraints, expressed by boolean equations over $F_B(A)$ (the free boolean algebra generated by A) stating which actions can (or have to) be executed in parallel.

Remark: In definition 1 we fix a model M for the system S (corresponding e.g. to the real world) and we allow as constraints formulas in the many-sorted language of the system.

Given a system S, if not otherwise specified, we will refer to its signature, set of variables, constraints on the values of the variables, model, set of actions and constraints on actions by Σ_S, X_S, Γ_S, M_S, A_S, C_S respectively. If a family of systems $\{S_i\}_{i \in I}$ is given, if not otherwise specified we will refer to the signature, set of variables, constraints on the values of the variables, model, set of actions and constraints on actions of S_i by Σ_i, X_i, Γ_i, M_i, A_i, C_i, for every $i \in I$.

Systems often arise in relationship with other systems. The corresponding relationships between systems are expressed by morphisms. Obviously, there is some choice in how to define appropriate morphisms, depending on the extent of the relationship between systems we want to express.

For instance, a *category of systems* SYS can be defined, having as objects systems. A morphism f in SYS from a system S_1 to a system S_2 consist of a "translation" of the language (resp. actions) of S_1 into the the language (resp. actions) of S_2 (i.e. a family of mappings $f_\Sigma : \Sigma_1 \to \Sigma_2$, $f_X : X_1 \to X_2$, $f_A : A_1 \to A_2$) such that for every $\phi \in \Gamma_1$ (resp. in C_1) the "translation" of ϕ to the language of S_2 via f_Σ, f_X (resp. f_A) is in Γ_2 (resp. in C_2); and transitions in S_1 are mapped into transitions of S_2.

Sometimes, however, the systems are not able to communicate using such "translations": important is that the systems have common subsystems by means of which the communication is done. In what follows, we will focus on this last aspect.

Definition 2. Let S_1 and S_2 be two systems. We say that S_1 *is a subsystem of* S_2 if and only if $\Sigma_1 \subseteq \Sigma_2$, $X_1 \subseteq X_2$, $A_1 \subseteq A_2$, and the constraints in Γ_1 (resp. C_1) are consequences of the constraints in Γ_2 (resp. C_2), and the model M_1 is the restriction of the model M_2 to the signature of S_1 $(M_1 = U_{\Sigma_1}^{\Sigma_2} M_2)$.

In concrete applications we usually are interested only in some subcategory Sys of SYS, having as objects those systems relevant for the given application (in the example in section 2 the relevant systems are $S_0, \ldots, S_3$ together with their common subsystems, and the systems obtained by interconnecting them), and a morphism between a system S_1 and S_2 if and only if S_1 is a subsystem of S_2.

In what follows we will assume given a family InSys of interacting systems. To enforce the compatibility of models on common sorts, we may assume that all these systems are subsystems of a "universal system" S_U. We further assume that the family InSys is closed under intersections (pullbacks of subsystems of S_U) i.e. it contains all those subsystems by means of which intercommunication is done. The elements of InSys are the "building blocks" from whose interconnection larger systems arise.

Definition 3. The *category of systems*, Sys, has as objects all the systems that can be obtained by interconnecting elements in In-Sys, and a morphism from S_1 to S_2 if and only if S_1 is a subsystem of S_2.

It is easy to see that Sys has pullbacks, namely for every $S_1, S_2 \hookrightarrow S$, $S_1 \times_S S_2 = (\Sigma_1 \cap \Sigma_2, X_1 \cap X_2, \Gamma_1 \cap \Gamma_2, M_{12}, A_1 \cap A_2, C_1 \cap C_2)$, where $M_{12} = U_{\Sigma_1 \cap \Sigma_2}^{\Sigma} M$.

Definition 4. A *covering family* for a system $S = (\Sigma, X, \Gamma, M, A, C)$ is a family $\{S_i \hookrightarrow S \mid i \in I\}$ of subsystems of S, with the property that $\bigcup_i \Sigma_i = \Sigma$, $\bigcup_i X_i = X$, $\bigcup_i A_i = A$, and such that the constraints in Γ(resp. C) are consequences of the constraints in $\bigcup_i \Gamma_i$ (resp. in $\bigcup_i C_i$).

Note that in general the distributivity condition in the definition of a Grothendieck topology is not satisfied, as shown by the following example (for the sake of simplicity we do not consider the actions):

Example 1. Let $\Sigma = \{0, 1, \leq\}$ and $M = \{0, 1\}$ (with $0 \leq 1$). Consider the following systems: $S_1 = (\Sigma, \{a, b, d\}, \{a \leq b\}, M)$, $S_2 = (\Sigma, \{b, c, d\}, \{b \leq c, c \leq d\}, M)$, $T = (\Sigma, \{a, c\}, \{a \leq c\}, M)$. Let S the system obtained by interconnecting S_1 and S_2. $S = (\Sigma, \{a, b, c, d\}, \{a \leq b, b \leq c, c \leq d\}, M)$. It is easy to see that in S also $a \leq d$ is satisfied (it is a consequence of the constraints $\{a \leq b, b \leq c, c \leq d\}$. Hence in S_1 seen as a "part" of S, also $a \leq d$ has to be satisfied (by interconnecting systems new constraints may arise). The family $\{S_1, S_2, S_1 \cap S_2\}$ is a cover for S. T is a subsystem of S, but $S_1 \cap T = \{\Sigma, \{a\}, \emptyset, M)$ and $S_2 \cap T = \{\Sigma, \{c\}, \emptyset, M)$, hence $\{S_1 \cap T, S_2 \cap T, S_1 \cap S_2 \cap T\}$ does not cover T.

Therefore, we additionally impose the condition that for every system $S \in$ InSys, the set Γ_S of constraints on the values of the control variables (respectively the set C of constraints that show which parallel actions are admissible), are the restriction of the set Γ_{S_U} (resp. C_{S_U}) to the language of S.

Lemma 5. *The function K assigning for every system S the set $K(S)$ of all covering families for S is a basis (for a Grothendieck topology) on* Sys.

We will denote by J the Grothendieck topology generated by the basis K.

3.1 States and Actions

The states of a system $S = (\Sigma, X, \Gamma, M, A, C)$ are those interpretations $s : X \to M$ that satisfy all the formulas in Γ.

One can see that if $\{S_i \hookrightarrow S \mid i \in I\}$ is a covering site for S, then for every family $\{s_i : X_i \to M_i\}_{i \in I}$ such that for every $i \in I$, $s_i \models \Gamma_i$ and for every $i, j \in I$, $s_{i|X_i \cap X_j} = s_{j|X_i \cap X_j}$ there is a unique $s : X \to M$ such that $s_{|X_i} = s_i$. If, additionally, the covering relation defines a Grothendieck topology on Sys, the following theorem holds.

Proposition 6. *The functor $St : \text{Sys}^{op} \to$ Sets, defined for every object $S = (\Sigma, X, \Gamma, M, A, C)$ in* Sys *by $St(S) = \{s : X \to M \mid s$ satisfies $\Gamma\}$, and for every morphism $S_1 \hookrightarrow S_2$ by $St(S_1 \hookrightarrow S_2)(s) = s_{|P_{S_1}}$ is a sheaf for J.*

The set of admissible actions of the system $S = (\Sigma, X, \Gamma, M, A, C)$ will be the set $Act(S) = \{f : A \to \{0, 1\} \mid f$ satisfies $C\}$. We assume that the compatibility of the actions is expressed by the following fact: If $f \in Act(S)$ and if $s \in St(S)$ such that for every $a \in A$ with $f(a) = 1$ there is a $s'_a \in St(S)_{|X_a}$ such that $(s_{|X_a}, s'_a) \in Tr_a$, then the new local states "agree on intersections", i.e. for every $x \in X_{a_1} \cap X_{a_2}$, $s'_{a_1}(x) = s'_{a_2}(x)$. Then we can associate a transition relation to f, that shows how the state of the system changes after the action is performed, namely: $Tr_f \subseteq St(S) \times St(S)$, $Tr_f = \{(s_1, s_2) \mid (s_{1|X_a}, s_{2|X_a}) \in Tr_a$ for every a such that $f(a) = 1$, and $s_1(x) = s_2(x)$ if $x \notin \bigcup_{a, f(a)=1} X_a\}$.

As pointed out before, in what follows we will assume that all the constraints imposed on the actions can be expressed by boolean equations (equations in the boolean algebra freely generated by A).

As in the case of states, it is easy to see that if $\{S_i \hookrightarrow S \mid i \in I\}$ is a covering site for S, then for every family $\{f_i : A_i \to \{0, 1\}\}_{i \in I}$ such that for every $i \in I$, $f_i \models C_i$ and for every $i, j \in I$, $f_{i|A_i \cap A_j} = f_{j|A_i \cap A_j}$ there is a unique $f : A \to \{0, 1\}$ such that $f_{|A_i} = f_i$. If, additionally, the covering relation defines a Grothendieck topology on Sys, the following theorem holds.

Proposition 7. *The functor Act* : $\mathsf{Sys}^{op} \to \mathsf{Sets}$, *defined for every object* $S = (\Sigma, X, \Gamma, M, A, C)$ *in* Sys *by* $Act(S) = \{f : A \to \{0, 1\} \mid f$ *satisfies* $C\}$, *and for every morphism* $S_1 \hookrightarrow S_2$ *by* $Act(S_1 \hookrightarrow S_2)(f) = s_{|A_{S_1}}$, *is a sheaf.*

The restriction imposed on the constraints on actions in particular subsumes also the situation in which a dependence relation on the set of actions is given. This approach leads to the study of traces [Die90]. We can model this case by defining the set of constraints as $\{a_1 a_2 = 0 \mid (a_1, a_2) \in D, a_1 \neq a_2\}$, where D is the dependence relation on A.

Note again that not every type of constraints leads to a sheaf condition. Consider for example the situation when $Act(S) = \{f : A \to \{0, 1\} \mid f^{-1}(1)$ is finite$\}$. It can be immediately seen that in this case the sheaf condition is not satisfied for infinite covers.

3.2 Applications

Category theory is a very powerful tool, and its language makes it possible to express things in a uniform way. Many important constructions on systems can for instance be modeled by limits or colimits in the category of systems. We will briefly show what meaning the standard limit and colimit constructions have in the category Sys.

Limits: The *product* of two systems is their largest common subsystem; the *pullback* is the largest common subsystem of two systems that are known to be contained in a larger system. Namely, if for $i = 1, 2$ $S_i = (\Sigma_i, X_i, \Gamma_i, M_i, A_i, C_i) \hookrightarrow S = (\Sigma, X, \Gamma, M, A, C)$, then $S_1 \times_S S_2 = (\Sigma_1 \cap \Sigma_2, X_1 \cap X_2, \Gamma_1 \cap \Gamma_2, M_{12}, A_1 \cap A_2, C_1 \cap C_2)$ where M_{12} is the restriction of M to the signature $\Sigma_1 \cap \Sigma_2$. The terminal object of Sys (if exists) is a system that contains all the systems in Sys.

Colimits: Colimits play a special rôle in the study of systems. Actually the system obtained by interconnecting a given family (diagram) of systems can be obtained computing the colimit of the given diagram.

The *coproduct* $S_1 \amalg S_2$ of two systems S_1, S_2 is the smallest system that contains both of them (and such that the two systems are assumed to be independent, i.e. they have no common subsystems); it is easy to see that if

S_1 and S_2 are independent systems then $St(S_1 \amalg S_2) = St(S_1) \times St(S_2)$, and $Act(S_1 \amalg S_2) = Act(S_1) \times Act(S_2)$. The *push-out* $S_1 \amalg_S S_2$ (see diagram 1) represents the system obtained by interconnecting S_1 and S_2 "gluing" them via their common subsystem S.

$$
\begin{array}{ccc}
S & \longrightarrow & S_1 \\
\downarrow & & \downarrow \\
S_2 & \longrightarrow & S_1 \amalg_S S_2
\end{array}
\tag{1}
$$

It is easy to see that if S_1 and S_2 have S as a largest common subsystem, then $St(S_1 \amalg_S S_2) = St(S_1) \times_{St(S)} St(S_2)$ and $Act(S_1 \amalg_S S_2) = Act(S_1) \times_{Act(S)} Act(S_2)$. More generally, we have:

Lemma 8. *Let* $\{S_i \mid i \in I\}$ *be a covering site for* S. *Then* $S = \varinjlim\{S_i \mid i \in I\}$, $St(S) = \varprojlim\{St(S_i) \mid i \in I\}$ *and* $Act(S) = \varprojlim\{Act(S_i) \mid i \in I\}$.

Note that the stability property required in the definition of a Grothendieck topology is in fact a "distributivity property" of intersections of systems w.r.t. arbitrary joins. If we assume that the (partially ordered) category of systems has a Heyting algebra structure, two more operations can be described, namely Heyting implication and negation:

Let S_1 and S_2 be two systems. The system $S_1 \Rightarrow S_2$ is the largest element in the family of systems S with the property that $S_1 \cap S$ is a subsystem of S_2.

Let S be a system. The complement of S (denoted by $\neg S$) is the largest system that is independent of S_1 (in the sense that they have no common subsystems). It is easy to see that $S \cap \neg S = \emptyset$ and that in general $\neg\neg S \neq S$.

We finish by briefly pointing out a possible direction of future work. The fact that states as well as actions can be expressed as sheaves on the category of systems w.r.t. a suitable Grothendieck topology J suggests that also transitions might be defined in a "generic" way as natural transformations between appropriate sheaves. Namely, we can define a natural transformation $tr : Act \to \Omega^{St \times St}$ as follows: For every $S \in$ Sys let $tr(S) : Act(S) \to \Omega^{St \times St}(S)$ be defined by $tr(S)(f) = tr_f : y(S) \times (St \times St) \to \Omega$, where for every $S' \hookrightarrow S$, $tr_f(S') : St(S') \times St(S') \to \Omega(S')$, with $tr_f(S')(s_1, s_2) = \{S'' \hookrightarrow S' \mid (s_{1|P_{f|S''}}, s_{2|P_{f|S''}}) \in Tr_{f|S''}\}$.

In [Sof96] it is proved that tr_f is a natural transformation for every system S and every action $f \in Act(S)$ and that the functor $tr : Act \to \Omega^{St \times St}$ is a natural transformation. Hence, one can associate with the category of systems Sys a "generic transition system" consisting of (St, Act, tr). It seems (see [Sof96]) that this fact can be exploited in order to study which properties of a system are "inherited" by the building blocks (seen as parts of the interconnection) and which properties of the "building blocks" that compose a system are preserved after their interconnection (possibly using the internal logic in a topos, e.g. Sh(Sys, J)). In order to achieve this, it is necessary to translate properties of systems (as for example the existence of deadlock, deadlock freedom, determinism) in the internal logic of $\mathsf{Sh}(\mathsf{Sys}, \mathsf{J})$.

4 Behavior

The starting point of our approach to deal with temporal behavior is the formalism developed by J. Goguen in [Gog92]. He starts with the assumption that every system can be described by a set of attributes X, each attribute $x \in X$ having a prescribed set of values V_x. Assume that time is considered to be discrete. Then in [Gog92] the behavior of a given system S in time is modeled by a functor $F : \mathcal{T}^{op} \to Set$, where $\mathcal{T}$ is the basis for the topology on $\mathbb{N}$ consisting of all the sets $\{0, 1, \ldots, n\}, n \in \mathbb{N}$. Intuitively, for every open set $U = \{0, 1, \ldots, n\}$, $F(U)$ represents the "observations" in the interval of time U. Formally, the functor F is defined on objects by $F(U) = \{h : U \to \prod_{p \in P} V_p \mid K(h)\}$ where $K(h)$ represents a set of conditions that have to be satisfied by h – usually some prescribed rules indicating how the states of the system can change, reflecting the pre- and postconditions of some actions. It is defined on morphisms by $F(\iota_V^U)(h) = h_{|V}$ for every $\iota_V^U : V \hookrightarrow U$ and every $h \in F(U)$. In order to study the behavior of a system consisting of several subsystems the intercommunication between the subsystems is taken into account. A system is seen as a diagram of subsystems, where the morphisms represent inheritance. Goguen shows that the behavior of the system can be described by $F(U) = \{\{h_i \mid i \in I\} \mid h_i \in F_i(U) \text{ and if } \phi_e : S_i \to S_{i'} \text{ then } \phi_e(h_i) = h_{i'}\}$, where for every $i \in I$, F_i is a sheaf that describes the behavior of the system S_i. Therefore, the behavior of a system is the limit of the behaviors of its subsystems (for details see [Gog92]).

In what follows we will develop this idea, modifying the definition for the behavior of a system slightly, by also taking into account the actions that are performed at every step. We will assume that all actions need one unit of time. In future work the more realistic case where actions can have different durations will be considered.

Definition 9. Let S be a system in Sys. The behavior of S is a functor $B_S : \mathcal{T}^{op} \to$ Sets defined for every $U \in \mathcal{T}$ by $B_S(U) = \{h : U \to St(S) \times Act(S) \mid K(h, U)\}$, where $K(h, U)$ can be expressed by "for every n, if $n, n + 1 \in U$ and $h(n) = (s, f)$, then $h(n + 1) = (s', f')$ such that $(s, s') \in Tr_f$ and f' can be executed at s'", and for every $\iota : V \subseteq U$, $B_S(\iota) : B_S(U) \to B_S(V)$ is the restriction to V.

Thanks to the particular form of the open sets of $\mathcal{T}$ (all $\{0, 1, \ldots, n\}$ for some n), it can easily be shown that B_S is a sheaf.

Assumption 1 *We assume that if S_1 is a subsystem of S_2, then for every action $a \in A_{S_1}$, if the set $X_a \subseteq X_2$ of variables a depends upon is such that $X_a \cap X_1 \neq \emptyset$, then $a \in A_{S_1}$, and Tr_a^1 is the restriction of Tr_a^2 to variables in $X_a \cap X_1$.*

Assumption 1 says that every action in S_2 that changes some control variables in a subsystem S_1 of S_2 is "known" in S_1, and that its transition relation is the restriction of the transition relation in S_2 to the variables in S_1. In what follows we will assume that Assumption 1 is satisfied.

Let $\iota : S_1 \hookrightarrow S_2$ in Sys. We define $\rho_{S_1}^{S_2} : B_{S_2} \to B_{S_1}$ by $\rho_{S_1}^{S_2}(U) : B_{S_2}(U) \to B_{S_1}(U)$ for every $U \in \mathcal{T}$, where for every $h : U \to St(S_2) \times Act(S_2)$, $\rho_{S_1}^{S_2}(U)(h) = \langle St(\iota), Act(\iota) \rangle \circ h : U \to St(S_1) \times Act(S_1)$ (with $St(\iota)(s) = s_{|P_1}$ for every $s \in St(S_2)$ and $Act(\iota)(f) = f_{|A_1}$). In what follows, for every $U \in \mathcal{T}$ and every $h \in B_{S_1}(U)$, we will abbreviate $\rho_{S_1}^{S_2}(U)(h)$ by $h_{|S_1}$.

Lemma 10. *Let $U \in \mathcal{T}$ be arbitrary but fixed. Let $B'_U : \mathrm{Sys}^{op} \to \mathrm{Sets}$ be defined for every object $S \in \mathrm{Sys}$ by $B'_U(S) = B_S(U)$ and for every morphism $\iota : S_1 \hookrightarrow S_2$ by $B'_U(\iota) = \rho_{S_1}^{S_2} : B_{S_2}(U) \to B_{S_1}(U)$. Then B'_U is a sheaf.*

Proposition 11. *Let $B : \mathrm{Sys}^{op} \to \mathrm{Sh}(\mathcal{T})$ be defined for every object S of Sys by $B(S) = B_S : \mathcal{T}^{op} \to \mathrm{Sets}$, and for every morphism $\iota : S_1 \hookrightarrow S_2$ by $B(i) = \rho_{S_1}^{S_2} : B(S_1) \to B(S_1)$, and let $B' : \mathcal{T}^{op} \to \mathrm{Sh}(\mathrm{Sys})$ be defined for every $U \in \mathcal{T}$ by $B'(U) : \mathrm{Sys}^{op} \to \mathrm{Sets}$, $B'(U)(S) = B_S(U)$. Then B and B' are functors.*

Proof: By Assumption 1 it follows that for every $h \in B_{S_2}(U)$, $\rho_{S_1}^{S_2}(h) \in B_{S_1}(U)$, i.e. that $\rho_{S_1}^{S_2}$ is well-defined. It is easy to see that $\rho_{S_1}^{S_2} : B_{S_2} \to B_{S_1}$ is a natural transformation. Let $V \subseteq U$, and let i_V^U be the inclusion of V in U. Then the following diagram commutes:

$$
\begin{array}{ccc}
B_{S_2}(U) & \xrightarrow{\rho_{S_1}^{S_2}(U)} & B_{S_1}(U) \\
\Big\downarrow{\scriptstyle B_{S_2}(i_V^U)} & & \Big\downarrow{\scriptstyle B_{S_1}(i_V^U)} \\
B_{S_2}(V) & \xrightarrow[\rho_{S_1}^{S_2}(V)]{} & B_{S_1}(V)
\end{array}
\tag{2}
$$

Hence B is a functor. In order to show that B' is a functor, note first that from Lemma 10 it follows that B' is well-defined on objects. Let $i : V \hookrightarrow U$ be the inclusion between the open sets $U, V \in \mathcal{T}$. Let us define $B'(i) : B'(U) \to B'(V)$ by $B'(i)(S) : B_S(U) \to B_S(V)$ by $B'(i)(S)(h) = h_U$ for every $h : U \to St(S) \times Act(S) \in B_S(U)$. $B'(i)$ is a natural transformation between the sheaves $B'(U)$ and $B'(V)$. This follows from the commutativity of diagram 2. $\square$

5 Conclusions and Prospects of Future Work

In this paper we made the first steps towards a sheaf-theoretic approach to the study of cooperative systems. We illustrated on a simple example the main ideas that motivated our study. Starting from this example we proposed a definition for systems and we defined a category of systems. We endowed this category with a Grothendieck topology and we showed that in certain (non-restrictive) conditions the states and the (parallel) actions are sheaves with respect to this topology. We used these results for studying the behavior of systems in time.

Of course, this work is only a beginning. One of our next goals is to see which properties can be expressed using results of topos theory: in particular to use geometric logic and to exploit the internal logic in topoi. This is part of ongoing work, see [Sof96]. Another direction for future research is to consider space- and time-dependent formulas. In [PS95] one section is devoted to a concept of space and time dependent formula handling, intended for applications in robotics where the state space of an agent can change with space and time. This makes it possible to model logically a variety of practical situations in the case of cooperating agents working in a geometrically (or topologically) modeled environment X.

Acknowledgments: I thank J. Pfalzgraf for encouraging me to pursue this research and for many stimulating discussions and helpful suggestions. I also thank K. Stokkermans and U. Sigmund for the interesting discussions that made me gain more insight in the field of modeling and planning for cooperating agents.

References

[Die90]   V. Diekert. *Combinatorics on Traces*, volume 454 of *Lecture Notes in Computer Science*. Springer Verlag, 1990.

[FS79]   M. P. Fourman and D. S. Scott. Sheaves and Logic. In M. Fourman, editor, *Durham Proceedings (1977). Applications of Sheaves*, volume 753 of *Lecture Notes in Mathematics*, pages 302–401. Springer Verlag, 1979.

[Gog75]   J. Goguen. Objects. *International Journal of General Systems*, 1:237–243, 1975.

[Gog92]   J. Goguen. Sheaf Semantics for Concurrent Interacting Objects. *Mathematical Structures in Computer Science*, 11:159–191, 1992.

[KC79]   P. H. Krauss and D. M. Clark. Global Subdirect Products. *Memoirs of the American Mathematical Society*, 17(210):1–109, January 1979.

[Lil93]   J. Lilius. A Sheaf Semantics for Petri Nets. Technical Report A23, Dept. of Computer Science, Helsinki University of Technology, 1993.

[Mal94]   G. Malcolm. Interconnections of Object Specifications. In R. Wieringa and R. Feenstra, editors, *Working Papers of the International Workshop on Information Systems – Correctness and Reusablity*, 1994. Appeared as internal report IR-357 of the Vrije Universiteit Amsterdam.

[MLM92]   S. Mac Lane and I. Moerdijk. *Sheaves in Geometry and Logic*. Universitext. Springer Verlag, 1992.

[MP86]   L. Monteiro and F. Pereira. A Sheaf Theoretic Model for Concurrency. *Proc. Logic in Computer Science (LICS'86)*, 1986.

[Pfa91]   J. Pfalzgraf. Logical Fiberings and Polycontextural Systems. In P. Jorrand and J. Kelemen, editors, *Proc. Fundamentals of Artificial Intelligence Research*, volume 535 of *Lecture Notes in Computer Science (subseries LNAI)*, pages 170–184. Springer Verlag, 1991.

[Pfa93]   J. Pfalzgraf. On Mathematical Modeling in Robotics. In J. Calmet and J.A. Campbell, editors, *AI and Symbolic Mathematical Computing. Proceedings*

AISMC-1, volume 737 of *Lecture Notes in Computer Science*, pages 116–132. Springer Verlag, 1993.

[Pfa95] J. Pfalzgraf. On Geometric and Topological Reasoning in Robotics. *Annals of Mathematics and AI, special issue on AI and Symbolic Mathematical Computing*, 1995. To appear.

[PS95] J. Pfalzgraf and K. Stokkermans. On Robotics Scenarios and Modeling with Fibered Structures. In J. Pfalzgraf and D. Wang, editors, Springer Series Texts and Monographs in Symbolic Computation, Automated Practical Reasoning: Algebraic Approaches, pages 53–80. Springer Verlag, 1995.

[PSS95] Jochen Pfalzgraf, Ute Cornelia Sigmund, and Karel Stokkermans. Modeling cooperative agents scenarios by deductive planning methods and logical fiberings. In Jacques Calmet and John A. Campbell, editors, *2nd Workshop on Artificial Intelligence and Symbolic Mathematical Computing*, volume 958 of *Lecture Notes in Computer Science*, pages 167–190. Springer-Verlag, 1995.

[PSS96] J. Pfalzgraf, U. Sigmund, and K. Stokkermans. Towards a General Approach for Modeling Actions and Change in Cooperating Agents Scenarios. *IGPL (Journal of the Interest Group in Pure and Applied Logics)*, 4(3):445–472, 1996.

[Sof96] V. Sofronie. A sheaf theoretic approach to cooperating agents scenarios. Manuscript in preparation, 1996.

[Sri93] Y.V. Srinivas. A Sheaf-theoretic Approach to Pattern Matching and Related Problems. *Theoretical Computer Science*, 112:53–97, 1993.

Data Types in Subdefinite Models *

Vitaly Telerman, Dmitry Ushakov

Institute of Informatics Systems,
Russian Academy of Science, Siberian Division
email: telerman@iis.nsk.su

Abstract. We consider the mechanism of subdefinite models and the problem of representing data types in such models. A justification of the method of subdefinite models is given; various kinds of subdefinite extensions of data types are presented. We also investigate their efficiency in the solution of various problems.

Introduction

The concept of *subdefiniteness* and data-flow computations on models with subdefinite values, proposed by A.S.Narin'yani in the early 80s [1-3] was implemented in various software products developed at the Russian Research Institute of Artificial Intelligence and at the Institute of Informatics Systems in Novosibirsk [4-7]. The principal idea of subdefiniteness is that with each object we associate not only one exact value, but a subset of the set of admissible values. The principle of indefiniteness regards the state of being partially defined as a solution to the problem which is possible at the given level of knowledge about the problem. Such a subdefinite solution can practically satisfy the user's requirements, and can also stimulate gathering additional information. When passing from the more general space of admissible values to a more narrow one, it is possible that other (for example, more specialized) solution methods become applicable, those which could not be applied in the original setting. The specification of a problem under subdefiniteness is called a *subdefinite model* (an SD-model).

The search for a solution under the assumption of subdefiniteness has much in common with the *constraint satisfaction problem* [7-10]. In both approaches, the main notions are objects associated with some values and constraints narrowing the domains of admissible values for these objects. We need to find the values of the objects which satisfy all of the constraints simultaneously.

Generality of the approach that is the basis of SD-models makes it possible to use them for the solution of problems which are traditionally placed in different classes. For example, the apparatus of SD-models can be applied to numerical problems (systems of linear and nonlinear equations, or inequalities over integer and real variables), to logical and combinatorial problems, to problems

---

* This work was partially supported by the foundation "Informatization of Russia", grant N 285.78

on sets, etc. The most remarkable fact is that all these problems can be solved simultaneously within a single SD-model, because each of them is solved using the same constraint satisfaction algorithm. Obviously, the special methods are more efficient than the universal one, but they can solve problems only from the specific class.

The main goal of our research is to perfect the basic computational mechanism so as to minimize the necessity to utilize specialized tools for solving problems. This is particularly important since for most problems such tools do not exist at all (considering the breadth of the area of application for SD-models).

One should note one more fact which has significant impact on the organization of computations in SD-models. With each traditional class of objects (type of a variable) we can associate several subdefinite extensions which differ in their expressive power as well as in the computational resources they require. In the original statement of the problem it is not always possible to point out the "most powerful" (from the standpoint of the quality of the results obtained) subdefinite extensions for all objects. For instance, the resources of the available computers may be insufficient.

In the present paper we will show the requirements which a subdefinite extension should satisfy; we will also demonstrate how we can associate with one object several subdefinite extensions. The paper is structured as follows. Section 1 defines the main notions of sorted algebras which are used to provide a formal representation of the apparatus of subdefinite models. In Section 2 we introduce the notion of a subdefinite extension for a sorted Σ-algebra, and present an algorithm for transforming an existential conjunction of atoms to a certain normalized form. We introduce also the notion of a *subdefinite witness* (SD-witness), which is the largest common fixed point of the system of operators of the model (by inclusion), and show that a subdefinite witness constructed with the help of a more powerful subdefinite extension will be more informative. In Section 3 we consider the operational semantics and describe a constraint satisfaction algorithm on the so-called generalized computational models, which generalize the well-known notion of an SD-model. Section 4 presents various kinds of subdefinite extensions and the relationships between them. Section 5 discusses the issues of the efficiency of using various kinds of subdefinite extensions in the solution of several problems. We summarize the paper in the conclusion.

1 Preliminaries

The traditional problem of SD-models is finding all solutions of an existential conjunction in a sorted model. Such sorted model completely corresponds to models from [11]. In the present paper we will use the terminology from [11]. We must specify a subdefinite extension of the corresponding sorted algebra to describe the denotational semantics of this problem. The operational semantics of the solving process is described in terms of the so-called generalized computational models [12].

Let us suppose that we have a signature (S, Σ), where

- S is the set of sorts $w = s_1, \ldots, s_n$, and λ is the empty sort;
- $\Sigma = \{\sigma_{w,s} | w \in S^*, s \in S\}$ is the set of function symbols.

A Σ-algebra A is the set of supports $A = \{A_s | s \in S\}$ and the set of operations $A_\sigma : A_w \to A_s$, where $\sigma \in \Sigma_{w,s}, A_w = A_{s_1} \times \cdots \times A_{s_n}$ for $w = s_1 \cdots s_n$ and $A_\lambda = \{*\}$.

A Σ-homomorphism $h : A \to B$ of one Σ-algebra into another is an s-indexed family of mappings $\{h_s : A_s \to B_s | s \in S\}$ satisfying the substitution condition:

$$h_s(A_\sigma(a_1, \ldots, a_n)) = B_\sigma(h_{s1}(a_1), \ldots, h_{sn}(a_n))$$

for all $\sigma \in \Sigma_{w,s}, w = s_1 \ldots s_n \in S^*, < a_1, \ldots, a_n > \in A_w$.

A Σ-algebra I is called initial in the class of all Σ-algebras if there is a Σ-homomorphism from I into any other Σ-algebra.

Σ-terms are the smallest s-indexed set T_σ such that

1) $\lambda_s \subseteq T_{\Sigma,s}$ for all $s \in S$;

2) if $\sigma \in \Sigma w, s$ and $t_i \in T_{\Sigma, s_i}$ for $w = s_1, \ldots, s_n \neq \lambda$, then the string $\sigma(t_1 \ldots t_n)$ belongs to $T_{\Sigma,s}$.

In addition, we define $T_\Sigma : T_w \to T_s$ as a mapping of $t_1, \ldots, t_n$ into the string $\sigma(t_1 \ldots t_n)$.

It is clear that the set of all Σ-terms constitutes a Σ-algebra which is initial in the class of all Σ-algebras.

Let us extend the signature Σ by additional constant symbols which will play the role of variables. Let X be an s-sorted set of variables such that $X_s \cap X_{s'} = \emptyset$ for $s \neq s'$. Thus, we obtain the extended signature $\Sigma(X)$ and the algebra $T_{\Sigma(X)}$, which we will denote by $T_\Sigma(X)$ also.

Let A be a Σ-algebra, and let X be some s-sorted set of variables (which does not intersect Σ). We call an *estimate* any s-sorted function $f : X \to A$. Since the Σ-algebra T_Σ is initial, there exists a unique Σ-homomorphism $f^* : T_\Sigma(X) \to A$ which extends f.

A sorted signature with predicates is a triple (S, Σ, Π), where (S, Σ) is a sorted signature and $\Pi = \{\Pi_w | w \in S^+\}$ is a family of predicate symbols containing the predicate symbols of equality $= \in \Pi_{ss}$ for each sort $s \in S$.

Let (S, Σ, Π) be a sorted algebra with predicates. A Σ, Π-model is a Σ, Π-algebra M in which each $P \in \Pi_w$ is associated with some subset $M_P^w \subseteq M_w$, with $(M_=^{ss}) = \{(a, a) | a \in M_s\}$.

Σ, Π-atoms are expressions of the form

$$P(t_1, \ldots, t_n),$$

where $t_i \in T_{\Sigma(X), s_i}, P \in \Pi_w, w = s_1, \ldots, s_n$.

Given a Σ, Π-signature with predicates, we define a Σ, Π-existential conjunction (of atoms) to be an existential formula of the form

$$(\exists X) A_1, \ldots, A_n, \tag{1}$$

where X is an s-sorted set of variables which contains all the variables occurring in the terms of Σ, Π-atoms $A_1, \ldots, A_n$. We say that a Σ, Π-model M realizes this conjunction (or that this conjunction is realized in M) if there exists an estimator $a : X \to M$ (called a witness) such that

$$(a^*(t_{i1}), \ldots, a^*(t_{in_i})) \in M_{P_i}^{wi}, i = 1, \ldots, n$$

(assuming that $A_i = P_i(t_{i1}, \ldots, t_{in_i})$, where $P_i \in \Pi_w, w = s_{i1} \ldots s_{in_i}$ and t_{ij} has sort s_{ij}). An SD-witness is a family of functions $\{f_s : X_s \to 2^{M_s} | s \in S\}$ such that, for any withness $a : X \to M$, we have $a_s(x) \in f_s(x)$ for all $x \in X_s$, $s \in S$. We say that an SD-withness f is *more informative* than an *SD-withness* g ($f \geq g$) if for all $s \in S$ and $x \in X_s$ we have $f_s(x) \subseteq g_s(x)$. It is clear that there always exists a trivial (the least informative) SD-witness $f(x) = M_s$ for all $x \in X_s, s \in S$. We will study below one way of obtaining the most informative SD-witnesses.

2 Subdefinite Extensions

The SD-extension of a Σ-algebra A is a Σ-algebra $^*A$ in which every witness satisfies the following conditions:

1. $^*A_s$ is a finite family of subsets A_s;
2. $\emptyset \in {}^*A_s$;
3. $A_s \in {}^*A_s$;
4. If $\alpha \in {}^*A_s$ and $\beta \in {}^*A_s$, then $\alpha \cap \beta \in {}^*A_s$.

The property (4) from this definition guarantees that for any $\xi \subseteq A_s$ its *unique representation* (which will be denoted by $^*[\xi]$) in the SD-extension $^*A_s$, namely:

$$^*[\xi] = \bigcap_{\xi \subseteq \alpha \in {}^*A_s} \alpha.$$

Thus, the representation of the set ξ in the system $^*A_s$ is the minimal subset of the SD-extension $^*A_s$ which contains ξ.

In addition, the operations of the algebra $^*A$ are defined as follows:

$$^*A_\sigma(\alpha_1, \ldots, \alpha_n) = {}^*[\{A_\sigma(a_1, \ldots, a_n) | a_i \in \alpha_i, i = 1, \ldots, n\}]$$

for $\sigma \in \Sigma_{w,s}$. We introduce also some auxiliary operations. Namely, for each $\sigma \in \Sigma_{w,s}$ we define

$$^*A_\sigma^i : {}^*A_{w'} \to {}^*A_{s_i},$$

where $w' = s_1 \ldots s_{i-1} s_{i+1} \ldots s_{n_s}$,

$$^*A_\sigma^i(\alpha_1, \ldots, \alpha_{i-1}, \alpha_{i+1}, \ldots, \alpha_n, \alpha) =$$

$$= {}^*[\{a_i | a = A_\sigma(a_1, \ldots, a_n), a_i \in \alpha_i, i = 1, \ldots, n, a \in \alpha\}].$$

Suppose that we have a Σ, Π-model and an existential conjunction (1), where $A_i = P_i(t_{i1}, \ldots, t_{in_i})$. This conjunction can be transformed to the following normal form:

$$(\exists (X \cup Y)) B_1, \ldots, B_{n+k}. \tag{2}$$

Here B_i have one of the following forms:

$$P(x_1, \ldots, x_n), \quad \text{where} \quad x_i \in Y_{si}, P \in \Pi_w, w = s_1, \ldots, s_n;$$
$$y = \sigma(x_1 \ldots x_m), \quad \text{where} \quad y \in Y_s, x_i \in Y_{s_i}, \sigma \in \Sigma_{w,s}, w = s_1, \ldots, s_n;$$
$$x = \sigma, \quad \text{where} \quad y \in Y_s, \sigma \in \Sigma_{\lambda,s} \cup X_s.$$

The meaning of this transformation is to replace correctly the complex atoms of the initial model by a conjunction of simpler and, thus, easier realizable, atoms.

Reduction to the normal form is performed by means of a recursive term transformation. Let A be the transforming function. Then the atom

$$P(t_1, t_2, \ldots, t_n),$$

where $P \in \Pi_w, w = s_1, \ldots, s_n$ is transformed into the conjunction of the atoms

$$P(x_1, x_2, \ldots, x_n),$$
$$x_1 = A(t_1),$$
$$\vdots$$
$$x_n = A(t_n),$$

with the new variables $x_i (i = 1, 2, \ldots, n)$ added to the sets Y_{s_i}. The term $\sigma(t_1, t_2, \ldots t_m)$, where $\sigma \in \Sigma_{w,s}, w = s_1, s_2, \ldots, s_n$ is transformed into the conjunction of the atoms

$$y = \sigma(x_1, x_2, \ldots, x_m),$$
$$x_1 = A(t_1),$$
$$\vdots$$
$$x_m = A(t_m),$$

where y and $x_i (i = 1, \ldots, n)$ are the new variables which are added to the sets Y_s and Y_{s_i}, respectively. The term $\sigma \in \Sigma_{\lambda,s} \cup X_s$ remains unchanged.

The conjunction of atoms resulting from this transformation is equivalent to the original one in the following sense: for any estimate

$$f : X \to M$$

there exists a unique Σ-homomorphism

$$f^* : T_\Sigma(X) \to M$$

extending M and, therefore, there exists a unique estimate

$$f^{**} : X \cup Y \to M$$

which is the extension of f (since each variable in Y is recursively related by the equality relation to some term in $T_\Sigma(X)$). Thus, every witness

$$a : X \to M$$

of the realization of the initial existential conjunction is uniquely extended to a witness

$$a^{**} : X \cup Y \to M$$

of realization of the conjunction resulting from recursive term reduction.

Suppose that we have some SD-extension $^*M$ of the Σ-algebra M, a Σ, Π-model and an existential conjunction (1), which have been transformed to the normal form (2). Each atom B_i is a mapping

$$\mathcal{T}_i : {}^*M \to {}^*M$$

such that

$$\mathcal{T}_i(\xi_1, \ldots, \xi_m) = (\zeta_1, \ldots, \zeta_m).$$

The formula by means of which $\zeta_j (j = 1, \ldots, m)$ are computed depends on the form of the corresponding atom B_i. The following four cases are possible:

1. $\zeta_j = {}^*[M_{P_i}^{w_i} \cap (\xi_{i_1} \times \cdots \times \xi_{i_{n(i)}})|_{\zeta_j}]$, if $B_i \equiv P_i(x_1, \ldots, x_{n(i)})$;
2. $\zeta_j = {}^*M_\sigma^j(\xi_{j_1}, \ldots, \xi_{j_{m(j)}}) \cap \xi_j$, if $B_i \equiv x_j = \sigma_j(x_1, \ldots, x_{m(j)})$;
3. $\zeta_j = {}^*M_{\sigma_k} \cap \xi_l$, if $B_i \equiv x_l = \sigma_k$;
4. $\zeta_j = \xi_k \cap \xi_l$, if $B_i \equiv x_l = x_k$.

Each of these mappings has the following two remarkable properties:

1. $\mathcal{T}_i(x) \subseteq x$, and
2. $x \subseteq y$ implies that $\mathcal{T}_i(x) \subseteq \mathcal{T}_i(y)$.

These properties, together with finiteness of $^*M$, are sufficient for the following assertion: *the system of operators $\mathcal{T}_i$ has the greatest (by inclusion) common fixed point, which can be found in finitely many steps by the method of sequential approximations.*

We will call such a point the SD-witness corresponding to the SD-extension $^*M$, denoting the function that yields this point by

$$f_{*M} : X \cup Y \to {}^*M.$$

It is clear that its restriction to X will by an SD-witness of the validity of the initial conjunction in the model M, since for any witness $a : X \to M$ of the validity of the initial conjunction in the model M and its extension

$$a^{**} : X \cup Y \to M$$

the point $^*[a^{**}(x)]$ will be a fixed point for all operators $\mathcal{T}_i$.

Suppose that we have a model M, an existential conjunction, and two SD-extensions of M: $^*M$ and $^{**}M$ such that $^*M_s \subseteq {}^{**}M_s$ for all $s \in S$. Then $f_{**M} \geq f_{*M}$, that is, the SD-witness obtained via a more powerful SD-extension is more informative.

3   Operational semantics

Let us return to arbitrary sorted algebras. Extend the algebra $< A, F, V >$ by a set C of elementary constraints. An elementary constraint C is a triple

$$C =< f, In, Out >, \qquad (3)$$

where

$$f \in A_{s_1} \times \cdots \times A_{s_m} \to A_{s_{m+1}} \times \cdots \times A_{s_{m+n}},$$
$$In = \{v_i \in V_{s_i} | i = 1, \ldots, m\} \subseteq V,$$
$$Out = \{u_j \in V_{s_{m+j}} | j = 1, \ldots, n\} \subseteq V,$$
$$In \cap Out = \emptyset.$$

We say that the constraint C connects the variables $v =< v_1, \ldots, v_m >$ and $u =< u_1, \ldots, u_n >$, writing this down as

$$f_j(v_1, \ldots, v_k) = u_j, j = 1, \ldots, n, \quad \text{or} \quad f(v) = u.$$

In (3) the set In is called the set of *input variables* of the constraint, Out is the set of its *output variables*, and the mapping f is the *interpreting function*.

Definition. A *generalized computational model* (GCM) [12,13] over an algebra $< A, F, V, C >$ is the pair $M =< X, R >$, where $X = \{v_i \in V_{s_i} | i = 1, \ldots, k\}$ is the set of variables, and $R = \{c_j \in C | j = 1, \ldots, l\}$ is the set of elementary constraints on the variables from X. To each variable $v_i \in V_{s_i} (i = 1, \ldots, k)$ we assign:

- an initial value $\alpha_i \in A_{s_i}$,
- an assignment function $w_i : A_{s_i} \times A_{s_i} \to A_{s_i}$,
- a correctness check function $corr_i : A_{s_i} \to \{true, false\}$.

A constraint satisfaction algorithm in a GCM $M = (X, R)$ is defined as follows ('$\leftarrow$' is the assignment operation):

algorithm *SD_CONSTRAINT_SATISFACTION*
var $u_i \in V_{s_i}, i = 1, \ldots, k$ are auxiliary variables,
$\quad\quad Q$ is the set of active constraints.
begin
$\quad\quad \forall v_i \in X : \quad v_i \leftarrow \alpha_i$
$\quad\quad Q \leftarrow R$
$\quad\quad$**while** $Q \neq \emptyset$
$\quad\quad\quad$**Choose** $c \in Q$
$\quad\quad\quad Q \leftarrow Q \setminus \{c\}$
$\quad\quad$(* Let $c =< f, In = \{v_1, \ldots, v_m\}, Out = \{v_{m+1}, \ldots, v_{m+n}\} > $ *)
$\quad\quad\quad \forall v_{m+i} \in Out$
$\quad\quad\quad u_{m+i} \leftarrow v_{m+i}$ (* save the old value *)
$\quad\quad\quad v_{m+i} \leftarrow w_{m+i}(v_{m+i}, f^i(v_1, \ldots, v_m))$
$\quad\quad\quad$**if** $u_{m+i} \neq v_{m+i}$ **then**
$\quad\quad\quad$**begin**

            if $corr_{m+i}(v_{m+i})$ **then** $Q \leftarrow Q \cup \{c' | v_{m+i} \in In_{c'}\}$
            else return *FAILURE*
        end
      end while
      return *SUCCESS*
    end *algorithm.*

Such algorithm can be implemented as a virtual data-flow machine [13]. The result of the algorithm is either SUCCESS, in which case the current values of the variables will point to possible values of objects of the model, or FAILURE, which signals that during the execution of the algorithm the value of one of the variables became invalid. The operator *Choose* assumes a certain strategy of choosing the next constraint from the set Q of active constraints. In what follows, we will consider only those GCM over the SD-extended algebras $< {}^*A, {}^*F, {}^*V, {}^*C >$ in which

- Each elementary constraint $< f, In, Out >\in {}^*C$ is a monotone mapping, that is,

$$\forall \xi_i, \zeta_i \in {}^*A_{s_i}, i = 1, \ldots, m \quad \xi_1 \times \cdots \times \xi_m \subseteq \zeta_1 \times \cdots \times \zeta_m \Rightarrow$$
$$\Rightarrow f(\xi_1, \ldots, \xi_m) \subseteq f(zeta_1, \ldots, zeta_m),$$

- the assignment function computes the intersection of SD-values, that is,

$$w_i(\xi^{old}, \xi^{new}) = \xi^{old} \cap \xi^{new},$$

- The correctness check function tests whether an SD-value is nonempty, that is,

$$corr_i(\xi) = \textbf{if } \xi \neq \emptyset \textbf{ then } true \textbf{ else } false \textbf{ fi}.$$

We will call these computational models *subdefinite models*, or SD-models. Note that the notion of an SD-model that was used in the previous papers [7,12,13] was less formal and, therefore, admits a broader interpretation. Henceforth we consider only the SD-models defined above.

In this case, the following assertions are valid:

1. The constraint satisfaction algorithm in SD-models terminates in finitely many steps.
2. The result of the algorithm (that is, FAILURE or SUCCESS) is determined by the input data (the initial SD-values of variables and constraints) and does not depend on the specific strategy of choosing the next constraint to be interpreted.
3. If the algorithm SUCCEEDS, then the output values of SD-variables depend only on their input values but not on the concrete strategy selecting the next constraint to be interpreted.

The proofs of these assertions can be found in [14].

Independently of our approach the similar results in the case of the interval representation of real, integer and boolean variables have been proved in [10].

4   Types of SD-extensions

Let us turn back to the definition of an SD-extension and consider some versions of them, which have been successfully used in SD-models for a long time. To simplify the notation, denote the set of values A_s of sort s of the algebra A by X, and call this set the universe. The class of all possible SD-extension of the universe X will be denoted by SD(X).

1) The simplest SD-extension of a universe X is the SD-extension *Single*, which has the following form:

$$X^{Single} = \{\emptyset\} \cup \{X\} \cup \{\{x\}|x \in X\}.$$

It is clear that X^{Single} conforms to the definition of an SD-extension: it is obtained by adding to the set X two special elements, UNDEFINED ($\{X\}$) and CONTRADICTION ($\emptyset$).

2) The maximal SD-extension of X, which we denote by X^{Enum}, is the set of all subsets of X, that is, $X^{Enum} = 2^X$. Clearly X^{Enum} is the maximal *X, satisfying the definition of the SD-extension. Of course, on a real computer it is not possible, to specify all elements of 2^X for large X. Therefore, in practice it is reasonable to restrict X^{Enum} somehow. For instance, we can define the SD-extension $X^{Enum(N)}$:

$$X^{Enum(N)} = \{x \in X|\#x \le N\} \cup \{X\}$$

The SD-extension $X^{Enum(N)}$ is the set of all subsets of X whose cardinality does not exceed some positive integer N, plus X itself. It is obvious that $X^{Enum(1)} = X^{Single}$, and $X^{Enum(M)} \subseteq X^{Enum(N)}$ for $M \le N$. If X is finite, then $X^{Enum(N)} = X^{Enum}$, if $N \ge \#X - 1$.

Note, that the requirement that the system *X must be finite, implies for the SD-extensions X^{Single} and X^{Enum} finiteness of the initial universe X. However, after a small modification of these definitions they can be used for some infinite and even continuous sets. If an infinite discrete set X is linearly ordered, then X^{Single} can be represented as

$$\{\emptyset\} \cup \{X\} \cup \{\{x\}|X_{min} \le x \le X_{max}\} \cup \{x|x < X_{min}\} \cup \{x|x > X_{max}\},$$

where $X_{min} < X_{max}$ are two distinguished elements of X. In the case of continuous sets, all one-element sets are replaced by some neighborhoods, whose disjoint union is equal to the entire universe. For example, if X is equal to $[a, b) \subset \mathbf{R}$ then

$$X^{Single} = \{\emptyset\} \cup \{X\} \cup \{[a + (k - 1)\epsilon, a + k\epsilon)|k = 1, \dots, (b - a)/\epsilon\},$$

where $\epsilon = (b - a)/N$ for some positive integer N.

In the case when X is a lattice (a set with two associative and commutative operations $\vee$ and $\wedge$ obeying the absorption law and the idempotent law) it is possible to define such types of SD-extensions of X as *intervals* [3,7,12,14,15] and *multi-intervals* [18,19].

3) SD-extension by intervals

$$X^{Interval} = \{[x^{Lo}, x^{Up}] | x^{Lo}, x^{Up} \in X\}.$$

Here $[x^{Lo}, x^{Up}] = \{x \in X | x^{Lo} \wedge x = x^{Lo}$ and $x \vee x^{Up} = x^{Up}\}$, x^{Lo} is called the lower bound of the interval, and x^{Up} is the upper one. It is obvious that $[x^{Lo}, x^{Up}] \cap [y^{Lo}, y^{Up}] = [x^{Lo} \vee y^{Lo}, x^{Up} \wedge y^{Up}]$. Here the empty set is represented by any interval $[x^{Lo}, x^{Up}]$, where $x^{Lo} \vee x^{Up} \neq x^{Up}$. The entire universe X is represented by the interval $[inf X, sup X]$, and a single element by $\{x\} = [x, x]$.

4) SD-extension by multi-intervals

$$X^{Multiinterval} = \{\alpha | \alpha = \cup \alpha_k, \alpha_k \in X^{Interval}, k = 1, 2, \ldots\}.$$

It is obvious that the properties 1–4 hold, and $\emptyset$, X, and $\{x\}$ are represented exactly as in the case of intervals. It is evident that $\alpha \cap \beta = \{\alpha_i \cap \beta_i | i = 1, 2, \ldots, j = 1, 2, \ldots\}$. Note that formally the systems X^{Enum} and $X^{Multiinterval}$ coincide, but the representation by multi-intervals is often better from the standpoint of efficiency of representation and computations.

It is obvious that interval and multi-interval SD-extensions can be applied to continuous universes. For example, the interval extension of the set $\mathbf{R}$ of real numbers may look as follows:

$$\mathbf{R}^{Interval} = \{[x^{Lo}, x^{Up}] | x^{Lo}, x^{Up} \in R_0\} \cup$$
$$\cup \{[x^{Lo}, +\infty) | x^{Lo} \in R_0\} \cup \{(-\infty, x^{Up}] | x^{Up} \in R_0\} \cup \{(-\infty, +\infty)\},$$

where $R0 = [m, M] \subset \mathbf{R}$ for arbitrary real $m < M$.

Intervals are of interest not only for linearly ordered universes, like real numbers (this is the domain where they are actively used, see [7,15,16]), but also for such peculiar universes as sets. Indeed, let us consider $X = 2^U$ for some set U and two elements $x, y \in X$. Let us define

$$x \wedge y \equiv x \cap y, \quad x \vee y \equiv x \cup y.$$

(that is, we define the natural order on the sets). What is then the semantics of the interval $[x^{Lo}, x^{Up}]$? Obviously $x^{Lo} \subseteq U$ includes those elements of U about which we have already learned in computations that they belong to the set modeled by the interval. Elements of $x^{Up} \subseteq U$ can potentially belong to the set. If we prove that some element cannot belong, then we must delete it from x^{Up}, that is, decrease the upper boundary of the interval. The cardinality of the set is modeled by the integer interval $[\#x^{Lo}, \#x^{Up}]$. Such sets were first proposed by A.S.Narin'yani in [2]. In this paper, the author proposed to store, instead of the set x^{Up}, the upper boundary of the interval, i.e. its complement. The semantics of the set $U \setminus x^{Up}$ is evident: it contains those elements about which we already know that they belong to the set.

General properties of the algebra of intervals in an arbitrary Boolean algebra were studied in [17].

The notion of a multi-interval was proposed independently in [18] and [19].

5) *Structural SD-extension.* Consider now the universe defined as a Cartesian product of sets: $X = X_1 \times \cdots \times X_n$. Just as to any other sets, we can apply to it the SD-extensions X^{Single} and X^{Enum}. Moreover, if each of X_i is a lattice, then X is a lattice also (as a Cartesian product of lattices), and so we can apply to X the SD-extensions $X^{Interval}$ and $X^{Multiinterval}$ as well. However, one more SD-extension of X can be proposed.

Let $^*X_i \in SD(X_i), i = 1, \ldots, n$. Then the system $^*X_1 \times \cdots \times {^*X_n}$ will satisfy conditions 1–4 too, that is, $^*X_1 \times \cdots \times {^*X_n} \in SD(X_1 \times \cdots \times X_n)$. The following question is of interest: if the form of an SD-extension applicable to X_i as well as to X is fixed, then what will be the relationship between the set systems $^*(X_1 \times \cdots \times X_n)$ and $^*X_1 \times \cdots \times {^*X_n}$? Consider these extensions for each type of SD-extensions considered above.

$$(X_1 \times \cdots \times X_n)^{Single} \subseteq X_1^{Single} \times \cdots \times X_n^{Single}.$$
$$(X_1 \times \cdots \times X_n)^{Enum} \supseteq X_1^{Enum} \times \cdots \times X_n^{Enum}.$$
$$(X_1 \times \cdots \times X_n)^{Enum(N)} \supseteq X_1^{Enum(N_1)} \times \cdots \times X_n^{Enum(N_n)},$$
$$\text{if } N_i \geq \#X_i - 1 \text{ and } N \geq N_1 * \cdots * N_n.$$
$$(X_1 \times \cdots \times X_n)^{Interval} = X_1^{Interval} \times \cdots \times X_n^{Interval}.$$
$$(X_1 \times \cdots \times X_n)^{Multiinterval} \supseteq X_1^{Multiinterval} \times \cdots \times X_n^{Multiinterval}.$$

We thus see that only for intervals the choice of the SD-extension in the form $^*(X_1 \times \cdots \times X_n)$ or $^*X_1 \times \cdots \times {^*X_n}$ is unimportant. In other case, this choice is essential. Note only that in the case of enumerations ($Enum$) and multi-intervals the choice of a "richer" representation (i. e. $^*(X_1 \times \cdots \times X_n)$) increases considerably resources necessary to store these subdefinite expressions.

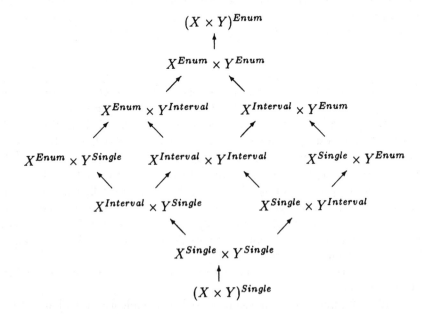

Consider the relationship between the SD-extensions introduced above, using as an example the Cartesian product of two universes $X \times Y$ (we do not consider multi-intervals since they are equal to enumerations).

We can see, that the more powerful SD-extensions are at the top of the picture, and the less powerful ones are at the bottom, with $X^{Interval} \times Y^{Interval} = (X \times Y)^{Interval}$, as mentioned above.

It should be noted that in order to simplify the exposition we consider only simple SD-extensions of sorted algebras, in which there is a unique SD-extension for each sort. It is clear that in the general case we can associate several SD-extensions within a single algebra with each sort. In this case, the above definition of an SD-extension of an algebra becomes more complex.

Example. Suppose that the signature Σ^1 over the sort *integer* has an operation $\sqrt[2]{}, \tau(\sqrt[2]{}) = integer \rightarrow integer$. In the Σ^1-algebra A^1 it corresponds to the function computing the square root of an integer, which calculates not only the arithmetic value of the root, but also its negative analog. The domain of definition of this function is the set

$$Dom(\sqrt[2]{}) = \{x \in A^1_{integer} | (\exists y \in A^1_{integer}) x^2 = y\} = \{0, 1, 4, 9, 16, 25, \ldots\}.$$

Let us examine some types of SD-extensions of the operation $\sqrt[2]{}$ in the case when the argument is the interval $[5, 35]$. We see that information is practically

SD-extension	Value of the root
Single	$A^1_{integer}$ (fully indefinite)
Enum	{ -5, -4, -3, 3, 4, 5 }
Interval	[-5, 5]
Multiinterval	{ [-5, -3], [3, 5] }

lost in the transition to the exact value; the interval extension is better than the exact one, but still is worse than the other two (since it contains the subinterval $[-2, 2]$) that does not belong to the value); the enumeration and multi-interval extensions are equivalent in their power, but the multi-interval representation is more compact.

Note that in the general case the superposition of the subdefinite extensions of two functions is not equal to the subdefinite extension of their superposition. Let $f : X \rightarrow Y$ and $g : Y \rightarrow Z$. Then the above definition implies only the following inclusion (not an equality!):

$$(\forall^* X \in SD(X))(\forall^* Y \in SD(Y))(\forall^* Z \in SD(Z))$$

$$(\forall \xi \in {}^*X)^*(f \circ g)(\xi) \subseteq ({}^*f \circ {}^*g)(\xi).$$

Example. Consider the signature $\Sigma^2 = \{+, -\}$ over the sort *integer*, where

$$\tau(+) = \tau(-) = integer \times integer \rightarrow integer.$$

In the Σ^2-algebra A^2, the set $A^2_{integer} = \mathbf{Z}$, and $+$ and $-$ are addition and subtraction of integers, respectively. Consider one of its possible SD-extensions, $^*A^2$. Let $^*A^2_{integer} = \mathbf{Z}^{Interval}$. Hence, following the definition of an interval SD-extension, we can easily obtain the well-known formulas of arithmetic operations of interval analysis:

$$[x^{Lo}, x^{Up}] + [y^{Lo}, y^{Up}] = [x^{Lo} + y^{Lo}, x^{Up} + y^{Up}],$$

$$[x^{Lo}, x^{Up}] - [y^{Lo}, y^{Up}] = [x^{Lo} - y^{Up}, x^{Up} - y^{Lo}].$$

5 Some Applications of the Framework

As we already noted above, a single object can have several SD-extensions differing in their expressive power and requirements for computing resources. Due to implementation restrictions, very often a problem cannot be specified by associating each object with the most powerful SD-extension. Sometimes, acceptable results can be obtained using an SD-model which is not the best one (in the sense of the power of SD-extensions).

For example, we can solve the system of two equations

$$x + y = 12;$$
$$2x = y;$$

in the weakest SD-extension *Interval*, obtaining a precise solution $x = 4$; $y = 8$. This example was studied in detail in [7].

It is obvious that we cannot obtain a precise solution in the SD-extension *Interval* if the problem has several solutions. Suppose, for example, that we must solve the following equation

$$sin(x) + sin(2x) + sin(3x) + sin(4x) + sin(5x) + sin(6x) = 0;$$

in the interval $-\pi/3 < x < \pi/3$.

If the variable x in the model is associated with the SD-extension *Interval*, then after computations we obtain the interval $x = [-1.04719, 1.0472]$ containing all solutions to the equation. If we pass to the SD-extension *Multiinterval*, then we will obtain the following five solutions:

$$x_1 = [-1.0471974, -1.0471976]; \quad x_2 = [-0.8975978, -0.8975980]; \quad x_3 = [0.0, 0.0];$$

$$x_4 = [0.8975978, 0.8975980]; \quad x_5 = [1.0471974, 1.0471976].$$

When solving problems in integer variables, we can also use the SD-extension *Enum*. Consider the following example:

$$x^2 + 3x + 2 = y;$$
$$2 - x = y;$$
$$y \geq 2;$$

Of course, in this setting we cannot assign to the integer variables x and y values in the SD-extension *Enum*, because the set of admissible values is too large. Therefore, we can do the following. First we solve this problem by using the SD-extension *Interval*. We obtain as a result $x = [-4, 0]$ and $y = [2, 6]$. Now we can pass either to *Multiinterval* or to *Enum*. In both cases we obtain the results $x_1 = 0$, $y_1 = 2$ and $x_2 = -4$, $y_2 = 6$.

Thus, we see that in the general case quality of the result depends on the SD-extension chosen for the specification of the corresponding SD-model. Since it is not always possible to specify a problem in terms of most powerful SD-extensions in the initial setting, we face the problem of choosing the most efficient representation of objects in the current state of the SD-model. Analysis of such states and the search of the most efficient transition from one SD-extension to another will be studied in the near future.

6 Conclusion

In the present paper we considered the apparatus of SD-models, which is a special kind of constraint programming. SD-models are traditionally used to solve problems which can be stated as finding all solutions of an existential conjunction in a sorted model analogous to the sorted model of Goguen. To describe the denotational semantics of this problem we specified an SD-extension of the sorted algebra corresponding to the model. Operational semantics of the process of finding the solutions was described in terms of generalized computational models.

We attempted to state the requriements for SD-extensions of values of variables and for SD-extensions of operations of the sorted algebra. We considered several versions of the most common SD-extensions and defined the relationship between them.

Since different SD-extensions can be associated with a single model, the problem of finding the most acceptable one arises. We believe that the most acceptable approach is that in which we first choose the SD-extensions that are the simplest for implementation. If the values of objects change when the computations in the corresponding SD-model are completed, then a transition to more complex and efficient SD-extensions may become possible.

A more detailed study of the problem of passing from one SD-extension to another, as well as the problem of automating this transition, are the research goals of the nearest future.

References

[1] Narin'yani A.S.: *Subdefinite Set - a Formal Model of Uncompletely Specified Aggregate*, Proc. of the Symp. on Fuzzy Sets and Possibility Theory, Acapulco, Mexico, (1980).

[2] Narin'yani A.S.: *Subdefinite Sets - New Data Type for Knowledge Representation*, Preprint USSR Acad. Sci., Siberian Division, Computer Center, **232**, Novosibirsk, (1980) (in Russian).

[3] Narin'yani A.S.: *Subdefiniteness, Overdefiniteness and Absurdity in Knowledge Representation (some Algebraic Aspects)*, Proc. of the II Conf. on AI Application, Miamy Beach, Dec. 9-13, (1985).

[4] Narin'yani A.S.: *Subdefinite Models: a Big Jump in Knowledge Proccesing Technology*, Proceeding of East-West AI Conf.: from theory to practice, Moscow, September, (1993).

[5] Babichev A.B., et al.: *UniCalc - an intelligent solver for mathematical problems*, Ibid, 257 – 260.

[6] Borde S.B., et al.: *Subdefiniteness and Calendar Scheduling*, Ibidem, 315 – 318.

[7] Telerman V.V.: *Propagation of Mathematical Constraints in Subdefinite Models*, In: J.Calmet and J.A.Campbell (Eds.), Integrating Symbolic Mathematical Computation and Artificial Intelligence, Lect. Notes in Comp. Sci., Vol. 958, Springer, (1995), 191 – 208.

[8] Mayoh B., Tyugu E., Uustalu T.: *Constraint Satisfaction and Constraint Programming: A Brief Lead-In*. Constraint Programming. - Springer-Verlag Berlin Heidelberg (1994), 1 – 16.

[9] Hentenryck P. van: *Constraint Satisfaction Using Constraint Logic Programming*, Artificial Intelligence, Vol. 58, (1992), 113 – 159.

[10] Benhamou F., Older W.J.: *Applying Interval Arithmetic to Real, Integer and Boolean Constraints*, Journal of Logic Programming, (1996). To appear.

[11] Goguen J.A., Meseguer J.: *Models and Equality for Logical Programming*, Lect. Notes in Comp. Sci., Vol. 250, Springer, (1987), p. 1 – 22.

[12] Telerman V.V.: *Active Data Types*, Preprint USSR Acad. Sci., Siberian Division, Computer Center, **792**, Novosibirsk, (1988), 30 p. (in Russian).

[13] Narin'yani A.S., Telerman V.V., Dmitriev V.E.: *Virtual Data-Flow Machine as Vehicle of Inference/Computations in Knowledge Bases*, In: Ph. Jorrand, V. Sgurev (Eds.) Artificial Intelligence II: Methodology, Systems, Application, North-Holland, (1987), 149 – 154.

[14] Telerman V.V., Ushakov D.M.: *Subdefinite Models: Formalisation and Perspectives*, In: Knowledge Processing Based on Subdefiniteness, RRIAI, Novosibirsk-Moscow, (1996), (in Russian).

[15] Alefeld G., Herzberger Ju.: *Introduction in Interval Computations*, Academic Press, New York, 1983.

[16] Hyvonen E.: *Constraint reasoning based on interval arithmetic: the tolerance propagation approach*, Artificial Intelligence, **58**, (1992), 71 – 112.

[17] Nechepurenko M.I.: *Elements of Boolean Interval Analisys*, In: System Simulation in Informatics, SS-11, Novosibirsk, (1985), 37 – 61. (In Russian).

[18] Yakovlev A.G.: *Computer Arithmetic of Multiintervals*, Problems of Cybernetics. Problem-oriented computational systems, (1987), 66 – 81. (In Russian).

[19] Telerman V.V.: *Using Multiintervals in Subdefinite Models*, In: Parallel programming and supercomputers: methods of knowledge representation in information technologies: Proc.of X All-Union Conf., Ufa, 19-26 June 1990 - Kiev, (1990), 128 – 129. (in Russian)

On Theorem-Proving in Horn Theories with Built-in Algebras

Nirina Andrianarivelo[1], Wadoud Bousdira[1], Jean-Marc Talbot[2]

[1] LIFO, Dépt. Informatique, Université d'Orléans,
Rue Léonard de Vinci, BP 6759, 45067 Orléans Cedex 02 (Fr.)
[2] LIFL, Université des Sciences et Technologies de Lille ,
Bâtiment M3, 59655 Villeneuve D'Ascq Cedex (Fr.)
e-mail: {andria, bousdira}@lifo.univ-orleans.fr, talbot@lifl.fr ***

Abstract. *We present a semi-decision procedure to prove ground theorems in Horn theories with built-in algebras. This is a maximal-unit-strategy based method, i.e in all our inference rules at least one of the premises clauses is an unit one. As in [4], constraint formalism is used as well; but more general specifications are studied. To limit the search space, an rpo-like ordering is used. Neither unification nor matching modulo the predefined algebra is needed. As a result, thanks to available constraint solvers on finite domains, naturals, integers, finite sets, ... our method is easy to implement and it is actually efficient to prove ground theorems.*

1 Introduction

Predefined algebras (natural numbers, real numbers, boolean values, ...) are available in all conventional programming languages. For a long time, one of the most important handicaps in rewriting had been the absence of such built-in structures. For example, to manipulate natural numbers, we had to specify them by equational theories. First, this is not natural at all; second a lot was lost in efficiency.

Over the last years, efforts have been developped to solve this problem. Among the most significant are Vorobyov's in 1989 [11] and Avenhaus and Becker's in 1994 [4].

Vorobyov has proved theorems in conditional specifications with built-in algebras but he has proposed very restrictive hypotheses. On one hand, restrictions on the specification are made: all its conditions have to be constraints in the predefined algebra and not any new function symbol defined in the conditional specification is of built-in sort; on the other hand, an algorithm to prove validity of all theorems in the predefined structure does exist.

Avenhaus and Becker do not have the above limitations. As far as we know, they are the first to have defined conditional rewriting modulo predefined algebras. They have to compute critical pairs between conditional rules and if all

---

*** Thanks to Klaus Becker for his fruitful comments on this work.

of them are *joinable* the studied system of rules is confluent. Test of ground confluence is the main purpose of their work. They need unification and match algorithm modulo the predefined algebra. As a result, problems of decidability may be encountered.

Our approach although it is quite close to Avenhaus and Becker's, has the following main characteristics:

- Unlike them, instead of testing the confluence of sets of Horn clauses, we only have in mind to prove ground theorems in conditional specifications with built-in algebras.
- We only rewrite by equations. Thus we need not find an ordering such that the heads of conditional rules are bigger than their tails.
- We compute only critical pairs between equations. This causes the number of deduced consequences to decrease tremendously.
- In all our inference rules, at least one of the used clauses is an equation. As a result, the search space is reduced tremendously.
- Furthermore, we do not use either unification or match algorithms *modulo the built-in algebra*. All we need is unification in hierarchical specifications and constraints solvers in the predefined algebras (regarded as *black boxes* in this paper).

Our method is an extension of the maximal-unit-strategy [6, 1]. As in [2], the formalism of constraints is used. Let us deal with a simple example on *stacks of naturals* as an illustration of our method.

Example 1. Consider the following specification:

$$sum(empty\_stack) = 0 \ . \tag{1}$$

$$y < 0 \land p = push(y, p') \Rightarrow sum(p) = sum(p') - y \ . \tag{2}$$

$$y \geq 0 \land p = push(y, p') \Rightarrow sum(p) = sum(p') + y \ . \tag{3}$$

Here, p, p' are variables of the type *stack*, and x, y variables on *natural numbers*. $sum(p)$ denotes the sum of the absolute values of numbers in p. Let us prove that:

$$sum(push(2, push(-4, empty\_stack))) = 6 \ . \tag{0}$$

The main steps of the proof are the following:

1. We deny the equation to be proved:

$$sum(push(2, push(-4, empty\_stack))) \neq 6 \ . \tag{0}$$

2. By *I-specialization* of (3) we obtain:

$$y \geq 0 \Rightarrow sum(push(y, p')) = sum(p') + y \ . \tag{4}$$

3. Simplifying (0) by (4) provides:

$$sum(push(-4, empty\_stack)) + 2 \neq 6 \ . \tag{5}$$

4. By *I-specialization* of (2) we obtain:

$$y < 0 \Rightarrow sum(push(y, p')) = sum(p') - y \ . \tag{6}$$

5. Simplifying (5) by (6) gives:

$$sum(empty\_stack) + 6 \neq 6 \ . \tag{7}$$

6. Simplifying (7) by (1) gives:

$$6 \neq 6, REFUTATION \ \bullet \tag{8}$$

This paper is organized as follows: in Section 2 we give precisions on our notations and give definitions, in Section 3 the semantical aspect of the problem is dealt with, in Section 4 we give the inference rules of our method. Correctness and completeness problems are dealt with in Section 5. A few words on the strategy of the implementation are said in Section 6, and in the last Section we speak about our conclusions and perspectives.

2 Syntax and framework

2.1 Hierarchical signatures

Let us precise all notations here developped are the usual ones.

Before introducing definitions, let us state our motivation shortly. Our proving process works in (user)-Horn clauses specifications based on predefined algebras. Therefore, two kinds of objects are used, built-in objects that are described in a built-in language given by a signature Σ_0. The second kind of objects -syntactic objects- are introduced by a signature extension - noted as usual $\Sigma_0 + \Sigma_1$ -, captured by the Horn clauses specification.

Definition 1. A signature $\Sigma = (S, F, D)$ is a set of sorts S, a set of function symbols F and a set D that includes function declarations $f : \langle \omega, s \rangle$ where $f \in F$, $\omega \in S^*$, $s \in S$ and sub-sort declarations $s_1 \lhd s_2$ ($s_1, s_2 \in S$) where $\lhd$ denotes a strict ordering relation between sorts. A signature $\Sigma = (S, F, D)$ is said to be flat iff D does not contain any sub-sort declaration and if for all $f \in F$ there is a unique function declaration of f in D

Definition 2. A signature extension $\Sigma_0 + \Sigma_1$ is a flat signature where $\Sigma_0 = (S_0, F_0, D_0)$ and $\Sigma_1 = (S_1, F_1, D_1)$ such that $S_0 \cap S_1 = \emptyset$, $F_0 \cap F_1 = \emptyset$

Let us note $F_0^{(=0)}$ the set of constants - functions with 0 arity - in F_0 and $F_0^{(\geq 1)}$ all the other symbols of F_0.

Definition 3. Let $\Sigma_0 + \Sigma_1$ be a flat signature extension. $\Sigma = (S, F, D)$ is said to be the hierarchical signature with respect to the sorted signature induced by Σ_0 and Σ_1, written $\Sigma = \Sigma_0 \oplus \Sigma_1$ iff

- $S = S_0 \cup S_1 \cup S^\wedge$ where $S^\wedge = \{s^\wedge \mid s \in S_0\}$

- $F = F_0 \cup F_1$
- $D = D_0 \cup D_{S_1}^\wedge \cup D_{S_0}^\wedge \cup D_{sort}$ where

  $D_{S_1}^\wedge = \{f : \langle \omega^\wedge, s \rangle \mid f \in F_1, f : \langle \omega, s \rangle \in (D_0 \cup D_1), \omega \in (S_0 \cup S_1)^* $ and $s \in S_1\}$

  where $\epsilon^\wedge = \epsilon, (s.\omega)^\wedge = s.\omega^\wedge$ if $s \in S_1$ and $(s.\omega)^\wedge = s^\wedge.\omega^\wedge$ if $s \in S_0$

- $D_{S_0}^\wedge = \{f : \langle \omega^\wedge, s^\wedge \rangle \mid f \in (F_0^{(\geq 1)} \cup F_1), f : \langle \omega, s \rangle \in (D_0 \cup D_1), \omega \in (S_0 \cup S_1)^+ $ and $s \in S_0\}$

  where $\omega^\wedge$ is defined as above.

- $D_{sort} = \{s \lhd s^\wedge \mid s \in S_0\}$.

As in [4], the elements of S_0 are called sorts of *low* type and the elements of $S^\wedge \cup S_1$ sorts of *high* type. Note that the function symbols from the built-in language with an arity greater or equal than 1 are declared twice, a declaration for built-ins and an additional one for "mixed" objects.

Sorts in $S^\wedge$ are called *junk sorts*. They are necessary to define "properly" *mixed terms* as for example the *term*: $1 + gcd(4,2)$, where $+$ stands for the addition in natural numbers and gcd is a function in F_1 : $gcd(x,y)$ denotes the greatest common divisor of x and y. $+$ being defined on naturals, it can not be applied to the term $1 + gcd(4,2)$ since $gcd(4,2)$ is a "term" in the high hierarchy - a user's term. So, we create another declaration for the operator $+$. That is $+ : \langle nat^\wedge, nat^\wedge, nat^\wedge \rangle$, that ensures that $1 + gcd(4,2)$ is a correct term - a well formed one. Note that this new declaration gives to $+$ a syntactical connotation; whereas on natural numbers, $+$ has a semantical meaning.

Example 2. We first define suitable built-in signatures to describe booleans and natural numbers. Let $\Sigma_0 = (S_0, F_0, D_0)$ with

$S_0 = \{bool, nat\}$

$F_0 = \{true, false, not, and, or, +, *, 0, 1, 2, \ldots\}$

$D_0 = \{true : \langle \epsilon, bool \rangle, false : \langle \epsilon, bool \rangle, not : \langle bool, bool \rangle,$
    $and : \langle bool, bool, bool \rangle, or : \langle bool, bool, bool \rangle,$
    $+ : \langle nat, nat, nat \rangle, * : \langle nat, nat, nat \rangle, 0 : \langle \epsilon, nat \rangle, 1 : \langle \epsilon, nat \rangle, \ldots\}$

then we define the high hierarchy signature $\Sigma_1 = (S_1, F_1, D_1)$:

$S_1 = \{stack\}$

$F_1 = \{empty\_stack, push, top, empty?\}$

$D_1 = \{empty\_stack : \langle \epsilon, stack \rangle, push : \langle nat, stack, stack \rangle,$
    $top : \langle stack, nat \rangle, empty? : \langle stack, bool \rangle\}$

To obtain $\Sigma = \Sigma_0 \oplus \Sigma_1$, let the following:

$S = S_0 \cup S_1 \cup \{nat^\wedge, bool^\wedge\}$

$F = F_0 \cup F_1$

$D = D_0 \cup D_{S_1}^\wedge \cup D_{S_0}^\wedge \cup D_{sort}$ where

$\bullet D_{S_1}^\wedge = \{empty\_stack : \langle \epsilon, stack \rangle, push : \langle nat^\wedge, stack, stack \rangle\}$

$\bullet D_{S_0}^\wedge = \{not : \langle bool^\wedge, bool^\wedge \rangle, and : \langle bool^\wedge, bool^\wedge, bool^\wedge \rangle,$
    $or : \langle bool^\wedge, bool^\wedge, bool^\wedge \rangle, + : \langle nat^\wedge, nat^\wedge, nat^\wedge \rangle,$
    $* : \langle nat^\wedge, nat^\wedge, nat^\wedge \rangle, top : \langle stack, nat^\wedge \rangle, empty? : \langle stack, bool^\wedge \rangle\}$

$\bullet D_{sort} = \{nat \lhd nat^\wedge, bool \lhd bool^\wedge\}$.

For the rest of this paper, let $\Sigma = \Sigma_0 \oplus \Sigma_1$. Note $\mathcal{V}_s$ the set of variables of sort s. $\mathcal{V}$ is defined as $\mathcal{V}_0 \cup \mathcal{V}_1$ where $\mathcal{V}_0 = \cup_{s \in S_0} \mathcal{V}_s$ and $\mathcal{V}_1 = \cup_{s \in S_1} \mathcal{V}_s$. Sets of variables are supposed to be countably infinite as usual.

In all the following, we suppose that each sort in S is non empty, so there exists at least one constant $c \in F$ such that $c : \rightarrow s, \forall s \in S$.

We can then define terms as follows:

Definition 4. The set $T_s(F, \mathcal{V})$ of Σ-terms of sort s is inductively defined as:

- if $x \in \mathcal{V}_s$ and $s \trianglelefteq s' \in D$ then $x \in T_{s'}(F, \mathcal{V})$
- if $f : \rightarrow s \in D$ and $s \trianglelefteq s'$ then $f \in T_{s'}(F, \mathcal{V})$
- if $f : \langle s_1, \ldots, s_n, s \rangle \in D$, $s \trianglelefteq s'$ and $t_i \in T_{s_i}(F, \mathcal{V}), i = 1, \ldots, n$ then $f(t_1, \ldots, t_n) \in T_{s'}(F, \mathcal{V})$.

Let $T(F, \mathcal{V}) = \cup_{s \in S} T_s(F, \mathcal{V})$. Two terms s and t are said to be sort compatible if there is a sort s in S such that s and t are terms of sort s.

We can distinguish the built-in terms - the ones only formed by function symbols from F_0 then whose sort is in S_0 -, from the user's terms - the ones that are in the high level. The latter contain at least one function symbol from F_1 while their sort may be in S_1 or $S^\wedge$.

$T(F_0)$ denotes the set of all ground terms containing only F_0-symbols, $T_0(F, \mathcal{V})$ the set of all terms made only with F_0 and $\mathcal{V}_0$, $T^\wedge(F, \mathcal{V})$ the set of all user's terms - terms containing at least one F_1-symbol, possibly mixed terms, $T^\wedge(F)$ the set af all ground user's terms, and $T(F)$ the set of all ground terms in $T(F, \mathcal{V})$.

A kind of hierarchy appears between built-in terms and user's terms. User's terms are said to be greater than the built-in ones. As we will see later, this hierarchy will be used to define orderings to compare terms.

As usual, the set of variables of a term t is noted $\mathcal{V}(t)$. We define a substitution σ as an assignment from $\mathcal{V}_1$ into $T^\wedge(F, \mathcal{V})$ such that $sort(\sigma x) = sort(x)$ and $\{\sigma(x) \neq x\}$ is finite.

2.2 Constraints

Our Horn clauses contain constraints - the notation of which are the same as in [10, 8]. We suppose readers are familiar with such definitions and notations. The notions in these papers are more general than ours, thus let us precise the particular form of constraints here developped.

We note $\overline{\Sigma} = (\Sigma, P)$ a first-order signature - defined by a many sorted signature $\Sigma = (S, F)$ and a set P of predicate symbols - and $\mathcal{F}(\overline{\Sigma}, \mathcal{V})$ the set of all formulæ built with $\overline{\Sigma}$ and the set $\mathcal{V}$ of variables. In this paper, a constraint C is the form of $s_1 = t_1 \wedge \ldots \wedge s_n = t_n$. Let us precise, when $s = t$ is in a constraint, $s = t$ may be either an equation in the predefined algebra - when the both terms are predefined ones - or an unification equation otherwise. A constraint C is satisfiable if there is one interpretation in which C has a solution. As usual, one kind of constraints to be solved is unification. Here, because of hierarchy, usual algorithms of unification can not be used. They need to be adapted. We will propose our inference system to unify both terms of an equation in a constraint. *fail* stands for a failure to find a substitution σ for an equation $s = t$ such that $\sigma s = \sigma t$. As in [8], $Sol_{I_A}(\mathcal{C})$ denotes the set of all the valuations of $\mathcal{C}$ in the interpretation $\mathcal{I}_A$.

For the sake of convenience, we note $\mathcal{C}$ as $\mathcal{Q} \wedge s = t$

− *Delete*

$$\frac{\mathcal{Q} \wedge s = s}{\mathcal{Q}}$$

− *Decompose*

$$\frac{\mathcal{Q} \wedge f(s_1, \ldots, s_n) = f(t_1, \ldots, t_n)}{\mathcal{Q} \wedge s_1 = t_1 \wedge \ldots \wedge s_n = t_n}$$

if $f(s_1, \ldots, s_n), f(t_1, \ldots, t_n) \in T^{\wedge}(F, V)$ and $f \in F_1$ and both terms have the same sort.

− *Instance*

$$\frac{\mathcal{Q} \wedge x = s}{\mathcal{Q}[x \leftarrow s]}$$

if $sort(x) = sort(s)$, $x \in V_1(\mathcal{Q})$, $x \notin V_1(s)$. Here $\mathcal{Q}[x \leftarrow s]$ stands for the constraint in which every instance of x is replaced by s.

− *Conflict*

$$\frac{\mathcal{Q} \wedge f(s_1, \ldots, s_n) = g(t_1, \ldots, t_m)}{fail}$$

if $f(s_1, \ldots, s_n), g(t_1, \ldots, t_m) \in T^{\wedge}(F, V)$ and $f, g \in F_1$ and both terms have the same sort.

− *Sort fail*

$$\frac{\mathcal{Q} \wedge s = t}{fail}$$

if $sort(s) \neq sort(t)$

− *Hierarchy fail*

$$\frac{\mathcal{Q} \wedge s = t}{fail}$$

if s (resp. t) $\in T_0(F, V)$ and t (resp. s) $\in T^{\wedge}(F, V)$

− *Occur check*

$$\frac{\mathcal{Q} \wedge x = s}{fail}$$

if $sort(x) = sort(s) \in S_1$, $x \in V(s)$, $s \notin V_1$

− *A-unsatisfiability*

$$\frac{\mathcal{Q} \wedge s = t}{fail}$$

if $s, t \in T_0(F, V)$ and $sort(s) = sort(t) = sort$ and $\mathcal{Q} \wedge s =_{sort} t$ is unsatisfiable.

Now let us define Horn clauses specifications with built-in algebra. The built-in objects are described by a Σ_0-algebra $\mathcal{A}$. Terms in $\mathcal{F}(\overline{\Sigma}_0, \mathcal{V})$ are constrained formulæ of the built-in algebra $\mathcal{A}$.

Definition 5. Let $\overline{\Sigma} = (\Sigma, \mathcal{P})$, $\Sigma = \Sigma_0 \oplus \Sigma_1$ be a first order signature. A Horn clause is a conditional equation defined over $\overline{\Sigma}$, of form $\mathcal{C} \wedge s_1 = t_1 \wedge \ldots \wedge s_n = t_n \Rightarrow A = B$ where the constraint $\mathcal{C}$ is a formula in $\mathcal{F}(\overline{\Sigma}_0, \mathcal{V})$ and $s_i, t_i, (i = 1 \ldots n), A, B \in T(F, \mathcal{V})$. Furthermore, $\forall s_i = t_i$, $s_i \in T^{\wedge}(F, \mathcal{V})$ or $t_i \in T^{\wedge}(F, \mathcal{V})$, $i = 1, .., n$. This condition is of $A = B$ too.

We have introduced here another "=" - induced by the user's specification - than this of $\mathcal{A}$, and the syntactic equality on terms. To avoid cumbersome notation, we use the same notation for all the three types of "=". The condition that A or B at least has to contain a new symbol is quite natural. Indeed, if we interpret built-in function symbols as constructors and if we consider all the built-in structure as an "imported" black box, this only allows to define user's axioms and then, axioms where at least a symbol from F_1 does appear.

A Horn clause is also written $\mathcal{C} \wedge \mathcal{L} \Rightarrow A = B$ where the literal $\mathcal{L}$ stands for the conjunction of equalities $s_i = t_i$ of definition 5.

By convention, we call equation a Horn clause without conditional part, i.e. a Horn clause of form $\mathcal{C} \Rightarrow A = B$ - the $\mathcal{L}$ part does not exist.

At least one member of the ground equational theorem to prove has to be in the high level hierarchy, i.e. theorem of form $\mathcal{C} \Rightarrow A = B$, such that i) at least A or B is in $T^{\wedge}(F, \mathcal{V})$, ii) all the variables of A and B are in the low level hierarchy, and iii) for all interpretation $\mathcal{I}_A$ there exists only one valuation ν such that $\nu \in Sol_{I_A}(\mathcal{C})$.

Let us give some examples of ground equations which could be proved.

Example 3. $gcd(4, 2) = 2$.
$x + y = 6 \wedge x - y = 2 \Rightarrow gcd(x, y) = y$.

While $xy = 8 \wedge x + y = 6 \Rightarrow A(x, y) = B(x)$, where $A, B \in F_1$, is not a ground equation. Its constraint has two solutions $(x = 4, y = 2)$ or $(x = 2, y = 4)$.

As usual, during proof process we use an ordering to minimize our search space. Our approach of ordering is as follows.

First, we suppose given a total precedence $<_{F_1}$ on function symbols of the high hierarchy i.e. on F_1, and we extend it to the whole set of function symbols F as follows:

- if $f, g \in F_1$ then $f <_F g \Leftrightarrow f <_{F_1} g$
- if $f \in F_0$ and $g \in F_1$ then $f <_F g$

Then, a complete *rpo-like* ordering [5] is built on the set of ground terms $T(F)$. Hereunder is its definition.

Definition 6. Let $<_F$ be as defined above, then ground terms in $\mathcal{F}(\overline{\Sigma}) \cup T(F)$ are compared as follows:

$$s = f(s_1, \ldots, s_n) \prec g(t_1, \ldots, t_m) = t \text{ if}$$

- either $f <_F g$ and $\forall i = 1, \ldots n,\ s_i \prec t$
- or $\exists i \in 1 \ldots m,\ s \prec t_i$ or $s = t_i$
- or $f = g$, $(s_1, \ldots, s_n) \prec_s (t_1, \ldots, t_n)$ and $\forall i = 1, \ldots n,\ s_i \prec t$

where $\prec_s$ is the extension obtained via the status s (multiset or lexicographic).

At last, to compare two equations $s = t$ and $u = v$, we use the multiset extension of $\prec$ ($s = t \prec u = v$ iff $\{s, t\} \prec\!\!\prec \{u, v\}$).

Remark. The definition of the ordering can lead to compare built-in terms. In this case, these terms are compared using the built-in ordering of $\mathcal{A}$ (keep in mind that $\mathcal{A}$ is regarded as a black box where we assume that everything - especially orderings - is given in advance). Before comparing users' terms, built-in terms they contain are always "$\mathcal{A}$-normalized", par exemple $f(1 + 0)$ is normalized in $f(1)$. Incidentally, a complete simplification ordering is built on $T(F)$.

For a Horn clause $C \wedge \mathcal{L} \Rightarrow A = B$, we say indifferently that $C \wedge \mathcal{L} \Rightarrow A = B$ is a conditional rule or a conditional equation.

3   Semantics

Our aim in this paper is to prove ground theorems automatically from specifications defined over built-in structures; namely to prove that a ground equation is logically deduced from a set of axioms. Keep in mind that we are only interested in proving *user's* ground theorems. Then we develop an operational process that is correct and complete with respect to the usual notion of models.

In this section, for the sake of readability, a ground equation is noted A = B.

From now on, let us note $(\overline{\Sigma}, \mathbf{H})$ a Horn clause specification where $\overline{\Sigma} = (\Sigma, \mathcal{P})$ is as defined above and $\mathbf{H}$ a set of Horn clauses. Note that $\overline{\Sigma}$ contains a built-in structure and then, the algebras of interest contain a built-in algebra $\mathcal{A}$ which is the interpretation of the Σ_0-objects, i.e. objects of the low hierarchy level of the signature $\Sigma (= \Sigma_0 \oplus \Sigma_1)$. Furthermore, these algebras have to satisfy the conditional equations from $\mathbf{H}$. Obviously, the models that capture the model-theoretical meaning of the Horn clause specification are the $\overline{\Sigma}$-algebras that are models of the whole specification (i.e. the built-in specification completed with $\mathbf{H}$). We assume that such models exist and they are consistent w.r.t the built-in algebra $\mathcal{A}$.

We say that an equation $A = B$ is valid in an algebra $\mathcal{B}$ iff $\mathcal{B} \models A = B$ ($\models$ stands for the usual notion of validity in the models). A set of Horn clauses $\mathbf{H}$ is valid in $\mathcal{B}$ iff every conditional equation in $\mathbf{H}$ is valid in $\mathcal{B}$. Then $\mathcal{B}$ is said to be a model of $\mathbf{H}$.

To take into account the built-in algebra $\mathcal{A}$, let $\mathbf{H}_\mathcal{A} = \{u = v \mid \mathcal{A} \models u = v;\ u, v \in T_0(F, \mathcal{V})\}$ be the set of F_0-equations induced by $\mathcal{A}$. Note $\sim_\mathcal{A}$ the congruence relation on $T(F_0)$. Let us extend it to $T(F)$ as follows:

Definition 7. Let $s,\ t \in T(F)$, $s \sim_\mathcal{A} t$ iff

- s, $t \in T(F_0)$ and $\mathcal{A} \models s = t$ (then $s = t \in \mathbf{H}_{\mathcal{A}}$)
- s, $t \in T^{\wedge}(F)$, s is of the form $f(s_1, \ldots, s_n)$ and t is of the form $f(t_1, \ldots, t_n)$ then $\forall i = 1, \ldots, n$, $s_i \sim_{\mathcal{A}} t_i$.

According to the idea that built-in equivalences are given in advance by $\mathcal{A}$, the inductive definition starts with $\sim_{\mathcal{A}}$:

Definition 8. Let A, B be two ground terms in $T(F)$, $=_{op(\mathbf{H})}$ is inductively defined as follows:

- $A =^0_{op(\mathbf{H})} B$ iff $A \sim_{\mathcal{A}} B$
- $A =^{i+1}_{op(\mathbf{H})} B$ iff $A =^i_{op(\mathbf{H})} B$ or there exists a conditional equation $C \wedge u_1 = v_1 \wedge \ldots \wedge u_n = v_n \Rightarrow u = v \in \mathbf{H}$, an occurrence p of A such that for all valuation ν in $Sol_{I_{\mathcal{A}}}(C)$, for all ground substitution σ, $\sigma \nu u_j =^i_{op(\mathbf{H})} \sigma \nu v_j$, $\forall j = 1, \ldots, n$ and $A_{/p} = \sigma \nu u$ and $B = A[p \leftarrow \sigma \nu v]$ or $A_{/p} = \sigma \nu v$ and $B = A[p \leftarrow \sigma \nu u]$.
  A and B play a symetrical role.

$=_{op(\mathbf{H})}$ is a congruence relation on $T(F)$. Let us note $\mathrm{Th}_g(\mathbf{H})$ the set $\{A = B | \exists i$ such that $A =^i_{op(\mathbf{H})} B\}$.

Finally in the following, like in Birkhoff's theorem, we prove that the operational validity is correct and complete with respect to the model semantics.

Theorem 9. Let $(\overline{\Sigma}, \mathbf{H})$ be a Horn clause specification and let $A = B$ be a ground equation then,

$$\mathbf{H} \models A = B \text{ iff } A =_{op(\mathbf{H})} B.$$

Proof. We need only to prove that $\mathbf{H} \models A = B \Rightarrow A =_{op(\mathbf{H})} B$. We only give the principal steps of the proof as it is similar to Becker's in [3].

Let $\mathbf{H} \models A = B$ and let $T_{(\mathbf{H}, \mathcal{A})}$ be the initial term algebra with carrier $T(\overline{\Sigma})_{/\sim_{\mathbf{H}, \mathcal{A}}}$ defined as usual. It is sufficient to prove that $T_{(\mathbf{H}, \mathcal{A})}$ is a model of $\mathbf{H} \cup \mathbf{H}_{\mathcal{A}}$.

If $u = v \in \mathbf{H}_{\mathcal{A}}$. In order to prove that $T_{(\mathbf{H}, \mathcal{A})} \models u = v$ we have to show that $\sigma u =_{op(\mathbf{H})} \sigma v$ for any ground substitution σ. This is trivial since $\sigma u = \sigma v \in \mathbf{H}_{\mathcal{A}}$, thus $\sigma u =^0_{op(\mathbf{H})} \sigma v$ and then $\sigma u =_{op(\mathbf{H})} \sigma v$.

Otherwise, let a conditional equation $C \wedge u_1 = v_1 \wedge \ldots \wedge u_n = v_n \Rightarrow u = v \in \mathbf{H}$ and let ν be a valuation that satisfies the constraint C and let a ground substitution τ such that $T_{(\mathbf{H}, \mathcal{A})} \models \tau \nu u_i = \tau \nu v_i$ for $i = 1, \ldots, n$. Then $\forall i = 1, \ldots, n$, $\tau \nu u_i =_{op(\mathbf{H})} \tau \nu v_i$. For an appropriate k, we have $\tau \nu u_i =^k_{op(\mathbf{H})} \tau \nu v_i$, $i = 1, \ldots, n$. We get $\tau \nu u =^{k+1}_{op(\mathbf{H})} \tau \nu v$ by definition of $=_{op(\mathbf{H})}$. Let $\sigma = \tau \nu$ then $\sigma u =_{op(\mathbf{H})} \sigma v$. $\square$

Now, since we have an operational proof process in Horn clause specifications, here is a formulation of this process by means of proof trees. This formulation has become classical and one can find such approaches in [7]. Let us give a formal definition of a ground proof.

Let $\mathbf{H}_E$ be the subset of $\mathbf{H}$ containing only the equations of $\mathbf{H}$ and let A, B be two ground terms in $T(F)$ such that at least A or B is in $T^\wedge(F)$;

Definition 10. A ground proof of $A == B$ in $\mathbf{H}$ is a $\mathbf{H}_E$-string of finite length from A to B. When the length of the string is 1, we say that it is an elementary proof.

Formally, $A == B$ is an elementary proof if there is an equation $C \Rightarrow g = d \in \mathbf{H}$, a valuation ν and a substitution σ such that $\nu \in Sol_{I_A}(C)$, $A = A[\sigma\nu g] == A[\sigma\nu d] = B$. An elementary proof is said to be a proof tree of height 0.

Let us define non-elementary proof trees:

Definition 11. A proof tree $\prod$ of $A == B$ in $\mathbf{H}$ is a structure of form:

$$\frac{\{\{\prod_{11} \prod_{12} \cdots \prod_{1r_1}\} \cdots \{\prod_{n1} \prod_{n2} \cdots \prod_{nr_n}\}\}}{A = s_0 == s_1 == \ldots == s_{n-1} == s_n = B}$$

such that, for each j, $1 \leq j \leq n$:
- s_j is a ground term,
  - there exists a conditional rule in $\mathbf{H}$ of the form

$$C_j \wedge (t_{j1} = t'_{j1}) \wedge \ldots \wedge (t_{jr_j} = t'_{jr_j}) \Rightarrow g_j = d_j$$

- an occurrence p_j of s_{j-1}, a substitution σ_j and a valuation ν_j on the variables of C_j such that
$s_{j-1}{}_{/p_j} = \sigma_j\nu_j g_j$, and $s_{j-1}[p_j \leftarrow \sigma_j\nu_j d_j] = s_j$,
- and $\nu_j \in Sol_{I_A}(C_j)$
- and for each k, $1 \leq k \leq r_j$, $\prod_{jk}$ is a proof tree of $\sigma_j\nu_j t_{jk} == \sigma_j\nu_j t'_{jk}$ in $\mathbf{H}$.
(We say that $s_{j-1} == s_j$ is a conditional step in $\prod$).

If $\prod$ is a proof tree of $A == B$ in $\mathbf{H}$, then we say that $\prod$ is a $\mathbf{H}$-proof tree or a $\mathbf{H}$-proof of $A == B$.

We note by $=$ the equational proof steps and by $==$ the conditional ones.

Example 4. Let us consider the following Horn set:

$$f(x,y) = g(x) \wedge \phi(x,y) = \psi(y) \Rightarrow A(x,y) = B(x,y) \ . \tag{9}$$

$$h(x) = k(y) \Rightarrow f(x,y) = g(x) \ . \tag{10}$$

$$\lambda(x) = \gamma(y) \Rightarrow \phi(x,y) = \psi(y) \ . \tag{11}$$

$$a = b \ . \tag{12}$$

$$h(a) = k(a) \ . \tag{13}$$

$$\lambda(a) = \gamma(b) \ . \tag{14}$$

Note that in this first example there is no constraints. As a result all function symbols are in F_1.
A proof tree of $A(a,b) == B(a,a)$ is:

$$\frac{h(a) =_{13} k(a) \qquad \lambda(a) =_{14} \gamma(b) =_{12} \gamma(a)}{\dfrac{f(a,a) ==_{10} g(a) \qquad \phi(a,a) ==_{11} \psi(a)}{A(a,b) =_{12} A(a,a) ==_9 B(a,a)}}$$

Example 5. Let us consider the following Horn set based on the built-in natural numbers specification (x, y are naturals, and all others function symbols are F_1-symbols):

$$x > 1 \;\Rightarrow\; a(x) = b(x) \;. \tag{15}$$

$$x > 1 \;\Rightarrow\; a(x) = c(x) \;. \tag{16}$$

$$x > 0 \wedge b(x) = c(x) \;\Rightarrow\; e(x) = d(x) \;. \tag{17}$$

$$x > 0 \wedge f(x) = g(x) \;\Rightarrow\; e(x) = h(x) \;. \tag{18}$$

$$x > 1 \;\Rightarrow\; a'(x) = f(x) \;. \tag{19}$$

$$x > 0 \;\Rightarrow\; a'(x) = g(x) \;. \tag{20}$$

A proof tree of $d(2) == h(2)$ is:

$$\frac{\dfrac{2>1 \quad 2>1}{2>0 \wedge b(2) =_{15} a(2) =_{16} c(2)} \qquad \dfrac{2>1 \quad 2>0}{2>0 \wedge f(2) =_{19} a'(2) =_{20} g(2)}}{\dfrac{d(2) ==_{17} e(2) \qquad ==_{18} h(2)}{}}$$

Now let us state one of the most important theorems of this work, which means that to each $A =_{op(\mathbf{H})} B$ is associated a proof tree where both extremity terms at its root are A and B.

Theorem 12. *Let $(\overline{\Sigma}, \mathbf{H})$ be a Horn clause specification and let $A = B$ be a ground equation then, $A =_{op(\mathbf{H})} B$ iff there exists a proof tree $\prod$ such that $A == B$ is the root of $\prod$.*

Proof. for '$\Rightarrow$', the proof is by induction on i as $=_{op(\mathbf{H})}$ is defined by induction. We do not consider the case '$i=0$' since if $A =^0_{op(\mathbf{H})} B$ then $A \sim_A B$ so A, B are in the built-in structure whereas we impose that the theorem has to be in the high hierarchy. If $A =^1_{op(\mathbf{H})} B$ then there is an equation $C \Rightarrow u = v \in \mathbf{H}_E$, σ, ν such that $\nu \in Sol_{I_A}(C)$ and $A = A[\sigma\nu u]$, $B = A[\sigma\nu v]$. Therefore, $A == B$ is an elementary proof said to be a proof tree of height 0. Else, let us suppose that for each $k \leq j$, if $A =^k_{op(\mathbf{H})} B$ then there is a proof tree whose root is $A == B$. If $A =^{j+1}_{op(\mathbf{H})} B$ then the case when $A =^j_{op(\mathbf{H})} B$ is trivial. Let $A =^{j+1}_{op(\mathbf{H})} B$ using a conditional equation $C \wedge \bigwedge_{i=1}^n u_i = v_i \Rightarrow u = v$, since $\sigma\nu u_i =_{op(\mathbf{H})} \sigma\nu v_i, \forall i$ then $\forall i = 1, \ldots, n$ there is a proof tree $\prod_i$ such that $\sigma\nu u_i == \sigma\nu v_i$ is the root of $\prod_i$. Therefore, we can build $\prod$ of the form $\dfrac{\prod_1, \ldots, \prod_n}{A == B}$.
For '$\Leftarrow$', first suppose that $A == B$ has a proof tree with an height equal to 1. Then $A == B$ is an elementary proof such that there is an equation $C \Rightarrow g = d$, a substitution σ and a valuation ν such that νC is satisfiable and $A = A[\sigma\nu g]$, $B = A[\sigma\nu d]$. Then $A =^1_{op(\mathbf{H})} B$. Else suppose that the proof tree of $A == B$ has an height equal to $k(> 1)$ and suppose that for each ground equality $s == t$ such

that there exists a proof tree of height $< k$, then $s =_{op(\mathbf{H})} t$. By definition of proof trees, the proof tree of each $\sigma_j \nu_j t_{jk} == \sigma_j \nu_j t'_{jk}, \forall j$ is of height $< k$, then $\forall j$, for an appropriate i, $\sigma_j \nu_j t_{jk} =^i_{op(\mathbf{H})} \sigma_j \nu_j t'_{jk}$. By definition, $\forall j$, $s_{j-1} =^{i+1}_{op(\mathbf{H})} s_j$ and then $s_{j-1} =_{op(\mathbf{H})} s_j$. Therefore, $A = s_0 =_{op(\mathbf{H})} s_1 =_{op\mathbf{H})} \cdots =_{op(\mathbf{H})} s_n = B$.

$\square$

4   The Inference System

From now on, we suppose that from a semantical point of view, the Horn clause specification is correctly defined. Thus double definitions of symbols and junk sorts are declared when necessary. For the operational point of view, this is not essential, and we only make distinction of built-in terms, in $T_0(F, \mathcal{V})$, from user's terms, in $T^\wedge(F, \mathcal{V})$. From the operational point of view, we assume all users' terms are always "$\mathcal{A}$-normalized", i.e all their built-in subterms are "$\mathcal{A}$-normalized". For the following, L denotes a conjunction (possibly empty) of equations in $T^\wedge(F, \mathcal{V})$; L can be viewed as a list of equations as well. In our approach we need computing rules more particularly via conditional-narrowing.

Definition 13 Conditional-narrowing. If (1) $C \wedge s[u] = t \wedge L \Rightarrow A = B$ is a conditional rule, u a non-variable subterm of s in the high hierarchy, and (2) $C' \Rightarrow v = w$ an equation, then $C \wedge C' \wedge C'' \wedge u = v \wedge s[w] = t \wedge L \Rightarrow A = B$ is a conditional-narrowing consequence from (1) and (2). C'' is $s = t \not\prec a = b$, $\forall a = b \in L$ and $C \wedge C' \wedge C'' \wedge u = v$ is satisfiable.

Example 6. Consider again our declaration of the type *stacks* in Example 1. Suppose on F_1 the precedence : $empty\_stack \prec push \prec sum$. Consider the two following rules.

$$sum(empty\_stack) = 0 \ . \tag{21}$$

$$x \geq 0 \wedge \ p = push(x, p') \wedge sum(p') = y \Rightarrow sum(p) = x + y \tag{22}$$

By conditional-narrowing, after solving constraints, we obtain :

$$x \geq 0 \wedge p = push(x, empty\_stack) \Rightarrow sum(p) = x \ . \tag{23}$$

We need to compute critical pairs between equations as well.

Definition 14 Critical pair. Let $C \Rightarrow s[u] = t$ and $C' \Rightarrow v = w$ be two equations, such that u is a non-variable subterm of s in the high hierarchy. Then $C'' \wedge C \wedge C' \wedge u = v \Rightarrow s[w] = t$ is a critical pair from the firsts. C'' is $s \not\prec t \wedge v \not\prec w$, and $C'' \wedge C \wedge C' \wedge u = v$ is satisfiable.

Example 7. On *stacks*, with the same precedence as above, consider the two equations :
(i) $sum(push(x, empty\_stack)) = x$ and
(ii) $y \geq 0 \Rightarrow sum(push(y, p)) = sum(p) + y$.
Then $x \geq 0 \Rightarrow sum(empty\_stack) + x = x$ is a critical pair from (i) and (ii).

Now, we are ready to give all the inference rules of our approach.

4.1 The inference system $\mathcal{SH}$

If $\mathbf{H}$ is a set of Horn clauses, we note $\mathbf{H}_E$ its equational part, (i.e the set of all its equational clauses), $CP(\mathbf{H}_E)$ is the set of all critical pairs of equations in $\mathbf{H}_E$, $CN(\mathbf{H})$ the set of all rules deduced by *conditional-narrowing* from the rules of $\mathbf{H}$.

Let $\mathbf{H}$ be the initial set of Horn clauses and $C \Rightarrow s = t$ the ground deductive theorem to prove. Then we start with $\mathbf{H}_0 = \mathbf{H} \cup \{C \Rightarrow s \neq t\}$. As in the UKB of Bachmair in 87, this is just to make distinction between trivialities to delete, and the end of the proof.

i) *Deduce*:
$$\frac{\mathbf{H}}{\mathbf{H} \cup \{C\}} \qquad \text{if } C \in CP(\mathbf{H}_E)$$

ii) *I-Superpose*:
$$\frac{\mathbf{H}}{\mathbf{H} \cup \{C\}} \qquad \text{if } C \in CN(\mathbf{H})$$

iii) *Refute*:
$$\frac{\mathbf{H} \cup \{C \Rightarrow s \neq t\}}{Refutation} \qquad \begin{array}{l} \text{if } C \text{ is satisfiable, } s,\, t \in T(F, \mathcal{V}), \\ \text{and } C \wedge s \neq t \text{ is unsatisfiable.} \end{array}$$

iv)a) *Specialization*:
$$\frac{\mathbf{H} \cup \{C \wedge (x = t) \wedge L \Rightarrow l = r\}}{\mathbf{H} \cup \{C \wedge \sigma L \Rightarrow \sigma l = \sigma r\}} \qquad \begin{array}{l} \text{if } x \in \mathcal{V}_1,\, t \in T^\wedge(F, \mathcal{V}), \\ x \notin V(t),\, \sigma = \{x \leftarrow t\} \end{array}$$

iv)b) *Evaluation*:
$$\frac{\mathbf{H} \cup \{C \wedge (x = t) \wedge L \Rightarrow l = r\}}{\mathbf{H} \cup \{\nu C \wedge \nu L \Rightarrow \nu l = \nu r\}} \qquad \begin{array}{l} \text{if } x \in \mathcal{V}_0,\, t \in T(F_0), \\ \nu = \{x \leftarrow t\} \end{array}$$

v)a) *Suppress1*:
$$\frac{\mathbf{H} \cup \{C \Rightarrow s = s\}}{\mathbf{H}} \qquad \text{if } C \text{ is satisfiable, } s \in T(F, \mathcal{V})$$

v)b) *Suppress2*:
$$\frac{\mathbf{H} \cup \{C \Rightarrow s = t\}}{\mathbf{H}} \qquad \text{if } C \text{ is unsatisfiable}$$

vi)a) *Simplify1*:

$$\frac{\mathbf{H} \cup \{C \Rightarrow s = t,\ C' \wedge s_1[s'] = t_1 \wedge L \Rightarrow u = v\}}{\mathbf{H} \cup \{C \Rightarrow s = t,\ C'' \wedge C' \wedge s_1[t] = t_1 \wedge L \Rightarrow u = v,\ \neg C'' \wedge C' \wedge s_1[s'] = t_1 \wedge L \Rightarrow u = v\}}$$

where s' is a non-variable subterm of the high hierarchy and C'' is $s \not< t \wedge s = \overline{s'} \wedge C$. Here $s = \overline{s'}$ means one has to unify by assuming that all variables of s' are constant symbols, i.e one has to compute match from s to s'.

vi)b) *Simplify2*:

$$\frac{\mathbf{H} \cup \{C \Rightarrow s = t,\ C' \wedge L \Rightarrow u[s'] = v\}}{\mathbf{H} \cup \{C \Rightarrow s = t,\ C'' \wedge C' \wedge L \Rightarrow u[t] = v,\ \neg C'' \wedge C' \wedge L \Rightarrow u[s'] = v\}}$$

where s' is a non-variable subterm of the high hierarchy and C'' is $s \not< t \wedge s = \overline{s'} \wedge C$. Here $s = \overline{s'}$ means one has to unify by supposing that all variables of s' are constant symbols, i.e one has to compute match from s to s'.

vii) *I-Reflexion*:

$$\frac{\mathbf{H} \cup \{C \wedge (g = d) \wedge L \Rightarrow l = r\}}{\mathbf{H} \cup \{C \wedge (g = d) \wedge L \Rightarrow l = r, C \wedge \sigma L \Rightarrow \sigma l = \sigma r\}}$$

if σ is a ground substitution in $T^\wedge(F, V)$ such that $\sigma g = \sigma d$, and $(\sigma g = \sigma d) \not< (\sigma s = \sigma t) \; \forall s = t \in L$

viii) *Trivialization*:

$$\frac{\mathbf{H} \cup \{C \wedge s = t \wedge L \Rightarrow s = t\}}{\mathbf{H}}$$

4.2 Remarks

1. *Simplify2* is used to reduce the theorem to prove as well.

2. As soon as a rule is deduced, the first task to do is to put its constraint in *normal form* (see [8] for details on this notion). Then its satisfiability should be evaluated. If so it is, solving contraints in the predefined algebra must always be attempted.

3. For the sake of efficiency, iv)a) and iv)b) have to be applied whenever it is possible before applying other rules. Roughly speaking iv)a) and iv)b) correspond to *propagations* of [8], naturally here adapted.

4. A *Subsumption* inference rule may well be defined here. It is similar of this in [8] with minor modifications.

5 Correctness and Completeness of $\mathcal{SH}$

5.1 Correctness

Theorem 15. $\mathcal{SH}$ *is correct.*

The proof of this theorem is trivial since $\mathcal{SH}$ conserves $\mathrm{Th}_g(\mathbf{H}_0)$.

5.2 Completeness

Let us see through examples how $\mathcal{SH}$ does destroy progressively the leaves of trees to prove theorems at their roots. Doing so $\mathcal{SH}$ leads to equational proofs. Recall equations are conditional rules with constraints only in predefined algebras and they do not have conditions on terms in the high hierarchy.

Examples

Example 8. First, let us consider the proof tree of Example 1. For the sake of readability, note *sum s, push pu, empty_stack e.* In the initial system we have:

$$\frac{2 \geq 0 \wedge pu(2, pu(-4, e)) = pu(2, pu(-4, e))}{s(pu(2, pu(-4, e))) \quad ==_3 \quad s(pu(-4, e)) + 2} \qquad \frac{-4 < 0 \wedge pu(-4, e)) = pu(-4, e))}{==_2 \quad s(e) + 4 + 2} \quad =_1 0{+}4{+}2$$

Specialization of 3, provides:

$$y \geq 0 \Rightarrow s(pu(y, p')) = s(p') + y \ . \tag{24}$$

In the new system, we have:

$$\frac{2 \geq 0}{s(pu(2, pu(-4, e))) \quad =_{24} \quad s(pu(-4, e)) + 2} \qquad \frac{-4 < 0 \wedge pu(-4, e)) = pu(-4, e))}{=_2 \quad s(e) + 4 + 2} \quad =_1 0{+}4{+}2$$

Specialization of 2, provides:

$$y < 0 \Rightarrow sum(pu(y, p')) = s(p') - y \ . \tag{25}$$

In the new system, we have:

$$\frac{2 \geq 0}{s(pu(2, pu(-4, e))) \quad =_{24} \quad s(pu(-4, e)) + 2} \quad \frac{-4 < 0}{=_{25} \quad s(e) + 4 + 2} \quad =_1 0{+}4{+}2$$

The latter is a constrained-equational proof whose two ends are the terms of the theorem to prove. From here, we apply standard reasoning. After three steps of simplification we deduce a triviality.

Example 9. Let us consider the following specification on *sets of naturals.*

$$add(x, add(y, e)) = add(y, add(x, e)) \ . \tag{26}$$

$$add(x, add(x, e)) = add(x, e) \ . \tag{27}$$

$$pres(x, empty) = false \ . \tag{28}$$

$$x = y \Rightarrow pres(y, add(x, e)) = true \ . \tag{29}$$

$$x \neq y \Rightarrow pres(y, add(x, e)) = pres(y, e) \ . \tag{30}$$

$$incl(empty, e) = true \ . \tag{31}$$

$$pres(x, e') = true \Rightarrow incl(add(x, e), e') = incl(e, e') \ . \tag{32}$$

$$pres(x, e') = false \Rightarrow incl(add(x, e), e') = false \ . \tag{33}$$

$$del(x, empty) = empty \ . \tag{34}$$

$$x = y \Rightarrow del(y, add(x, e)) = e \ . \tag{35}$$

$$x \neq y \Rightarrow del(y, add(x, e)) = add(x, del(y, e)) \ . \tag{36}$$

x, y are variables on naturals; e, e' on sets. *empty* denotes the empty set; $add(x, e)$ denotes $\{x\} \cup e$; $del(x, e)$ denotes $e - \{x\}$; $incl(e, e') = true$ means $e \subseteq e'$; $incl(e, e') = false$ means $e \not\subseteq e'$; $pres(x, e) = true$ means $x \in e$; $pres(x, e) = false$ means $x \notin e$.

Prove the theorem:

$$incl(add(2, empty), del(4, add(1, add(2, add(1+3, empty))))) = true \qquad (0)$$

The proof-tree of it in $\mathbf{H_0}$ - because of a lack of space, we have to cut the tree into several parts, and use small font.

$$\frac{4 \neq 1}{incl(add(2,empty), del(4, add(1, add(2, add(1+3, empty))))) =_{36} incl(add(2, empty), add(1, del(4, add(2, add(1+3, empty}$$

$$\frac{4 \neq 2}{incl(add(2,empty), add(1, del(4, add(2, add(1+3, empty)))) =_{36} incl(add(2, empty), add(1, add(2, del(4, add(1+3, empty))}$$

$$\frac{4 = 4}{incl(add(2,empty), add(1, add(2, del(4, add(1+3, empty)))) =_{35} incl(add(2, empty), add(1, add(2, empty)))}$$

$$\frac{\dfrac{2 \neq 1}{pres(2, add(1, add(2, empty)))=_{30}} \quad \dfrac{2 = 2}{pres(2, add(2, empty)) =_{29} \; true}}{incl(add(2,empty), \; add(1, add(2, empty))) ==_{32} incl(empty, add(1, add(2, empty))) =_{31} \; true}$$

By three steps of simplification in the equation to prove, we obtain the last subtree.

Rename variables of rules (30) and (32):

$$x_1 \neq y_1 \Rightarrow pres(y_1, add(x_1, e_1)) = pres(y_1, e_1) \; . \qquad (37)$$

$$pres(x_2, e'_2) = true \Rightarrow incl(add(x_2, e_2), e'_2) = incl(e_2, e'_2) \; . \qquad (38)$$

By narrowing-superposition of (30) in (32), we get:

$$x_1 \neq y_1 \wedge y_1 = x_2 \wedge e'_2 = add(x_1, e_1) \wedge pres(y_1, e_1) = true \Rightarrow incl(add(x_2, e_2), e'_2) = incl(e_2, e'_2). \qquad (39)$$

Thanks to the third remark of 4.2, we obtain:

$$x_1 \neq y_1 \wedge y_1 = x_2 \wedge pres(y_1, e_1) = true \Rightarrow incl(add(x_2, e_2), add(x_1, e_1)) = incl(e_2, add(x_1, e_1)) \; .$$

In this new set, the last proof-tree becomes:

$$\frac{2 \neq 1 \wedge \; \dfrac{2 = 2}{y_1 = 2 \wedge \; pres(2, add(2, empty)))=_{29} true}}{incl(add(2,empty), \; add(1, add(2, empty))) ==_{40} incl(empty, add(1, add(2, empty)))) =_{31} \; true}$$

Renamimg rule (29):

$$x_3 = y_3 \Rightarrow pres(y_3, add(x_3, e_3)) = true \; . \qquad (41)$$

By narrowing-superposition of (29) into (40), we get:

$$x_1 \neq y_1 \wedge y_1 = x_2 \wedge x_3 = y_3 \wedge y_1 = y_3 \wedge e_1 = add(x_3, e_3) \Rightarrow incl(add(x_2, e_2), add(x_1, e_1)) = incl(e_2, add(x_1, e_1)) \ .$$
(42)

which is, according to the precedent remark:

$$x_1 \neq y_1 \wedge y_1 = x_2 \wedge x_3 = y_3 \wedge y_1 = y_3 \Rightarrow incl(add(x_2, e_2), add(x_1, add(x_3, e_3))) = incl(e_2, add(x_1, add(x_3, e_3))) \ .$$

In the new system of rules, we have the following proof-tree:

$$\dfrac{2 \neq 1 \wedge y_1 = 2 \wedge x_3 = y_3 \wedge y_1 = y_3}{incl(add(2, empty),\ add(1, add(2, empty))) =_{43} incl(empty,\ add(1, add(2, empty)))) =_{31} true}$$

From here, we can apply standard reasoning. Simplifications with (43) and (31) yield triviality.

After observing those two examples, we can state the main theorem. Fairness notion in our case is quite close to that in [1], or [6].

Theorem 16. Let **H** be a set of Horn clauses, and $C \Rightarrow A = B$ a ground theorem in **H**. Applying $\mathcal{SH}$ on $\mathbf{H} \cup \{C \Rightarrow A \neq B\}$ - respecting the fairness hypothesis - yields a refutation.

Thanks to minor modifications on the complexities of proof-trees, the proof of this theorem is very close of those in [1] or [6]. The main steps of the proof are: 1) we define the complexities of proof-trees, 2) we prove that the inference rules decrease the complexity of the minimal proof-tree.

6 On the implementation

As it is mentioned in section 2.2, two main kinds of constraints are encountered here. On terms of the high level hierarchy, ordering and unification constraints are to be considered. By means of algorithms based on inferences of section 2.2, such constraints are "normalized" possibly leading to constraints on predefined terms. If so it is, "constraint solvers" in the predefined algebras are used as those of CLP(FD), CHIP.

Neither algorithm of unification nor matching modulo predefined algebras is used in our approach. Only syntactic unification based on inferences in section 2.2. is performed. Equational constraints in predefined algebras would be generated which would be dealt with by means of usual constraint solvers.

Theoretically, a nœtherian ordering on ground terms is needed. Thus, we suppose the used orderings in predefined algebras are suitable to be extended naturally on all ground terms. In practice, such hypothesis is often satisfied. Bounded intervals of integers, sets of finite elements (like *boolean, binary values*,...) are kinds of sets worth considering.

In our approach, neither unit equations, nor equations which are heads of Horn clauses are oriented. Thus both sides of unit equations are studied in simplifications. One may wonder about the effectiveness of such choice. In front of the theorem to be proved, such approach is quite suitable. In fact, as it is always ground, only one of the side of the simplifying equation runs. Works on the other side are bound to yield unsatisfiability. Anyway, orientations made by constraints should deliver a lot of unused rules as well ! Both cases have their own problems.

7 Conclusion and perspectives

We have proposed a new method to prove theorems in Horn theories with built-in algebras. As in Kounalis and Rusinowitch's work [9], let us note our method can be generalized to prove ground Horn clauses - of form $L \Rightarrow A = B$ - as well. For that purpose, we add all the equations of L and $A \neq B$ to the initial system of clauses.

Our method does not deal with confluence problem at all, even on ground terms. It is to be used only to prove ground theorems. Incidentally, unlike Becker and Avenhaus, we need neither unification nor matching algorithms modulo predefined algebras. We cumulate constraints until they can be solved. As we are treating ground problems, steps of interesting cumulation are bound to be met, allowing us to conclude.

From the theorem proving point of view, we can say our method is more effective than Becker and Avenhaus's. The main reason for such efficiency is : in all our inference rules at least one of the manipulated clauses is an unit one. The implementation of the method here described is under way.

In the future, one application of the approach developped here is program or circuit validation. For that purpose, we will work to prove 'inductive' theorems. This is one of the next steps of our study.

The interesting aspect of such topic, among others, consists also in computing the ground confluent system of Horn clauses corresponding to the initial set. To our knowlegde, Avenhaus and Becker have proposed a method to check the ground confluence of systems; they do not say how to compute them really. To replace "programming" only by "writing some conditional rules", strategies to compute ground canonical Horn theories are to be found. As a result this makes up one aspect of our future works as well.

References

1. Anantharaman, S., Andrianarivelo, N.: A semi-decision procedure in Horn Theories. technical report, Laboratoire d'Informatique Fondamentale d'Orléans-Laboratoire de Recherche en Informatique d'Orsay, (1991)
2. Bousdira, W., Andrianarivelo, N.: A rewrite-based strategy for theorem proving in first order logic with equality and ordering constraints. technical report, Laboratoire d'Informatique Fondamentale d'Orléans, (1994), 94-4
3. Avenhaus, J., Becker, K.: Conditional Rewriting modulo a Built-in Algebra. technical report Number 11, Universitat Kaiserslautern, West Germany, (1992)
4. Avenhaus, J., Becker, K.: Operational Specifications with Built-Ins. 7th Symposium on Theoretical Aspects of Computer Science, Caen, France, (1994), 263-274

5. Dershowitz, N.: Corrigendum to Termination of Rewriting. In Journal of Symbolic Computation, Vol. 4, 409–410
6. Dershowitz, N.: A maximal-literal strategy for Horn Clauses. 2nd International Workshop on Conditional Term Rewriting Systems, Montreal, Canada, (1990), 143–154
7. Ganzinger, H.: Ground Term Confluence in Parametric Conditional Equational Specifications. 4th annual Sympoisum on Theoretical Aspects of Computer Science, Passau, RFA, (1987), Vol. 247, 286–298
8. Kirchner, C. and Kirchner, H. and Rusinowitch, M.: Deduction with Symbolic Constraints. Revue Française d'Intelligence Artificielle, (1990), Vol. 4, Number. 3, 9–52
9. Kounalis, E., Rusinowitch, M.: Mechanizing Inductive Reasoning. EATCS Bulletin, Vol. 41, (1990), 216–226
10. Smolka, G.: Logic Programming over Polymorphically Order-Sorted Types. PHD Thesis, Universitat Kaiserslautern, West Germany, (1989)
11. Vorobyov, S.G.: Conditional Rewrite Rules Systems with Built-in Arithmetic and Induction. 3rd International Conference on Rewriting Techniques and Applications, Chapel Hill, North California, (1989), 492–512

Backward Reasoning in Systems with Cut

Elmar Eder

Institut für Computerwissenschaften Salzburg,
5020 Salzburg, Austria

Abstract

In backward reasoning, a proof of a theorem is constructed by trying
to obtain a set of axioms from the theorem by backward application of inference rules. Backward reasoning can be automatized if, for each instance
of a rule, the premises are uniquely determined by the conclusion. The cut
rule does not have this property. On the contrary, for every formula which
is the conclusion of an instance of the cut rule, there are infinitely many
pairs of premises. Hence, the known efficient techniques of automated
backward reasoning cannot be used in systems with cut.

In this paper, an approach to the problem is presented using *formula
schemes* rather than formulas. A formula scheme is like a formula except
it can contain *metavariables* instead of some subformulas or subterms. For
example, backward application of modus ponens to a formula F yields the
two formula schemes Φ and $\Phi \to F$ where Φ is a metavariable for a formula. Rules just consist of a number of formula schemes (premises and
conclusion) and of a number of conditions under which the rule is applicable. Two or more rules can be composed into a new rule. This allows
an arbitrary mixture of forward and backward reasoning. The composite
rule is computed by a sort of unification on the set of formulas. Since a
unifiable pair of formulas in general does not have a single most general
unifier, unification is only done partially (*pseudounification*) leaving an
equation between formulas as a new condition of the resulting rule. Since
these equations are always solvable, it is guaranteed that an inference step
on the level of formula schemes fails whenever there is no corresponding
inference on the level of formulas. So, considering formula schemes rather
than formulas really reduces the search tree.

1   Introduction

In existing automated theorem proving systems, a number of different inference
mechanisms are being used. The most common of them is resolution. Another
class of common inference mechanisms are calculi derived from backward reasoning in cut-free sequent calculi. Among them are the connection method, and
tableau calculi with built-in unification. The proof process in these calculi starts
from the theorem as a goal. A proof step can be viewed as a backward application of a rule of the sequent calculus. In a proof step, a goal which is an instance

of the conclusion of the rule is replaced by a number of new goals (*subgoals*), namely by the corresponding instances of the premises of the rule. Once there are no more goals left, the theorem is proven. The rules of the cut-free sequent calculus have the *subformula property* which states that each formula occurring in one of the premises is essentially[1] a subformula of a formula occurring in the conclusion. The premises are obtained from the conclusion by splitting the conclusion into subformulas or, as we say, by analysing the conclusion, and then reassembling some of the resulting subformulas. A calculus based on such an analysis mechanism is called an *analytic calculus*. The connection method and the tableau calculus are such analytic calculi. An analytic calculus allows a proof search which does not involve infinite branching at any node of the search tree. In fact, the search tree is comparatively narrow. So, efficient systems can be and, in fact, have been built with these calculi.

Sequent calculi were first introduced by Gentzen in [5]. They try to formalize the way mathematicians reason. The full sequent calculus contains the cut rule whose most simple version (modus ponens) states that from A and $A \to B$ we can infer B. Such an inference occurs often in human reasoning. A problem with applying this rule in backward reasoning in an automated theorem proving system is that it does not have the subformula property. Namely, if we want to prove a goal B, we obtain two subgoals A and $A \to B$ where A is an arbitrary formula. There is an infinite number of possible choices for such a formula A. Thus, an infinite branching of the search tree occurs, which is a catastrophy for an automated system.

This is the reason why existing automated theorem proving systems based on backward reasoning in sequent calculi use sequent calculi without a cut rule rather than the full sequent calculus. In a cut-free sequent calculus the number of possible choices for a set of subgoals for a given goal may still be infinite. For example, each formula $F(t)$ (where t is a ground term) is a possible choice of a subgoal[2] for a goal $\exists x F(x)$. The way this problem is overcome in existing automated theorem proving systems is to generate a subgoal $F(x)$ instaed of choosing a subgoal $F(t)$. Here the variable x plays the role of a metavariable standing for ground terms and its value is determined only in later proof steps through the mechanism of unification. But it is conceptually clearer to actually introduce a metavariable τ standing for ground terms, called a *term variable*, and to generate the subgoal $F(\tau)$. The value of τ is then determined in later proof steps through unification on the term variables. The subformula property of the cut-free sequent calculus assures that the number of possible choices of subgoals for any given goal, and thus the branching factor of the search tree, is always finite. Since the cut rule does not have the subformula property, it is not possible to apply this trick to the full sequent calculus in order to get an inference system which has a finitely branching search tree and which incorporates the cut rule.

---

[1]More precisely: it is obtained from such a subformula by application of a substitution of variables.

[2]identifying a seqent with an empty antecedent and a succedent consisting of just one formula with that formula

By Gentzen's cut elimination theorem [5], every application of the cut rule can be eliminated from a proof in the full sequent calculus. So, every valid formula can be proved in the cut-free sequent calculus. However, it has been shown by Statman [6] that the elimination of applications of the cut rule from a proof may increase the length of a proof dramatically (essentially from n to $f(n)$ where $f(0) := 1$ and $f(k+1) := 2^{f(k)}$). In these cases, even such a seemingly stupid brute force search method as enumerating all strings and checking them one after the other whether they are proofs of the formula in the full sequent calculus, would be more efficient than the most clever proof search in a cut-free sequent calculus.

So, inference mechanisms derived from cut-free sequent calculi are efficient tools for automated theorem proving only for formulas admitting short cut-free proofs. For formulas whose proofs make essential use of the cut rule, they are unfeasible. Thus it would be desirable to have a calculus which is derived from backward reasoning in the full sequent calculus or in an equivalent calculus but which still has a finitely branching search tree.

There are a number of calculi which are equivalent to the full sequent calculus in the sense that a proof of a formula in any one of these calculi can be simulated in any of the other calculi at polynomial expense. Among them are the Frege systems and Gentzen's calculus of natural deduction. For more details see [4, 1]. Frege systems have a simpler syntactical structure than sequent calculi since they work directly with formulas rather than with sequents of formulas. Therefore it is easier to develope techniques for narrowing the search tree in Frege systems than it is in sequent calculi. For this reason the investigations done in this paper are based on Frege systems rather than on sequent calculi. Although sequent calculi seem to be much more adequate than Frege systems for implementations of theorem provers, the equivalence of the calculi assures that the techniques and results of our investigation carry over to the sequent calculus and can be applied as well to construct more efficient backward reasoning systems based on the full sequent calculus.

In this paper a way is shown how to construct a calculus which is based on backward reasoning in a Frege system and which has a finitely branching search tree. In order to see how this works, remember the trick used in conventional theorem proving systems to make the branching factor of the search tree finite. There, the use of metavariables for terms allows to code a whole infinite family of formulas as a finite string. We shall call such a string a *formula scheme*. Thus the formula scheme $F(\tau)$ codes the family of all formulas $F(t)$ such that t is a ground term. Since the cut rule does not have the subformula property, backward reasoning in systems with cut (such as the full sequent calculus or Frege systems) makes it necessary to introduce metavariables for formulas, for constants, and for object variables in addition to the metavariables for terms. A *formula scheme* will be a string which is similar to a formula but which may contain metavariables. Formula schemes can then be used to write rules in sequent calculi or in Frege systems as strings. For example, the modus ponens can be written as $\Gamma, \Gamma \to \Delta \vdash \Delta$ where Γ and Δ are formula variables.

In this paper we show how two rules can be composed into a new rule whose application has the same effect as an application of the first rule followed by an application of the second rule. Thus, each Frege system gives rise to a calculus for generating rules by composition of the rules of the Frege system. Let us call a rule sound if the conclusion of an instance of this rule is valid provided the premises are valid. Then if the rules of a Frege system are sound then any composition of these rules is sound and the calculus generating compositions of rules can be used to do any mixture of forward and backward reasoning. In backward reasoning, the conclusion of the last of the composed rules must be instantiated to the goal. In Section 5 this instantiation which involves a unification process is discussed.

At first glance it might seem that it would not be necessary to go into all this trouble introducing so many different types of metavariables and defining concepts of unification for formula schemes containing these metavariables. It is known that there is a simple and efficient transformation to normal form , namely the structure preserving transformation to definitional form [3, 4, 1]. Also it is known that for any short proof of a formula in the full sequent calculus or in a Frege system, there is also a short proof of its definitional form in extended resolution [4]. Extended resolution has a finite branching factor at each node of the search tree. So it might seem that a simple way to obtain a short proof for a formula whose proof makes essential use of the cut rule might be to transform the formula to its definitional form and then search for a proof in extended resolution. However, the transformation only transforms a node of the search tree at which inifinite branching occurs to an infinite number of nodes at each of which a fininite branching occurs. It does not help with the search since none of the branches of the search tree have been eliminated.

We denote formulas of first order predicate logic with function symbols by capital Latin characters $F, G, H, \ldots$, variables by $x, y, z, \ldots$, function symbols by $f, g, h, \ldots$, and constants by $a, b, c, \ldots$. If F is a formula, x is a variable, and t is a ground term, then F_t^x denotes the result of replacing each free occurrence of x in F with t.

2 Frege Systems

Frege systems are finite sets of rules such as the following system S.

$$\vdash \quad \Gamma \vee \neg\Gamma \tag{1}$$

$$\vdash \quad \Gamma \to \Gamma \tag{2}$$

$$\vdash \quad \Gamma \to \Delta \to \Gamma \tag{3}$$

$$\vdash \quad (\Gamma \to \Gamma \to \Delta) \to \Gamma \to \Delta \tag{4}$$

$$\vdash \quad (\Gamma \to \Delta \to \Theta) \to \Delta \to \Gamma \to \Theta \tag{5}$$

$$\vdash \quad (\Gamma \to \Delta) \to (\Delta \to \Theta) \to \Gamma \to \Theta \tag{6}$$

$$\vdash \quad \Gamma \wedge \Delta \to \Gamma \tag{7}$$

$$\vdash \quad \Gamma \wedge \Delta \to \Delta \tag{8}$$

$$\vdash \quad (\Gamma \to \Delta) \to (\Gamma \to \Theta) \to \Gamma \to \Delta \wedge \Theta \tag{9}$$

$$\vdash \quad \Gamma \to \Gamma \vee \Delta \tag{10}$$

$$\vdash \quad \Delta \to \Gamma \vee \Delta \tag{11}$$

$$\vdash \quad (\Gamma \to \Theta) \to (\Delta \to \Theta) \to \Gamma \vee \Delta \to \Theta \tag{12}$$

$$\vdash \quad (\Gamma \to \Delta) \to (\Gamma \to \neg\Delta) \to \neg\Gamma \tag{13}$$

$$\vdash \quad \neg\Gamma \to \Gamma \to \Delta \tag{14}$$

$$\vdash \quad \forall\xi\,\Gamma \to \Gamma^\xi_\tau \tag{15}$$

$$\vdash \quad \Gamma^\xi_\tau \to \exists\xi\,\Gamma \tag{16}$$

$$\Gamma, \quad \Gamma \to \Delta \quad \vdash \quad \Delta \tag{17}$$

$$\alpha \notin \Gamma\Delta: \quad \Gamma \to \Delta^\xi_\alpha \quad \vdash \quad \Gamma \to \forall\xi\,\Delta \tag{18}$$

$$\alpha \notin \Gamma\Delta: \quad \Gamma^\xi_\alpha \to \Delta \quad \vdash \quad \exists\xi\,\Gamma \to \Delta \tag{19}$$

Each of the rules has the following syntactic structure:

$$\mathcal{C}: \quad \Phi_1, \quad \ldots, \Phi_n \quad \vdash \quad \Psi$$

where $\mathcal{C}$ is a sequence of *conditions* and $\Phi_1, \ldots, \Phi_n, \Psi$ are *formula schemes*. Both of these concepts will be defined later. If the sequence of conditions is empty—such as in rules (1)–(17)—, its terminating colon is omitted. The formula schemes $\Phi_1, \ldots, \Phi_n$ are the *premises* and the formula scheme Ψ is the *conclusion* of the rule. Rules which have no premises—such as rules (1)–(16) are called *axiom schemes*. In a formula scheme, metavariables are used for formulas, for ground terms, for constants, and for object variables. These metavariables are called *formula variables*, *term variables*, *constant variables*, and *variable variables*, respectively. The symbols used for these metavariables are capital Greek characters $\Gamma, \Delta, \Theta, \ldots$ for formula variables, and small Greek characters $\rho, \sigma, \tau, \ldots$ for term variables, $\alpha, \beta, \gamma, \ldots$ for constant variables, and $\xi, \eta, \zeta, \ldots$ for variable variables.

An *atomic formula scheme* is defined inductively as follows.

- Each formula variable is an atomic formula scheme.

- If Φ is an atomic formula scheme, ξ is a variable variable, and μ is a constant variable or a term variable, then Φ_μ^ξ is an atomic formula scheme.

A *formula scheme* and its atoms are defined inductively as follows.

- If Φ is an atomic formula scheme then Φ is a formula scheme whose only atom is Φ.

- If Φ is a formula scheme then $\neg\Phi$ is a formula scheme whose atoms are the atoms of Φ.

- If Φ and Ψ are formula schemes then $\Phi\wedge\Psi$, $\Phi\vee\Psi$, and $\Phi\to\Psi$ are formula schemes whose atoms are the atoms of Φ and the atoms of Ψ.

- If Φ is a formula scheme and ξ is a variable variable then $\forall\xi\,\Phi$ and $\exists\xi\,\Phi$ are formula schemes whose atoms are the atoms of Φ.

Appropriate use of parentheses is assumed where it is necessary for determination of precedence of operators. For brevity, we omit explicit mentioning of parentheses throughout this paper where this does not lead to ambiguities. In a sense, the atoms of a formula scheme are its maximal atomic sub-formulaschemes.

We do not allow an expression Φ_μ^ξ as a formula scheme or as a subexpression of a formula scheme unless Φ is atomic. However, as we did for formulas, we define the formula scheme which such an expression *denotes*. In the following definition, Υ and Ω are expressions, ξ and η are two distinct variable variables, and μ is either a constant variable or a term variable.

- An atomic formula scheme denotes itself.

- If expression Υ denotes the formula scheme Φ then $\neg\Upsilon$ denotes $\neg\Phi$.

- If Υ denotes Φ and Ω denotes Ψ then $\Upsilon\wedge\Omega$ denotes $\Phi\wedge\Psi$, $\Upsilon\vee\Omega$ denotes $\Phi\vee\Psi$, and $\Upsilon\to\Omega$ denotes $\Phi\to\Psi$.

- If Υ denotes Φ then $\forall\xi\Upsilon$ denotes $\forall\xi\Phi$ and $\exists\xi\Upsilon$ denotes $\exists\xi\Phi$.

- If Υ denotes $\neg\Phi$ then Υ_μ^ξ denotes $\neg\Phi_\mu^\xi$.

- If Υ denotes $\Phi\wedge\Psi$ then Υ_μ^ξ denotes $\Phi_\mu^\xi\wedge\Psi_\mu^\xi$.

- If Υ denotes $\Phi\vee\Psi$ then Υ_μ^ξ denotes $\Phi_\mu^\xi\vee\Psi_\mu^\xi$.

- If Υ denotes $\Phi\to\Psi$ then Υ_μ^ξ denotes $\Phi_\mu^\xi\to\Psi_\mu^\xi$.

- If Υ denotes $\forall\xi\Phi$ then Υ_μ^ξ denotes $\forall\xi\Phi$ and Υ_μ^η denotes $\forall\xi\Psi$ where Φ_μ^η denotes Ψ.

- If Υ denotes $\exists\xi\Phi$ then Υ_μ^ξ denotes $\exists\xi\Phi$ and Υ_μ^η denotes $\exists\xi\Psi$ where Φ_μ^η denotes Ψ.

A *condition* is either an expression of the form $\alpha \notin \Phi_1 \ldots \Phi_n$ where α is a constant variable and where $\Phi_1, \ldots, \Phi_n$ are atomic formula schemes, or it is an equation of the form $\Phi = \Psi$ where Φ and Ψ are atomic formula schemes. Intuitively speaking, the former kind of conditions is used to express the requirement that an eigenconstant in a rule application must not occur in the conclusion, while the latter kind of conditions is necessary when we compose two rules into a new rule using the idea of unification of formula schemes (rather than of terms as in usual first order theorem proving). Since such a unification is in general only possible down to the level of atomic formula schemes, we are left with equations between atomic formula schemes which have to remain ununified in the form of conditions.

An *instantiation* is a function mapping each formula variable to a formula, each term variable to a ground term, each constant variable to a constant, and each variable variable to an object variable, which is injective on the constant variables and on the variable variables. We use postfix notation for instantiations. If ι is an instantiation and Φ is a formula scheme then the *result* $\Phi\iota$ *of applying* ι to Φ is the formula denoted by the expression which is obtained from Φ by replacing each metavariable μ with $\mu\iota$. The formula $\Phi\iota$ is then said to be an *instance* of the formula scheme Φ. A condition $\alpha \notin \Phi_1 \ldots \Phi_n$ *holds* for an instantiation ι if the constant $\alpha\iota$ does not occur in any of the formulas $\Phi_1\iota, \ldots, \Phi_n\iota$. A condition $\Phi = \Psi$ *holds* for an instantiation ι if $\Phi\iota = \Psi\iota$. An *instance* of a rule $C: \Phi_1, \ldots, \Phi_n \vdash \Psi$ is a string $\Phi_1\iota, \ldots, \Phi_n\iota \vdash \Psi\iota$ where the condition C holds for the instantiation ι. A *derivation* of a formula G from a set S of formulas in a Frege system $\mathcal{F}$ is a finite sequence of formulas ending with G such that for each formula H of the sequence it holds that $H \in S$ or there are formulas $H_1, \ldots, H_n$ preceeding H in the sequence such that $H_1, \ldots, H_n \vdash H$ is an instance of a rule of $\mathcal{F}$.

A *propositional Frege system* is a Frege system in propositional logic in which no conditions, term variables, constant variables or variable variables occur. An example is the set of rules (1)–(14) and (17) of system S. For propositional Frege systems a backward application of rules using unification is easy. You just consider the propositional connectives as function symbols in the term algebra of propositional formulas and use unification in this term algebra. A logic program written in Prolog syntax for the system consisting of rules (1)–(14) and (17) looks like this.

```
prove(Gamma or (non Gamma)).
...
prove((non Gamma) implies (Gamma implies Delta)).
prove(Delta) :- prove(Gamma), prove(Gamma implies Delta).
```

A logic programming system providing an occur check and reasonable search strategies, would find a proof of a formula F when started with this program and with the query ?- prove(F). It is interesting to note that if we resolve two clauses of this program then we obtain a clause which corresponds to a new rule of a possible Frege system. This new rule is the composition of the two old rules in the sense that applying the new rule is equivalent to applying the two old rules one after the other. Any sound rule of propositional logic can be obtained in this way by composition of one or more rules of any complete propositional Frege system containing the modus ponens rule (17).

3 Composition of Rules in First order Logic

Matters are more complicated in first order logic. In analogy to propositional logic we should be able to compose rules. In systems without cut such as the cut-free sequent calculus, such a general composition of rules is not necessary, since there in backward reasoning you always can compute each formula (except for necessary unification of terms). But when you have a cut rule such as modus ponens (17) then this does not hold. If you want to prove a formula G by (17) then you have to prove some formula F and the formula $F \to G$. If you take the next step backward from F (which is of course usually not a good idea) then you do not even have a clue about the conclusion F. If you take the other alternative first proving $F \to G$ then, by the time you want to prove F you might still not have determined F completely. So we should try to answer the question what is the composition of two (or more) rules. In order to do this we must decide whether there is a common instance of two formula schemes (namely of the conclusion of the first rule and of one of the premises of the second rule).

A *unifying instance* of two formula schemes Φ and Ψ is a formula $\Phi\iota$ where ι is an instantiation such that $\Phi\iota = \Psi\iota$. Two formula schemes or rules are said to be *disjoint* if they have no metavariables in common. Obviously, for disjoint formula schemes the concepts of a common instance and of a unifying instance are equivalent. Since we are interested in the composition of rules (not only in the composition of instances of rules), we need a concept of substitution for formula schemes.

A *substitution* is a function mapping each formula variable to a formula scheme, each term variable to a term variable, each constant variable to a constant variable, and each variable variable to a variable variable. We use postfix notation for substitutions. The *domain* of a substitution ς is the set of metavariables μ such that $\mu\varsigma \neq \mu$. If ς is a substitution whose domain is (a subset of) $\{\mu_1, \ldots, \mu_n\}$ and if $\mu_i\varsigma = \nu_i$ for $i = 1, \ldots, n$ then the substitution ς is also denoted by $[\begin{smallmatrix} \mu_1 \ldots \mu_n \\ \nu_1 \ldots \nu_n \end{smallmatrix}]$. If Φ is a formula scheme and ς is a substitution which is injective on the constant variables of Φ and on the variable variables[3] of Φ then the result $\Phi\varsigma$ of *applying* ς to Φ is defined as the formula scheme denoted by the expression obtained from Φ by replacing each metavariable μ with $\mu\varsigma$. Otherwise the substitution ς is *not applicable* to the formula scheme Φ. The result of *applying* a substitution ς to a condition $\alpha \notin \Phi_1 \ldots \Phi_n$ is the condition $\alpha\varsigma \notin \Psi_1 \ldots \Psi_r$ where $\Psi_1, \ldots \Psi_r$ are the atoms of the formula schemes

---

[3]Injectivity on variable variables assures that, for example, that an application of a substitution to $\forall\xi\forall\eta\Gamma$ does not yield $\forall\xi\forall\xi\Gamma$. Injectivity on constant variables is expresses the fact that, in backward reasoning in Frege systems, constants appear through backward application of a critical rule and are required to be new constants not already appearing in the conclusion (see the condition on α in rules (18) and (19) of system $\mathcal{S}$). Therefore, two distinct constant variables should not be unified.

$\Phi_1\varsigma, \ldots, \Phi_n\varsigma$. The result of applying ς to a condition $\Phi = \Psi$ is the condition $\Phi\varsigma = \Psi\varsigma$. The result of *applying* a substitution ς to a rule $\mathcal{C}: \Phi_1, \ldots, \Phi_n \vdash \Psi$ is the rule $\mathcal{C}\varsigma: \Phi_1\varsigma, \ldots, \Phi_n\varsigma \vdash \Psi\varsigma$. If R is a rule then the set of instances of $R\varsigma$ is always a subset of the set of instances of R. So, application of a substitution to a sound rule yields again a sound rule.

The concept of two formula schemes Φ and Ψ *having the same logical structure*, in symbols $\Phi \simeq \Psi$ is defined inductively as follows.

- If Φ and Ψ are atomic formula schemes then $\Phi \simeq \Psi$.

- If $\Phi \simeq \Psi$ then $\neg\Phi \simeq \neg\Psi$.

- If $\Phi_1 \simeq \Psi_1$ and $\Phi_2 \simeq \Psi_2$ then $\Phi_1 \wedge \Phi_2 \simeq \Psi_1 \wedge \Psi_2$, $\Phi_1 \vee \Phi_2 \simeq \Psi_1 \vee \Psi_2$, and $\Phi_1 \rightarrow \Phi_2 \simeq \Psi_1 \rightarrow \Psi_2$.

- If $\Phi \simeq \Psi$ then $\forall\xi\Phi \simeq \forall\xi\Psi$ and $\exists\xi\Phi \simeq \exists\xi\Psi$.

If two formula schemes have the same logical structure, then to each occurrence of an atom in one of them corresponds exactly one occurrence of an atom in the other formula scheme. These two occurrences occupy 'identical places' in the two structure trees.

Proposition 1 *If $\Phi \simeq \Psi$ then Φ and Ψ have a unifying instance.*

Proof: Let U be a propositional variable and let ι be an instantiation defined by $\Gamma\iota = U$ for all formula variables Γ and defined arbitrarily for all other metavariables. Then $\Phi\iota = \Psi\iota$.   *q.e.d.*

A *unifier* of two formula schemes Φ and Ψ is a substitution ς such that $\Phi\varsigma = \Psi\varsigma$. Unfortunately, the existence of a unifying instantiation of two formula schemes does not imply the existence of a unifier of these formula schemes. Therefore, this concept of a unifier is not very useful for our purposes.

A *pseudounifier* of two formula schemes Φ and Ψ is a substitution ς such that $\Phi\varsigma \simeq \Psi\varsigma$. The *unifying equations* for Φ, Ψ, and ς are the equations $\Phi'\varsigma = \Psi'\varsigma$ where Φ' is an atom of Φ and Ψ' is the corresponding atom of Ψ. By a *most general pseudounifier (mgpu)* of two formula schemes Φ and Ψ we mean a pseudounifier ς of Φ and Ψ such that each unifying instance of Φ and Ψ is a unifying instance of $\Phi\varsigma$ and $\Psi\varsigma$.

Proposition 2 *Two formula schemes Φ and Ψ have a unifying instance iff they have a pseudounifier.*

Proof: Let Φ and Ψ have a unifying instance $\Phi\iota = \Psi\iota$ and let Γ be a formula variable. Now let ς denote the action of replacing each atomic formula with Γ and replacing consistently (injectively) each object variable with a variable variable. Build the composition, taking the instantiation ι followed by the action ς. This composition is a pseudounifier of Φ and Ψ.

On the other hand, assume that Φ and Ψ have a pseudounifier ς. Then $\Phi\varsigma \simeq \Psi\varsigma$. By Proposition 1 there is an instantiation ι such that $\Phi\varsigma\iota = \Psi\varsigma\iota$. This is the wanted unifying instance.   *q.e.d.*

Proposition 3 *If two formula schemes Φ and Ψ have a pseudounifier then they have a most general pseudounifier.*

Proof: Let Φ' and Ψ' be the formula schemes obtained from Φ and Ψ, respectively, by replacing in them each atom with the formula variable ocurring in that atom. Consider Φ' and Ψ' as terms of a term algebra whose variables are the formula variables. Then Φ' and Ψ' must have a unifier ς in this term algebra. For each formula variable Γ take the formula scheme $\Gamma\varsigma$ and replace in it each occurrence of a formula variable with a new formula variable. Call the result $\Gamma\kappa$. This defines a substitution κ whose domain is the set of formula variables. Each formula variable occurs in at most one of the formula schemes $\Gamma\kappa$, and there also only once. Then the substitution κ is a most general pseudounifier of Φ and Ψ. $\hspace{2cm}$ q.e.d.

If ς is a most general pseudounifier of Φ and Ψ then, by definition, every unifying instance of Φ and Ψ is also a unifying instance $\Phi\varsigma\iota = \Psi\varsigma\iota$ of $\Phi\varsigma$ and $\Psi\varsigma$ with some instantiation ι. Moreover, $E\iota$ holds for each unifying equation E for Φ, Ψ, and ς.

Obviously, a substitution is a unifier of two formula schemes iff it is a pseudounifier of these formula schemes and the unifying equations hold. If two formula schemes have a unifying instance then they have a most general pseudounifier (mgpu).

Let $F_1,\ldots,F_m,F,G_1,\ldots,G_n,G$ be formulas and assume that $1 \leq j \leq n$ and that $F = G_j$. Then the string $G_1,\ldots,G_{j-1},F_1,\ldots,F_m,G_{j+1},\ldots,G_n \vdash G$ is said to be a *composition* of the two strings $F_1,\ldots,F_m \vdash F$ and $G_1,\ldots,G_n \vdash G$. In particular, if we have two instances of rules and the conclusion of the first instance equals one of the premises of the second, then a composition of the two instances exists. In a complete Frege system, every string $\vdash F$ where F is valid formula, can be obtained by repeated composition of instances of rules of the Frege system.

A *composition* of two disjoint rules $\mathcal{C}\colon \Phi_1,\ldots,\Phi_m \vdash \Phi$ and $\mathcal{D}\colon \Psi_1,\ldots,\Psi_n \vdash \Psi$ is a rule

$$\mathcal{C}\varsigma,\mathcal{D}\varsigma,\mathcal{E}\colon \Psi_1\varsigma,\ldots,\Psi_{j-1}\varsigma,\Phi_1\varsigma,\ldots,\Phi_m\varsigma,\Psi_{j+1}\varsigma,\ldots,\Psi_n\varsigma \vdash \Psi\varsigma$$

where ς is an mgpu of Φ and Ψ_j, and $\mathcal{E}$ is a sequence consisting of the corresponding unifying equations. If the two rules are not disjoint then the metavariables in one of the rules have to be renamed before the pseudounification in the same way as it is done in resolution (separating metavariables apart). Any conditions of the form $\Phi = \Phi$ (tautological equations) may be omitted. Similarly, if an equation $\Gamma = \Phi$ occurs as a condition in a rule (where Γ is a formula variable and Φ is an atomic formula scheme) then the rule can be simplified by omitting the equation and replacing each occurrence of Γ with Φ.

Proposition 4 *Let $\mathcal{R}$ and $\mathcal{R}'$ be two rules. Then a string is a composition of an instance of $\mathcal{R}$ and an instance of $\mathcal{R}'$ if and only if it is an instance of a composition of $\mathcal{R}$ and $\mathcal{R}'$.*

This proposition guarantees that rules can be composed in any order in the process of theorem proving. In particular, backward reasoning is possible despite the presence of a cut rule. In the following two sections the mechanism of composition of rules involved in the process of theorem proving is shown with an example.

4 An Example

As an example of a composition of rules consider a derivation of

$$\exists x \, \forall u \, Pxu \to \forall y \, \exists z \, Pzy$$

in the Frege system $\mathcal{S}$:

$$
\begin{array}{lll}
Pab \to \exists z \, Pzb & (20) & \text{by (16)} \\
\forall u \, Pau \to Pab & (21) & \text{by (15)} \\
(\forall u \, Pau \to Pab) & & \\
\quad \to (Pab \to \exists z \, Pzb) \to \forall u \, Pau \to \exists z \, Pzb & (22) & \text{by (6)} \\
(Pab \to \exists z \, Pzb) \to \forall u \, Pau \to \exists z \, Pzb & (23) & \text{by (17) from (21,22)} \\
\forall u \, Pau \to \exists z \, Pzb & (24) & \text{by (17) from (20,23)} \\
\exists x \, \forall u \, Pxu \to \exists z \, Pzb & (25) & \text{by (19) from (24)} \\
\exists x \, \forall u \, Pxu \to \forall y \, \exists z \, Pzy & (26) & \text{by (18) from (25)}
\end{array}
$$

This proof involves a composition of 7 instances of rules. But the rules themselves can also be composed in the same way without considering any instances. More precisely, let us call the sequence of lines

$$
\begin{array}{ll}
(20) & \text{by (16)} \\
(21) & \text{by (15)} \\
(22) & \text{by (6)} \\
(23) & \text{by (17) from (21,22)} \\
(24) & \text{by (17) from (20,23)} \\
(25) & \text{by (19) from (24)} \\
(26) & \text{by (18) from (25)}
\end{array}
$$

the *skeleton* of the derivation. Then a rule can be constructed such that every derivation which has this skeleton, has the same premises (in our example none) and the same conclusion as a suitable instance of this rule. We shall construct this rule which is displayed below as (32).

Let us start from the bottom by composing first the rules (19) and (18). By renaming the metavariables in (18) we obtain the rule

$$\beta \notin \Theta\Lambda: \quad \Theta \to \Lambda_\beta^\eta \quad \vdash \quad \Theta \to \forall \eta \, \Lambda.$$

A most general pseudounifier of the conlusion of rule (19) and of the premise of this rule is the substitution $[\begin{smallmatrix}\Theta & \Delta \\ \exists \xi \Gamma & \Lambda_\beta^\eta\end{smallmatrix}]$. The composition of the two rules using this mgpu is the rule

$$\alpha \notin \Gamma\Lambda_\beta^\eta, \ \beta \notin \Gamma\Lambda: \quad \Gamma_\alpha^\xi \to \Lambda_\beta^\eta \quad \vdash \quad \exists \xi \Gamma \to \forall \eta \Lambda. \tag{27}$$

Here we have omitted the tautological equations $\Gamma = \Gamma$ and $\Lambda_\beta^\eta = \Lambda_\beta^\eta$. If the pseudounifier is also a unifier—as is the case here—then all equations introduced newly through the composition are tautological and can be omitted. Only equations which already occur in the parent clauses have to be kept as conditions of the composed rule. In a similar way, a composition of rules (17) and (27) yields the rule

$$\alpha \notin \Gamma\Lambda_\beta^\eta,\ \beta \notin \Gamma\Lambda: \quad \Delta,\quad \Delta \to \Gamma_\alpha^\xi \to \Lambda_\beta^\eta \quad \vdash \quad \exists\xi\Gamma \to \forall\eta\Lambda. \tag{28}$$

Our next step is composing rules (16) and (28). By renaming of metavariables we obtain from (16) the rule $\vdash \Theta_\sigma^\varsigma \to \exists\varsigma\Theta$. A most general pseudounifier of the conclusion of this rule and of the first premise of rule (28) is $[^\Delta_{\Theta_\sigma^\varsigma \to \exists\varsigma\Theta}]$. Thus the composition of (16) and (28) is

$$\alpha \notin \Gamma\Lambda_\beta^\eta,\ \beta \notin \Gamma\Lambda: \quad (\Theta_\sigma^\varsigma \to \exists\varsigma\Theta) \to \Gamma_\alpha^\xi \to \Lambda_\beta^\eta \quad \vdash \quad \exists\xi\Gamma \to \forall\eta\Lambda. \tag{29}$$

The composition of (17) and (29) is

$$\alpha \notin \Gamma\Lambda_\beta^\eta,\ \beta \notin \Gamma\Lambda: \quad \Delta,\quad \Delta \to (\Theta_\sigma^\varsigma \to \exists\varsigma\Theta) \to \Gamma_\alpha^\xi \to \Lambda_\beta^\eta \quad \vdash \quad \exists\xi\Gamma \to \forall\eta\Lambda. \tag{30}$$

Now let us compose (15) and (30). By renaming of metavariables we obtain from (15) the rule $\vdash \forall\theta\Xi \to \Xi_\tau^\theta$. A most general pseudounifier of this rule and of the first premise of rule (30) is $[^\Delta_{\forall\theta\Xi \to \Xi_\tau^\theta}]$. Thus the composition of (15) and (30) is

$$\alpha \notin \Gamma\Lambda_\beta^\eta,\ \beta \notin \Gamma\Lambda: \quad (\forall\theta\Xi \to \Xi_\tau^\theta) \to (\Theta_\sigma^\varsigma \to \exists\varsigma\Theta) \to \Gamma_\alpha^\xi \to \Lambda_\beta^\eta \quad \vdash \quad \exists\xi\Gamma \to \forall\eta\Lambda. \tag{31}$$

Finally we compose rules (6) and (31). By renaming metavariables we obtain from (6) the rule $\vdash (\Delta \to \Pi) \to (\Pi \to \Sigma) \to \Delta \to \Sigma$. A most general pseudounifier of the conclusion of this rule and of the premise of rule (31) is $[^{\Delta\ \ \Pi\ \ \Sigma\ \ \Gamma\ \ \Lambda}_{\forall\theta\Xi\Xi_\tau^\theta\exists\varsigma\Theta\forall\theta\Upsilon\exists\varsigma\Omega}]$. This is, however, not a unifier. So the composition of rules (6) and (31) contains equations as conditions:

$$\alpha \notin \Upsilon\Omega_\beta^\eta,\ \beta \notin \Upsilon\Omega,\ \Xi_\tau^\theta = \Theta_\sigma^\varsigma,\ \Xi = \Upsilon_\alpha^\xi,\ \Theta = \Omega_\beta^\eta: \quad \vdash \quad \exists\xi\forall\theta\Upsilon \to \forall\eta\exists\varsigma\Omega$$

This rule can be simplified by replacing Ξ with Υ_α^ξ and Θ with Ω_β^η. The resulting rule is the following.

$$\alpha \notin \Upsilon\Omega_\beta^\eta,\ \beta \notin \Upsilon\Omega,\ \Upsilon_{\alpha\tau}^{\xi\theta} = \Omega_{\beta\sigma}^{\eta\varsigma}: \quad \vdash \quad \exists\xi\forall\theta\Upsilon \to \forall\eta\exists\varsigma\Omega \tag{32}$$

It seems that a further simplification is impossible, and in particular that an explicit unification of the equation $\Upsilon_{\alpha\tau}^{\xi\theta} = \Omega_{\beta\sigma}^{\eta\varsigma}$ cannot be done. For example, if you make the ansatz $\Upsilon = \Psi_{\beta\sigma}^{\eta\varsigma}$ and $\Omega = \Psi_{\alpha\tau}^{\xi\theta}$ and then try to match the right hand side of equation (32) with the formula $\exists x \forall u\, Pxu \to \forall y \exists z\, Pzy$, then you find out that you have to instantiate σ by x and τ by y which is not allowed since the terms by which term variables are instantiated have to be ground.

5 Backward Reasoning

In order for compositions of rules to be useful for backward reasoning, it is necessary that an instance of a given rule be found such that this instance has a given formula as its conclusion. In our example let us assume that we want to prove the formula

$$\exists x \, \forall u \, Pxu \to \forall y \, \exists z \, Pzy$$

using the skeleton shown in the previous section. We want to find out to which extent we can determine the premises of the derivation. The skeleton is equivalent to the rule (32). It follows that it is known for all metavariables on the right hand side of rule (32) how they are to be instantiated, namely ξ by x, θ by u, η by y, ζ by z, Υ by Pxu, and Ω by Pzy. With these instantiations plus some instantiations of the metavariables which do not occur on the right-hand side of (32), the conditions of rule (32) have to be satisfied. The constant variables α and β just have to be instantiated by two distinct constants, say a and b. In our example, the conditions $\alpha \notin \Upsilon\Omega_\beta^\eta$ and $\beta \notin \Upsilon\Omega$ are automatically fulfilled if a and b are distinct. The remaining condition $\Upsilon_{\alpha\tau}^{\xi\theta} = \Omega_{\beta\sigma}^{\eta\zeta}$ is fulfilled if σ is replaced by a ground term s and τ by a ground term t such that $Pat = Psb$. This is a unification problem which can be solved efficiently by one of the well-known unification algorithms for first order predicate logic. In our example, the ground term s has to be chosen as a, and t has to be chosen as b.

If the formula which has to be proved contains constants, such as the formula

$$\exists x \, \forall u \, P(x, u) \to \forall y \, \exists z \, P(z, f(y, c)),$$

then the first two conditions of rule (32) are not automatically fulfilled. In this case the constants a and b must be distinct from c which, however poses no problems. It is interesting to note that this formula can be proved by the same skeleton although it has quite a different syntactic structure, and it seems (although the author has not yet proved it) that there is no sound rule (axiom scheme) which has both formulas as instances and which has no equations as conditions. So it seems that equations are essential.

A problem occurs, however, if there is a condition such as $\alpha \notin \Gamma_\tau^\xi$. If α is instantiated by a constant a, Γ by a formula F, ξ by a variable x, and τ by a ground term t, then this means—unless x does not occur freely in F—that the constant a must not occur in the ground term t. Deciding whether there is an instance of a rule containing such conditions amounts to checking a graph for cycles. This is similar and closely related to the way unification is treated in non-normal form versions of W. Bibel's connection method [2]. In fact, the skeletons introduced here are similar to Bibel's skeletons. Bibel goes somewhat further in the abstraction in that his skeletons contain only partial information about the order of proof steps. For pairs of steps whose order is irrelevant, in his setting both possible orders result in the same skeleton whereas they result in different skeletons in the setting presented here. On the other hand, Bibel does not have the cut rule. Perhaps further research will show whether the two approaches can be combined.

6 Conclusion

The concept of a *rule* of a Frege system and of the composition of two rules has been defined. If an application of one rule followed by an application of another rule is possible for some formulas then the corresponding composition of the two rules exists and its application is equivalent to the application of the first rule followed by the application of the second rule. Backward reasoning or, more generally, any combination of forward and backward reasoning can be considered as a special case of repeated composition of rules. Since the composition of two rules is uniquely determined through pseudounification, even a backward application of the cut rule does not introduce an infinite branching of the search tree. Branching of the search tree occurs at each point where a choice is made which rule to apply to which formulas. The branching factor is always finite.

In this paper the Frege systems have been investigated because they have a simple syntax. The ideas presented here can be applied to the sequent calculus as well as to Frege systems. The sequent calculus has the great advantage over the Frege systems that most of its rules are analytic and the only rule which is not analytic, namely the cut rule, can be eliminated. For an efficient implementation of a backward reasoning system it is necessary to control the search in such a way that most of the time only analytic rules are applied. A non-analytic rule should be applied only when there is a strong reason to do so. Since proofs in Frege systems are essentially based on application non-analytic rules, Frege systems seem to be inadequate for backward reasoning. For an implementation, I would suggest the sequent calculus together with a strategy that tries to avoid the cut rule unless it has strong evidence that an application of the cut rule is likely to shorten the proof considerably.

References

[1] W. Bibel, M. Davis, E. Eder, N. Eisinger, M. Fitting, W. Hodges, D.J. Israel, H.J. Ohlbach, and D.A. Plaisted. *Handbook of Logic in Artificial Intelligence and Logic Programming*, volume 1. Clarendon Press, Oxford, 1993. (Dov M. Gabbay, C.J. Hogger, and J.A. Robinson, editors).

[2] Wolfgang Bibel. *Automated Theorem Proving*. Artificial Intelligence. Vieweg, Braunschweig/Wiesbaden, second edition, 1987.

[3] Elmar Eder. An implementation of a theorem prover based on the connection method. In W. Bibel and B. Petkoff, editors, *Artificial Intelligence, Methodology, Systems, Applications (AIMSA'84), Varna, Bulgaria (Sept. 1984)*, pages 121–128, Amsterdam, New York, Oxford, 1985. European Coordinating Committee for Artificial Intelligence, North-Holland.

[4] Elmar Eder. *Relative Complexities of First Order Calculi*. Artificial Intelligence. Vieweg, Wiesbaden, 1992. (Wolfgang Bibel and Walther von Hahn, editors).

[5] Gerhard Gentzen. Untersuchungen über das logische Schließen. *Mathematische Zeitschrift*, 39:176–210, 405–431, 1935. Engl. transl. in [7], pp. 68–131.

[6] R. Statman. Lower bounds on herbrand's theorem. *Proc. AMS*, 75, 1979.

[7] M. E. Szabo. *The Collected Papers of Gerhard Gentzen*. Studies in Logic and the Foundations of Mathematics. North-Holland, Amsterdam, 1969.

Soundness and Completeness versus Lifting Property*

Jan A. Plaza

University of Miami
Department of Mathematics and Computer Science
P.O. Box 249085
Coral Gables, Florida 33124
U.S.A.

janplaza@math.cs.miami.edu
http://www.cs.miami.edu/~janplaza

Abstract. We give new formulations of the property of soundness and completeness of a resolution system and of the lifting lemma, and we discuss their relationship.

The discussion points out why certain resolution systems are not complete, and that there is a simple method for showing that a resolution system is "absolutely incomplete" – that there is no notion of program completion and no logic which could give soundness and completeness. The method is demonstrated on the case of basic variants of the SLDNF-resolution. The same approach can be used to analyze other resolution systems.

Keywords: soundness and completeness, lifting lemma, program completion, SLDNF-resolution.

1 Introduction

One of the ideas underlying the development of logic programming is that programming should be to a greater extent declarative. Declarativeness of logic programming is characterized in a formal way, by so called, soundness and completeness properties. Many sound resolution systems have been proposed as possible bases for logic programming; [AB94] gives a broad overview. For most of these systems it has been shown that they do not satisfy a particular completeness property for the full class of normal programs. Still, one could ask whose fault it is: is it a deficiency of the resolution system or a deficiency of a particular formulation of the soundness and completeness property. This paper presents a simple method which can be used to put the blame entirely on the resolution system.

---

* This research has been partially supported from the NSF grant IRI 9308970.

In *The Craft of Prolog* [O'Keefe90] wrote: "Elegance is not optional," and then repeated: "Elegance is not optional." Let us add that elegance is not optional not only in programming but also in the presentations of a theory. It is our intention to make in this paper a step in the direction of presenting issues of completeness in an elegant way. For this purpose we will modify the definition of Herbrand universes; calling the resulting structures omega-Herbrand universes. The concept of the omega-Herbrand universe is *not* a generalzation of that of Herbrand universe. The change in the definition is not a mere technicality, and will have far reaching consequences: without it the universe would not have desired strong compactness properties. These compactness properties, in turn, will be crucial for showing equivalence of conventional and new formulations of soundness and completeness, and lifting properties. References to other non-technical properties of omeaga-Herbrand universes will be mentioned in Conclusion. Unlike conventional formulations of lifting, which provided just technical tools to be used in proofs of other theorems, the lifting property in our new formulation will have a clear philosophical reading. These ideas will be presented using a notation which allows to think of answers returned by a resolution system as subsets of a universe.

In the paper we assume that the reader is familiar with basic notions in foundations of logic programming, as presented, for instance, in chapters 1-3 of [Lloyd87]. Following conventions of the set theory, the set of natural numbers $\{0, 1, 2, \ldots$

2 Terms and Substitutions as Subsets of the omega-Herbrand Universe

A resolution system, such as the SLDNF, when given a program and a query, returns substitutions as answers. In order to discuss resolution systems conveniently, we will need to introduce notation which will allow us to think of such answers as of subsets of (a power of) a universe. The universe will be somewhat different from the Herbrand universe:

Definition 1 (omega-Herbrand universe).
Let $\mathcal{L}$ be a first-order language and let $\mathcal{L}^{\mathcal{K}}$ result by adding a countable set $\mathcal{K} = \{k_i \mid i < \omega\}$ of new individual constants to the alphabet of $\mathcal{L}$. By the *omega-Herbrand universe* $U_{\mathcal{L}}^{\omega}$ *for* $\mathcal{L}$ we understand the set of all ground terms of the language $\mathcal{L}^{\mathcal{K}}$. We refer to members of $U_{\mathcal{L}}^{\omega}$ as *elements*. Members of the set $\mathcal{K}$ will be called *free elements*. □

Notice that free elements of the omega-Herbrand universe for language $\mathcal{L}$ do not have names in $\mathcal{L}$. So the difference between the *ω-Herbrand universe* and the conventional Herbrand universe results from the presence of these unnamed elements. The reason for such a definition will be explained in the next section.

Now, with any term t we will associate the set $[\![t]\!]$ of all its instances in $U_{\mathcal{L}}^{\omega}$. For instance: $[\![f(c_1, g(c_2))]\!] = \{f(c_1, g(c_2))\}$, $[\![x]\!] = U_{\mathcal{L}}^{\omega}$ and $[\![g(x)]\!] = \{g(e) \mid e \in U_{\mathcal{L}}^{\omega}\}$.

As any substitution can be represented as a sequence of terms, not only terms but also substitutions can be represented using the same idea. Before we give a formal definition, a few words about substitutions and valuations.

By a *substitution* we understand any partial function from the set $Var = \{x_i \mid i < \omega\}$ of individual variables of a language $\mathcal{L}$, into the set of terms of $\mathcal{L}$. Substitutions are written as sets of elementary substitutions, e.g. $\theta = \{t_i/x_i \mid i \in I\}$ where I is a set of natural numbers. Any substitution $\{t_i/x_i \mid i \in I\}$ which is not total, is identified with the following total substitution: $\{t_i/x_i \mid i \in I\} \cup \{x_i/x_i \mid i \in (\omega - I)\}$. The symbol ϵ stands for the *identity substitution*: $\epsilon = \{x_i/x_i \mid i < \omega\}$. Notice that, by the convention above, the empty substitution (i.e. Prolog's YES) is identified with the identity substitution.

Given an interpretation of a language $\mathcal{L}$, by a *valuation* we always understand a total function from the set Var of variables of $\mathcal{L}$, into the domain of the interpretation.

Definition 2 (Terms and substitutions as subsets of $U_{\mathcal{L}}^{\omega}$).

1. Let t be a term of a first-order language $\mathcal{L}$. We define:
   $[\![t]\!] = \{t[\nu] \in U_{\mathcal{L}}^{\omega} \mid \nu : Var \longrightarrow U_{\mathcal{L}}^{\omega}\}$. Always $[\![t]\!] \subseteq U_{\mathcal{L}}^{\omega}$.
2. Let $x_i : i < \omega$ be a fixed enumeration of all the variables of $\mathcal{L}$, let $\theta = \{t_i/x_i \mid i < \omega\}$ be a substitution in $\mathcal{L}$, and let $\tilde{\theta} = \langle t_0, t_1, t_2, \ldots \rangle$ be the sequence of terms substituted for consecutive variables. We define: $[\![\theta]\!] = \{\tilde{\theta}[\nu] \in (U_{\mathcal{L}}^{\omega})^{\omega} \mid \nu : Var \longrightarrow U_{\mathcal{L}}^{\omega}\}$. Always $[\![\theta]\!] \subseteq (U_{\mathcal{L}}^{\omega})^{\omega}$. □

The meaning of $[\![t]\!]$ and $[\![\theta]\!]$ depends on the chosen language $\mathcal{L}$. Notice however that for every language $\mathcal{L}$ we have $[\![c]\!] \neq [\![x]\!]$. (For a language with a single individual constant c and no other function symbols, there wouldn't be any difference between $[\![c]\!]$ and $[\![x]\!]$ if one defined these notions using conventional Herbrand universes.) Notice also that we may write $[\![\theta_1]\!] \subseteq [\![\theta_2]\!]$ without specifying the language of the substitutions: If θ_1, θ_2 are substitutions in language $\mathcal{L}$, and if $\mathcal{L} \subseteq \mathcal{L}'$, then the relation $[\![\theta_1]\!] \subseteq [\![\theta_2]\!]$ does not depend on whether the substitutions are interpreted as subsets of $U_{\mathcal{L}}^{\omega}$ or of $U_{\mathcal{L}'}^{\omega}$.

The following proposition shows that the notation with $[\![t]\!]$ and $[\![\theta]\!]$ can concisely express basic notions related to unification.

Proposition 3.
1. $[\![t_1]\!] \subseteq [\![t_2]\!]$ *iff there exists a substitution* θ *such that* $t_1 = t_2\theta$.
2. $[\![t_1]\!] \cap [\![t_2]\!] = [\![t]\!]$ *iff* t *is the result of a most general unification of* t_1 *and* t_2.
3. $[\![\theta_1]\!] \subseteq [\![\theta_2]\!]$ *iff there exists a substitution* θ *such that* $\theta_1 = \theta_2\theta$ *(which means that* θ_2 *is a more general substitution than* θ_1.)

The proof is straightforward and will be omitted.

In topology, the compactness property says that every (open) cover contains a finite subcover. In the theory of logics intermediate between the classical and intuitionistic logic, one defines strongly compact pseudo-Boolean algebras as those in which every cover contains a single covering element. By analogy, we will call the properties listed in the following theorem a strong compactness.

Theorem 4. (Strong compactness)

    1. If $[\![t]\!] \subseteq \bigcup_{i \in I} [\![t_i]\!]$ then there exists $i_0 \in I$ such that $[\![t]\!] \subseteq [\![t_{i_0}]\!]$.

    2. If $[\![\theta]\!] \subseteq \bigcup_{i \in I} [\![\theta_i]\!]$ then there exists $i_0 \in I$ such that $[\![\theta]\!] \subseteq [\![\theta_{i_0}]\!]$.

Proof.

1. Assume $[\![t]\!] \subseteq \bigcup_{i \in I} [\![t_i]\!]$. Consider valuation $\iota = \{k_i/x_i \mid i < \omega\}$. Now, $t[\iota]$ is a ground term and we have $t[\iota] \in \bigcup_{i \in I} [\![t_i]\!]$. Take i_0 such that $t[\iota] \in [\![t_{i_0}]\!]$. There exists valuation ν such that $t[\iota] = t_{i_0}[\nu]$ and there exists substitution θ' such that $\nu = \theta'[\iota]$. Thus $t = t_{i_0}\theta'$ and $[\![t]\!] \subseteq [\![t_{i_0}]\!]$.

2. Assume $[\![\theta]\!] \subseteq \bigcup_{i \in I} [\![\theta_i]\!]$. Consider valuation $\iota = \{k_i/x_i \mid i < \omega\}$. Now, $\tilde{\theta}[\iota]$ is a sequence of ground terms and we have $\tilde{\theta}[\iota] \in \bigcup_{i \in I} [\![\theta_i]\!]$. Take i_0 such that $\tilde{\theta}[\iota] \in [\![\theta_{i_0}]\!]$. There exists valuation ν such that $\tilde{\theta}[\iota] = \tilde{\theta}_{i_0}[\nu]$ and there exists substitution θ' such that $\nu = \theta'[\iota]$. Thus $\tilde{\theta} = \tilde{\theta}_{i_0}(\theta')$ and $[\![\theta]\!] \subseteq [\![\theta_{i_0}]\!]$.   □

Remark. The choice of omega-Herbrand universes instead of conventional Herbrand universes in Definition 2 is essential. The definition has been constructed so that it guarantees that Theorem 4 on strong compactness holds. In [Kunen87] a notion analogous to $[\![t]\!]$ is defined using the conventional Herbrand universe $U_{\mathcal{L}}$; this change causes that the statement analogous to Theorem 4 is not true, even under Kunen's assumption that the language $\mathcal{L}$ contains infinitely many function symbols or constants. Indeed, for every such language $\mathcal{L}$ we would have $[\![x]\!] \subseteq \bigcup_{t \in U_{\mathcal{L}}} [\![t]\!]$ but there would be no single term $t_0 \in U_{\mathcal{L}}$ such that $[\![x]\!] \subseteq [\![t_0]\!]$. Lemma 1 in [Kunen87] states a property much weaker than 4. (One could also notice that having *infinitely* many free elements in the universe is essential for Theorem 4.)   □

Remark. In constraint logic programming over a domain of finite terms one can consider substitutions constrained by formulas involving no predicates except equality. For instance,

$$\{c/x_1,\ f(x_5)/x_2,\ x_4/x_3\} \text{ WHERE } (x_5 \neq x_4 \wedge \forall_y x_5 \neq f(y)).$$

Also such substitutions can be interpreted as subsets of the omega-Herbrand universe:

$$[\![\theta \text{ WHERE } S]\!] = \{\tilde{\theta}[\nu] \in (U_{\mathcal{L}}^{\omega})^{\omega} \mid U_{\mathcal{L}}^{\omega} \models S[\nu]\}.$$

but the compactness property does not hold. Indeed,

$$[\![\epsilon]\!] = \bigcup_{i < \omega} [\![\{f^i(x_1)/x_0\} \text{ WHERE } (\forall_y x_1 \neq f(y))]\!]$$

but there is no finite $I \subseteq \omega$ such that

$$[\![\epsilon]\!] = \bigcup_{i \in I} [\![\{f^i(x_1)/x_0\} \text{ WHERE } (\forall_y x_1 \neq f(y))]\!].$$
   □

3 Soundness and Completeness versus Lifting Lemma

Now we are ready to start analyzing resolution systems. For SLD-resolution, definite program P and goal $\leftarrow B$, we could define the *SLD-answer set* as: $SLD(P, B) = \bigcup_{i \in I} [\![\theta_i]\!]$, where $\theta_i : i \in I$ is the collection of all SLD-computed answers for P and $\leftarrow B$. Similarly, for SLDNF-resolution, normal program P and goal $\leftarrow B$, the *SLDNF-answer set* is: $SLDNF(P, B) = \bigcup_{i \in I} [\![\theta_i]\!]$, where $\theta_i : i \in I$

is the collection of all SLDNF-computed answers for P and $\leftarrow B$. The next definition generalizes these notions.

Definition 5. Let *Resolution* be a resolution system which, if given a program P and a goal $\leftarrow B$, returns a sequence of computed answer substitutions $\theta_i : i \in I$. Then, by the *Resolution-answer set* for P and $\leftarrow B$ we understand the set: $Resolution(P, B) = \bigcup_{i \in I}[\![\theta_i]\!]$. □

Soundness and completeness of SLD-resolution for the class of definite programs and goals are usually[2] formulated as two separate implications, cf. [Clark79], [Lloyd87]:

Soundness: If θ is an SLD-computed answer for P and $\leftarrow B$, then $Comp(P) \vdash B\theta$.
Completeness: If $Comp(P) \vdash B\theta$ then there exists a substitution θ', more general than θ, which is an SLD-computed answer for P and $\leftarrow B$.

These statements could not be combined into one equivalence because of a technical difficulty with the "substitution θ' that is more general than θ". However if we use answer sets from Definition 5 we may conveniently formulate soundness and completeness as a single equivalence:
$$SLD(P, B) \supseteq [\![\theta]\!] \quad \text{iff} \quad Comp(P) \vdash B\theta.$$
This is explained in the following proposition.

Proposition 6. *The following formulations of the soundness and completeness property are equivalent:*

1. *If θ is a Resolution-computed answer for P and $\leftarrow B$*
   then $Completion(P) \vdash B\theta$;
   If $Completion(P) \vdash B\theta$ then there exists a substitution θ' more general than θ, which is a Resolution-computed answer for P and $\leftarrow B$.
2. *$Resolution(P, B) \supseteq [\![\theta]\!] \quad$ iff $\quad Completion(P) \vdash B\theta$*

The reader should notice that that this proposition critically depends on the Theorem 4 on strong compactness.

Proof.

$1 \Rightarrow 2$. For the proof of the implication to the right assume that $Resolution(P, B) \supseteq [\![\theta]\!]$. This means that $[\![\theta]\!]$ is covered by the *Resolution*-computed answers for P and $\leftarrow B$. By Theorem 4 on strong compactness there exists a single *Resolution*-computed answer $[\![\theta']\!]$ such that $[\![\theta']\!] \supseteq [\![\theta]\!]$. Then, by 1, $Completion(P) \vdash B\theta'$ and of course $Completion(P) \vdash B\theta$.
For the proof of the implication to the left assume that $Completion(P) \vdash B\theta$. Then, by 1, *Resolution* computes an answer θ' more general than θ. We have $Resolution(P, B) \supseteq [\![\theta']\!] \supseteq [\![\theta]\!]$.

---

[2] perhaps using just P instead of Clark's program completion $Comp(P)$.

$2 \Rightarrow 1$. For the proof of the first implication assume that θ is a *Resolution*-computed answer for P and $\leftarrow B$. Thus $Resolution(P, B) \supseteq [\![\theta]\!]$ and by 2, $Completion(P) \vdash B\theta$.

For the proof of the second implication assume that $Completion(P) \vdash B\theta$. Then, by 2, $Resolution(P, B) \supseteq [\![\theta]\!]$ and by Theorem 4 on strong compactness there exists a single *Resolution*-computed answer $[\![\theta']\!]$ such that $[\![\theta']\!] \supseteq [\![\theta]\!]$ which means that θ' is more general than θ. □

Another important property of a resolution system is related to the lifting lemma. The lifting lemma, in its many variants, has been used as a technical tool in the proofs of completeness of resolution systems: it allows to "lift" ground derivations to the general derivations involving variables, cf. [Robinson65], [Clark79], [Lloyd87] and [Doets94]. In our new formulation the lifting lemma reads:

$$Resolution(P, B) \supseteq [\![\theta]\!] \quad \text{iff} \quad Resolution(P, B\theta) \supseteq [\![\epsilon]\!].$$

This can be understood as:

"*Resolution* is capable of computing answer θ, exactly when it accepts θ as appropriate".

Indeed, by Theorem 4 on strong compactness, $Resolution(P, B) \supseteq [\![\theta]\!]$ means that θ can be computed, and $Resolution(P, B\theta) \supseteq [\![\epsilon]\!]$ means that given P and $\leftarrow B\theta$ the resolution returns the identity substitution, i.e. *YES*, thus accepting θ as appropriate. As our formulation has a clear intuitive meaning we will no longer call it a lemma, referring to it as to a *lifting property*. We believe that the lifting property in this formulation is interesting for its own sake, independently of its applications in proofs of other theorems. The next proposition explains how this version is related to other versions.

Proposition 7. *Consider the following formulations of lifting properties.*

1. *If θ' is a Resolution-computed answer for P and $\leftarrow B\theta$, then there exists a substitution θ'' more general than $\theta\theta'$, such that θ'' is a Resolution-computed answer for P and $\leftarrow B$.*
   (This is the conventional formulation, cf. [Lloyd87], lm. 8.2 p. 47.)
2. *$Resolution(P, B\theta) \supseteq [\![\theta_0]\!]$ implies $Resolution(P, B) \supseteq [\![\theta\theta_0]\!]$.*
3. *$Resolution(P, B) \supseteq [\![\theta]\!]$ iff $Resolution(P, B\theta) \supseteq [\![\epsilon]\!]$*

Formulations 1 and 2 are equivalent. Formulation 3 implies each of 1 and 2.

The reader should notice that that this proposition critically depends on the Theorem 4 on strong compactness.

Proof.

$1 \Rightarrow 2$. Assume that $Resolution(P, B\theta) \supseteq [\![\theta_0]\!]$. Thus by Theorem 4 on strong compactness *Resolution* with P and $\leftarrow B\theta$ computes θ' such that $[\![\theta']\!] \supseteq [\![\theta_0]\!]$. By 1, there exists θ'' such that $[\![\theta'']\!] \supseteq [\![\theta\theta']\!]$ and $Resolution(P, B) \supseteq [\![\theta'']\!]$. Thus $Resolution(P, B) \supseteq [\![\theta'']\!] \supseteq [\![\theta\theta']\!] \supseteq [\![\theta\theta_0]\!]$.

$2 \Rightarrow 1$. Assume that *Resolution* with P and $\leftarrow B\theta$ computes θ'. Then, by 2, *Resolution*$(P, B) \supseteq [\![\theta\theta']\!]$. By Theorem 4 on strong compactness *Resolution* with P and $\leftarrow B\theta$ computes θ'' which is more general than $\theta\theta'$.

$3 \Rightarrow 2$. Assume that *Resolution*$(P, B\theta) \supseteq [\![\theta_0]\!]$. By 3, *Resolution*$(P, B\theta\theta_0) \supseteq [\![\epsilon]\!]$. By 3 again (using the other implication), *Resolution*$(P, B) \supseteq [\![\theta\theta_0]\!]$. $\square$

The following theorem offers a method for showing that a resolution system is "absolutely incomplete."

Theorem 8. (the test of lifting)

If a resolution system does not satisfy the lifting property:

$(*)$ *Resolution*$(P, B) \supseteq [\![\theta]\!]$ *iff*   *Resolution*$(P, B\theta) \supseteq [\![\epsilon]\!]$

then there is no notion of program completion Completion and no logic **L** *for which the soundness and completeness property:*

$(**)$ *Resolution*$(P, B) \supseteq [\![\theta]\!]$ *iff*   *Completion*$(P) \vdash_{\mathbf{L}} B\theta$

could hold.

Proof. It is enough to notice that the soundness and completeness $(**)$ implies the lifting property $(*)$. Indeed:

$$
\begin{aligned}
&\textit{Resolution}(P, B) \supseteq [\![\theta]\!] && \text{iff (by } (**)) \\
&\textit{Completion}(P) \vdash_{\mathbf{L}} B\theta && \text{iff} \\
&\textit{Completion}(P) \vdash_{\mathbf{L}} (B\theta)\epsilon && \text{iff (by } (**)) \\
&\textit{Resolution}(P, B\theta) \supseteq [\![\epsilon]\!]. &&
\end{aligned}
$$
$\square$

Theorem 8 can be paraphrased as: lifting $(*)$ is a necessary condition for soundness and completeness $(**)$. The question whether it is a sufficient condition cannot be considered unless one defines what constitutes a program completion or a logic and what doesn't. Considering this issue is not in the scope of this paper.

Remark. The proof above does not use the strong compactness property, so Theorem 8, on the test of lifting, generalizes to the case of resolution systems for constraint logic programming over universes of terms. $\square$

As a consequence of the soundness and completeness of SLD-resolution, from Theorem 8 we obtain the following expected corollary:

Corollary 9. *For definite programs and goals, the SLD-resolution satisfies the lifting property:*

$$SLD(P, B) \supseteq [\![\theta]\!] \quad \textit{iff} \quad SLD(P, B\theta) \supseteq [\![\epsilon]\!]$$

4   "Absolute Incompleteness" of the SLDNF-resolution

The SLD-resolution is sound and complete for definite programs and goals cf. [Clark79], [Lloyd87]. We will formulate this in the following two ways:

$$
\begin{aligned}
SLD(P, B) \supseteq [\![\theta]\!] &\quad \text{iff} \quad P \vdash B\theta \\
SLD(P, B) \supseteq [\![\theta]\!] &\quad \text{iff} \quad Comp(P) \vdash B\theta
\end{aligned}
$$

where *Comp* is Clark's program completion. Now let us consider basic generalizations of SLD-resolution which can handle programs with negation.

In [Clark78] SLDNF-resolution was introduced and proved sound for normal programs and goals. We will formulate this as:

$$SLDNF(P, B) \supseteq [\![\theta]\!] \quad \text{implies} \quad Comp(P) \vdash B\theta.$$

It is well known that for normal programs and goals the SLDNF-resolution is not complete with respect to Clark's program completion and inferences in classical logic. The reason for this lack of completeness of has been identified as related to the safeness condition.

Let us recall that the safeness condition disallows selecting a non-ground negative literal for processing. It is a reasonable idea to relax this condition. We paraphrase the description from [Lloyd87], (p. 94):

It is possible to weaken the safeness condition a little and still maintain soundness. Consider the following weaker safeness condition. Non-ground negative literals can be selected for processing. If the subgoal $\neg A$ is selected, a "lemma" is created with a single subgoal A. If A fails, then $\neg A$ succeeds. However, if A succeeds, a check is made to make sure no bindings were made to any variables in the top-level goal of the corresponding refutation. If no such binding was made, then $\neg A$ fails. But, if such a binding was made, then a different literal is selected and $\neg A$ is delayed in the hope that more of its variables will be bound later. If we are left with only negative literals, neither of which can be processed, the system halts without returning an answer – i.e. flounders.

Let us call such a resolution SLDNF$^+$. SLDNF$^+$ is a generalization of SLDNF-resolution: $SLDNF(P, B) \subseteq SLDNF^+(P, B)$. It is known that for normal programs and goals SLDNF$^+$ resolution is sound with respect to Clark's program completion and inferences in classical logic:

$$SLDNF^+(P, B) \supseteq [\![\theta]\!] \quad \text{implies} \quad Comp(P) \vdash B\theta$$

but the converse implication still does not hold.

Finally let us consider the version of SLDNF in which we drop the safeness condition entirely. This is an idealization of Prolog's computational mechanism and let us call it PROLOG$^+$. Notice however that in Prolog not only the safeness condition is dropped but also, for pragmatic reasons, unification is done without the occurs check, and a deterministic depth first search is implemented. PROLOG$^+$ doesn't make these further concessions — its unification is correct and selection of subgoals and input clauses in derivations is non-deterministic.

It is known that for normal programs and goals, PROLOG$^+$ is neither sound nor complete with respect to Clark's program completion and inferences in classical logic.

The reader already expects that the variants of SLDNF-resolution above are "absolutely incomplete". This term, which was also used in the title of this section, is surrounded by quotes, in order to emphasize that it is not precise. A more precise term would be: absolutely not sound-and-complete for a specific class of programs. A precise formulation is given in the following theorem.

Theorem 10. *For each of the following resolution systems:*
$$SLDNF, \; SLDNF^+, \; PROLOG^+$$
there is no notion of program completion Completion and no logic **L** *for which the soundness and completeness property:*
$$Resolution(P, B) \supseteq [\![\theta]\!] \quad \textit{iff} \quad Completion(P) \vdash_{\mathbf{L}} B\theta$$
could hold for the class of normal programs and goals.

Proof. By Theorem 8 it is enough to show that the resolution systems listed above do not satisfy the lifting property $(*)$. Consider the following normal program:
$$p(x) \leftarrow \neg q(x)$$
$$q(a) \leftarrow false$$
$$q(b) \leftarrow true$$
In the case of SLDNF, the derivation for P and $\leftarrow p(x)$ encounters subgoal $\neg q(x)$ and flounders (i.e. halts, due to the safeness condition, without returning any answer.) So $SLDNF(P, p(x)) = \emptyset$. On the other hand $SLDNF(P, p(a)) = [\![\epsilon]\!] = YES = (U_{\mathcal{L}}^\omega)^\omega$. So the lifting property does not hold.

In the case of SLDNF$^+$, the derivation for P and $\leftarrow p(x)$ encounters subgoal $\neg q(x)$ and finds an illegal substitution b/x which it discards, and then flounders. So $SLDNF^+(P, p(x)) = \emptyset$. On the other hand $SLDNF^+(P, p(a)) = [\![\epsilon]\!] = YES = (U_{\mathcal{L}}^\omega)^\omega$. So the lifting property does not hold.

In the case of PROLOG$^+$, the derivation for P and $\leftarrow p(x)$ encounters subgoal $\neg q(x)$ and finds a substitution b/x which causes the subgoal to fail. So $PROLOG^+(P, p(x)) = \emptyset = NO$ and $PROLOG^+(P, p(a)) = [\![\epsilon]\!] = YES$. So the lifting property does not hold. $\square$

Remark. It is known that there exist subclasses of the class of normal programs and goals, for which SLDNF-resolution is sound and complete with respect to Clark's program completion. [Kunen89] contains such a result for allowed programs and inferences in Kleene's three-valued logic. [Plaza90, Plaza91] contains such a result for propositional normal programs, and inferences in classical logic. [DM91] contains such a result for allowed programs and inferences in classical logic. $\square$

¿From the results mentioned in the remark above, by Theorem 8 we obtain the following corollary.

Corollary 11. *For allowed programs and goals, the SLDNF-resolution satisfies the lifting property:*
$$SLDNF(P, B) \supseteq [\![\theta]\!] \quad \textit{iff} \quad SLDNF(P, B\theta) \supseteq [\![\epsilon]\!]$$

One source of the lack of completeness of SLDNF for the full class of normal programs and goals is the lack of symmetry in the way it treats positive and negative subgoals; SLDNF makes substitutions in positive but not in negative subgoals. So to obtain completeness one need a resolution system which never flounders, which can process negative subgoals making substitutions, and when it stops, always returns an answer. Many such resolution systems have been considered, see for instance [Chan88, Chan89], [Przymusiński89], [Plaza90, Plaza92f] and [Drabent95] which consider various forms of constructive negation. For an overview of other directions see [AB94]. In this paper we cannot argue whether these systems satisfy the lifting property (∗), but for the researchers working on these systems it should be straightforward to produce appropriate argument.

5 Conclusion

We presented omega-Herbrand universes and discussed one of their advantages over conventional Herbrand universes. Further advantages are demonstrated in [Plaza92o] and [Plaza92f].

We gave a formulation (∗∗) of the soundness and completeness property which allows for a convenient analysis of this notion.

We gave a non-technical formulation (∗) of the lifting property which, we believe, makes this property interesting in itself, not only as a tool in proofs of other theorems. The informal reading of the property is: a resolution system is capable of computing an answer exactly when it accepts this answer as correct.

We formulated the test of lifting (Theorem 8) – a condition sufficient for an "absolute incompleteness" of a resolution system.

We showed that three basic variants of the SLDNF-resolution are "absolutely incomplete" (Theorem 10): there is no notion of program completion and no logic which could yield the soundness and completeness property for the class of normal programs and goals.

Acknowledgments
Special thanks to Professor Melvin Fitting.

References

[AB94]   K. R. Apt and R. N. Bol, Logic Programming And Negation: A Survey, *Journal of Logic Programming* Vol.19/20, pp. 9-71, 1994.

[Chan88]   D. Chan, Constructive Negation Based on Completed Database, in [LPconf88], pp. 111-125.

[Chan89]   D. Chan, An extension of constructive negation and its application in corouting, in [LPconfNA89], pp. 477-493.

[Clark78]   K. L. Clark, Negation as Failure, in *Logic and Databases*, edited by H. Gallaire and J. Minker, Plenum Press, New York, 1978, pp. 193-322.

[Clark79]   K. L. Clark, Predicate Logic as a Computational Formalism, *Research Report DOC 79/59*, Dept. of Computing, Imperial College, 1979.

[Doets94]   K. Doets, *From Logic to Logic Programming* MIT Press, 1994.

[DM91]   W. Drabent and M.Martelli, Strict Completion of Logic Programs, *New Generation Computing* 9, pp. 69-79, 1991.

[Drabent95]   What is Failure? An Approach to Constructive Negation, *Acta Informatica*, to appear.

[Kunen87]   K. Kunen, Answer Sets and Negation as Failure, in [LPconf87].

[Kunen89]   K. Kunen, Signed Data Dependencies in Logic Programs, *Journal of Logic Programming*, 1989, vol. 7, no. 3, pp.231-247.

[Lloyd87]   J. W. Lloyd, *Foundations of Logic Programming*, Second extended edition, Springer Verlag, 1987.

[LPconf87]   J-L. Lassez (ed.), *Logic Programming, Proceedings of the Fourth International Conference*, MIT Press, 1987.

[LPconf88]   R. A. Kowalski and K. A. Bowen (eds.), *Logic Programming, Proceedings of the Fifth International Conference and Symposium*, MIT Press, 1988.

[LPconfNA89]   E. L. Lusk and R. A. Overbeek (eds.), *Logic Programming, Proceedings of the North American Conference 1989*, MIT Press, 1989.

[O'Keefe90]   R. A. O'Keefe, *The Craft of Prolog*, MIT Press, 1990.

[Plaza90]   J. A. Plaza, *Fully Declarative Programming with Logic – Mathematical Foundations* Ph.D. Dissertation, City University of New York, July 1990.

[Plaza91]   J. A. Plaza, Completeness for propositional logic programs with negation, in: Methodologies for Intelligent Systems, 6th International Symposium, edited by Z. W. Ras and M. Zemankova, Springer Verlag, LNAI 542, 1991, pp. 600-609.

[Plaza92f]   J. A. Plaza, Fully Declarative Logic Programming, in *Programming Language Implementation and Logic Programming, 4th International Symposium*, edited by M. Bruynooghe and M. Wirsing, Springer Verlag, LNCS 631, 1992, pp. 415-427.

[Plaza92o]   J. A. Plaza, Operators on Lattices of omega-Herbrand Interpretations, in: A. Nerode and M. Taitslin (eds.) *Logical Foundations of Computer Science – Tver '92, Second International Symposium, Tver, Russia, July 1992, Proceedings*, Springer Verlag, LNCS 620, 1992 pp. 358-369.

[Przymusiński89] T. C. Przymusiński, On Constructive Negation in Logic Programming, in [LPconfNA89], addendum.

[Robinson65]   J. A. Robinson, A machine-oriented logic based on the resolution principle, *J. ACM* 12, 1965, pp. 23-41.

Reasoning with Preorders and Dynamic Sorts Using Free Variable Tableaux

A. Gavilanes, J. Leach, P.J. Martín, S. Nieva

Dep. de Informática y Automática. Universidad Complutense de Madrid
e-mail: {agav, leach, pjmartin, nieva}@dia.ucm.es *

Abstract. In this paper we present a three valued many sorted logic for dealing with preorders, incorporating subsort relations into the syntax of the language, and where formulas taking the third boolean value as interpretation contain a term or a predicate which is not well-sorted w.r.t. the signature. For this logic a ground tableau-based deduction method and a free variable extension version are proposed, proving their completeness.

1 Introduction

The study of efficient methods for dealing with equality has been traditionally considered an important workline in different areas of theoretical computer science. More recently has emerged the need of extending this study in order to cover non-symmetric relations different from those of equivalence. This is the case, for instance, of using ordered sorts in equational and object-oriented programming [GM 92], or constraints in declarative programming languages [JM 94]. In these cases, the logical framework includes a specific handling for some aspect of the involved relation, like sorted unification [Wei], constraint solvers [JM 94], or bi-rewriting [LA 93]. In the field of automated deduction this situation has resulted in the development of provers with additional rules expressing the properties of the relation. This line was followed for example in [BKS 85], and, more recently, in [BG 95], by applying rewriting-based techniques. These and other attempts take place considering resolution as proof method.

On the other hand sorts are commonly argued as a great applied benefit and, specifically, ordered sorts are considered as a simple and natural way for incorporating partial functions, multiple representation and constructor-selectors in structured data [GM 92]. In automated deduction, we add to these considerations the fact that sorts significantly reduce the search space. When reasoning in an order-sorted setting, it is usual to have the sort information apart from data; then there is no interaction between the statically declared information about sorts and the formulas. Examples of this framework are the resolution calculi in [Wal 87] based on a many sorted logic, and the tableau method of [SW 90] for an order sorted logic. However in several situations it is necessary reasoning under assumptions about sort relations [Wei 91]. In this case formal mechanisms must be incorporated to the language for expressing relations about sorts at the same level than formulas; it is said that sorts behave dynamically. For instance the formula $s \leq s' \to (\forall x^{s'} P(x^{s'}) \to \forall x^s P(x^s))$ expressing that predicate P holds for every s-element, whenever it holds for every s'-element, if s is a subsort of s'. But this kind of logics are forced to be unsorted since static sort declarations can not be given when sorts are dynamic. In fact, if the declared sort of P is s', a formula as the previous one would be false because $P(x^s)$ has no sense. We propose a way for merging a (many) sorted logic with dynamic sorts that consists of making the

---

*This paper has been supported by Proyecto Precompetitivo PR 219/94 5564.

logic three valued, where undefinedness is produced by not well-sorted formulas, that is, formulas including terms or predicates not respecting the restrictions imposed by the sort declarations of the signature.

In this paper a three-valued many sorted predicate logic with preorders and dynamic sorts is defined, allowing a uniform treatment of relations about sorts and non-symmetric relations about data, where functions and predicates behave monotonically or antimonotonically in their arguments. For this logic we present sound and complete tableau-based deduction systems. Although semantic tableaux are not the most efficient tool for automated theorem proving, they possess some characteristics which make them suitable for mechanization (they can be extended to many nonclassical logics used in AI research [Fit 88],[GL 90],[LN 93], they easily allow the introduction of heuristics and human interaction [OS 88], they do not require conversion to canonical forms, and bring the possibility of generating counterexamples for non-theorems).

Our deduction system is based on the free variable semantic tableaux proposed by Fitting [Fit 96] for adding equality to classical tableaux, where some specific extension rules are defined, expressing equality as a congruence relation and, at the same time, restricting the search space. Other specific free variable tableau versions have been proposed for handling equalities [HS 94], [BH 92]. In a previous work, [GLN 96], we defined tableaux methods for dealing with data inequalities in a classical framework (only two boolean values). In this paper we have significantly improved the expressive power of the logic by allowing dynamic sorts.

The organization of the paper is as follows. In Section 2 our logic is outlined and its basic properties are shown. Section 3 extends first order ground semantic tableaux to cope with preorders, formulating additional tableau expansion rules dealing with inequalities between terms –for the monotonic and antimonotonic cases– and sorts. Section 4 presents a more efficient system, such that the inequality expansion rules are defined in presence of γ and δ rules that use free variables instead of ground terms. The completeness of this system is shown by means of a Lifting Lemma. Examples of applications of the method are presented in Section 5. We finish commenting some related works.

2   The Logic LPDS

LPDS stands for *Logic with Preorders using Dynamic Sorts*, a many-sorted three-valued predicate logic with data inequalities, where operations behave monotonically or antimonotonically in their arguments, and subsort information incorporated within the language.

A preorder is a pair $(D, \sqsubseteq_D)$ where D is a nonempty set and $\sqsubseteq_D$ is a reflexive and transitive binary relation on D. In particular a partial order is an antisymmetric preorder, and an equivalence (e.g. equality) is a symmetric preorder. Given $(D_i, \sqsubseteq_{D_i})$, $1 \leq i \leq n$, and $(D, \sqsubseteq_D)$ preorders, a mapping $f : D_1 \times \ldots \times D_n \to D$ is monotonic (resp. antimonotonic) in the i-th argument, if for every $d_i, d'_i \in D_i$, $d_i \sqsubseteq_{D_i} d'_i \Rightarrow$ $f(d_1, \ldots, d_i, \ldots, d_n) \sqsubseteq_D f(d_1, \ldots, d'_i, \ldots, d_n)$ (resp. the inverse order $\sqsupseteq_D$).

A signature Σ for *LPDS* consists of a finite set S of sorts s, and sorted sets of constant $C^s (s \in S)$, function $F^{s_1, \ldots, s_l \to s}$ $(s_1, \ldots, s_l, s \in S)$, and predicate symbols $P^{s_1, \ldots, s_r}$ $(s_1, \ldots, s_r \in S)$. In addition to this, for every function symbol f of arity l, Σ contains a mapping $m(f) : \{1, \ldots, l\} \to \{0, 1\}$ expressing in which arguments f will be interpreted as monotonic $(m(f)(i) = 0)$ or antimonotonic $(m(f)(i) = 1)$. Analogously a mapping $m(P)$ is supplied for every predicate symbol P. Given a signature Σ and a sorted family of variables $X = (X^s)_{s \in S}$, terms and formulas are defined as follows.

Definition 1 *The sets $T(\Sigma)$ of Σ-terms and $F(\Sigma)$ of Σ-formulas are given by the following formation rules:*

$$t ::= x^s (\in X^s) \mid c^s \ (\in C^s) \mid f(t_1, \ldots, t_n) \ (f \in F^{s_1, \ldots, s_n \to s}; \ t_i \in T(\Sigma)).$$
$$\varphi ::= t \sqsubseteq t' \mid s \sqsubseteq_0 s' \mid P(t_1, \ldots, t_n) \ (P \in P^{s_1, \ldots, s_n}; t_i \in T(\Sigma)) \mid \neg\varphi \mid \varphi \wedge \varphi' \mid \exists x^s \varphi.$$

$\forall x^s \varphi, \ \varphi \vee \varphi', \varphi \to \varphi'$ will stand for their classical abbreviations $\neg \exists x^s \neg \varphi$, $\neg(\neg\varphi \wedge \neg\varphi')$ and $\neg\varphi \vee \varphi'$, respectively. We omit superscripts for sorts where they are not relevant.

Observe that we are allowing the construction of terms and predicates not well-sorted w.r.t. the signature. The reason for this apparent misuse is to allow the logic to express properties about data, that are conditioned to assumptions about sorts. When making reasonings, we expect that the combination of the subsort information provided by formulas and the sorted information given in the signature will reduce the search space significantly.

We generalize the set of terms of an order sorted logic defining sets of well-sorted terms w.r.t. the information about subsort relations contained in a set of formulas.

Definition 2 *Given a set of Σ-formulas Φ, the sorted family of Φ-well-sorted sets of Σ-terms $\mathcal{T}_\Sigma^\Phi(s)$ is the least sorted family of sets Y^s of Σ-terms such that:*
1. $C^s \subset Y^s$.   2. $X^s \subset Y^s$.   3. If $s' \sqsubseteq_0 s \in \Phi$ then $Y^{s'} \subseteq Y^s$.
4. If $f \in F^{s_1, \ldots, s_n \to s}$ and $t_i \in Y^{s_i}, i = 1, \ldots, n$ then $f(t_1, \ldots, t_n) \in Y^s$.

For terms $t \in T(\Sigma)$, we use $sort(t)$ for representing the sort of t deduced from the signature Σ; it is defined by $sort(x^s) = sort(c^s) = s$ and $sort(f(t_1, \ldots, t_n)) = s$, if $f \in F^{s_1, \ldots, s_n \to s}$. When needed we will use $T(\Sigma, s)$ to represent $\{t \in T(\Sigma) \mid sort(t) = s\}$.

Substitutions in *LPDS* are mappings $\tau : X \to T(\Sigma)$ such that $\tau(x^s) \in T(\Sigma, s)$. The result of substituting a term t for a variable x^s in a term t' or a formula φ is defined as usual and respectively denoted by $t'[t/x^s]$ and $\varphi[t/x^s]$. Remark that substitutions keep the sort of the substituted variable, so when unification were demanded we will face to syntactic rather than sorted unification problems.

As we have said, LPDS allows constructions which are not well-sorted w.r.t. the signature. Nevertheless, these construcctions may be well-sorted w.r.t. the subsort information provided by a set of formulas in the sense of the following definition.

Definition 3 *The well-sortedness of a formula φ w.r.t. a set of Σ-formulas Φ, written $WS(\varphi, \Phi)$, is a property defined by structural induction on φ as follows:*

- $WS(s \sqsubseteq_0 s', \Phi) \Leftrightarrow_{def} s \sqsubseteq_0 s' \in \Phi$.

- $WS(t_1 \sqsubseteq t_2, \Phi) \Leftrightarrow_{def} WS(t_i, \Phi)$ *and there is $s \in S$ ($sort(t_i) \sqsubseteq_0 s \in \Phi, i = 1, 2$).*

  Where the well-sortedness of a term t w.r.t. Φ, written $WS(t, \Phi)$, is a property defined by induction on t as follows:

  — $WS(t, \Phi)$, *if t is a constant or a variable.*

  — $WS(f(t_1, \ldots, t_n), \Phi) \Leftrightarrow_{def} WS(t_i, \Phi)$, $sort(t_i) \sqsubseteq_0 s_i \in \Phi, 1 \leq i \leq n, f \in F^{s_1 \ldots s_n \to s}$.

- $WS(P(t_1, \ldots, t_n), \Phi) \Leftrightarrow_{def} WS(t_i, \Phi)$, $sort(t_i) \sqsubseteq_0 s_i \in \Phi, 1 \leq i \leq n, P \in P^{s_1 \ldots s_n}$.

- $WS(\neg\varphi, \Phi) \Leftrightarrow_{def} WS(\varphi, \Phi)$. • $WS(\varphi_1 \wedge \varphi_2, \Phi) \Leftrightarrow_{def} WS(\varphi_i, \Phi), i = 1, 2$.

- $WS(\exists x^s \varphi, \Phi) \Leftrightarrow_{def} WS(\varphi[c/x^s], \Phi)$, *where $c \in C^s$.*

For a set of formulas Φ' we define $WS(\Phi', \Phi) \Leftrightarrow_{def} (WS(\varphi, \Phi)$, for every $\varphi \in \Phi)$.
A substitution τ is well-sorted w.r.t. a set of Σ-formulas Φ, written $WS(\tau, \Phi)$, if $WS(\tau(x), \Phi)$, for every variable x in the domain of τ.

Remark that whenever Φ is closed w.r.t. transitivity in subsort relations, Def. 3 concerning to terms coincides with Def. 2, in the sense that: $WS(t, \Phi) \Leftrightarrow$ there is $s \in S$ such that $t \in \mathcal{T}_{\Sigma}^{\Phi}(s)$.

Lemma 4 *Given terms $t, t' \in T(\Sigma, s)$, a set of Σ-formulas $\Phi \cup \{\varphi\}$ and a substitution τ, the following conditions hold:*
(1) $WS(t\tau, \Phi) \Rightarrow WS(t, \Phi)$ and $WS(\varphi\tau, \Phi) \Rightarrow WS(\varphi, \Phi)$.
If $WS(\tau, \Phi)$ then $(WS(t, \Phi) \Rightarrow WS(t\tau, \Phi))$ and $(WS(\varphi, \Phi) \Rightarrow WS(\varphi\tau, \Phi))$.[1]
(2) If σ is the most general unifier of t, t', then $WS(t, \Phi), WS(t', \Phi) \Rightarrow WS(\sigma, \Phi)$.

For interpreting terms and formulas, Σ-structures supply preordered domains for each sort and a special element $\perp$ for representing the value of non well-sorted terms. Observe that we do not have partial functions and that undefinedness does not express here partiality. If a term is interpreted as undefined then it will contain a non well-sorted subterm. Of course, the opposite direction is in general false, because the semantic value of a term depends on the inclusion relation between the domains of the structure.

Definition 5 *A Σ-structure $\mathcal{D}$ is composed by a system $\{(D^s, \sqsubseteq_s^{\mathcal{D}}) | s \in S\} \cup \{\perp\}$ and interpretations for constants $\{c^{\mathcal{D}} \in D^s | c \in C^s\}$, function symbols $\{f^{\mathcal{D}} : D^{s_1} \times \ldots \times D^{s_l} \to D^s | f \in F^{s_1, \ldots, s_l \to s}\}$, and predicate symbols $\{P^{\mathcal{D}} : D^{s_1} \times \ldots \times D^{s_r} \to \{\underline{f}, \underline{t}\} | P \in P^{s_1, \ldots, s_r}\}$ such that:*

1. *$\{(D^s, \sqsubseteq_s^{\mathcal{D}}) | s \in S\}$ is a family of preorders that is transitive in the following sense: if $d \sqsubseteq_s^{\mathcal{D}} d'$, $d' \sqsubseteq_{s'}^{\mathcal{D}} d''$, and $d, d'' \in D^{s''}$ then $d \sqsubseteq_{s''}^{\mathcal{D}} d''$.*

2. *If $m(f)(i) = 0$ (resp. 1) then $f^{\mathcal{D}}$ is monotonic (resp. antimonotonic) in the i-th argument. Analogously for predicate symbols.[2]*

A valuation ρ for $\mathcal{D}$ is a sorted set of applications $\rho = \{\rho^s | s \in S\}$, where for each sort $s \in S$, $\rho^s : X^s \to D^s$ is a valuation of the set X^s. A Σ-interpretation is a pair $\langle \mathcal{D}, \rho \rangle$ such that $\mathcal{D}$ is a Σ-structure and ρ is a valuation for $\mathcal{D}$.

Transitivity is demanded in order to avoid some pathological structures. In particular with this condition we assure that the following property holds for every sorts s, s'. If $d, d' \in D^s \cap D^{s'}$, and $d \sqsubseteq_s^{\mathcal{D}} d'$ then $d \sqsubseteq_{s'}^{\mathcal{D}} d'$.

It is evident that a functional term or predicate formula behaves monotonically whenever we increase the value of a subterm which is on the scope of an even number of antimonotonic arguments. As we will see later, some deduction rules will be founded on this fact. This is precised through a concept of (anti)monotonic position, presented by using a notation usual in the context of rewrite term systems. Given a term $t \in T(\Sigma)$ we denote by $Pos(t)$ the set of positions of t, recursively defined by $\{\varepsilon\}$, if t is a variable or a constant, and $\{\varepsilon\} \cup \{i.p | p \in Pos(t_i), 1 \leq i \leq n\}$, if t is $f(t_1, \ldots, t_n)$. We denote by $root(t)$ the root of t viewed as a tree. For two positions p, q, we say that $p \leq q$ if there is a position r such that $p.r = q$; in that case $q - p$ will be r. Given a position $p \in Pos(t)$, $t|_p$ is the subterm of t at the position p, and $t[t']_p$ the result of substituting t' for $t|_p$ in t.

---

[1] Note that $WS(\varphi, \Phi) \Leftrightarrow WS(\varphi, \Phi\tau)$, because sort inequalities are not affected by substitutions.
[2] Monotonicity (resp. antimonotonicity) of $P^{\mathcal{D}}$ in the i-th argument means that if $d_i \sqsubseteq_{s_i}^{\mathcal{D}} d_i'$, $P^{\mathcal{D}}(\ldots d_i \ldots) = \underline{t} \Rightarrow P^{\mathcal{D}}(\ldots d_i' \ldots) = \underline{t}$ (resp. $P^{\mathcal{D}}(\ldots d_i' \ldots) = \underline{t} \Rightarrow P^{\mathcal{D}}(\ldots d_i \ldots) = \underline{t}$).

Definition 6 *Given* $t \in T(\Sigma)$ *and* $p \in Pos(t)$, *the function* $\mathcal{F}_{t,p} : \{q \in Pos(t) | q < p\} \to \mathbb{N}$ *is defined by* $\mathcal{F}_{t,p}(q) = $ *first digit of* $p - q$.

Note that $q \in dom(\mathcal{F}_{t,p})$ if $t|_p$ is subterm of $t|_q$, and that $\mathcal{F}_{t,p}(q)$ is the ordinal of the argument of $t|_q$ in which $t|_p$ occurs as subterm. For simplicity, we will write $\mathcal{F}$ instead of $\mathcal{F}_{t,p}$.

Definition 7 *Given* $t \in T(\Sigma)$ *and* $p \in Pos(t)$, *if* $\sum_{q<p} m(root(t|_q))(\mathcal{F}(q))$ *is even, we say that* p *is a monotonic position of* t. *Otherwise we say that* p *is an antimonotonic position of* t. ε *is a monotonic position of* t.
Given $P(t_1, \ldots, t_n)(\equiv A)$ *and a position* $p \in Pos(t_i)$, *for some* $1 \leq i \leq n$, *we say that* p *is a monotonic position of* A *at* i, *if* $\sum_{q<p} m(root(t_i|_q))(\mathcal{F}(q)) + m(P)(i)$ *is even. Otherwise we say that* p *is an antimonotonic position of* A *at* i.

In the next definition we express that functions are strict. Predicates are also strict (boolean) functions, hence their semantics is three-valued.

Definition 8 *The semantic value of* $t \in T(\Sigma, s)$ *in a* Σ-*interpretation* $\langle \mathcal{D}, \rho \rangle$ *is an element* $[\![t]\!]_\rho^{\mathcal{D}}$ *of* $D^s \cup \{\bot\}$. *In particular, given* $f \in F^{s_1,\ldots,s_n \to s}$, *it is defined by:*

$$- [\![f(t_1, \ldots, t_n)]\!]_\rho^{\mathcal{D}} = \begin{cases} f^{\mathcal{D}}([\![t_1]\!]_\rho^{\mathcal{D}}, \ldots, [\![t_n]\!]_\rho^{\mathcal{D}}) & \text{if } [\![t_i]\!]_\rho^{\mathcal{D}} \in D^{s_i}, i = 1, \ldots, n \\ \bot & \text{otherwise.} \end{cases}$$

The truth value of a Σ-*formula* φ *in a* Σ-*interpretation* $\langle \mathcal{D}, \rho \rangle$ *will be denoted by* $[\![\varphi]\!]_\rho^{\mathcal{D}}(\in \{\underline{u}, \underline{f}, \underline{t}\})$[3] *and defined by the following rules:*

$$- [\![s \sqsubseteq_{\mathbf{0}} s']\!]_\rho^{\mathcal{D}} = \begin{cases} \underline{t} & \text{if } D^s \subseteq D^{s'} \\ \underline{f} & \text{otherwise.} \end{cases}$$

$$- [\![t_1 \sqsubseteq t_2]\!]_\rho^{\mathcal{D}} = \begin{cases} \underline{t} & \text{if there is } s \text{ such that } ([\![t_i]\!]_\rho^{\mathcal{D}} \in \mathcal{D}^s, i = 1, 2, \text{ and } [\![t_1]\!]_\rho^{\mathcal{D}} \sqsubseteq_s^{\mathcal{D}} [\![t_2]\!]_\rho^{\mathcal{D}}) \\ \underline{u} & \text{if } [\![t_i]\!]_\rho^{\mathcal{D}} = \bot \text{ for some } i = 1, 2, \text{ or there is not } s ([\![t_1]\!]_\rho^{\mathcal{D}}, [\![t_2]\!]_\rho^{\mathcal{D}} \in \mathcal{D}^s) \\ \underline{f} & \text{otherwise.} \end{cases}$$

$$- [\![P(t_1, \ldots, t_n)]\!]_\rho^{\mathcal{D}} = \begin{cases} P^{\mathcal{D}}([\![t_1]\!]_\rho^{\mathcal{D}}, \ldots, [\![t_n]\!]_\rho^{\mathcal{D}}) & \text{if } [\![t_i]\!]_\rho^{\mathcal{D}} \in \mathcal{D}^{s_i}, i = 1, \ldots, n \\ \underline{u} & \text{otherwise.} \end{cases} \quad P \in P^{s_1,\ldots,s_n}$$

φ	$\underline{t}$	$\underline{f}$	$\underline{u}$
$\neg\varphi$	$\underline{f}$	$\underline{t}$	$\underline{u}$

φ	$\underline{t}$	$\underline{t}$	$\underline{t}$	$\underline{f}$	$\underline{f}$	$\underline{f}$	$\underline{u}$	$\underline{u}$	$\underline{u}$
ψ	$\underline{t}$	$\underline{f}$	$\underline{u}$	$\underline{t}$	$\underline{f}$	$\underline{u}$	$\underline{t}$	$\underline{f}$	$\underline{u}$
$\varphi \wedge \psi$	$\underline{t}$	$\underline{f}$	$\underline{u}$	$\underline{f}$	$\underline{f}$	$\underline{f}$	$\underline{u}$	$\underline{f}$	$\underline{u}$

$$- [\![\exists x^s \varphi]\!]_\rho^{\mathcal{D}} = \begin{cases} \underline{t} & \text{if there is } d \in D^s \text{ such that } [\![\varphi]\!]_{\rho[d/x^s]}^{\mathcal{D}} = \underline{t} \\ \underline{f} & \text{if } [\![\varphi]\!]_{\rho[d/x^s]}^{\mathcal{D}} = \underline{f}, \text{ for all } d \in D^s \\ \underline{u} & \text{otherwise.} \end{cases}$$

In the context of a third boolean value we can define, at least, two kinds of models. An interpretation $\langle \mathcal{D}, \rho \rangle$ is a *strong model*, or *s-model*, of a formula φ (resp. a set of formulas Φ), written $\langle \mathcal{D}, \rho \rangle \models_s \varphi$, if $[\![\varphi]\!]_\rho^{\mathcal{D}} = \underline{t}$ (resp. for every $\psi \in \Phi$); and it is a *weak model*, or ω-*model*, of φ, written $\langle \mathcal{D}, \rho \rangle \models_\omega \varphi$, if $[\![\varphi]\!]_\rho^{\mathcal{D}} \neq \underline{f}$. Also we will say that φ is *s-satisfiable* or ω-*satisfiable*, respectively. Then φ is a *logical consequence* of Φ, written $\Phi \models \varphi$, if every weak model of Φ is a strong model of φ, which means that true properties are concluded from not false hypothesis. Note that this logical consequence can be expressed in terms of ω-satisfiability as follows: $\Phi \models \varphi \Leftrightarrow \Phi \cup \{\neg\varphi\}$ is not ω-satisfiable.

---

[3]The set of boolean values $\{\underline{u}, \underline{f}, \underline{t}\}$ is ordered by setting $\underline{u} \sqsubseteq_b \underline{t}, \underline{f}$, and $\underline{f}, \underline{t}$ incomparable.

Lemma 9 *(i) Substitution Lemma for terms and formulas.* $[t[t'/x]]^{\mathcal{D}}_{\rho} = [t]^{\mathcal{D}}_{\rho[[t']^{\mathcal{D}}_{\rho}/x]}$ *and* $[\varphi[t'/x]]^{\mathcal{D}}_{\rho} = [\varphi]^{\mathcal{D}}_{\rho[[t']^{\mathcal{D}}_{\rho}/x]}$, *provided that* $[t']^{\mathcal{D}}_{\rho} \neq \perp$.

(ii) Monotonicity and antimonotonicity of terms. Let $p \in Pos(t)$ a monotonic (resp. antimonotonic) position of t such that $t|_p \equiv t_1$. If $\langle \mathcal{D}, \rho \rangle \models_s t_1 \sqsubseteq t_2$ and there is a sort s such that $[t[t_i]_p]^{\mathcal{D}}_{\rho} \in D^s$, $i = 1, 2$, then $[t[t_1]_p]^{\mathcal{D}}_{\rho} \sqsubseteq^{\mathcal{D}}_s [t[t_2]_p]^{\mathcal{D}}_{\rho}$ (resp. $[t[t_2]_p]^{\mathcal{D}}_{\rho} \sqsubseteq^{\mathcal{D}}_s [t[t_1]_p]^{\mathcal{D}}_{\rho}$).

(iii) Monotonicity and antimonotonicity of atomic formulas. Let $P(t_1, \ldots, t_n)(\equiv A)$ and $p \in Pos(t_i)$ a monotonic (resp. antimonotonic) position of A at i. If $\langle \mathcal{D}, \rho \rangle \models_s t^1 \sqsubseteq t^2$ and $[t_i[t^j]_p]^{\mathcal{D}}_{\rho} \in D^{s_i}$, $j = 1, 2$, then $[P(t_1, \ldots, t_i[t^1]_p, \ldots, t_n)]^{\mathcal{D}}_{\rho} = \underline{t} \Rightarrow [P(t_1, \ldots, t_i[t^2]_p, \ldots, t_n)]^{\mathcal{D}}_{\rho} = \underline{t}$ (resp. $[P(t_1, \ldots, t_i[t^2]_p, \ldots, t_n)]^{\mathcal{D}}_{\rho} = \underline{t} \Rightarrow [P(t_1, \ldots, t_i[t^1]_p, \ldots, t_n)]^{\mathcal{D}}_{\rho} = \underline{t}$).

(iv) If $p \in Pos(t)$ and $[t']^{\mathcal{D}}_{\rho} = \perp$ then $[t[t']_p]^{\mathcal{D}}_{\rho} = \perp$.

(v) If $[t]^{\mathcal{D}}_{\rho} = \perp$ and $[t']^{\mathcal{D}}_{\rho} \neq \perp$, then $[\varphi[t/x]]^{\mathcal{D}}_{\rho} \sqsubseteq_b [\varphi[t'/x]]^{\mathcal{D}}_{\rho}$.

The fact that we are using the third boolean value for dealing with not well-sorted formulas is expressed in the following result, where we prove that in any ω-model of Φ, every (negated) atomic $\varphi \in \Phi$ is interpreted as true if it is well-sorted w.r.t. Φ.

Theorem 10 (Well-sortedness) *Given a set of Σ-formulas Φ, if $\langle \mathcal{D}, \rho \rangle \models_\omega \Phi$ then:*
(1) $WS(t, \Phi) \Rightarrow [t]^{\mathcal{D}}_{\rho} \in D^{sort(t)}$, for every Σ-term t.
(2) $WS(\varphi, \Phi) \Rightarrow [\varphi]^{\mathcal{D}}_{\rho} \neq \underline{u}$, for every (negated) atomic formula φ.

Proof. (1). By induction on t. In the basic case, if $t \equiv c$ then $[t]^{\mathcal{D}}_{\rho} = c^{\mathcal{D}} \in D^{sort(c)}$, while if $t \equiv x$ then $[t]^{\mathcal{D}}_{\rho} = \rho(x) \in D^{sort(x)}$. For the induction step, suppose that $t \equiv f(t_1, \ldots, t_n)$, where $f \in F^{s_1, \ldots, s_n \to s}$, and that $WS(t, \Phi)$. By Def. 3 we have $WS(t_i, \Phi)$ and $sort(t_i) \sqsubseteq_0 s_i \in \Phi$, $i = 1, \ldots, n$. So, using induction hypothesis and the fact that $\langle \mathcal{D}, \rho \rangle \models_\omega \Phi$, we obtain $[t_i]^{\mathcal{D}}_{\rho} \in D^{sort(t_i)} \subseteq D^{s_i}$, $i = 1, \ldots, n$. Then $[t]^{\mathcal{D}}_{\rho} = f^{\mathcal{D}}([t_1]^{\mathcal{D}}_{\rho}, \ldots, [t_n]^{\mathcal{D}}_{\rho}) \in D^s$, by Def. 8.

(2). Let φ be an atomic formula, then φ is of the form $s \sqsubseteq_0 s'$, $t \sqsubseteq t'$ or $P(t_1, \ldots, t_n)$. In the first case, the theorem is trivial by Def. 8. In the second one, by hypothesis we have $WS(\varphi, \Phi)$ and then $WS(t, \Phi)$, $WS(t', \Phi)$ and there exists s such that $sort(t) \sqsubseteq_0 s, sort(t') \sqsubseteq_0 s \in \Phi$. By the previous result and using that $\langle \mathcal{D}, \rho \rangle \models_\omega \Phi$, it can be proved that there exists s such that $[t]^{\mathcal{D}}_{\rho}, [t']^{\mathcal{D}}_{\rho} \in D^s$ and so $[\varphi]^{\mathcal{D}}_{\rho} \neq \underline{u}$, from Def. 8. The other cases are proved similarly. ∎

3 A Ground Tableau System

This section presents an extension of first order semantic tableaux [Fit 96] to deal with preordered terms and sorts. Attempts of adding the axioms characterising a preorder structure to the initial set of formulas do not lead to satisfying results, because of the huge increasing of the search space. It is better to formulate additional tableau expansion rules to capture the essential properties of preorders.

In the definition of the method we need to extend the signature Σ by countably infinite sets AC^s, $s \in S$, of new auxiliary constants; the new signature is denoted $\overline{\Sigma}$. The partition of $\overline{\Sigma}$-formulas in *Alpha, Beta, Gamma* and *Delta* classes is the usual in classical logic. The *Basic* class is the set of atomic $\overline{\Sigma}$-formulas, $P(t_1, \ldots, t_n)$, $t \sqsubseteq t'$, $s \sqsubseteq_0 s'$, and their complementaries $\neg P(t_1, \ldots, t_n)$, $\neg t \sqsubseteq t'$, $\neg s \sqsubseteq_0 s'$.

Definition 11 *A tableau sequence for a non-empty set Φ of Σ-sentences is any sequence $\mathcal{T}_0, \mathcal{T}_1, \mathcal{T}_2, \ldots$ where $\mathcal{T}_0$ is a linear tree with a branch labeled by some sentences $\{\varphi_1, \ldots, \varphi_n\} \subseteq \Phi$. And $\mathcal{T}_{k+1}$ arises from $\mathcal{T}_k$ through one of the tableau expansion rules α—ξ_2 below. An infinite tableau for a set of Σ-sentences Φ is defined as the limit of some tableau sequence. α and β are the usual tableau rules, α applied onto $\varphi \wedge \psi$ and $\neg\neg\varphi$ formulas and β applied onto $\neg(\varphi \wedge \psi)$. If $\mathcal{T}$ is a finite tableau for Φ and B is a branch of $\mathcal{T}$, the other rules are:*

  γ *If $\neg\exists x^s \varphi, s' \sqsubseteq_\mathbf{0} s \in B$[4], then enlarge B with a node labeled by $\neg\varphi[t/x^s]$, where t is ground and $t \in T(\overline{\Sigma}, s')$.*

  δ *If $\exists x^s \varphi \in B$, then enlarge B with a node labeled by $\varphi[c/x^s]$, $c \in AC^s$ new for B.*

  In Enlarge B with a new node labeled by $\varphi \in \Phi$, where $\varphi \notin B$.

  Ref Enlarge B with a new node labeled by $t \sqsubseteq t$, where t is a ground $\overline{\Sigma}$-term.

  Ref$_\mathbf{0}$ Enlarge B with a new node labeled by $s \sqsubseteq_\mathbf{0} s$, where $s \in S$.

  Tr$_\mathbf{0}$ If $s \sqsubseteq_\mathbf{0} s'$, $s' \sqsubseteq_\mathbf{0} s'' \in B$, then enlarge B with a new node labeled by $s \sqsubseteq_\mathbf{0} s''$.

  ζ_1, ζ_2 *If $t \sqsubseteq t'$ ($t' \sqsubseteq t$, for ζ_2), $t_1 \sqsubseteq t_2[t]_p \in B$, being both well-sorted w.r.t. B, and p is a monotonic (antimonotonic, for ζ_2) position of t_2, then extend B with a node labeled by $t_1 \sqsubseteq t_2[t']_p$.*

  ξ_1, ξ_2 *If $P(t_1, \ldots, t_i[t]_p, \ldots, t_n)(\equiv A)$, $t \sqsubseteq t'$ ($t' \sqsubseteq t$, for ξ_2) $\in B$, being both well-sorted w.r.t. B, and p is a monotonic (antimonotonic, for ξ_2) position of A at i, then extend B with $P(t_1, \ldots, t_i[t']_p, \ldots, t_n)$.*

α–δ are classical first order expansion rules and are applied to sentences in classes *Alpha*, *Beta*, *Gamma* and *Delta*, respectively. *Ref*, ζ_1, ζ_2, ξ_1, ξ_2 are the inequality extension rules. *Ref$_\mathbf{0}$, Tr$_\mathbf{0}$* are the order sorted extension rules.

Definition 12 *We say that a branch B in a tableau is closed, and then it is neither enlarged nor split anymore, when an atomic contradiction is detected at its labels. That means, φ and $\neg\varphi$ (φ atomic) are on B and are well-sorted w.r.t. B. Otherwise B is open. A tableau is closed if all its branches are closed.*

3.1 Soundness and Completeness

Soundness of the tableau method expresses that not ω-satisfiability of a set of formulas is consequence from the existence of a closed tableau for it. The proof is based on the fact that ω-satisfiability of a tableau (i.e., that of some of its branches) is preserved when tableaux are extended by applying the expansion rules.

Lemma 13 *The ω-satisfiability of a tableau is preserved by the expansion rules.*

Proof. We see some cases. Let $\langle \mathcal{D}, \rho \rangle$ be an ω-model of a branch B, then:
  γ. Let $\neg\exists x^s \varphi, s' \sqsubseteq_\mathbf{0} s \in B$, $t \in T(\overline{\Sigma}, s')$. If $[t]_\rho^\mathcal{D} \in D^{s'}$ then $[t]_\rho^\mathcal{D} \in D^s$ because $s' \sqsubseteq_\mathbf{0} s \in B$. By Lemma 9.(i), $[\neg\varphi[t/x]]_\rho^\mathcal{D} \neq \underline{f}$. The same result applies if $[t]_\rho^\mathcal{D} = \bot$, by Lemma 9.(v).
  ζ_1. Let $\varphi_1 \equiv t \sqsubseteq t'$, $\varphi_2 \equiv t_1 \sqsubseteq t_2[t]_p$, where p is a monotonic position of t_2, and $WS(\varphi_i, B), i = 1, 2$; then, by Theorem 10.(2), we have $[\varphi_1]_\rho^\mathcal{D}, [\varphi_2]_\rho^\mathcal{D} = \underline{t}$. Let

---

[4]We identify a branch with the set of formulas labeling its nodes.

$\psi \equiv t_1 \sqsubseteq t_2[t']_p$. We can suppose that there exists a sort s such that $[t_1]_\rho^{\mathcal{D}}, [t_2[t']_p]_\rho^{\mathcal{D}} \in D^s$, otherwise $[\psi]_\rho^{\mathcal{D}} = \underline{u}$. Then it can be proved that there exists a sort s' such that $[t_2[t]_p]_\rho^{\mathcal{D}}, [t_2[t']_p]_\rho^{\mathcal{D}} \in D^{s'}$; if $p = \varepsilon$, use that φ_1 is true, otherwise use $sort(t_2)$, since φ_2 is true. Now by Lemma 9.(ii), $[t_2[t]_p]_\rho^{\mathcal{D}} \sqsubseteq_{s'}^{\mathcal{D}} [t_2[t']_p]_\rho^{\mathcal{D}}$. Hence $[\psi]_\rho^{\mathcal{D}} = \underline{t}$, by transitivity of Σ-structures. ∎

Theorem 14 (Soundness) *For every set Φ of Σ-sentences, if Φ has a closed tableau, then Φ is not ω-satisfiable.*

Now we show completeness, proving that if a set of formulas Φ does not have a closed tableau, then we can systematically build a tableau with an open branch having special saturation properties, that allows to define an ω-model of Φ. These properties can be defined in a *Hintikka's* way.

Definition 15 (Hintikka set) *A set of $\overline{\Sigma}$-sentences $\mathbf{H}$ is a Hintikka set if it satisfies the following conditions:*

(1) If $\varphi_1 \wedge \varphi_2 \in \mathbf{H}$ then $\varphi_1, \varphi_2 \in \mathbf{H}$.

(2) If $\neg(\varphi_1 \wedge \varphi_2) \in \mathbf{H}$ then $\neg\varphi_1 \in \mathbf{H}$ or $\neg\varphi_2 \in \mathbf{H}$.

(3) If $\neg\neg\varphi \in \mathbf{H}$ then $\varphi \in \mathbf{H}$.

(4) If $\exists x^s \varphi \in \mathbf{H}$ then there exists a constant $c \in AC^s$ such that $\varphi[c/x] \in \mathbf{H}$.

(5) If $\neg\exists x^s \varphi \in \mathbf{H}$ and $s' \sqsubseteq_{\mathbf{Q}} s \in \mathbf{H}$ then $\neg\varphi[t/x] \in \mathbf{H}$, for all ground $t \in T(\overline{\Sigma}, s')$.

(6) For all ground $t \in T(\overline{\Sigma})$, $t \sqsubseteq t \in \mathbf{H}$, and for all $s \in S$, $s \sqsubseteq_{\mathbf{Q}} s \in \mathbf{H}$.

(7) Let $t, t', t_1, t_2 \in T(\overline{\Sigma})$ and $s, s', s'' \in S$:

    (i) if $t \sqsubseteq t'$ (resp. $t' \sqsubseteq t$), $t_1 \sqsubseteq t_2[t]_p \in \mathbf{H}$, $WS(t \sqsubseteq t', \mathbf{H})$, $WS(t_1 \sqsubseteq t_2[t]_p, \mathbf{H})$ and p is a monotonic (resp. antimonotonic) position of t_2, then $t_1 \sqsubseteq t_2[t']_p \in \mathbf{H}$,

    (ii) if $t \sqsubseteq t'$ (resp. $t' \sqsubseteq t$), $P(t_1, .., t_i[t]_p, .., t_n)(\equiv A) \in \mathbf{H}$, $WS(t \sqsubseteq t', \mathbf{H})$, $WS(A, \mathbf{H})$, p is a monotonic (resp. antim.) position of A at i, then $P(t_1, .., t_i[t']_p, .., t_n) \in \mathbf{H}$,

    (iii) if $s \sqsubseteq_{\mathbf{Q}} s', s' \sqsubseteq_{\mathbf{Q}} s'' \in \mathbf{H}$, then $s \sqsubseteq_{\mathbf{Q}} s'' \in \mathbf{H}$.

(8) $\mathbf{H}$ is a coherent set of sentences, i.e. no atomic formula well-sorted w.r.t. $\mathbf{H}$ belongs to $\mathbf{H}$ together with its negation.

Lemma 16 *Given $t \in T(\overline{\Sigma})$ and $s \in S$:*
(1) If $t \in \mathcal{T}_{\overline{\Sigma}}^{\mathbf{H}}(s)$ then $sort(t) \sqsubseteq_{\mathbf{Q}} s \in \mathbf{H}$. (2) If $t \in \mathcal{T}_{\overline{\Sigma}}^{\mathbf{H}}(s)$ then $WS(t, \mathbf{H})$.

Lemma 17 *The system $\mathcal{D} = \langle \{(D^s, \sqsubseteq_s^{\mathcal{D}}) | s \in S\} \cup \{\bot\}, \{c^{\mathcal{D}}, f^{\mathcal{D}}, P^{\mathcal{D}} | c, f, P \in \overline{\Sigma}\}\rangle$ where: $D^s = \{t | t \in \mathcal{T}_{\overline{\Sigma}}^{\mathbf{H}}(s), t\ ground\}$. For all $t, t' \in D^s$, $t \sqsubseteq_s^{\mathcal{D}} t'$ iff $t \sqsubseteq t' \in \mathbf{H}$. $f^{\mathcal{D}}$: $D^{s_1} \times \ldots \times D^{s_n} \to D^s$, $f^{\mathcal{D}}(t_1, \ldots, t_n) = f(t_1, \ldots, t_n)$.[5] $P^{\mathcal{D}}: D^{s_1} \times \ldots \times D^{s_n} \to \{\underline{t}, \underline{f}\}$,*
$$P^{\mathcal{D}}(t_1, \ldots, t_n) = \begin{cases} \underline{t} & if\ P(t_1, \ldots, t_n) \in \mathbf{H} \\ \underline{f} & otherwise \end{cases}\ is\ a\ \Sigma\text{-}structure.$$

---

[5] If f is a constant c then $c^{\mathcal{D}} = c$.

Proof. (1). First of all, note that D^s is not empty because every sort is inhabited. Reflexivity: $t \sqsubseteq t \in \mathbf{H}$ by Def. 15.(6), then $t \sqsubseteq_s^{\mathcal{D}} t$ for any $t \in D^s$. Transitivity: if $t \sqsubseteq_s^{\mathcal{D}} t'$ and $t' \sqsubseteq_s^{\mathcal{D}} t''$, then $t \sqsubseteq t'$, $t' \sqsubseteq t'' \in \mathbf{H}$, being both well-sorted w.r.t. $\mathbf{H}$ by Lemma 16.(1) and (2), so $t \sqsubseteq t'' \in \mathbf{H}$ by Def. 15.(7)(i); therefore $t \sqsubseteq_s^{\mathcal{D}} t''$. Transitivity of the structure: let $t, t'' \in D^{s''}$, $t \sqsubseteq_s^{\mathcal{D}} t', t' \sqsubseteq_{s'}^{\mathcal{D}} t''$, then $t \sqsubseteq t'$, $t' \sqsubseteq t'' \in \mathbf{H}$, being both well-sorted w.r.t $\mathbf{H}$ by Lemma 16.(1) and (2), so from Def. 15.(7)(i), $t \sqsubseteq t'' \in \mathbf{H}$ and then $t \sqsubseteq_{s''}^{\mathcal{D}} t''$.

(2). Suppose $m(f)(i) = 0$. If $t_i \sqsubseteq_{s_i}^{\mathcal{D}} t_i'$ then $t_i \sqsubseteq t_i' \in \mathbf{H}$, being well-sorted w.r.t $\mathbf{H}$ by Lemma 16.(1) and (2). So if $t_j \in D^{s_j}, j \neq i$, from Def. 15.(6), (7)(i) and Lemma 16.(1), (2) we can deduce $f(t_1, \ldots, t_i, \ldots, t_n) \sqsubseteq f(t_1, \ldots, t_i', \ldots, t_n) \in \mathbf{H}$, therefore $f^{\mathcal{D}}(t_1, \ldots, t_i, \ldots, t_n) \sqsubseteq_s^{\mathcal{D}} f^{\mathcal{D}}(t_1, \ldots, t_i', \ldots, t_n)$. We would reason analogously for $m(f)(i) = 1$ and $P^{\mathcal{D}}$. ∎

Theorem 18 *Every Hintikka set $\mathbf{H}$ has an ω-model.*

Proof. Let $\mathcal{D}$ be the $\overline{\Sigma}$-structure just defined. It can be proved that $WS(t, \mathbf{H}) \Rightarrow \llbracket t \rrbracket^{\mathcal{D}} = t$, for any $t \in T(\overline{\Sigma})$ ground (†). The theorem is proved by induction on $\varphi \in \mathbf{H}^6$, for instance:

- $\varphi \equiv t_1 \sqsubseteq t_2$. Suppose that there exists a sort s such that $\llbracket t_i \rrbracket^{\mathcal{D}} \in D^s, i = 1, 2$; otherwise $\llbracket \varphi \rrbracket^{\mathcal{D}} = \underline{u}$. Then by Lemma 16.(2) and (†) we obtain $\llbracket t_i \rrbracket^{\mathcal{D}} = t_i, i = 1, 2$; so $\llbracket t_1 \rrbracket^{\mathcal{D}} \sqsubseteq_s^{\mathcal{D}} \llbracket t_2 \rrbracket^{\mathcal{D}}$ because $\varphi \in \mathbf{H}$. Then $\llbracket \varphi \rrbracket^{\mathcal{D}} = \underline{t}$.
- $\varphi \equiv \neg \exists x^s \psi$. Let t be any term of D^s, then by Lemma 16.(1),(2) and (†) it can be proved that $sort(t) \sqsubseteq_{\mathbf{0}} s \in \mathbf{H}$ and $\llbracket t \rrbracket^{\mathcal{D}} = t$. From the former we obtain that $\neg \psi[t/x] \in \mathbf{H}$, so by induction hypothesis $\llbracket \neg \psi[t/x] \rrbracket^{\mathcal{D}} \neq \underline{f}$. From the latter and Lemma 9.(i), $\llbracket \neg \psi \rrbracket_{[t/x]}^{\mathcal{D}} = \llbracket \neg \psi[t/x] \rrbracket^{\mathcal{D}}$. So $\llbracket \varphi \rrbracket^{\mathcal{D}} \neq \underline{f}$ by the semantics of quantifiers.
- $\varphi \equiv \neg(s \sqsubseteq_{\mathbf{0}} s')$. From Def. 15.(8), it is impossible that $s \sqsubseteq_{\mathbf{0}} s' \in \mathbf{H}$. Let $c \in C^s$ then $\llbracket c \rrbracket^{\mathcal{D}} = c \in D^s$, so $c \notin D^{s'}$ because $s \sqsubseteq_{\mathbf{0}} s' \notin \mathbf{H}$. Then $\llbracket \varphi \rrbracket^{\mathcal{D}} = \underline{t}$. ∎

A ground tableau construction rule is a procedure $\mathcal{R}$ that given a finite tableau $\mathcal{T}$ produces as answer either a refusal if no continuation is possible, or a new tableau $\mathcal{T}'$ that results from the application of a ground tableau expansion rule to $\mathcal{T}$. A ground tableau construction rule $\mathcal{R}$ is fair if, for any tableau sequence for a set of Σ-sentences Φ constructed according to $\mathcal{R}$, it verifies that every $\varphi \in \Phi$ is eventually introduced onto the open branches, Ref and $Ref_{\mathbf{0}}$ are applied arbitrarily often, and every occurrence of a sentence or pair of sentences in an open branch, for which an expansion rule can be used, eventually has the appropriate rule applied to it (arbitrarily often for the case of γ).

Theorem 19 (Completeness) *Let $\mathcal{R}$ be a fair ground tableau construction rule. If Φ is a not ω-satisfiable set of Σ-sentences then Φ has a closed tableau constructed according to $\mathcal{R}$.*

If Φ would be well-sorted w.r.t. $\mathbf{H}$, a s-model could be obtained for Φ. Hence we could restrict the hypothesis and formulate a new completeness theorem. Next, a syntactic condition is defined allowing to get completeness from not s-satisfiability.

Definition 20 *The positive sort static information of a formula φ, written $\mathcal{P}(\varphi)$, is defined recursively as follows:*

$\mathcal{P}(s \sqsubseteq_{\mathbf{0}} s') =_{def} \{s \sqsubseteq_{\mathbf{0}} s'\}$, $\mathcal{P}(\varphi) =_{def} \emptyset$, *for any other Basic formula* φ,
$\mathcal{P}(\neg\neg\varphi) =_{def} \mathcal{P}(\varphi)$, $\mathcal{P}(\varphi \land \psi) =_{def} \mathcal{P}(\varphi) \cup \mathcal{P}(\psi)$, $\mathcal{P}(\neg(\varphi \land \psi)) =_{def} \emptyset$,
$\mathcal{P}(\exists x \varphi) =_{def} \mathcal{P}(\varphi)$, $\mathcal{P}(\neg \exists x \varphi) =_{def} \mathcal{P}(\neg\varphi)$.
For a set of formulas Φ we define $\mathcal{P}(\Phi) = \cup\{\mathcal{P}(\varphi)|\varphi \in \Phi\}$.

---

[6] No valuation is needed because $\mathbf{H}$ is a set of sentences.

From the previous definition, it is obvious that $\mathcal{P}(\Phi)$ will be contained in every Hintikka branch of every tableau for Φ. Then, by Theorem 10, we can conclude the expected result.

Theorem 21 *Let $\mathcal{R}$ be a fair ground tableau construction rule. If Φ is a not s-satisfiable set of Σ-sentences and $WS(\Phi, \mathcal{P}(\Phi))$ then Φ has a closed tableau constructed according to $\mathcal{R}$.*

This theorem is a particular case of Theorem 19. It can be proved that if Φ is a not s-satisfiable set of Σ-sentences and $WS(\Phi, \mathcal{P}(\Phi))$ then Φ is not ω-satisfiable. However, the opposite result is not true. For example $\Phi \equiv \{s \sqsubseteq_{\mathbf{0}} s' \vee s' \sqsubseteq_{\mathbf{0}} s, \neg(s' \sqsubseteq_{\mathbf{0}} s), \forall x^{s'} \neg(x^{s'} \sqsubseteq c^s)\}$ has a closed tableau but it does not satisfy $WS(\Phi, \mathcal{P}(\Phi))$.

4 Free Variable Tableaux

Regarding implementation, the ground version of first order semantic tableaux is very inefficient, even more in presence of inequalities. A more efficient system may be defined using free variable quantifier rules, which introduce new free variables, instead of ground terms, substituting the quantified variable of a formula of class *Gamma*. These free variables will be instantiated on demand, either to close a tableau or applying inequality rules to expand a branch.

We will assume that the extended signature $\overline{\Sigma}$ also contains a collection of countable sets $SF^{s_1,\dots,s_l \to s}$, $s_1,\dots,s_l, s \in S$, of Skolem function symbols, which are monotonic in all their arguments. Some tableau expansion rules are redefined in the following sense. If $\mathcal{T}$ is a finite tableau for a set of Σ-sentences Φ and B is a branch of $\mathcal{T}$ then:

γ' If $\neg \exists x^s \varphi, s' \sqsubseteq_{\mathbf{0}} s \in B$, then enlarge B with a node labeled by $\neg\varphi[y^{s'}/x^s]$, where $y^{s'}$ is a new free variable, that is, it is not used for $\neg\exists x^s \varphi$ in B.

δ^+ If $\exists x^s \varphi \in B$, then enlarge B with a node labeled by $\varphi[f(x_1,\dots,x_n)/x^s]$, where $f \in SF^{s_1,\dots,s_n \to s}$ is not used before and $x_1^{s_1},\dots,x_n^{s_n}$ are the free variables in $\exists x^s \varphi$.

ζ_1', ζ_2' If $t \sqsubseteq t'$ ($t' \sqsubseteq t$, for ζ_2'), $t_1 \sqsubseteq t_2[t'']_p \in B$, being both well-sorted w.r.t. B, σ is the most general unifier of t and t'', and p is a monotonic (antimonotonic, for ζ_2') position of t_2, then apply σ to $\mathcal{T}$ and extend the branch $B\sigma$ with a node labeled by $(t_1 \sqsubseteq t_2[t']_p)\sigma$.

ξ_1', ξ_2' If $t \sqsubseteq t'$ ($t' \sqsubseteq t$, for ξ_2'), $P(t_1,\dots,t_i[t'']_p,\dots,t_n)(\equiv A) \in B$, being both well-sorted w.r.t. B, σ is the most general unifier of t and t'', and p is a monotonic (antimonotonic, for ξ_2') position of A at i, then apply σ to $\mathcal{T}$ and extend the branch $B\sigma$ with a node labeled by $P(t_1,\dots,t_i[t']_p,\dots,t_n)\sigma$.

Refv, Reff Enlarge some branch B with a new node labeled by $x \sqsubseteq x$, where x is new for B (by $f(x_1,\dots,x_n) \sqsubseteq f(x_1,\dots,x_n)$, where f is any function symbol of $\overline{\Sigma}$ and $x_1,\dots,x_n$ are new in B, for *Reff*). These rules replace *Ref*.

Definition 22 *A free variable tableau $\mathcal{T}$ with branches $B_1,\dots,B_k$ is closed iff there is a grounding substitution σ such that, for every $1 \le i \le k$, there are two basic formulas $\varphi_i, \psi_i \in B_i$, such that $\varphi_i\sigma$ and $\psi_i\sigma$ are well-sorted w.r.t. $B_i\sigma$ and complementary.*

The soundness of the new method can be established along the same lines as in the ground case, but using the arguments of Hähnle and Schmitt [HS 94] for the δ^+-rule. On the other hand, note that, by Lemma 4.(2), if a mgu σ is used in the application of

ζ_1', ζ_2', ξ_1' and ξ_2' to a branch B, then $WS(\sigma, B)$. So proving that these rules preserve ω-satisfiability is similar to the ground case, having into account that, according to Lemma 4.(1), if $WS(\varphi, B)$ then $WS(\varphi\sigma, B\sigma)$. For any other branch B', ω-satisfiability is preserved even in the case that for some $\varphi' \in B'$, not $WS(\varphi'\sigma, B'\sigma)$.

4.1 Completeness

The main result of this section is a Lifting Lemma that allows to show completeness of the free variable method leaning on the completeness of the ground case. We show that any extension of a closed ground tableau can be lifted to an extension of a closed free variable tableau.

Lemma 23 (Lifting Lemma) *Let $\mathcal{T}$ be a free variable tableau and τ be a grounding substitution whose domain is the set of free variables of $\mathcal{T}$. If $\mathcal{T}\tau$ can be closed applying only the inequality extension rules for basic tableaux, then $\mathcal{T}$ can be closed using the inequality extension rules for free variable tableaux.*

Proof. By induction on the number n of applications of inequality rules needed for closing $\mathcal{T}\tau$. If $n = 0$, then $\mathcal{T}\tau$ is closed and $\mathcal{T}$ is obviously also closed.

For the inductive step, let $B_1, \ldots, B_l$ be the branches of $\mathcal{T}$, and B^* the branch obtained from extending, say $B_i\tau$, with the first application of Ref, ζ_1, ζ_2, ξ_1 and ξ_2 rules. The idea of the proof is to extend $\mathcal{T}$ to a tableau $\mathcal{T}'$ with branches $B_1', \ldots, B_l'$, using the rules allowed in the lemma, and to extend τ to a grounding substitution τ' such that $B_j'\tau' = B_j\tau$, $1 \leq j \leq l$, $j \neq i$, and $B_i'\tau' \supseteq B^*$. Then the proof is completed by applying induction hypothesis to $\mathcal{T}'\tau'$.

$\mathcal{T}'$ is constructed according to the rule used for obtaining B^*. In the sequel, the subscript i is avoided in order to simplify the notation. Let us prove the case for ξ_1, being similar the proof of the other cases. We can assume that $B\tau$ is extended to B^* as follows:

$$
\begin{array}{ccc}
\underline{B} & \underline{B\tau} & \underline{B^*} \\
\vdots & \vdots & B\tau \\
t \sqsubseteq t' & t\tau \sqsubseteq t'\tau & \big| \\
P(t_1, \ldots, t_k) & P(t_1\tau, \ldots, t_k\tau) & P(t_1\tau, \ldots, t_i\tau[t'\tau]_p, \ldots, t_k\tau)
\end{array}
$$

where $t_i\tau|_p = t\tau$, p is a monotonic position of $P(t_1\tau, \ldots, t_k\tau)$ at i, and $WS(t\tau \sqsubseteq t'\tau, B)$, $WS(P(t_1\tau, \ldots, t_k\tau), B)$. As the next lemma states, there is $q \in Pos(t_i)$, $q \leq p$, such that B can be extended, using only $Reff$, ζ_1' and ζ_2' rules, to a branch B'' containing B and an inequality $a_q \sqsubseteq a_q'$, if q is a monotonic position of $P(t_1, \ldots, t_n)$ at i (resp. $a_q' \sqsubseteq a_q$, if q is antimonotonic), being $WS(a_q \sqsubseteq a_q', B)$, and there is a substitution τ' extending τ such that (i) $a_q'\tau' = (t_i\tau|_q)[t'\tau]_{p-q}$, and (ii) τ' unifies a_q and $t_i|_q$.

Therefore, we obtain the branch B' from B'' applying ξ_1' to $a_q \sqsubseteq a_q'$ (or ξ_2' to $a_q' \sqsubseteq a_q$) and $P(t_1, \ldots, t_k)$. Note that this is possible because, by Lemma 4.(1), $WS(P(t_1\tau, \ldots, t_k\tau), B)$ implies $WS(P(t_1, \ldots, t_k), B)$. So if σ is the mgu of a_q and $t_i|_q$ then $B' = B''\sigma \cup P(t_1\sigma, \ldots, (t_i[a_q']_q)\sigma, \ldots, t_k\sigma)$. Therefore $B'\tau' \supseteq B^*$ as we show:

- $B''\sigma\tau' = B''\tau'$ (by (ii)) $\supseteq B\tau = B\tau$ (because τ' agrees with τ in B).
- $P(t_1\sigma\tau', \ldots, (t_i[a_q']_q)\sigma\tau', \ldots, t_k\sigma\tau') = P(t_1\tau, \ldots, t_i\tau[a_q'\tau']_q, \ldots, t_k\tau)$
  (because $(t_i|_q)\tau = t_i\tau|_q$, τ' agrees with τ in B, and (ii))
  $= P(t_1\tau, \ldots, t_i\tau[(t_i\tau|_q)[t'\tau]_{p-q}]_q, \ldots, t_k\tau)$ (by (i)) $= P(t_1\tau, \ldots, t_i\tau[t'\tau]_p, \ldots, t_k\tau)$.

For the other branches $B_j' = B_j\sigma$ then $B_j'\tau' = B_j\tau$ trivially. ∎

The next lemma establishes that $\mathcal{T}\tau$ can be lifted even in the case that the inequality extension rule were applied to a new subterm t introduced by τ. This will be possible by previously building an inequality containing a subterm unifying with t, by successive applications of ζ_1', ζ_2' and $Reff$.

Lemma 24 *Using the notation and conditions of the preceding lemma, if $t_i\tau|_p = t\tau$ then there is a position q of t_i, $q \leq p$, such that $\mathcal{T}$ can be expanded using only $Reff$, ζ_1' and ζ_2' rules, in such a way that only B changes and is extended to a branch B' such that $B' \supseteq B$, and B' also contains, for each position r, $q \leq r \leq p$, an inequality $a_r \sqsubseteq a_r'$ ($a_r' \sqsubseteq a_r$) if r is a monotonic position of $P(t_1\tau, \ldots, t_k\tau)$ at i (resp. antimonotonic), being $WS(a_r \sqsubseteq a_r', B')$, and there is a substitution τ' extending τ such that:*

(i) $a_r\tau' = (t_i\tau|_r)[t\tau]_{p-r}$, $a_r'\tau' = (t_i\tau|_r)[t'\tau]_{p-r}$, and (ii) τ' is a unifier of a_q and $t_i|_q$.

In the following we will suppose that, for any tableau construction, we enumerate the set of variables X as $x_1, x_2, \ldots$. Applying γ' to a formula φ of class *Gamma* on a branch B, φ is instantiated with the first free variable x_i^s of the enumeration not used for φ in B, for an adequate sort s. Similarly, applying $Refv$ and $Reff$. Also we will consider a limit $l \geq 1$ for the number of applications of γ'-rule to every occurrence of a formula of class *Gamma* on a branch.

The notions of free variable tableau construction rule and fair free variable tableau construction rule are similar to the ground case, but considering the new expansion rules to open tableaux and respecting a limit for the application of γ'-rule.

Furthermore we are assuming that the set of sentences for which a tableau is constructed is finite. If not, a limit for the application of *In*-rule on each branch should be fixed. In that case, a completeness result for some such limit could be stated.

Theorem 25 (Completeness) *Let Φ be a not w-satisfiable finite set of Σ-sentences. There is a limit $l \geq 1$ for which, if $\mathcal{R}$ is a fair free variable tableau construction rule with respect to l such that: a) the applications of all first order expansion rules, order sorted extension rules and In come first, and b) all possible applications of inequality extension rules come next, then Φ has a closed free variable tableau constructed according to $\mathcal{R}$.*

Proof. From $\mathcal{R}$ it is possible to define a fair ground tableau construction rule $\mathcal{R}'$, similar to $\mathcal{R}$, not considering any limit for the application of γ-rule, and such that it uses an enumeration $t_1, t_2, \ldots$ of the ground $\overline{\Sigma}$-terms, in the same way that $\mathcal{R}$ uses an enumeration of the variables $x_1, x_2, \ldots$, and being $sort(t_i) = sort(x_i)$[7]. Moreover, $\mathcal{R}'$ will be such that if $\mathcal{T}'$ is constructed according to $\mathcal{R}'$, applying only first order expansion rules, order sorted extension rules or In, then there is a free variable tableau $\mathcal{T}$, constructed according to $\mathcal{R}$, such that $\mathcal{T}'=\mathcal{T}\sigma$, where σ is the substitution $[t_1/x_1, t_2/x_2, \ldots]$.

The completeness of the ground method assures the existence of a closed ground tableau $\mathcal{T}_g$ for Φ constructed according to $\mathcal{R}'$. Let l be the greatest number of γ-rule applications to a formula on a branch of $\mathcal{T}_g$. Extend first $\mathcal{T}_g$ by applying γ-rule to every occurrence in $\mathcal{T}_g$ of a formula of class *Gamma* until to reach the limit l, then move the applications of Ref, ζ_1, ζ_2, ξ_1 and ξ_2 rules to the end. This movement does not matter because the resulting formulas from these applications are *Basic* and do not participate in γ neither in Tr_0 rules. Let $\mathcal{T}_g'$ be the part of this tableau, before applying those inequality extension rules, and let $\mathcal{T}_v$ be a free variable tableau constructed according to $\mathcal{R}$ such that $\mathcal{T}_v\sigma = \mathcal{T}_g'$. By Lifting Lemma, $\mathcal{T}_v$ can be closed using inequality rules, since $\mathcal{T}_g'$ can be closed applying only Ref, ζ_1, ζ_2, ξ_1 and ξ_2 rules. ∎

---

[7]See [GLMN 96] for more details.

5 Examples

Example 26 Consider a signature with sorts $S = \{s_1, s_2, s_3, s_4\}$, a constant $d \in D^{s_4}$ and a binary predicate symbol $P \in P^{s_2 s_1}$, monotonic in both arguments. Let Φ be the following set of formulas $\{1 : s_1 \sqsubseteq_0 s_2 \rightarrow \forall x^{s_2} \forall x^{s_1} P(x^{s_2}, x^{s_1}), 2 : (s_4 \sqsubseteq_0 s_3) \vee (s_4 \sqsubseteq_0 s_1), 3 : (s_1 \sqsubseteq_0 s_2) \vee (s_3 \sqsubseteq_0 s_2), 4 : \forall x^{s_2} \neg P(x^{s_2}, d), 5 : \forall x^{s_3} P(x^{s_3}, d)\}$ and $\varphi \equiv s_4 \sqsubseteq_0 s_3$. We prove that $\Phi \models \varphi$ with the tableau below, where the first five formulas correspond to the formulas of Φ.

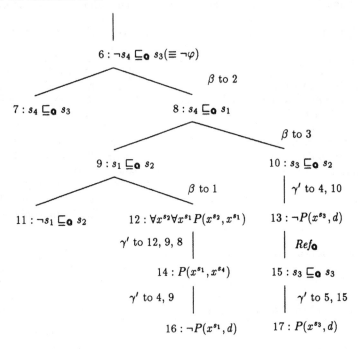

Enumerating branches B_i from left to right, it is easy to check that the substitution $\sigma = [c^{s_1}/x^{s_1}, c^{s_3}/x^{s_3}, d/x^{s_4}]$, where c^{s_1}, c^{s_3} are auxiliary constants of sort s_1 and s_3, respectively, closes B_3 and B_4. In fact, in B_3 for example, the literals $\neg P(c^{s_1}, d), P(c^{s_1}, d)$ are complementary and well-sorted w.r.t. the sort information $\{s_1 \sqsubseteq_0 s_2, s_4 \sqsubseteq_0 s_1\}$ contained in B_3. The branches B_1 and B_2 are obviously closed.

Example 27 The following example treated in [Wal 90] also with tableaux, can be easily solved in our system.

(i) No used-car dealer buys a used car. (ii) Some people who buy used cars are absolutely dishonest. Conclude that (iii) some absolutely dishonest people are not used-car dealers.

Let s_1, s_2 and s_3 be sorts representing, respectively, people who buy used cars, absolutely dishonest people and used-car dealers. The formulas below labeled by 1 and 2 axiomatize (ii), 3 axiomatizes the negation of (iii), and for (i) we establish the formula 4. A closed tableau is found for the set of formulas $\{1 : s_4 \sqsubseteq_0 s_1, 2 : s_4 \sqsubseteq_0 s_2, 3 : s_2 \sqsubseteq_0 s_3, 4 : \forall x^{s_3} \neg \exists y^{s_1}(x \approx y)^8\}$. The substitution $[c/x^{s_4}, c/y^{s_4}]$ $(c \in AC^{s_4})$ allows to close the whole tableau.

---

[8]$t \approx t'$ stands for $t \sqsubseteq t' \wedge t' \sqsubseteq t$.

$$\left|\ Ref_{\mathbf{o}}\right.$$
$$5 : s_4 \sqsubseteq_{\mathbf{o}} s_4$$
$$\left|\ Tr_{\mathbf{o}} \text{ to } 2,3\right.$$
$$6 : s_4 \sqsubseteq_{\mathbf{o}} s_3$$
$$\left|\ \gamma' \text{ to } 6,4\right.$$
$$7 : \neg \exists y^{s_1}(x^{s_4} \approx y^{s_1})$$
$$\left|\ \gamma' \text{ to } 1,7\right.$$
$$8 : \neg(x^{s_4} \approx y^{s_4})$$
$$\diagup\diagdown\ \beta \text{ to } 8$$
$$9 : \neg x^{s_4} \sqsubseteq y^{s_4} \qquad 10 : \neg y^{s_4} \sqsubseteq x^{s_4}$$
$$\left|\ Ref_v \qquad\qquad\qquad \right|\ Ref_v$$
$$11 : x^{s_4} \sqsubseteq x^{s_4} \qquad 12 : x^{s_4} \sqsubseteq x^{s_4}$$
$$\natural \text{ by } 9,11 \qquad\qquad \natural \text{ by } 10,12$$

6 Related Works

There has been many proposals for incorporating order-sorted relations to logic, defining resolution-based calculi. However concerning with tableaux, the number of attempts is fewer. [SW 90] defines a many sorted logic with static order-sorted relations, and extends tableaux by properly modifying the γ-rule. [Wei 91] proposes a first order logic $\mathcal{L}_S$ without equality and with dynamic sorts. Here the dynamism is achieved by means of the so-called *term declarations* [Sch 89] $t \lessdot s$, expressing that a term t has sort s. The use of these constructions as a new kind of formulas significantly increases the expressive power of the logic. Compared *LPDS* with $\mathcal{L}_S$, it can express some term declarations; for example, for $\forall x^{s'}(x^{s'} \lessdot s)$ we have the formula $s' \sqsubseteq_{\mathbf{o}} s$, and for constants declarations $c \lessdot s$, $c \lessdot s'$, we use the many-sortedness of our logic and assume there is $s'' \in S$, being $s'' \sqsubseteq_{\mathbf{o}} s$, $s'' \sqsubseteq_{\mathbf{o}} s'$ and $c \in C^{s''}$. Tableaux methods were obtained for a version of $\mathcal{L}_S$ in [Wei 95], by modifying γ and δ rules in such a way that a variable is instantiated with a term which is demanded —declared, for δ— to have the sort(s) of the variable; for free variable tableaux, sorted instead of syntactic unification is used. As an advantage, *LPDS* embodies the possibility to handle equality and partial orders as particular cases of preorders, and keeps the many-sortedness of the language; although an efficient implementation based in the replacement rules should incorporate specific strategies and heuristics. It is out of the scope of this paper to formally compare $\mathcal{L}_S$ and *LPDS*, above all because *LPDS* is three-valued. We leave for future work an adequate implementation of the tableaux that can be compared with the resolution and tableaux calculi based on $\mathcal{L}_S$.

References

[BG 95] L. Bachmair, H. Ganzinger. *Ordered Chaining Calculi for First-Order Theories of Binary Relations*. MPI-I-95-2-009, 1995.

[BH 92] B. Beckert, R. Hähnle. *An Improved Method for Adding Equality to Free Variable Semantic Tableaux*. Proc. CADE'10. LNAI 607, 507–521, 1992.

[BKS 85] W. W. Bledsoe, K. Kunen, R. Shostak. *Completeness Results for Inequality Provers*. Artificial Intelligence 27, 255–288, 1985.

[Fit 88] M. Fitting. *First-Order modal tableaux*. J. of Automated Reasoning 4, 191–213, 1988.

[Fit 96] M. Fitting. *First-Order Logic and Automated Theorem Proving*. Second edition. Springer, 1996.

[GL 90] A. Gavilanes-Franco, F. Lucio-Carrasco. *A first order logic for partial functions*. TCS 74, 37–69, 1990.

[GLN 96] A. Gavilanes, J. Leach, S. Nieva. *Free Variable Tableaux for a Many Sorted Logic with Preorders*. To appear in Proc. AMAST'96, Springer, 1996.

[GLMN 96] A. Gavilanes, J. Leach, P. J. Martín, S. Nieva. *Reasoning with Preorders and Dynamic Sorts using Free Variable Tableaux*. Technical Report DIA 34/96, Univ. Complutense de Madrid, 1996.

[GM 92] J. A. Goguen, J. Meseguer. *Order-sorted algebra I: Equational deduction for multiple inheritance, overloading, exceptions and partial operations*. TCS 105, 217–273, 1992.

[HS 94] R. Hähnle, P. H. Schmitt. *The liberalized δ-rule in free variable semantic tableaux*. J. of Automated Reasoning 13, 211-221, 1994.

[JM 94] J. Jaffar, M. J. Maher. *Constraint logic programming: A survey*. J. of Logic Programming 19/20, 503–582, 1994.

[LA 93] J. Levy, J. Agustí. *Bi-rewriting, a term rewriting technique for monotonic order relations*. Proc. RTA'93. LNCS 690, 17–31, 1993.

[LN 93] J. Leach, S. Nieva. *Foundations of a Theorem Prover for Functional and Mathematical Uses*. J. of Applied Non-Classical Logics 3(1), 7–38, 1993.

[OS 88] F. Oppacher, E. Suen. *HARP: A Tableau-Based Theorem Prover*. J. of Automated Reasoning 4, 69–100, 1988.

[Sch 89] M. Schmidt-Schauss. *Computational aspects of an order sorted logic with term declarations*. LNAI 395. Springer,1989.

[SW 90] P.H. Schmitt, W. Wernecke. *Tableau Calculus for Order Sorted Logic*. Proc. Workshop on Sorts and Types in Artificial Intelligence (1989). LNAI 418, 49–60, 1990.

[Wal 87] C. Walther. *A Many-sorted Calculus based on Resolution and Paramodulation*. Research Notes in Artificial Intelligence. Pitman, 1987.

[Wal 90] C. Walther. *Many Sorted Inferences in Automated Theorem Proving*. Proc. Workshop on Sorts and Types in Artificial Intelligence (1989). LNAI 418, 18–48, 1990.

[Wei 91] C. Weidenbach. *A sorted logic using dynamic sorts*. MPI-I-91-218, 1991.

[Wei 95] C. Weidenbach. *First-Order Tableaux with Sorts*. J. of the Interest Group in Pure and Applied Logics 3(6), 887–907, 1995.

[Wei] C. Weidenbach. *Unification in Sort Theories and its Applications*. Annals of Mathematics and Artificial Intelligence. To appear.

Author Index

This international journal publishes mathematically rigorous, original research papers reporting on algebraic methods and techniques relevant to all domains concerned with computers, intelligent systems and communications.

Its scope includes algebra, computational geometry, computational algebraic geometry, computational number theory, computational group theory, differential algebra, signal processing, signal theory, coding, error control techniques, cryptography, protocol specification, networks, system design, fault tolerance and dependability of systems, microelectronics including VLSI technology and chip design, algorithms, complexity, computer algebra, symbolic computation, programming languages, logic and functional programming, automated deduction, algebraic specification, term rewriting systems, theorem proving, graphics, modeling, knowledge engineering, expert systems, artificial intelligence methodology, vision, robotics.

Electronic edition in preparation

Applicable Algebra in Engineering, Communication and Computing (AAECC)

Managing Editor:
Jacques Calmet, Karlsruhe

ISSN 0938-1279   Title No. 200

Subscription information for 1997:
Volume 8, 6 issues
DM 448,–*

* plus carriage charges
In EU countries the local VAT is effective.

Springer

Please order by
Fax: +49 30 82787 448
e-mail: subscriptions@springer.de
or through your bookseller

Springer-Verlag, P. O. Box 31 13 40, D-10643 Berlin

Lecture Notes in Computer Science

For information about Vols. 1–1071

please contact your bookseller or Springer-Verlag